FRESHWATER FISH DISTRIBUTION

TIM M. BERRA

FRESHWATER FISH DISTRIBUTION

THE UNIVERSITY OF CHICAGO PRESS   Chicago and London

Cover image: Groombridge, B. and M. Jenkins. 2002. World Atlas of Biodiversity. University of California Press, Berkeley, CA.

To Joseph S. Nelson, Ph.D., and William N. Eschmeyer, Ph.D., whose *Fishes of the World* and *Catalog of Fishes*, respectively, make the lives of ichthyologists easier.

Tim M. Berra is professor emeritus of evolution, ecology, and organismal biology at the Ohio State University.

The University of Chicago Press, Chicago 60637
The University of Chicago Press, Ltd., London

16 15 14 13 12 11 10 09 08 07 1 2 3 4 5

ISBN-13: 978-0-226-04442-2 (paper)

ISBN-10: 0-226-04442-4   (paper)

Library of Congress Cataloging-in-Publication Data

Berra, Tim M., 1943–
    Freshwater fish distribution / Tim M. Berra.
      p. cm.
    Originally published: San Diego, Calif. : Academic Press, 2001. With new pref.
    Includes bibliographical references and index.
    ISBN-13: 978-0-226-04442-2 (pbk. : alk. paper)
    ISBN-10: 0-226-04442-4 (pbk. : alk. paper) 1. Freshwater fishes. 2. Freshwater
fishes—Classification. 3. Fishes—Geographical distribution I. Title.
    QL624.B47 2007
    597.176—dc22
                                                                    2006103499

♾ The paper used in this publication meets the minimum requirements of the American National Standard for Information Sciences—Permanence of Paper for Printed Library Materials, ANSI Z39.48-1992.

Contents

x CONTENTS

Preface to the Paperback Edition

THIS 2007 PAPERBACK EDITION OF *FRESHWATER FISH DIS-*
tribution is a reprint of the 2001 hardback edition. Typographical errors
were corrected and an alphabetical list of families was added to the rear
of the book.

I followed the third edition of Nelson (1994) for much of the classification used in the 2001 edition. Major exceptions include the Characiformes (Buckup, 1998) and Siluriformes (dePinna, 1998). The fourth
edition of Nelson (2006) reflected many changes in numbers of species
within each family and in some family arrangements. References and ra-
tionale for most of the changes formalized by Nelson (2006) were mentioned in Berra (2001). Below is a brief listing of how Nelson (2006) dif-
fered from the classification in this reprint of Berra (2001).

Living lungfishes were placed in a single order, Ceratodontiformes,
and treated at the rear of the book as sister group to Tetrapoda. Polypteridae was placed in the subclass Cladistia as sister to Chondrostei.
Pantodontidae was included in Osteoglossidae. Hiodontidae was put
in Hiodontiformes instead of Osteoglossiformes. Sundasalangidae was
placed in Clupeidae as a subfamily. Cyprinidae was divided into 11 subfamilies from the previous eight. Psilorhynchinae of Cyprinidae was elevated to family status. Catostomidae was divided into four subfamilies
from three.

Nelson (2006) listed 35 catfish families, I cited 32. He recognized
Pseudopimelodidae and Heptapteridae from Pimelodidae and Aucheno-
glandidae from Claroteidae. The electric eel was placed in the Gymnotidae instead of the Electrophoridae.

Salangidae was relegated to a tribe within Osmeridae, and Lepidogalaxiidae was treated as a subfamily within Galaxiidae. Within Paracanthopterygii, Nelson retained the orders and families but varied the sequence. Within Atherinomorpha, Bedotiidae, Pseudomugilidae, and Telmatherinidae were recognized as subfamilies within Melanotaenidae, and New World silversides, Atherinopsidae, were treated as a separate family from Old World silversides, Atherinidae. Nelson recognized the removal of African rivulines from Aplochilidae and their inclusion in the resurrected Nothobranchiidae. He placed Indostomidae in the suborder Gasterosteoidei (sticklebacks) rather than Syngnathoidei (pipefishes).

Lates and *Psammoperca* were removed from Centropomidae and placed in Latidae. The two species of Chilean *Percilia* were removed from Percichthyidae and placed in Perciliidae. Nelson retained *Lobotes* and *Datnioides* (*Coius*) in the same family but changed the family name from Coiidae to Lobotidae. Pristolepidae was made a subfamily within Nandidae.

Pseudaphritis urvillii was removed from Bovichthyidae (now called Bovichtidae) and placed in a monotypic Pseudaphritidae. Nelson reduced the suborder Anabantoidei from five to three families. *Luciocephalus pulcher* was removed from the monotypic Luciocephalidae and placed in a subfamily within Osphronemidae along with five other genera of gouramies. Belontiidae was treated as a subfamily of Osphronemidae.

Fish classification continues its slow but steady march toward stability.

Preface to the 2001 Edition

THIS BOOK IS A COMPLETE REVISION OF MY 1981 *AN ATLAS OF Distribution of the Freshwater Fish Families of the World*. It was written without reference to the first edition, and I did not check the 1981 book until after I had written the first draft of the new material. In a very few cases I discovered that I had said it more clearly in 1981, so I kept some of the wording. Much has happened since 1981. Families have been lumped, and families have been split. New species have been discovered, and known ranges have been extended. As in the first edition, the distribution maps are intended to give a general picture of where a family occurs. At this scale, the maps cannot pinpoint the precise distribution. Classification and family names more or less follow Nelson (1994). However, I have largely relied on Buckup (1998) and de Pinna (1998) for characiform and siluriform relationships. The numbers of genera and species in a family were taken from Nelson (1994) and updated from FISHBASE (*www.fishbase.org/search.cfm*), which incorporates Eschmeyer's (1998) catalog, or from recent primary literature.

I intend this book for a variety of users. It will be helpful as a ready reference and as an entrance into the literature for undergraduate students writing a term paper or preparing an oral presentation. Graduate students in ichthyology might use it as a starting point and source of ideas for selecting a thesis topic. Professors should find it helpful when preparing lectures on various groups. Ichthyologists and zoogeographers will find it a convenient summary of distribution patterns. In addition, tropical fish hobbyists and anglers may find it a source of information relevant to their pastimes.

By necessity, almost any book is a product of the author's experience. It seems natural that I have included more detail about some families than about others. Most of my career has been spent working with North American and Australian fishes so I have emphasized what I know best. I have included all primary and secondary division freshwater fish families and a sprinkling of peripheral or marine families that have some representatives in fresh water. There are other marine families that could have been included by virtue of their species that enter rivers throughout the world.

The maps in this book show native and not introduced distributions, and map references marked with an asterisk include a distribution map.

Acknowledgments

I AM REALLY TOUCHED TO BE PART OF A PROFESSION IN which colleagues are so willing to help one another. The following ichthyologists commented on the text and maps of various families, but any errors that remain are solely my own. *James S. Albert* (Sternopygidae, Rhamphichthyidae, Hypopomidae, Apteronotidae, Gymnotidae, Electrophoridae, and Adrianichthyidae); *Gerald R. Allen* (Plotosidae, Melanotaeniidae, Pseudomugilidae, Telmatherinidae, Chandidae, Apogonidae, Toxotidae, and Kuhliidae); *Clyde D. Barbour* (Atherinidae and Cichlidae); *Robert J. Behnke* (Salmonidae); *Michael A. Bell* (Gasterosteidae); *R. Bigorne* (Mormyridae); *Marcelo Brito* (Callichthyidae); *Paulo A. Buckup* (Distichodontidae, Citharinidae, Parodontidae, Curimatidae, Prochilodontidae, Anostomidae, Chilodontidae, Crenuchidae, Hemiodontidae, Alestidae, Gasteropelecidae, Characidae, Acestrorhynchidae, Cynodontidae, Erythrinidae, Lebiasinidae, Ctenoluciidae, and Hepsetidae); *Warren E. Burgess* (Callichthyidae, Ambassidae, Cichlidae, Luciocephalidae, Anabantidae, Helostomatidae, Belontiidae, and Osphronemidae); *Brooks M. Burr* (Ictaluridae); *Douglas M. Carlson* (Acipenseridae and Polyodontidae); *Robert C. Cashner* (Fundulidae and Centrarchidae); *José I. Castro* (Carcharhinidae); *Ning Labbish Chao* (Sciaenidae); *Francois Chapleau* (Achiridae, Soleidae, and Cynoglossidae); *Brian W. Coad* (Acipenseridae, Anguillidae, Cyprinidae, Balitoridae, Bagridae, Siluridae, Heteropneustidae, Sisoridae, Esocidae, Salmonidae, Gadidae, Mugilidae, Cyprinodontidae, Gasterosteidae, Mastacembelidae, Percidae, Cichlidae, Gobiidae, and Channidae); *Miles M. Coburn* (Cyprinidae); *Daniel M. Cohen* (Bythitidae and Gadidae); *Bruce B. Collette* (Batrachoididae, Belonidae, and Hemiramphidae); *John Edwin*

Cooper (Esocidae and Osmeridae); *Wilson J.E.M. Costa* (Aplocheilidae and Rivulidae); *Ed J. Crossman* (Esocidae and Umbridae); *Fabio di Dario* (Pristigasteridae); *Guido Dingerkus* (Carcharhinidae, Potamotrygonidae, Dasyatidae, Ceratodontidae, Lepidosirenidae, Protopteridae, Polypteridae, Acipenseridae, Polyodontidae, Lepisosteidae, Osteoglossidae, Pantodontidae, Hiodontidae, Notopteridae, Mormyridae, Gymnarchidae, Anguillidae, Gadidae, and Rhyacichthyidae); *Joseph T. Eastman* (Bovichthyidae); *Ross M. Feltes* (Polynemidae); *Carl J. Ferraris, Jr.* (Cetopsidae, Bagridae, Pimelodidae, Chacidae, Auchenipteridae, and Aspredinidae); *J. Michael Fitzsimons* (Goodidae, Eleotridae, and Gobiidae); *Byron J. Freeman* (Synbranchidae); *Lance Grande* (Acipenseridae, Polyodontidae, Lepisosteidae, and Amiidae); *Carl D. Hopkins* (Mormyridae, Gymnarchidae, and Hypopomidae); *Walter Ivantsoff* (Mugilidae, Pseudomugilidae, Atherinidae, and Telmatherinidae); *Robert E. Jenkins* (Catostomidae); *Jean M. P. Joss* (Ceratodontidae, Lepidosirenidae, and Protopteridae); *Maurice Kottelat* (Sundasalangidae, Cyprinidae, Gyrinocheilidae, Cobitidae, Balitoridae, Bagridae, Siluridae, Schilbeidae, Pangasiidae, Chacidae, Clariidae, Heteropneustidae, Amblycipitidae, Akysidae, Sisoridae, Telmatherinidae, Indostomidae, Chaudhuriidae, Mastacembelidae, Coiidae, Nandidae, Polycentridae, Pristolepidae, Odontobutidae, Eleotridae, Gobiidae, Scatophagidae, Luciocephalidae, Anabantidae, Helostomatidae, Belontiidae, Osphronemidae, Channidae, Soleidae, and Tetradontidae); *Sven O. Kullander* (Nandidae, Polycentridae, Pristolepididae, and Cichlidae); *Francisco Langeani* (Hemiodontidae); *Helen K. Larson* (Rhyacichthyidae, Odontobutidae, Eleotridae, Gobiidae, and Kurtidae); *Carlos A. S. Lucena* (Acestrorhynchidae and Cynodontidae); *Don E. McAllister* (Osmeridae); *John E. McCosker* (Anguillidae); *Robert M. McDowall* (Retropinnidae, Galaxiidae, and Cheimarrhichthyidae); *John D. McEachran* (Carcharhinidae, Potamotrygonidae, and Dasyatidae); *Peter B. Moyle* (Cottidae, Centrarchidae, and Embiotocidae); *Steven M. Norris* (Malapteruridae and Anabantidae); *Lawrence M. Page* (Petromyzontidae, Acipenseridae, Polyodontidae, Amiidae, Hiodontidae, Anguillidae, Clupeidae, Cyprinidae, Catostomidae, Characidae, Ictaluridae, Esocidae, Osmeridae, Salmonidae, Percopsidae, Aphredoderidae, Amblyopsidae, Gadidae, Mugilidae, Belonidae, Hemiramphidae, Aplocheilidae, Fundulidae, Anablepidae, Poeciliidae, Goodeidae, Cyprinodontidae, Gasterosteidae, Cottidae, Centropomidae, Moronidae, Centrarchidae, Percidae, Apogonidae, Sciaenidae, Elassomatidae, Cichlidae, Embiotocidae, Gobiidae, Achiridae, and Cynoglossidae); *Lynne R. Parenti* (Phallostethidae, Adrianichthyidae, Aplocheilidae, Rivulidae, Profundulidae, Fundulidae, Valencidae, Anablepidae, Poeciliidae, Goodeidae, and Cyprinodontidae); *D. Paugy* (Osteoglossidae, Kneriidae, Citharinidae, Alestidae, Hepsetidae, and Mochokidae); *Mario C. C. de Pinna* (Diplomystidae, Cetopsidae, Ictaluridae, Claroteidae,

Austroglanididae, Bagridae, Pimelodidae, Cranoglanididae, Siluridae, Schilbeidae, Pangasiidae, Chacidae, Plotosidae, Clariidae, Heteropneust-idae, Malapteruridae, Ariidae, Mochokidae, Doradidae, Auchenip-teridae, Amblycipitidae, Akysidae, Sisoridae, Erethistidae, Aspredinidae, Amphiliidae, Nematogenyidae, Trichomycteridae, Callichthyidae, Scolo-placidae, Loricariidae, and Astroplepidae); *Ian C. Potter* (Petromyzontidae, Geotriidae, and Mordacidae); *Roberto E. Reis* (Callichthyidae and Loricariidae); *Aldemaro Romero* (Characidae and Amblyopsidae); *Stuart J. Rowland* (Percichthyidae and Teraponidae); *Scott A. Schaefer* (Scoloplacidae, Loricariidae, and Astroblepidae); *Gerald R. Smith* (Catostomidae); *Jay R. Stauffer, Jr.* (Esocidae, Aphredoderidae, and Cichlidae); *Donald J. Stewart* (Pimelodidae and Aspredinidae); *Melanie L. J. Stiassny* (Bedotiidae and Cichlidae); *Guy G. Teugels* (Protopteridae, Polypteridae, Pantodontidae, Mormyridae, Denticipitidae, Citharinidae, Hepsetidae, Bagridae, Clariidae, Malapteruridae, Mochokidae, and Amphiliidae); *D. F. E. Thys van den Audenaerde* (Phractolaemidae, Channidae, Nandidae, and Cichlidae); *James C. Tyler* (Scatophagidae and Tetraodontidae); *Richard P. Vari* (Distichodontidae, Citharinidae, Parodontidae, Curimatidae, Prochilodontidae, Anostomidae, Chilodon-tidae, Crenuchidae, Hemiodontidae, Alestidae, Gasteropelecidae, Characidae, Acestrorhynchidae, Cynodontidae, Erythrinidae, Lebiasinidae, Ctenoluciidae, Hepsetidae, Cetopsidae, Auchenipteridae, Lobotidae, and Teraponidae); *Emmanuel Vreven* (Mastacembelidae); *John Waldman* (Moronidae, Centrarchidae, and Percidae); *Stephen J. Walsh* (Doradidae, Auchenipteridae, and Elassomatidae); *Ronald E. Watson* (Rhyacichthyidae, Odontobutidae, Eleotridae, and Gobiidae); *Marilyn Weitzman* (Lebiasinidae); *Stanley H. Weitzman* (Distichodontidae, Citharinidae, Parodontidae, Curimatidae, Prochilodontidae, Anostomidae, Chilodon-tidae, Crenuchidae, Hemiodontidae, Alestidae, Gasteropelecidae, Characidae, Acestrorhynchidae, Cynodontidae, Erythrinidae, Lebiasinidae, Ctenoluciidae, and Hepsetidae); *Mark V. H. Wilson* (Osteoglossidae, Pantodontidae, Hiodontidae, Notopteridae, Mormyridae, Gymnarchidae, Esocidae, Umbridae, and Osmeridae); *Robert M. Wood* (Cyprinidae and Percidae); and *Richard S. Wydoski* (Umbridae, Percopsidae, Cottidae, and Embiotocidae).

The drawings were selected by availability, aesthetic value, and infor-mation content, mainly from works published in the late 1800s and early 1900s. In a few cases, recent drawings had to be used. The following indi-viduals and institutions made drawings under their control available to me: Laurence Richardson, Figs. 6, 7; UNESCO, Fig. 8; Jean Joss, Fig. 13; Musee Royal De L'Afrique Centrale and ORSTOM, Figs. 36, 39, 90, and 272; California Academy of Sciences, Fig. 42; American Museum of Natural History, Figs. 54, 62, and 192; Sterling Publishing Co., Figs. 85, 87, and 88; The Natural History Museum, Fig. 86; T. F. H. Publishing Co., Figs. 91, 107, and 146; Paul Skelton and the J. L. B. Smith Institute of Ichthyology,

Fig. 99; Academy of Natural Sciences of Philadelphia, Fig. 134; James Albert, Fig. 159; Brad Pusey and Kluwer Academic Publishers, Fig. 169; Robert M. McDowall, Figs. 170, 174–176, and 295; Carnegie Institution of Washington, Fig. 188; Bruce B. Collette and the American Society of Ichthyologists and Herpetologists, Fig. 190; Melanie L. J. Stiassny and the American Museum of Natural History, Fig. 192; Museum of Comparative Zoology, Harvard University, Fig. 198; Hamlyn Publishing Group, Fig. 212; A. A. Balkema Publishers, Figs. 226 and 229; and E. J. Brill Publishing Co., Figs. 268, 274, 276, 277, 291, 292, and 300. Dave Dennis drew Figs. 1–3 and 324.

Elaine Bednar compiled much of the index, proofread an early version of the manuscript, checked the Bibliography, filed reprints and slides of drawings, searched FISHBASE for numbers of genera and species, printed sections of the text, and made photocopies. Dixie Barton located obscure articles and sources of drawings through interlibrary loans. My wife, Rita, proofread sections of the manuscript. I appreciate the support of John Riedl, Dean of The Ohio State University at Mansfield, for allowing me to treat my early retirement as a permanent sabbatical.

Introduction

What Is a Fish?

This question is not as simple as it sounds at first reading. What makes it a difficult question is that there are many exceptions for each character used to define "fish." For example,

Character: Fishes swim in water. Exception: The mudskipper, *Periophthalmus*, skips about on mangrove flats completely out of water; grunions, *Leuresthes*, spawn on beaches above the low tide mark; the walking catfish, *Clarias*, ambles overland from pond to pond.

Character: Fishes respire by means of gills. Exception: Lungfishes, *Protopterus*, can breathe atmospheric air via lungs and can even remain encysted in a dry mud cocoon; anabantids have a labyrinth organ in their head that provides a large, highly vascularized surface for oxygen uptake.

Character: Fishes have scales. Exception: Catfishes lack scales; some sticklebacks have an armor plating of modified scales; some fishes are only partially scaled; and some (eels) have such tiny, embedded scales that they are easily overlooked.

Character: Fishes have fins. Exception: *Gymnarchus* lacks pelvic, anal, and caudal fins, whereas *Gymnotus* lacks dorsal and pelvic fins; the swamp eels (Synbranchidae) lack both pectoral and pelvic fins, and the anal and dorsal fins exist only as ridges.

Character: Fishes are cold-blooded. Exception: Many fast-swimming species such as tunas and their relatives develop muscle temperatures in excess of ambient water temperatures.

If we allow room for these and other exceptions, we can define a fish as a poikilothermic, aquatic chordate with appendages (when present) developed as fins, whose chief respiratory organs are gills and whose body is usually covered with scales. This broad definition makes no mention of skeletal material and therefore includes the cartilaginous lampreys and elasmobranchs as well as the bony fishes.

Primary, Secondary, and Peripheral Division Fish Families

Myers (1938, 1949, 1951) developed a widely used classification of fishes in fresh water based on their tolerance to salt water. Of course, this system is ecological rather that taxonomic. As modified by Darlington (1957), the three divisions are primary, secondary, and peripheral. Primary division freshwater fish families are those whose members have little salt tolerance and are confined to fresh waters. Salt water is a major barrier for them, and their distribution has not depended on passage through the sea. The salt tolerance of individual species within a family does vary, however. Secondary division freshwater fish families are those whose members are usually confined to fresh water, but they have some salt tolerance and their distribution may reflect dispersal through coastal waters or across short distances of salt water. Some peripheral family members may be confined to fresh water, and others may spend a considerable portion of their life cycle in fresh water, but they both are derived from marine ancestors who used the oceans as dispersal routes. Other peripheral division families are basically marine groups, some of which enter fresh water.

The ecological designation followed in this book is taken from Darlington (1957), who wrote, "We may, if we wish, doubt the 'reality' of Myers' divisions, but how can we doubt their usefulness in zoogeography?" On the other hand, Rosen (1974) challenged the usefulness of such a salt tolerance classification scheme, and he viewed fishes as being either continental or oceanic. In the family accounts "1st," "2nd," and "Per" refer to primary, secondary, and peripheral divisions, respectively.

How Many Kinds of Fishes Are There?

Cohen (1970) attempted an analysis of this problem by surveying ichthyologists who were experts in the various taxonomic groups. His results indicated about 50 species of Agnatha (lampreys and hagfishes), 555 species of Chondrichthyes (sharks, skates, rays, and chimaerids), and about 20,980 species of Osteichthyes (bony fishes), excluding fossil forms. A similar estimate, 18,300, was made by Bailey (1971). Of the 20,000 bony fishes, approximately 33% (6650) are primary division freshwater species, and 93% of these are ostariophysan fishes. Secondary division freshwater fishes accounted for about 8% (1625) of the total Osteichthyes, and most of these are cichlids and cyprinodontoids. Thus, freshwater fishes make up about 41% (8275) of all bony fishes, and as Cohen (1970) pointed out, this surprisingly high percentage is a reflection of the degree of isolation and diversity of niches possible in the freshwater environment. The large numbers of species in fresh water become even more amazing if the volume of fresh water on Earth is compared to the volume of the oceans. The oceans account for 97% of all the water on Earth, whereas the fresh water in lakes and rivers is an almost negligible percentage, 0.0093% (Horn, 1972). (The rest of the water is tied up as ice, groundwater, atmospheric water, and so on.) Therefore, 41% of all fish species live in less than 0.01% of the world's water (Horn, 1972). Further data on productivity of marine and freshwater environments and calculations which show that there are about 113,000 km^3 of water for each marine species while only 15 km^3 for each freshwater species—an approximate 7500-fold difference—can be found in Horn (1972).

The rate at which new species have been named is interesting. The 10th edition of Linnaeus's *Systema Naturae* (1758) listed 478 species in 50 teleost genera. By 1870, Gunther put the total number of known teleosts at about 8700. An analysis of new names proposed for teleosts from 1869 to 1970 (Berra and Berra, 1977) yielded an estimate of 15,370 new names for the 100-year period. If this figure is added to Gunther's calculation, we arrive at a total of 24,070 teleost names, which, when synonymies are taken into account, agrees well with Cohen's data derived independently.

Berra and Berra (1977) found that the families receiving the most new names were the cyprinids, gobiids, characins, and cichlids. They also reported that of the newly named freshwater species 35% were from South America, 30% from Africa, and 23% from Asia. This agrees well with the diversity and endemism data shown in Tables 1 and 2 and with the figures given by Gilbert (1976).

I revisited this topic 20 years later (Berra, 1997) and found very similar results. The families Cichlidae, Cyprinidae, Characidae, Loricariidae, and Cyprinodontidae constituted about half of all the recently described new freshwater fish species named in the period 1978–1993. The most new freshwater teleosts names came from South America (39%), Africa (32%), and Asia (17%).

Nelson (1994) estimated a total of 24,618 fish species compared to 23,550 tetrapod species, which is 51% of the 48,170 recognized living vertebrate species. This number of fish species is in close agreement with Cohen's (1970) maximum estimate of 21,585 agnathans, chondrichthyians, and osteichthyians considering the 24 years that have elapsed between the two publications and the rate of new species descriptions suggested by Berra and Berra (1977) and Berra (1997).

Today, we have the computerized advantage of Eschmeyer's (1998) *Catalog of Fishes*. That database yields an estimate of 27,300 valid fish species. Approximately 250 new species are described each year, but some

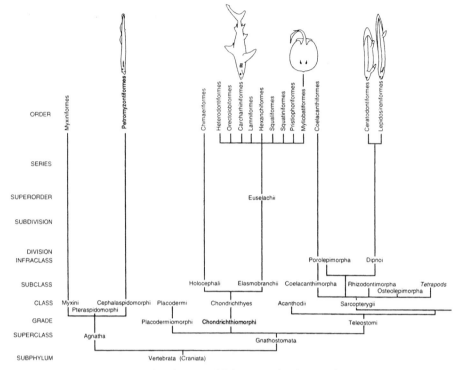

FIGURE 1. Classification of fishes. (See also front end papers.)

areas of the world are nearly completely cataloged, such as North America and Europe. Likewise, some groups of fishes are almost completely known, such as the large sharks, tunas, and billfishes. Eschmeyer estimates that when all is said and done there will be about 31,500 valid species of fishes (W. Eschmeyer, personal communication).

Wallace's Line and Zoogeographical Realms

The foundations of biogeography can be traced to the writings of Georges Louis Leclerc, comte de Buffon, and Augustin Pyramus de Candolle (Nelson, 1978). However, it was Alfred Russel Wallace (1823–1913), a British naturalist and codiscoverer with Charles Darwin of evolution via natural selection, who elaborated the concept of zoogeographic regions with the publication of his classic work, *The Geographical Distribution of Animals*, in 1876. He recognized six zoogeographical realms, which he called Nearctic (North America except tropical Mexico), Neotropical (South and Central America with tropical Mexico), Palearctic (nontropical Eurasia and north tip of Africa), Ethiopian (Africa and southern Arabia), Oriental (tropical Asia and nearby islands), and Australian (Celebes and

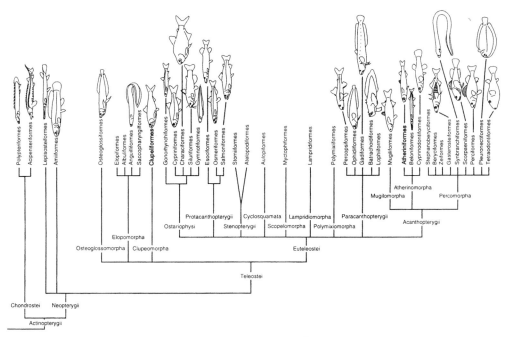

FIGURE 1. *Continued*

nearby small islands to the east, Australia, New Guinea, and New Zealand) (Fig. 2). Matthews (1998) conducted an extensive and very interesting analysis of freshwater fish distribution based on the maps from the first edition of this book (Berra, 1981). His analysis reflected the reality of Wallace's six zoogeographic realms. Darlington (1957) presented a detailed discussion of continental patterns of the distribution of vertebrates.

In 1860 Wallace proposed a hypothetical boundary between the Oriental and Australian faunas. This line passes between Bali and Lombok, through the Makassar Strait between Borneo and the Celebes, and south of the Philippines (Fig. 3). A few years later, T. H. Huxley named the boundary Wallace's line. He also modified it on the basis of bird studies to place the Philippines east of the line within the Australian realm. Mayr (1944) treated this controversy in depth. Camerini (1993) discussed the early history of Wallace's line, which is further illustrated in Daws and Fujita (1999). Quammen (1996) reviewed the history of Wallace's adventures and explored the theory of island biogeography. As the fauna of this region (Wallacea) became better known, many examples of animal groups crossing Wallace's line were noted, and most zoogeographers today do not take Wallace's line literally as an exact boundary between the Oriental and Australian faunas. However, the line does suggest a major faunal break separating the rich Oriental continental fauna from the depauperate Australian island fauna.

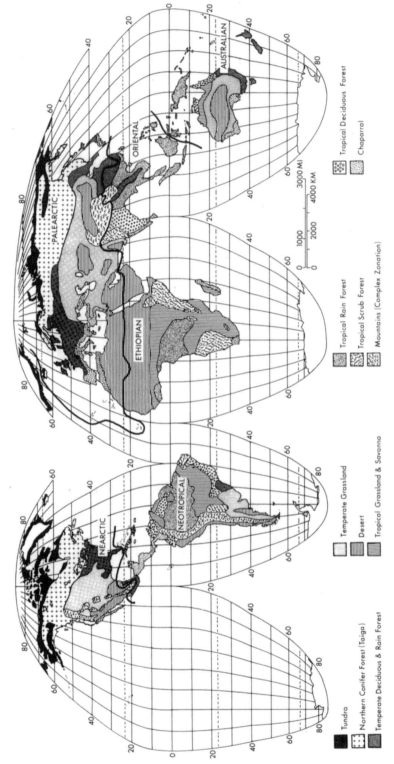

FIGURE 2. Wallace's six zoogeographic realms and the major biomes of the world.

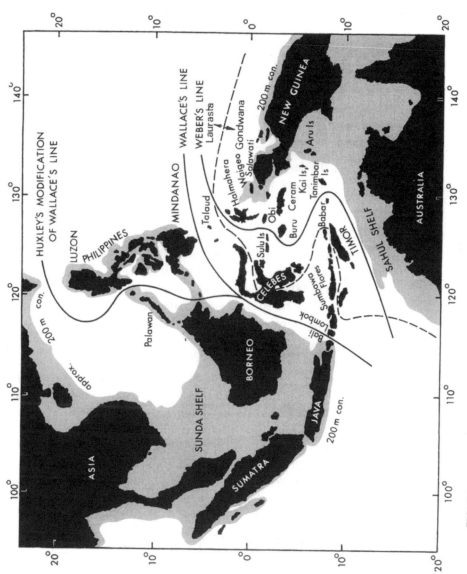

FIGURE 3. Lines of zoogeographic importance in the Malay archipelago (redrawn from Mayr, 1944).

Of the vertebrate groups, the freshwater fishes are most inhibited by a saltwater barrier and, therefore, most closely follow Wallace's line. There are 23 families of primary division freshwater fishes on Borneo. Only one family, Osteoglossidae, has managed to cross Wallace's line naturally. There are three species in three genera from 3 families (*Anabas, Ophicephalus,* and *Clarias*) that humans likely transported across the Makassar Strait to the Celebes, now called Sulawesi. A few salt-tolerant, secondary division oryziins have reached the Celebes, Lombok, and Timor, and the endemic secondary division Adrianichthyinae occurs in the Celebes. The Celebes rainbowfishes, Telmatherinidae, a peripheral division group, also only occur in Sulawesi, and the peripheral division Phallostethidae barely extends into Sulawesi. Only two cyprinid genera, *Puntius* and *Rasbora,* occur on both Bali and Lombok, with *Rasbora* reaching Sumbawa. These two cyprinid genera and a few others also occur in the Philippines along with an endemic silurid and clariid catfish (Darlington, 1957). With these few exceptions, the Asian and Australian freshwater fish faunas do not mix but rather end abruptly, which is unlike the situation in the New World, where the faunas of North and South America mingle in a Central American transitional or filter zone.

A second line, Weber's line, was proposed by Pelseneer in 1904. Weber's line represents the "line of faunal balance" that separates the islands with a majority of Oriental groups from those with an Australian majority. Mayr (1944) favored this line and provided more details. Weber's line (Fig. 3) is closer to Australia and reflects the fact that the Oriental fauna has made more of an intrusion to the east than has the Australian fauna to the west.

Wallacea, the island area between Wallace's and Weber's lines, is one of the most geologically complex areas of the world. Geophysicists suggest that the first collision between Australia–New Guinea and the Asian islands occurred about 15 million years ago in the mid-Miocene, near Sulawesi. By late Miocene or early Pliocene a land connection between Australia and Sulawesi was exposed (Audley-Charles, 1981). This could have provided a corridor for plant and animal migration. For purposes of discussion in this book, I consider Sulawesi (= Celebes) as part of the Australian realm since it is east of Wallace's line. However, evidence suggests that Sulawesi is of composite origin from both Laurasia and Gondwana (Fig. 3) (Parenti, 1991). Its biota, although predominantly Asian, is highly endemic but has elements of both regions. Modern consideration of Wallace's line should probably circumscribe Sulawesi, which could be considered a separate province. Whitmore (1981) provided further information on the relationship between Wallace's line and plate tectonics. Weber's line places Sulawesi within the Asian realm.

Which line, if any, one chooses to defend really depends on the taxonomic group in question and how many exceptions one is willing to tolerate. Furthermore, these lines are irrelevant to plant geographers, whose subject is much more directly influenced by climate.

Distribution Patterns

Freshwater fishes are one of the most important groups zoogeographically because they are more or less confined to drainage systems which can be thought of as dendritic islands of water surrounded by land, which is in turn bordered by a saltwater barrier. The freshwater fishes provide a relatively conservative system for examining patterns of distribution that may reflect continental changes. The family is the taxon that best reflects the evolution and dispersal of a group, and, in fact, zoogeographic patterns form strong evidence of evolution often cited in introductory biology texts.

The distributions of 139 primary, secondary, and peripheral division families are given in Table 2. The Neotropical realm has the largest number (35) of primary division families (Table 1), whereas the Ethiopian and Oriental regions are tied with 28 and 28 primary division families, respectively. The Nearctic and Palearctic regions have 15 primary division families each. The only primary division freshwater fishes in Australia are the lungfish, *Neoceratodus forsteri*, and two species of osteoglossids, *Scleropages*.

The Neotropical realm also has the greatest number of secondary division freshwater families (eight), whereas the Nearctic has seven division families, most of which are cyprinodontoids, and the Ethiopian region has five secondary families including the Cichlidae represented by an enormous number of endemic species with a great diversity of forms and habits (Table 1). In Australia and New Guinea, the Melanotaeniidae and the Pseudomugilidae occupy many cyprinodont and cyprinid niches. The Oriental family Adrianichthyidae in Sulawesi adds to the secondary division diversity in the Australian realm, as does one wide-ranging Asian species from the Aplocheilidae. The Oriental region has only three secondary families—the Adrianichthyidae, the Aplocheilidae, and the Cichlidae. The Valenciidae and cyprinodontids make up the secondary freshwater fish fauna of the Paleartic (Table 2).

The Neotropical and Oriental realms have the largest number of families in fresh water—51 and 42, respectively (Table 1). The Ethiopian, Nearctic, and Palearctic have 36, 31, and 30 families, respectively. The Australian freshwater fauna is depauperate, as expected of an island, and is dominated by peripheral groups. If the other marine families with freshwater representatives included in this book were taken into account, the proportion of peripheral representatives in Australian fresh waters would be even more lopsided.

By far the greatest number of endemic primary division families (32) is found in the Neotropical zone (Table 1). In fact, all but 3 primary families are endemic. This is a reflection of the extensive radiation of the characiforms and siluriforms (Otophysi). The Ethiopian and Oriental regions are a distant second and third place in the endemism race, with 17 and 14 families, respectively. The Nearctic region has 8 endemic primary division families. The only endemic primary division family in Australia is the Ceratodontidae.

TABLE 1
Number of Fish Families and Percentage Endemism in Each Biogeographical Realm, Based on 139 Families[a]

	Nearctic	Neotropical	Paleartic	Ethiopian	Oriental	Australian
No. primary families	15	35	15	28	28	2
No. secondary families	7	8	2	5	3	4
No. peripheral families	9	8	13	3	11	14
Total No. families	31	51	30	36	42	21[b]
No. endemic primary	8	32	0	17	14	1
No. endemic secondary	1	2	1	1	0	2
No. endemic peripheral	0	1	2	0	2	3
Total endemic families	9	35	3	18	16	7[b]
% Endemic Primary	26	63	0	47	33	5
% Endemic Secondary	3	4	3	3	0	10
% Endemic Peripheral	0	2	7	0	5	14
% Endemic Total	29	69	10	50	38	33[b]

[a]Widely distributed peripheral families excluded.
[b]The Lepidogalaxiidae is included in the total.

The Australian and Neotropical regions each have two endemic secondary division families, the Melanotaeniidae and Pseudomugilidae and the Profundulidae and Anablepidae, respectively. The Goodeidae of the Nearctic, the Valenciidae of the Palearctic, and the Bedotiidae of the Ethiopean realms are the only endemic secondary division families of those regions. The Oriental realm has no endemic secondary division families (Tables 1 and 2).

The Australian realm (including Sulawesi) has three endemic peripheral families (Retropinnidae, Telmatherinidae, and Cheimarrhichthyidae) which make up 14% of the freshwater families. The Palearctic (Comephoridae and Abyssocottidae) and Oriental (Indostomidae and Chaudhuriidae) regions each have two endemic peripheral families constituting 7 and 5% of the family totals, respectively (Table 1).

The highest percentage of endemism of freshwater fish families is in the Neotropical area, with 69% of the families found nowhere else (Table 1).

TABLE 2
Primary, Secondary, and Selected Peripheral Freshwater Fish Families and the Geographical Areas Where They Occur[a]

Family	Division	Nearctic	Neotropical	Palearctic	Ethiopian	Oriental	Australian
Petromyzontidae	Per	X		X			
Geotriidae	Per		X				X
Mordaciidae	Per		X				X
Potamotrygonidae	Per		X				
Ceratodontidae	1st						X
Lepidosirenidae	1st		X				
Protopteridae	1st				X		
Polypteridae	1st				X		
Acipenseridae	Per	X		X		X	
Polyodontidae	1st	X		X			
Lepisosteidae	2nd	X	X				
Amiidae	1st	X					
Osteoglossidae	1st		X		X	X	X
Pantodontidae	1st				X		
Hiodontidae	1st	X					
Notopteridae	1st				X	X	
Mormyridae	1st				X		
Gymnarchidae	1st				X		
Denticeptidae	1st				X		
Sundasalangidae	Per					X	
Kneriidae	1st				X		
Phractolaemidae	1st				X		
Cyprinidae	1st	X		X	X	X	
Gyrinocheilidae	1st					X	
Catostomidae	1st	X		X			
Cobitidae	1st			X		X	
Balitoridae	1st			X	X	X	
Distichodontidae	1st				X		
Citharinidae	1st				X		
Parodontidae	1st		X				
Curimatidae	1st		X				
Prochilodontidae	1st		X				
Anostomidae	1st		X				
Chilodontidae	1st		X				
Crenuchidae	1st		X				
Hemiodontidae	1st		X				
Alestidae	1st				X		
Gasteropelecidae	1st		X				
Characidae	1st	X	X				
Acestrorhynchidae	1st		X				
Cynodontidae	1st		X				
Erythrinidae	1st		X				
Lebiasinidae	1st		X				
Ctenoluciidae	1st		X				
Hepsetidae	1st				X		
Diplomystidae	1st		X				
Cetopsidae	1st		X				
Ictaluridae	1st	X					
Claroteidae	1st				X		
Austroglanididae	1st				X		

(*continues*)

TABLE 2 (*continued*)

Family	Division	Nearctic	Neotropical	Palearctic	Ethiopian	Oriental	Australian
Bargidae	1st			X	X	X	
Pimelodidae	1st		X				
Cranoglanididae	1st					X	
Siluridae	1st			X		X	
Schilbeidae	1st				X	X	
Pangasiidae	1st					X	
Chacidae	1st					X	
Clariidae	1st			X	X	X	
Heteropneustidae	1st					X	
Malapteruridae	1st				X		
Mochokidae	1st				X		
Doradidae	1st		X				
Auchenipteridae	1st		X				
Amblycipitidae	1st					X	
Akysidae	1st					X	
Sisoridae	1st			X		X	
Erethistidae	1st					X	
Aspredinidae	1st		X				
Amphiliidae	1st				X		
Nematogenyidae	1st		X				
Trichomycteridae	1st		X				
Callichthyidae	1st		X				
Scoloplacidae	1st		X				
Loricariidae	1st		X				
Astroblepidae	1st		X				
Sternopygidae	1st		X				
Rhamphichthyidae	1st		X				
Hypopomidae	1st		X				
Apteronotidae	1st		X				
Gymnotidae	1st		X				
Electrophoridae	1st		X				
Esocidae	1st	X		X			
Umbridae	1st	X		X			
Osmeridae	Per	X		X			
Salangidae	Per			X		X	
Retropinnidae	Per						X
Lepidogalaxiidae	?						X
Galaxiidae	Per		X		X		X
Salmonidae	Per	X		X			
Percopsidae	1st	X					
Aphredoderidae	1st	X					
Amblyopsidae	1st	X					
Bedotiidae	2nd				X		
Melanotaeniidae	2nd						X
Pseudomugilidae	2nd						X
Telmatherinidae	Per						X
Phallostethidae	Per					X	X
Adrianichthyidae	2nd					X	X
Aplocheilidae	2nd				X	X	X
Rivulidae	2nd	X	X				
Profundulidae	2nd		X				
Fundulidae	2nd	X	X				

(*continues*)

TABLE 2 (*continued*)

Family	Division	Nearctic	Neotropical	Palearctic	Ethiopian	Oriental	Australian
Valenciidae	2nd			X			
Anablepidae	2nd		X				
Poeciliidae	2nd	X	X		X		
Goodeidae	2nd	X					
Cyprinodontidae	2nd	X	X	X	X		
Gasterosteidae	Per	X		X			
Indostomidae	Per					X	
Synbranchidae	Per		X	X	X	X	X
Chaudhuriidae	1st					X	
Mastacembelidae	1st			X	X	X	
Cottidae	Per	X		X			
Comephoridae	Per			X			
Abyssocottidae	Per			X			
Moronidae	Per	X		X	X		
Percichthyidae	Per		X				X
Centrarchidae	1st	X					
Percidae	1st	X		X			
Coiidae	Per					X	X
Toxotidae	Per					X	X
Nandidae	1st					X	
Polycentridae	1st		X		X		
Pristolepididae	1st					X	
Elassomatidae	1st	X					
Cichlidae	2nd	X	X		X	X	
Embiotocidae	Per	X		X			
Bovichthyidae	Per		X				X
Cheimarrhichthyidae	Per						X
Rhyacichthyidae	Per					X	X
Odontobutidae	Per			X		X	
Kurtidae	Per					X	X
Luciocephalidae	1st					X	
Anabantidae	1st				X	X	
Helostomatidae	1st					X	
Belontiidae	1st			X		X	
Osphronemidae	1st					X	
Channidae	1st			X	X	X	
Achiridae	Per	X	X				

[a]The following widely distributed peripheral families (mostly marine) are omitted from this analysis although they are included in the text: Carcharhinidae, Dasyatidae, Anguillidae, Engrauillidae, Pristigasteridae, Clupeidae, Plotosidae, Ariidae, Bythitidae, Gadidae, Batrachoididae, Mugilidae, Atherinidae, Belonidae, Hemiramphidae, Syngnathidae, Centropomidae, Ambassidae, Apogonidae, Polynemidae, Sciaenidae, Monodactylidae, Terapontidae, Kuhlidae, Eleotridae, Gobiidae, Scatophagidae, Soleidae, Cynoglossidae, and Tetraodontidae.

Half (50%) of the families found in the fresh waters of the Ethiopian realm are endemic, and 38% of the families in Oriental fresh waters are endemic. The fish faunas, at the family level, are about 33% endemic in the fresh waters of the Australian realm and 29% endemic in the Nearctic realm. The endemic families of the Palearctic region constitute only 10% of the freshwater fish families (Table 1). Several families share a Nearctic and Palearctic distribution which may be considered a Holarctic pattern. It should be noted that some of the peripheral families excluded from Tables 1 and 2 have endemic freshwater species in the various biogeographical realms. This is not reflected in an analysis based on families.

Three recent ichthyology textbooks (Bond, 1996; Moyle and Cech, 1996; Helfman *et al.*, 1997) and W. Matthews (1998) provide discussions of the major biogeographic realms and their subdivisions. In addition, Lundberg *et al.* (2000) summarized the taxonomic composition and biogeography of freshwater ichthyofaunas. In North America we are fortunate to have such excellent reference works as Miller (1958), Gilbert (1976), Lee *et al.* (1980), Hocutt and Wiley (1986), and Mayden (1992) to summarize biogeographical data. Géry (1969) described eight faunistic regions of South America, and Lundberg *et al.* (1986) and Lundberg (1993) explained many patterns of distribution of Neotropical fishes. Roberts (1975), Greenwood (1983), and Lowe-McConnell (1987) analyzed African freshwater fish distribution. The biogeographic analyses of Berg (1948/1949) and Chereshnev (1990) covered much of Eurasia. Jayaram (1977) reviewed Indian freshwater fish distribution. Banarescu and Coad (1991) and Rainboth (1991) dealt with various aspects of Southeast Asian ichthyogeography. Zakaria-Ismail (1994) recognized five zoogeographic regions in Southeast Asia. Lake (1971), Roberts (1978), Merrick and Schmida (1984), Allen (1989, 1991), and Coates (1993) discussed the freshwater fish fauna and drainage regions of Australia and New Guinea, and McDowall (1990) comprehensively treated New Zealand fish distribution. On a worldwide scale deBeaufort (1951), Darlington (1957), Novacek and Marshall (1976), Briggs (1979, 1995), and Banarescu (1990, 1992, 1995), for example, summarized various aspects of freshwater zoogeography and provided an entrance into its literature.

Continental Drift

One of the most interesting aspects of zoogeography is the idea that the distribution of certain fish groups can be explained by the past movement of the continents. This is especially true for ancient groups such as the lungfishes and Osteoglossiformes, but continental drift also has implications for the distribution of the Otophysi. In 1915 Alfred Wegener published the first edition of *Die Entstehung der Kontinente und Ozeane (The Origin of Continents and Oceans)*, in which he called attention to the similarity of the east coast of South America and the west coast of Africa and suggested that they fit together like the pieces of a giant jigsaw puzzle. This concept was slow to be accepted, but in light of the modern geophysical evidence of plate tectonics and paleomagnetics, this hypothesis has become elevated to one of the basic theories in science. A geologic time chart is provided in Table 3.

TABLE 3
Geologic Time[a]

Eras, Periods, and Epochs	Span of time covered (in millions of years before present)		Emerging groups, or events
	From	To	
Cenozoic			
Quaternary			
Recent	0.01	Today	Retreat of last ice age (epoch covers just last 10,000 years)
Pleistocene	2.0	0.01	*Homo* (modern human)
Tertiary			
Pliocene	5.1	2.0	*Australopithecs* (ape–human)
Miocene	24.6	5.1	Apes
Oligocene	38	24.6	Monkeys
Eocene	54.9	38	Lemurs, lorises, tarsiers
Paleocene	65	54.9	Origin of primates
Mesozoic			
Cretaceous	144	65	First flowering plants (period closes with extinction of dinosaurs)
Jurassic	213	144	First birds; abundant dinosaurs
Triassic	248	213	First dinosaurs, mammal-like reptiles; origin of mammals
Paleozoic			
Permian	286	248	Radiation of reptiles
Carboniferous	360	286	First reptiles
Devonian	408	360	First amphibians
Silurian	438	408	First jawed fishes
Ordovician	505	438	Spread of jawless fishes
Cambrian	590	505	Trilobites and other hard-bodied invertebrates
Precambrian			
Ediacaran	4600	590	First multicellular fossils of jellyfish, sea pens, wormlike animals, and algae

[a]Source: Revised geological time scale based on Harland *et al.* (1982) from Berra (1990).

Continental drift theory states that all the continents were once part of a large landmass, Pangaea. This supercontinent became divided about 180 million years ago (Jurassic) into a northern portion, Laurasia (North America and Eurasia), and a southern portion, Gondwana (South America, Africa, India, Antarctica, Australia, and New Zealand). About 90 million years ago (Cretaceous), Gondwana broke apart into the separate southern continents [see references cited in Novcek and Marshall (1976) for the geologic timing of these events]. The propulsive forces for these movements, which are still going on today, are convection currents in the earth's mantle generated by the heat of radioactivity within the interior. These currents move the six major plates that make up the earth's surface and that float on the mantle. For historical perspective on this subject see the following: Berggren and Hollister (1974), Brown and Lomolino (1998), Brundin (1975), Colbert (1973), Cox (1974), Cox and Moore (1993), Cracraft (1974, 1975), Dietz and Holden (1970), Glen (1975), Howden (1974), Mayr (1952), Runcorn (1962), Smith and Briden (1977), Tarling and Tarling (1975), Udvardy (1969), Wegener (1966), Whitmore (1981), and the Scientific American book *Continents Adrift* (1973).

Dispersal

The relationship between continental drift and freshwater fish distribution has been discussed by Bond (1996), Moyle and Cech (1996), Helfman *et al.* (1997), Matthews (1989), and Lundberg *et al.* (2000). How the present patterns of fish distribution evolved is beyond the scope of this book and requires an understanding of the phylogeny and fossil history of the taxa under consideration as well as the geological history of the region. Lundberg (1993) provided an excellent model of such understanding. The "poster child" for vicariant vs dispersalist explanations of freshwater fish distribution is the Southern Hemisphere family Galaxiidae, and an extended discussion is presented in that section of this book based on Berra *et al.* (1996).

Some additional, and sometimes conflicting, explanations (but by no means an exhaustive list) of various aspects of ichthyogeography include those by Briggs (1974, 1979, 1984, 1986, 1987), Chardon (1967), Croizat *et al.* (1974), Darlington (1957), Gilbert (1976), Grande (1990), McDowall (1978), Miller (1958), Myers (1966, 1967), Nelson (1969), Novacek and Marshall (1976), Parenti (1981), Platnick and Nelson (1978), Rosen (1974, 1975, 1978), Rosen and Bailey (1963), Vari (1978), and Wiley (1988a,b). The papers by Wiley explain the basic premise of vicariance biogeography that depends on the congruence of distribution patterns of many unrelated monophyletic groups.

Naturally, major vicariant events such as the movement of landmasses will affect worldwide fish distribution. In addition to these rather dramatic movements, freshwater fishes may also disperse over a continent in a variety of ways. During times of flooding, drainage systems may overlap on a flood plain, allowing fishes access to other drainages. Stream capture, whether by headwater piracy or lowland meanders, may allow fishes to move from one river system to another and thereby spread across large continental areas. Freshwater fishes with some salt tolerance may swim from one river mouth through brackish or marine waters into another river mouth and then move upstream. During periods of glacial melt, drainage systems may be connected by overflow streams. Conversely, streams may be isolated in times of glacial advance because of lowered sea levels. Some fishes such as eels can actually wiggle across grassy areas from one stream to another. Other fishes may survive dispersal in the egg or other stages on mud attached to aquatic birds' feet or by waterspouts. The uplifting of mountain ranges may separate river systems, which can result in a similar fauna on both sides of a divide, which may then diverge when isolated. These are just a few of the considerations necessary for the understanding of the patterns of freshwater fish distribution.

Introduced Fishes

The maps in this book show native ranges only. This gives the most informative view of natural evolutionary patterns. Introductions of species by humans into areas outside the fishes' natural range are not shown. For example, no salmonids are shown in the Southern Hemisphere, although they have been introduced to South America, Africa, and Australia. The damage caused by exotic species to native fish faunas has been documented by Courtenay and Stauffer (1990), Lever (1996), and Fuller et al. (1999).

Rahel (2000) detailed the homogenization of fish faunas throughout the United States. He showed that fish faunas have become more similar through time due to widespread introductions. Eighty-nine pairs of states that formerly had no species in common now share an average of 25.2 species. States now share, on average, 15.4 more species than before European settlement. More than half the fish species in Nevada, Utah, and Arizona are not native to those states. Furthermore, introductions have been far more important than extripations in the homogenization of fish faunas. Fuller et al. (1999) provide maps showing the distribution of nonindigenous fishes in inland waters of the United States.

Pronunciation of Family Names

THE NAMES OF ZOOLOGICAL FAMILIES ARE, FOR THE MOST part, derived from Latin and Greek roots and always end in *-idae*. There are, of course, rules for the pronunciation of these names derived from the original language. Some of these rules are listed as follows:

1. All vowels are pronounced.
2. Diphthongs (two vowels written together) are pronounced as a single vowel.
3. *Ch* is pronounced as *k* if it is derived from a Greek word.
4. When *c* is followed by *ae, e, oe, i*, or *y*, it is pronounced as a soft *s* as in "cell"; however, when it is followed by *a, o, oi*, or *u*, it has the hard *k* sound as in "call."
5. *G*, when followed by *ae, e, i, oe*, or *y*, has the soft *j* sound as in "gel," but when followed by *a, o, oi*, or *u* it is pronounced as in "go."
6. Scientific names beginning with *ps, pt, ct, gn*, or *mn* are pronounced as if the first letter were not there; however, the first letter is sounded if these combinations appear in the middle of a word.
7. *X* is pronounced as *z* in the beginning of a word but as *ks* elsewhere.
8. In family names the major accent is on the antepenult (the third syllable from the end of the word)—the syllable before *-idae*. There may be a secondary accent on a syllable near the beginning of a very long word.
9. The vowel of the antepenult is short where it is followed by a consonant, except where the vowel is *u*.
10. The vowel of the antepenult is long where it is followed by another vowel.

For a detailed exposition of these and other rules and for a source of word roots, consult Borror (1960). Steyskal (1980) provided a list of family names that he considered to be in need of correction to bring them into compliance with the *Rules of Zoological Nomenclature* or Latin grammar.

These rules are not inflexible, however, and due consideration should be given to accepted usage as well as to the classically correct way. For example, many ichthyologists would pronounce the Amiidae as "ā'-mē-i-dē," which is closer to the generic name on which it is based, but Webster's *New International Dictionary* (2nd ed.) lists the classically correct "a-mī'-i-dē." The accent and the *a* and *i* sounds are different in the two pronunciations. A similar pronunciation difference exists with the Galaxiidae and most other family names which have an *i* before the *-idae* ending.

For the first edition of this book in 1981, George S. Myers was kind enough to send me his pronunciations of the families. He explained that his usage, as well as that of Carl L. Hubbs, was derived from David Starr Jordan, who was a noted classicist. Jordan, in turn, based his pronunciations on botanists who used an Americanized pronunciation. In the text I have given the classical pronunciation and syllabification in parentheses if listed in *Webster's*. If it is substantially the same as the Jordan–Hubbs–Myers version, only one pronunciation is listed. If there are significant differences in pronunciation, I list both, with the latter in brackets. If the family name or its root is not in *Webster's*, I give just the Jordan–Hubbs–Myers method in brackets. Where the Jordan–Hubbs–Myers pronunciation is different, it is usually an attempt to make the family name sound more like its generic base by placing the major accent on the fourth syllable from the end rather than on the antepenult.

Families and Maps

AGNATHA—Jawless Fishes

Superclass Agnatha
Class Cephalaspidomorphi

Order Petromyzontiformes
(Per) Family Petromyzontidae—**lampreys** (pet'-rō-mī-zon'-ti-dē)

THE HOLARCTIC LAMPREY FAMILY CONSISTS OF 34 SPECIES IN four genera: _Ichthyomyzon_, 6 species in fresh waters of eastern North America; _Petromyzon marinus_, anadromous in Atlantic drainages of North America, Iceland, Europe, and landlocked in Great Lakes; _Caspiomyzon wagneri_ in the Caspian Sea region; and _Lampetra_, 29 species, anadromous and freshwater, in North America and Eurasia.

FIGURE 4. _Lampetra spadicea_ (Meek, 1904, Fig. 1).

The holarctic distribution of lampreys, mostly north of 30°N, is probably related to temperature. The Petromyzontidae is generally found north of the 20°C isotherm, and an average lethal temperature for lampreys is approximately 28°C (Potter, 1980). The Mexican species _Lampetra spadicea_ (Fig. 4) and _L. geminis_ (subgenus _Tetrapleurodon_) are found as landlock relics on the Mesa Central in the Rio Lerma system (Miller and Smith, 1986). They occur further south than any other petromyzontid; however, the altitude of their habitat protects them from high temperatures (Potter, 1980). As a generalization, larger lamprey species have a greater geographical range (Potter, 1980).

Lampreys lack paired fins, scales, and bone. They have an elongate body with seven pairs of external gill openings and dorsal and caudal fins. The mouth is a circular disc. The skull and gill region are cartilaginous, as are the vertebrae, which lack centra. Lampreys present a good representation of what the earliest vertebrates were like: unarmored and lacking mineralized skeletons (Zimmer, 2000). There is a single nasal opening between the eyes. Chromosomes are mainly acrocentric and the diploid number of 164–168 in several different species is higher than that of any other vertebrate group except the Southern Hemisphere lamprey, _Geotria australia_ (Potter, 1980). (See the next family, Geotriidae.)

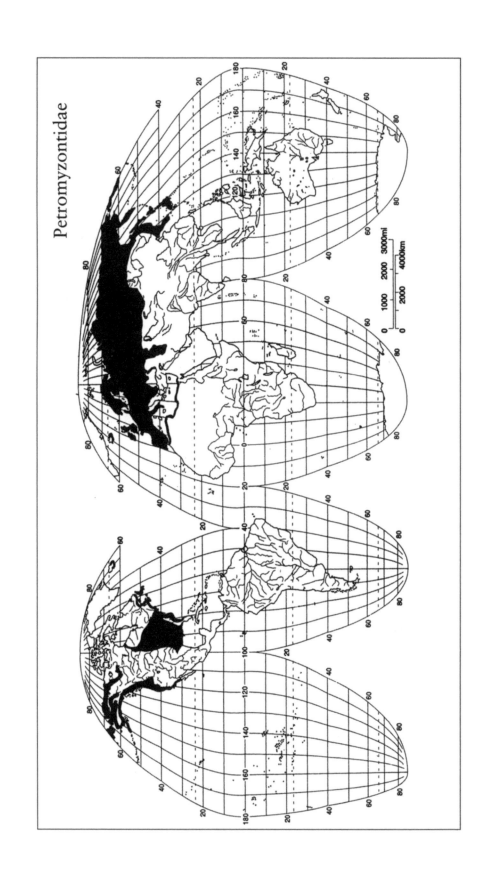

Some lampreys are parasitic, whereas others are free living. The nonparasitic lampreys are thought to be derived from parasitic ancestors. The terms "paired species" and "satellite species" are used for the parasitic ancestor and its nonparasitic descendant, but not all parasitic species have given rise to nonparasitic ones. Some lampreys have both a parasitic and nonparasitic phase in their life cycle.

Lampreys have an extended larval life. Small eggs are deposited by the female in gravel nests in stream bottoms in late winter–early spring. The eggs hatch in 2–4 weeks depending on water temperature. The larvae are known as ammocoetes. Ammocoetes have a very small dorsal fin fold, skin-covered eyes, and a well-developed pharynx. The ammocoetes burrow into the soft substrata in the slow-flowing regions of streams and rivers. There they filter plankton and detritus just off the bottom. The ammocoetes may remain in the mud for 3–7 years, then metamorphose into the adult lamprey. Parasitic adults attach themselves to fishes with their oral sucking disc and rasp away flesh, blood, and tissue fluids with horny teeth on their oral disc and piston-like tongue. Anadromous parasitic sea lampreys, *P. marinus* (Fig. 5), spend time growing at sea and may reach up to 1 m in length before ascending streams to spawn and die. Some parasitic lampreys reach far smaller sizes and remain in fresh water. The adult stage of the large lampreys may last 2 years.

FIGURE 5. *Petromyzon marinus* (Goode, 1884, Plate 251).

Nonparasitic lampreys, sometimes called "brook" lampreys, do not feed after metamorphosis. These small species, usually less than 20 cm, hide for a few months after metamorphosis while their gonads develop. Then they spawn. All lampreys die after spawning. Spawning females usually have a ventral fin fold and spawning males have a genital papilla.

The parasitic sea lamprey, *Petromyzon marinus*, has done tremendous damage to fish stocks in the Great Lakes. Important commercial species such as lake trout (*Salvelinus namaycush*) and lake whitefish (*Coregonus clupeaformis*) were nearly eliminated by sea lamprey predation combined with overfishing. The sea lamprey is not native to the Great Lakes. It arrived in Lake Erie through the Welland Canal in approximately 1921. From there it spread, reaching Lakes Huron and Michigan by 1936 and Lake Superior by 1946. This led to the collapse of several fisheries. After much intensive study to learn the lamprey life cycle (Applegate, 1950), a specific ammocoete poison was developed. Streams draining into the Great Lakes were treated with the lampricide, and the lamprey population has been controlled. Restocking has enabled some fisheries activity to resume, but the species mix has been altered forever. There is concern that native nonparasitic lamprey populations have been damaged by the control measures for the sea lamprey. Lake Erie is shallow and warm compared to the other Great Lakes and is not an ideal lamprey habitat.

Fossil lampreys date to the Carboniferous (Janvier and Lund, 1983). *Ichthyomyzon unicuspis*, which has a simple dentition, is considered close to the ancestral stock (Potter, 1980). Hubbs and Potter (1971) and Potter (1980, 1986) suggested that the differences between the Northern Hemisphere lampreys and the Southern Hemisphere lampreys (*Geotria* and *Mordacia*) were significant enough to warrant separate family status. Nelson (1994) considered the two Southern Hemisphere lamprey families as subfamilies within the Petromyzontidae. See Bailey (1980, 1982) and Vladykov and Kott (1982) for a discussion of lamprey systematics.

Berg (1948/1949), Hubbs and Potter (1971),* Lee *et al.* (1980),* Maitland (1977),* Nelson (1976),* Page and Burr (1991),* Rostlund (1952),* Scott and Crossman (1973),* Vladykov (1984)*

Superclass Agnatha
Class Cephalaspidomorphi

Order Petromyzontiformes
(Per) Family Geotriidae—**pouched lamprey**
(jē-ō-trī'-i-dē) [jē-ō'-trē-i-dē]

THIS MONOTYPIC FAMILY IS COMPOSED OF THE ANADROMOUS, parasitic *Geotria australis* found in southern Australia including Tasmania, New Zealand, Chile, Argentina, and the Falkland and South Georgia islands. Along with *Galaxias maculatus*, which has a similar distribution,

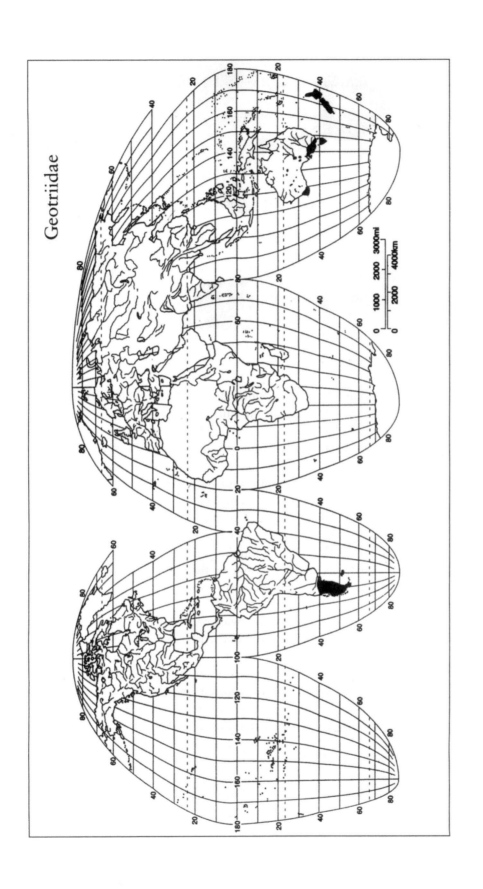

Geotriidae

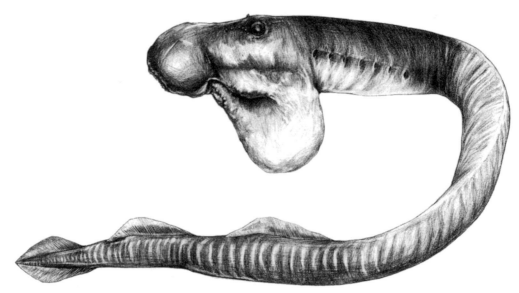

FIGURE 6. *Geotria australis*. Original drawing by Laurence Richardson.

G. australis is one of the most widely dispersed of all the species that are found in fresh water throughout the world (Fig. 6).

Mature males have a very large mouth and a well-developed gular pouch. Feeding adults have two prominent dorsolateral blue-green stripes that become less visible after the lamprey begins its upstream migration from the sea. The mouth is surrounded by fimbriae. Ammocoetes rarely exceed 120 mm, whereas adults may reach 500–700 mm when they ascend rivers. The lateral teeth of the oral disc are spatulate shaped rather than conical like those of other lampreys. The second dorsal fin is well separated from the caudal fin in the ammocoete. *Geotria* ammocoetes have very high levels of hemoglobin in their blood and tissues (Potter, 1986). This may allow the ammocoetes to survive in poorly oxygenated waters.

Potter (1996) developed the following natural history information from Western Australian populations of pouched lampreys. Spawning occurs in October and November following a 15- or 16-month spawning migration up river. Major morphological changes occur during this protracted migration, including modification of the dentition and development of the male's gular pouch. Metamorphosis begins in January or February (summer) when the ammocoetes are about 4 years old. It is complete by July (winter) when the downstream migration begins. It is not known what fish species *Geotria* feeds on while at sea (Potter, 1996).

Geotria has about 180 acrocentric chromosomes, a higher diploid number than any other vertebrate (Potter, 1986).

Map references: Allen (1989),* Arratia (1981),* Hubbs and Potter (1971),* McDowall (1990),* Potter (1996)*

Superclass Agnatha
Class Cephalaspidomorphi

Order Petromyzontiformes
(Per) Family Mordaciidae—**Southern Hemisphere lampreys**
(mōr-dā-sī'-i-dē) [mōr-dā'-sē-i-dē]

THERE ARE THREE SPECIES IN THIS FAMILY. *MORDACIA LAPI-cida*, a parasitic form, spawns in the rivers of southwestern South America. There are two Australian species. *Mordacia mordax* occurs in rivers in southeastern Australia and Tasmania and is parasitic (Fig. 7). Its nonparasitic derivative species, *M. praecox*, is found in two rivers within the northern part of *M. mordax*'s range.

FIGURE 7. *Mordacia mordax*. Original drawing by Laurence Richardson.

The males of *M. lapicida* develop a gular pouch at sexual maturity similar to *Geotria*, but *M. mordax* and *M. praecox* do not. *Mordacia* lack mucus-producing fimbriae around the suctorial disc, and their eyes are dorsolateral rather than lateral. These characteristics may be adaptations to the burrowing habit of adult *Mordacia* during migration (Potter, 1980). The ammocoetes of the two Australian species are indistinguishable. Adults of *M. mordax* typically reach 300–400 mm, whereas *M. praecox* rarely exceeds 170 mm (Potter, 1996).

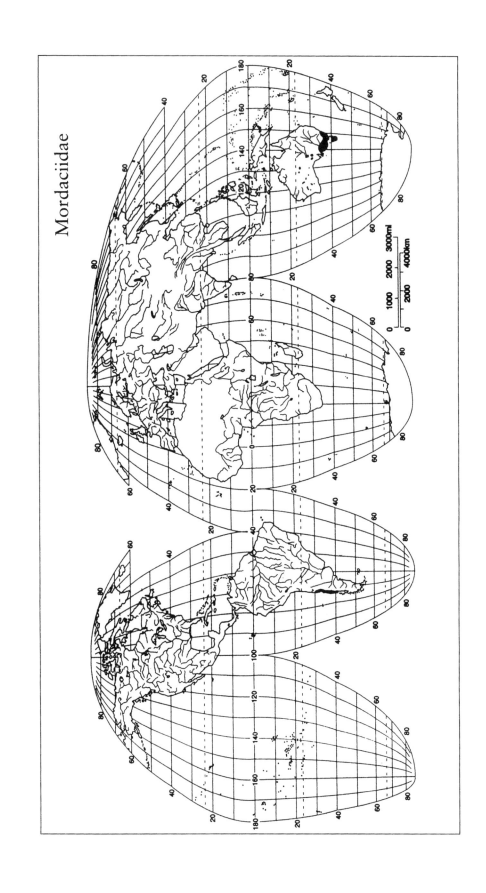

Mordaciidae

The chromosomes of *Mordacia* are predominantly meta- or submetacentric and number 76, perhaps as the result of fusion from a higher number as found in the Petromyzontidae and the Geotriidae (Potter, 1986).

Map references: Allen (1989),* Arratia (1981),* Hubbs and Potter (1971),* Potter (1996)*

GNATHOSTOMATA—Jawed Fishes
Chondrichthyes—Cartilaginous Fishes

Superclass Gnathostomata
Class Chondrichthyes
SubClass Elasmobranchii

Order Carcharhiniformes
(Per) Family Carcharhinidae—**requiem sharks** (kar'-ka-rī'-ni-dē)

THIS COSMOPOLITAN MARINE FAMILY OF ABOUT 58 SPECIES IN 13 genera has 7 species that enter fresh water. However, only the bull shark, *Carcharhinus leucas* (Fig. 8), and the rare Ganges shark, *Glyphis gangeticus*, make extensive movements into rivers and lakes. The map shows the distribution of these two species.

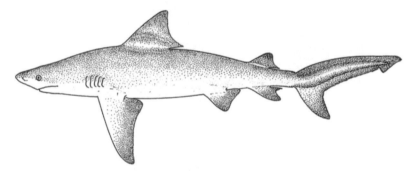

FIGURE 8. *Carcharhinus leucas* (Compagno, 1984, p. 478; reproduced with permission).

The bull shark is widespread along coastal regions of the tropics and warm temperate zones. There are five North American freshwater records (Burgess and Ross, 1980) including a specimen from the Mississippi River near St. Louis (Thomerson and Thorson, 1977). They may also occur up the Hudson River in New York (Smith, 1985). In Central America, Lake Nicaragua is infamous for its bull shark population that moves between the Caribbean Sea and the lake via the San Juan River (Thorson, 1976). Bull sharks have been taken 3700 km from the sea in the Amazon River as far as Peru, and in South Africa the Zambezi and Limpopo Rivers also house this species (Compagno, 1984). Both the bull shark and the Ganges shark are found in some Indian rivers. In Australia bull sharks, called river or freshwater whalers, have been reported from the Adelaide, Daly, East Alligator, Herbert, Brisbane, Clarence, and Swan Rivers (Last and Stevens, 1994). In New Guinea bull sharks have been taken in the Sepik and Ramu

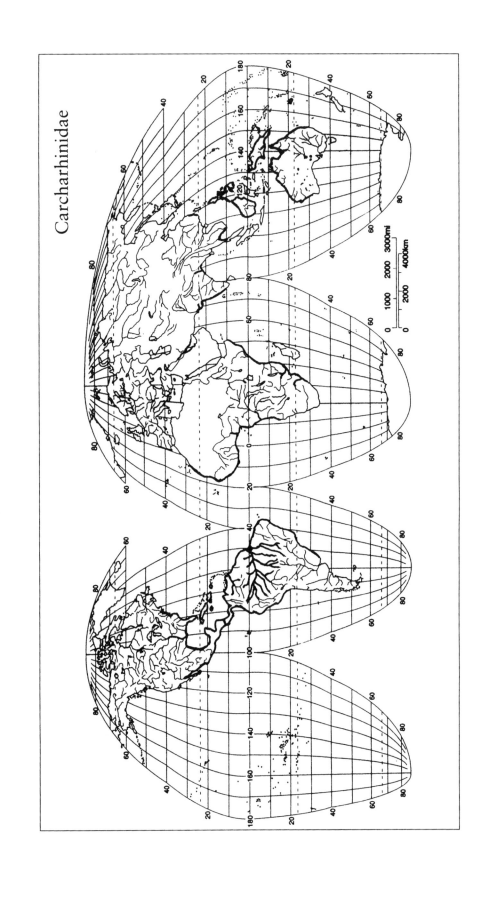

Carcharhinidae

Rivers and Lake Jamur, which is approximately 130 km inland from the Arafur Sea (Allen, 1991).

The bull shark is a gray, heavy-bodied shark with a short, broadly rounded snout, large and broad pectoral fins, and small eyes. The extremely short snout is much shorter than the width of the mouth (Castro, 1983). The first dorsal fin is large and triangular. The second dorsal is usually about a third of the height of the first dorsal. There are often trematode parasites or scars between the dorsal fins (Ellis, 1976). Upper teeth are broadly triangular and serrated. It normally feeds on bony fishes and other sharks. Because of its distribution near populated coasts and its aggressive attitude, large size, and formidable jaws and teeth, the bull shark is dangerous to humans. Along with the great white shark, *Carcharodon carcharias*, and the tiger shark, *Galeocerdo cuvieri*, the bull shark ranks in the top three in attacks on humans (Garrick and Schultz, 1963; Baldridge, 1974).

Bull sharks are viviparous, producing 1–13 pups per litter, usually in estuaries. Pups are between 56 and 81 cm at birth. Adults reach 3.4 m (Compagno, 1984). Reproduction in the fresh waters of Lake Nicaragua is not common (Jensen, 1976). Adult and neonatal bull sharks are broadly euryhaline, ranging from the hyperuremic condition of ocean-dwelling sharks to the hypouremic levels of sharks from fresh waters (Thorson and Gerst, 1972). They do not appear to do well in hypersaline waters (Compagno, 1984).

FIGURE 9. *Glyphis gangeticus* as *Carcharias gangeticus* (Day, 1878b, Plate CLXXXVII, Fig. 1).

The eyes of the Ganges shark are even smaller than those of the bull shark (Fig. 9). This may be an adaptation to the turbid conditions of tropical rivers and estuaries. The Ganges shark has a fierce reputation as a "man-eater," but because bull sharks also live in the Hooghly–Ganges River system, it is difficult to determine which species is responsible for the attacks (Compagno, 1984). Bull sharks are larger and their dentition is more robust. Bull sharks should probably receive most of the blame for human attacks in Indian rivers.

The biology of freshwater elasmobranchs was the subject of a symposium edited by Oetinger and Zorzi (1995) that forms a worthy extension of Thorson (1976).

Map references: Burgess and Ross (1980),* Bussing (1998),* Compagno (1984),* Dingerkus (1987),* Last and Stevens (1994)*

Class   Chondrichthyes
SubClass   Elasmobranchii

Order   Myliobatiformes
(Per) Family   Potamotrygonidae—**river stingrays**
(pot'-a-mō-trī-gon'-i-dē)

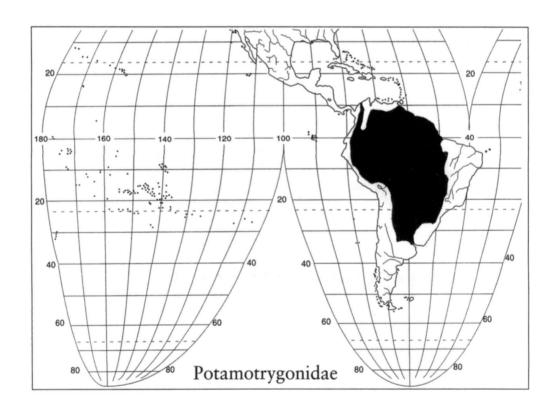

Potamotrygonidae

THIS GROUP IS TREATED AS A SUBFAMILY OF THE DASYATIDAE within the order Rajiformes by Nelson (1994). Compagno (1999a,b) considered the Potamotrygonidae as a family distinct from the Dasyatidae and included both in the order Myliobatiformes. I follow Compagno's scheme.

The river stingrays consist of about 20 species in three genera (Compagno, 1999a,b). They occur in fresh waters of the Atlantic drainages of South America, such as the Atrato, Maracaibo, Magdalena, Orinoco, Essequibo, Amazon, Paraná, and de la Plata Rivers.

They are restricted to fresh water and do not tolerate salt water (Dingerkus, 1995). Freshwater adaptations of the Potamotrygonidae include very little urea in body fluids and an atrophied rectal gland (Thorson *et al.*, 1967, 1978). They osmoregulate like a freshwater bony fish by excreting copious quantities of dilute urine.

Paratrygon aireba lives in the rivers of northern Bolivia, eastern Peru, and northern Brazil (Fig. 10). *Plesiotrygon iwamae* occurs in the upper and mid-Amazon River and its tributaries in Ecuador and Brazil (Compagno, 1999b). There are 18 described species of *Potamotrygon* and at least 2 undescribed species. *Potamotrygon* is widespread in rivers of Colombia,

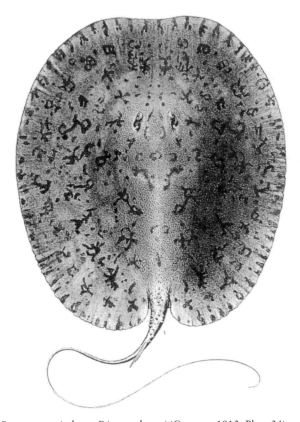

FIGURE 10. *Paratrygon aireba* as *Disceus thayeri* (Garman, 1913, Plate 34).

FIGURE 11. *Potamotrygon constellata* as *P. circularis* (Garman, 1913, Plate 31).

Venezuela, Bolivia, Guyana, French Guiana, Surinam, Peru, Brazil, Argentina, Uruguay, and Paraguay (Compagno, 1999b). *Potamotrygon constellata* is representative of the genus (Fig. 11).

Potamotrygonids have a circular or longitudinally oval pectoral disk and a bluntly rounded or truncate snout. The disk may be covered with small denticles or small to large thorns. They have an extended median prepelvic process directed anteriorly. This process is absent in the Dasyatidae. The tail may be much longer or shorter than the snout–vent length. The large sting is located behind the pelvic fins. The caudal fin is reduced to a fin fold. Adults are usually less than 1 m long, but some may reach 2 m (Compagno, 1999a).

Marine and freshwater stingrays bury themselves in sandy bottoms. Walking with a probing stick and shuffling one's feet will help dislodge hidden stingrays. If an unwary wader treads on its dorsal surface, the ray lashes out with its tail, driving its barbed and grooved spine into the leg of the victim and causing great pain. The stinging spine is actually modified from a placoid scale, and venom-producing cells lie in epidermal layers along the base of the spine (Halstead, 1978). The venomous sting of freshwater stingrays is extremely painful and produces a predominantly local symptomology with a torpid and chronic involvement of the affected parts, whereas marine rays cause less local symptoms and a more general response (Castex, 1967).

The initial wound from a freshwater stingray is a laceration or puncture which soon becomes necrotic and ulcerated. If left untreated the wound may become gangrenous and the leg may require amputation. First-aid treatment includes irrigation of the wound with cold water to remove surface venom and removal of any pieces of the sting and its sheath. The limb should then be soaked in hot water (50°C) for 30–90 minutes (Halstead, 1978). This helps stop the pain by denaturing the proteinaceous venom. Seek medical attention. Antitetanus, antibiotics, and sutures may be required.

The Potamotrygonidae was the subject of a classic study in biogeography utilizing host–parasite coevolution. Brooks *et al.* (1981) studied the phylogenies of helminth parasites infecting freshwater and marine stingrays. They concluded that the parasites of the potamotrygonids were more closely related to those of coastal Pacific rays of the genus *Urolophus* rather than to the rays in Atlantic drainages. Brooks *et al.* speculated that the Cretaceous–Miocene mountain building that resulted in the Andes blocked a Pacific-draining proto-Amazon River, thereby trapping marine rays of Pacific origin in what was to become the Atlantic-draining Amazon system. Lovejoy (1996, 1997) conducted a phylogenetic analysis of stingrays and produced findings that do not agree with those of Brooks *et al.* (1981). Lovejoy's results suggest that a clade of Pacific and Caribbean *Himantura* species is the sister group to the potamotrygonids. Nevertheless, Brooks *et al.*'s very clever idea has been extremely useful for thinking about Neotropical biogeography and has generated hypotheses that have stimulated further discovery. Lovejoy *et al.* (1998) proposed that the potamotrygonids diverged from their closest marine relative in the early Miocene as a by-product of massive movements of marine waters into the upper Amazon region about 15–23 million years ago.

Dingerkus (1995) reviewed the classification of the river stingrays which, at various times, have been placed within the Dasyatidae (= Trygonidae) or Urolophidae or have been considered a distinct family, Potamotrygonidae. He concluded that they should be included within the Myliobatidae based on a cladistic analysis of the visceral arch complex. McEachran *et al.* (1996) studied the interrelationships of batoid fishes and concluded that the classification should reflect separate families Potamotrygonidae and Dasyatidae within the order Myliobatiformes, distinct from the Rajiformes. They assigned the three species of the Eastern Hemisphere marine dasyatid genus *Taeniura* and two American species of the dasyatid genus *Himantura* to the Potamotrygonidae.

Map references: Castex (1967),* Compagno (1999b), McEachran and Capapé (1984),* Roberts (1989a)

Class Chondrichthyes
SubClass Elasmobranchii

Order Myliobatiformes
(Per) Family Dasyatidae—**whiptail stingrays** (das'-i-at'-i-dē)

THE DASYATIDAE IS CIRCUMGLOBAL IN ALL TEMPERATE AND tropical seas, and some species enter tropical and warm-temperate rivers and lakes (Compagno, 1999b). Nelson treated this group as a subfamily and included the South American river rays (Potamotrygonidae) within the same family. There are six genera and more than 62 species in the Dasyatidae as recognized by Compagno (1999b). Further studies are needed to clarify the taxonomic limits of the Dasyatidae (McEachran *et al.*, 1996).

Dasyatids are similar to potamotrygonids, with a circular or longitudinally oval disk and bluntly rounded or truncate snout. Like potamotrygonids, the dasyatids may have small denticles or small to large thorns covering the disk. The pelvic girdle lacks the prepelvic process found in potamotrygonids, or there may be a low, blunt process instead of the high, sharp process of the potomotrygonids. The tail of dasyatids ranges from short to greatly elongate and may be more than twice as long to somewhat shorter than the snout–vent length. The sting is located behind the pelvic fins but may be greatly reduced or absent in *Urogymnus*. The caudal fin may be reduced to a fin fold or completely lost. Adults vary from less than 1 m to more than 4 m long (Compagno, 1999a). Dasyatids, like most elasmobranchs, have a high level of urea as an osmoregulatory agent and a prominent rectal gland used for salt excretion.

Several species in the genera *Dasyatis* and *Himantura* enter rivers. For example, the Atlantic stingray, *D. sabina*, commonly enters fresh waters in southeastern North America although it does not penetrate far upstream (Ross and Burgess, 1980; Schmid *et al.*, 1988). Some tropical dasyatids have only been found in fresh water in Asia, Africa, New Guinea, and Australia. Endemic stingrays from African fresh waters such as *D. garouaensis* from the Niger basin, Lagos Lagoon in western Nigeria, and the Cross River in Cameroon were reported by Compagno and Roberts (1984). They also reported a second fresh water species, *D. ukpam*, from the Cross River in Nigeria, the Ogowe River, and the lower Congo River. Roberts and Karnasuta (1987) described *Dasyatis laosensis*, an endemic stingray, from the Mekong River on the border between Laos and Thailand.

Compagno and Roberts (1982) reported on *H. krempfi* from fresh water near Pnom Penh, Kampuchea, and *H. signifer* from rivers in western

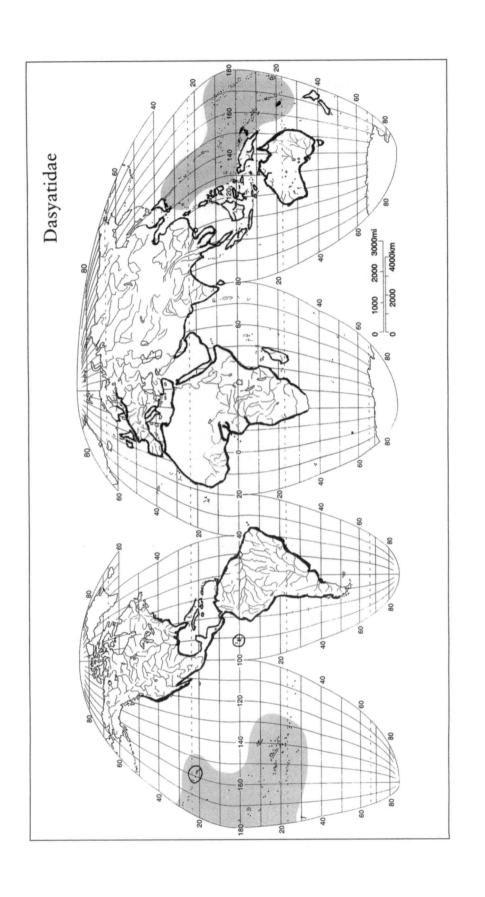

Dasyatidae

Borneo, Sumatra, western Malay Peninsula, and Thailand. *Himantura chaophraya* is a giant freshwater stingray found in the fresh waters of the Fly River, Papua New Guinea, the Mahakam basin of Borneo, several rivers of Thailand (Monkolprasit and Roberts, 1990), and in tropical Australian rivers such as the Gilbert (Queensland), Daly and South Alligator (Northern Territory), and the Ord and Pentecost Rivers (Western Australia) (Last and Stevens, 1994). It may reach 500 kg (Monkolprasit and Roberts, 1990).

Pastinachus sephan, referred to as *Hypolophus sephan* by Talwar and Jhingran (1992), is the most commonly reported stingray from fresh water in Southeast Asia (Fig. 12). This coastal, marine species reaches 1.2 m disc width and is consumed as food. Its tough skin is used for polishing wood (Talwar and Jhingran, 1992).

FIGURE 12.   *Pastinachus sephen* as *Trygon sephen* (Day, 1878b, Plate CXCV, Fig. 2).

Map references: Castex (1967),* Compagno (1999b), Compagno and Roberts (1982), Halstead (1978),* Last and Stevens (1994),* McEachran and Capapé (1984),* Monkolprasit and Roberts (1990), Roberts (1989), Roberts and Karnasuta (1987), Ross and Burgess (1980)*

GNATHOSTOMATA—Jawed Fishes
Sarcopterygii—Lungfishes

Class Sarcopterygii

Order Ceratodontiformes
(1st) Family Ceratodontidae—**Australian lungfish**
(ser'-a-tō-don'-ti-dē)

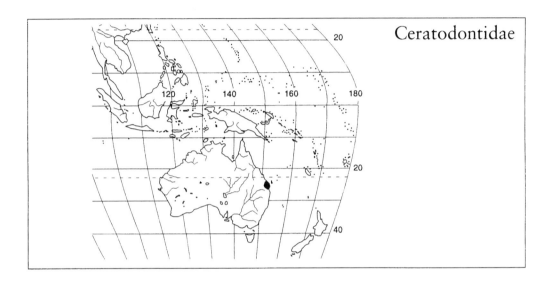

Ceratodontidae

THIS FAMILY HAS ONLY ONE LIVING SPECIES, *NEOCERATODUS forsteri*, which is restricted to the Mary and Burnett River systems in south-eastern Queensland (Fig. 13). They have been successfully introduced into the Brisbane, Albert, Coomer, and Stanly Rivers as well as into Enoggera Reservoir of Queensland (Merrick and Schmida, 1984).

FIGURE 13. *Neoceratodus forsteri* (original drawing by Josh Ferris provided by Jean Joss; reproduced with permission).

The family is known from the Lower Triassic, and extinct species in the genus *Ceratodus* were widespread around the world. *Neoceratodus forsteri* dates back to the Lower Cretaceous of New South Wales, making it one of

29

the oldest known vertebrate species (Kemp and Molnar, 1981; Marshall, 1987).

The Australian lungfish has an elongate body with large, bony, over-lapping scales. Its paired fins are paddle-like, and the pointed caudal fin is continuous with the dorsal and anal fins. There is only one lung, but it is divided internally into two distinct lobes that interconnect along its length. The lung is very well vascularized and compartmentalized to provide increased surface area for gaseous exchange. The lung opens into the esophagus almost ventrally via a duct that curves around the gut from the dorsal median point between the two lobes. There are five pairs of gill arches, all bearing gills. The fifth arch bears a hemibranch. The endocranium is cartilaginous.

This sluggish fish prefers still or slow-flowing waters with deep pools. It has the ability to gulp air at the surface in stagnant waters or when under-taking vigorous activity such as spawning, but normally it extracts oxygen from the water with its gills. It is a facultative air breather, not an obligate one. It will die if forced to depend on air breathing (Lenfant et al., 1970). It is unable to release CO_2 aerially. This makes its blood acidic, which reduces O_2 transport capacity of hemoglobin and thus reduces total O_2 consumption (Graham, 1997). It cannot survive the desiccation of its habitat like African or South American lungfishes, and it probably cannot survive for more than a few days out of the water, even if kept moist, because the gills need to be supported by surrounding water in order to function (J. Joss, personal communication).

Australian lungfish are omnivorous, feeding on aquatic invertebrates, but they also ingest some aquatic and terrestrial plant material (Merrick and Schmida, 1984). *Neoceratodus* browses among the detritus lying above the substrate of its habitat, using its electroreceptors to pick out hidden mollusks, worms, crustaceans, etc. They feed on native fruits fallen from trees overhanging the creeks, and they eat the aquatic weed, *Vallisneria* (J. Joss, personal communication). They can reach a length of more than 1.5 m and exceed 40 kg.

Spawning takes place in shallow waters from August to December when temperatures are between 10 and 25°C. Spawning occurs in aquatic plants, and the newly fertilized large eggs adhere to the weeds. Egg development is similar to that of an amphibian (Kemp, 1982). Larvae hatch in about 3–5 weeks depending on temperature and gradually develop the adult form over several months without an obvious metamorphosis (Kemp, 1981). Hatching can take place over a range of developmental stages (Joss et al., 1997). The young do not have external gills.

The lungfishes as a group (Australian plus the South American Lepidosirenidae and the African Protopteridae) are frequently referred to as a subclass or infraclass, Dipnoi (Fig. 1). The distribution of living Dipnoi presents a striking pattern on the Southern Hemisphere continents. This suggests continental drift as an explanation of their geographical range. However, their fossil distribution is very broad on all continents in both the Northern Hemisphere and Southern Hemisphere. Some Cretaceous lungfish were found in marine deposits and, therefore, may have been able to disperse through the sea (Marshall, 1987).

Lungfish biology and evolution have been summarized by Bemis *et al.* (1987) and included within their book is a bibliography of more than 2000 papers on lungfish biology (Conant, 1987). An area of active discussion today involves the relationship of lungfishes to tetrapods. Paleontologists and molecular systematists are debating the relative phylogenetic positions of coelacanths, lungfishes, and amphibians (Schultz, 1994; Meyer, 1995; Cloutier and Ahlberg, 1996; Zardoya and Meyer, 1997; Rassmussen *et al.*, 1998). The controversy is far from over.

Map references: Allen (1989),* Kemp (1987),* Merrick and Schmida (1984)

Class Sarcopterygii

Order Lepidosireniformes
(1st) Family Lepidosirenidae—**South American lungfish** (lep'-i-dō-sī-ren'-i-dē)

THERE IS ONLY ONE SPECIES OF SOUTH AMERICAN LUNGFISH, *Lepidosiren paradoxa* (Fig. 14). It is found in the freshwater swamps of Brazil and Paraguay throughout the Amazon and Paraná River systems (Lowe-McConnell, 1987). Darlington (1957) cited records for the lower Paraná at Resistencia, Argentina, and from below San Pedro, Argentina, in the Paraná delta, at about 34°S.

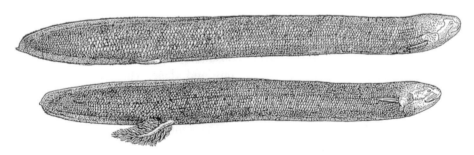

FIGURE 14   *Lepidosiren paradoxa* (top, female; bottom, male) (Cunningham, 1912, Fig. 25).

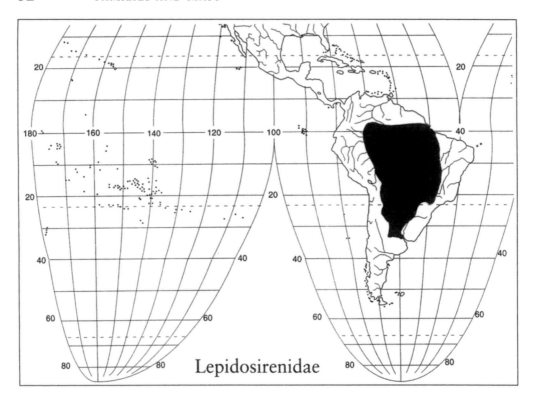

Lepidosirenidae

The body of *Lepidosiren* is very elongate with small scales. The paired fins are filamentous and have tactile and chemosensory capabilities (Graham, 1997). In breeding males, the pelvic fins develop highly vascularized, gill-like, feathery structures that may release oxygen from the bloodstream into the poorly oxygenated water surrounding the eggs or young (Cunningham and Reid, 1933). Foxon (1933a,b) considered the feathery fins to be auxiliary gills that benefit the adult male, not the eggs and young. The filaments atrophy after the end of the breeding season. Foxon (1933a) cited a 1908 report that described similar filament development for the pectoral fins. This species has two lungs and five gill arches. The connection of the lungs to the esophagus is ventral, just as it is in other air-breathing vertebrates (Norman and Greenwood, 1975).

South American lungfish are obligate air breathers and will drown if denied access to the surface (Johansen and Lenfant, 1967; Bone *et al.*, 1995). They are omnivorous, consuming aquatic vertebrates; invertebrates such as snails, clams, and shrimp; and algae. They can reach 1.25 m in length. South American lungfish can burrow into the bottom substrate as their swampy pools dry up and survive several months in an inactive metabolic state by breathing air. The young look like amphibian tadpoles with four external gills (Norman and Greenwood, 1975). The external gills are replaced by internal gills and lungs during metamorphosis. Very little is known about their natural history (Carter and Beadle, 1930). Kerr (1900) mentioned that *Lepidosiren* formed a cocoon similar to *Protopterus*. Pettit and Beitinger (1980) described thermal experiments, and Graham (1997) reviewed its air-breathing abilities and cited several early anatomical studies.

Map references: Bartholomew *et al.* (1911),* Bertin and Arambourg (1958),* Darlington (1957), Eigenmann (1909a),* Géry (1969), Lowe-McConnell (1987), Norman and Greenwood (1975)*

Class **Sarcopterygii**

Order **Lepidosireniformes**
(1st) Family **Protopteridae—African lungfishes**
(prō-top-ter'-i-dē)

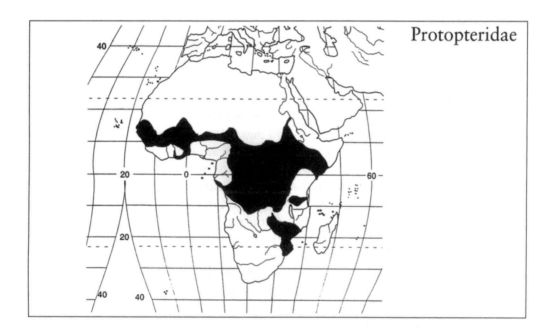

FOUR SPECIES OF *PROTOPTERUS* ARE FOUND IN THE FRESH waters of Africa: *P. annectens, P. aethiopicus, P. amphibius*, and *P. dolloi* (Gosse, 1984a). *Protopterus annectens* and *P. dolloi* are in west, central, and southern Africa and *P. aethiopicus* and *P. amphibius* are essentially in east and south-central Africa. African lungfishes occur in a wide variety of shallow habitats from standing (lentic) to running (lotic) water.

Like *Lepidosiren, Protopterus* is elongate with long, filamentous paired fins and small scales. African lungfishes have two lungs and are obligate air breathers. Although *Protopterus* has five gill arches, as does *Neoceratodus*, it does not bear such functional gills as the Australian lungfish. Two of the arches bear hemibranchs (2 and 4) and gill arch 3 has no gills at all (J. Joss, personal communication). They do not develop the vascularized pelvic fins of *Lepidosiren*. The largest species, *P. aethiopicus*, reaches more than 2 m, making it one of Africa's largest freshwater fishes (Greenwood, 1987a). The other species are smaller at approximately 1 m. Males are larger than females. The anus of lungfishes has an asymmetrical position offset from the midline posterior to the pelvic fins (Skelton, 1993).

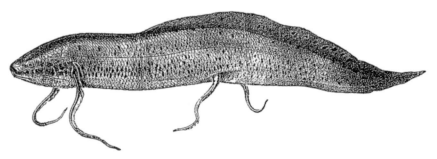

FIGURE 15. *Protopterus annectens* (Boulenger, 1909, Fig. 14).

The seminal study of lungfish estivation that has become a classic is the work of Johnels and Svensson (1954) on *P. annectens* (Fig. 15). The fish literally chews its way into the substrate ejecting mud out its gill openings. It may reach a depth of 30–250 mm below the bottom depending on the length of the fish. The lungfish wriggles around, thereby hollowing out a bulb-shaped chamber and coming to rest with its nose pointed upward. Lungfish breathe air at the mouth of the chamber's tube and sink back into the expanded part of the chamber. The fish secretes a great deal of mucus into the water-filled chamber. As the water table falls below the snout level of the fish and the bottom mud desiccates, the respiratory trips cease. The lungfish remains coiled in a U-shaped position and the mucus dries into a snug, thin-walled cocoon around the estivating lungfish. Air reaches the enclosed fish's mouth via the tube to the surface. The dry season usually lasts 8 or 9 months. When the rains come and the cocoon and bottom mud are softened, the lungfish wriggles free into the standing water. Under laboratory conditions, *P. aethiopicus* has remained in its cocoon for 4 years (Coates, 1937). Greenwood (1987a) summarized the strategies of the other lungfish species that involve estivation in water-filled burrows or retreat to permanent water.

The energy requirements during estivation are very low and are met by the metabolism of lungfish muscle tissue (Janssens and Cohen, 1968). Glycogen is actually stored during this time of protein destruction, and it provides energy for immediate use upon arousal when the rains come. Urea is a waste product of protein metabolism, and this normally toxic substance is allowed to build up in lungfish tissues (Fishman *et al.*, 1987). It is excreted when the habitat fills with water.

Lungfishes spawn when rainfall is plentiful. Several nest types are built by the various lungfish species. Males usually guard the eggs. The embryos hatch in 1 or 2 weeks with well-developed external gills. They are not obligate air breathers for at least a month after hatching (Greenwood, 1987a).

Lungfishes are omnivorous carnivores consuming a wide variety of prey, especially fishes, mollusks, crustaceans, and insects. Lungfishes are widely consumed by humans throughout their range and specialized fishing techniques have been developed for their capture during estivation.

Unlike the Ceratodontidae, there is no fossil evidence that the South American or African lungfishes ever occurred outside of their respective continents. They are also more closely related to each other than to the Australian lungfish. South America and Africa were broadly connected until the early Cretaceous, about 135 million years ago (MYA). Their final separation occurred in the mid-Cretaceous, approximately 95 MYA (Lundberg, 1993). *Lepidosiren* and *Protopterus*, as sister groups, provide convincing evidence of a lineage that has experienced South American–African vicariance (Lundberg, 1993).

J. Graham (1997) summarized air-breathing information, and Conant (1987) compiled a useful bibliography of 2209 papers published between 1811 and 1985 that deal with all aspects of lungfish biology.

Map references: Gosse (1984a), Greenwood (1987a),* Lévêque (1990a),* Poll (1973),* Skelton (1993)*

GNATHOSTOMATA—Jawed Fishes
Actinopterygians—Ray-Finned Fishes

Class Actinopterygii

Subclass Chondrostei

Order Polypteriformes

(1st) Family Polypteridae—**bichirs** (pol'-ip-ter'-i-dē)

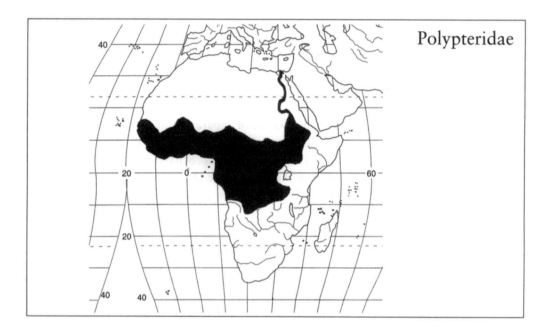

THIS FAMILY CONSISTS OF AT LEAST 10 LIVING SPECIES IN TWO genera from African freshwaters (Gosse, 1984b, 1988). The reedfish or ropefish, *Calamoichthys* (= *Erpetoichthys) calabaricus*, is confined to a narrow strip of coastline near the Gulf of Guinea (Fig. 16). It reaches 90 cm total length (TL) in nature but only about 15–25 cm in aquaria.

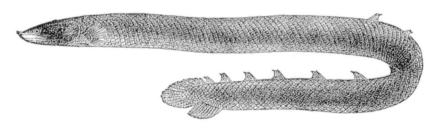

FIGURE 16. *Calamoichthys calabaricus* (Boulenger, 1909, Fig. 13).

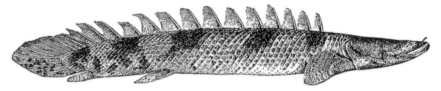

FIGURE 17. *Polypterus endlicheri* (Boulenger, 1909, Fig. 6).

It is eel-like and lacks pelvic fins in contrast to the 9 species of bichirs in the genus *Polypterus*. Bichirs inhabit shores and flood plains of rivers and lakes of central and west tropical Africa including the Nile. They do not occur in Indian Ocean drainages (Greenwood, 1984a). *Polypterus endlicheri* is among the largest bichirs at 750 mm TL (Fig. 17).

Some authorities classify the bichirs with the lungfishes in the Sarcopterygii. Others consider them separately in the subclass Cladistia or Brachiopterygii. Recent studies support their affinity to the actinopterygians (Fig. 1) (Gardiner and Schaeffer, 1989). The fossil record of this family extends from the Middle Cretaceous of Africa (Greenwood, 1974a). Late Cretaceous polypterids in Bolivia postdate the final separation of South America and Africa (Lundberg, 1993). Polypterids became extinct in South America sometime after the early Paleocene.

The name *Polypterus* refers to the 5–18 flag-like dorsal finlets set atop an elongate body. Each finlet is supported by a sharp spine that can make handling the fish painful. Bichirs have a suite of primitive characters such as diamond-shaped ganoid scales, spiracles, spiral valve, gular plate, maxillary fused to the skull, heterocercal tail, lungs with ventral attachment to the gut, and fleshy pectoral fins (Greenwood, 1984a). They range in size from 30 to 120 cm.

Bichirs are nocturnal predators feeding on fishes, amphibians, and aquatic invertebrates (Greenwood, 1984a). Because of their small left lung and larger right lung they can utilize atmospheric oxygen, which gives them the ability to inhabit poorly oxygenated swamps. They can remain out of water for several hours and have been reported to make overland excursions while feeding on terrestrial insects (Graham, 1997). Babiker (1984) and Pettit and Beitinger (1981, 1985) discussed nonobligatory air breathing by both genera. Graham (1997) summarized various airbreathing studies. Bichirs are egg layers, and the young have a pair of well-developed external gills until metamorphosis several weeks after hatching.

Map references: Gosse (1984b, 1990a),* Greenwood (1984a),* Poll (1973)*

Class Actinopterygii
Subclass Chondrostei

Order Acipenseriformes; Suborder Acipenseroidei
(Per) Family Acipenseridae—**sturgeons** (as-i-pen-ser'-i-dē)

THIS HOLARCTIC FAMILY OF ANADROMOUS AND FRESHWATER fishes is composed of four genera and 24 species (Nelson, 1994). *Acipenser* occurs throughout the family range. Five of its 16 species are North American. Two species of *Huso* are Eurasian from the Adriatic Sea to the Caspian Basin and Amur River. *Pseudoscaphirhynchus* consists of 3 species in the Aral Sea basin, and *Scaphirhynchus* has 3 species in the Mississippi River and Mobile basins. Fossils date to the Upper Cretaceous of western North America (Gardiner, 1966; Wilimovsky, 1956). Grande and Bemis (1996) and Bemis *et al.* (1997) reviewed the generic diversity of the Acipenseriformes and provided osteological characters to support the monophyly of the sturgeons.

The most striking feature of sturgeons is the five rows of bony scutes (derived from ganoid scales) along the body. Between these ridges the body is mostly naked except for small patches of scales over the postcleithrum and branchiostegals. Gardiner (1984) presented technical details of the skeletal system that define the family and order. Grande and Bemis (1991, 1996) provided an emended diagnosis of the family and order based on new and better preserved material. There are many more fin rays than there are basal supports. Sturgeons have small, protractile mouths on the underside of the head. A row of four barbels is located anterior to the mouth. Adults lack teeth, and the upper jaw does not articulate with the skull. The endoskeleton is mostly cartilaginous, but some dermal bone is present in the head. The vertebrae lack centra, and the tail is heterocercal with the persistent notochord extending into the upper lobe of the tail. The anus and urogenital opening are near the pelvic fins instead of the anal fin as in most actinopterygians. *Acipenser* and *Huso* have a spiracle. This structure is absent in the two genera of river sturgeons. An intestinal spiral valve is present.

The beluga sturgeon, *Huso huso*, of the Black and Caspian seas is the largest fish living in fresh water. From a larva just 13 mm may grow a giant that reaches 8 m and weighs 1300 kg with a life span of nearly 100 years. A female this size may contain more than 7 million eggs (Bond, 1996). Various other size records can be found in Wood (1972). One ounce (28 g) of beluga caviar sells for $55 in a New York restaurant (Matthews, 1998).

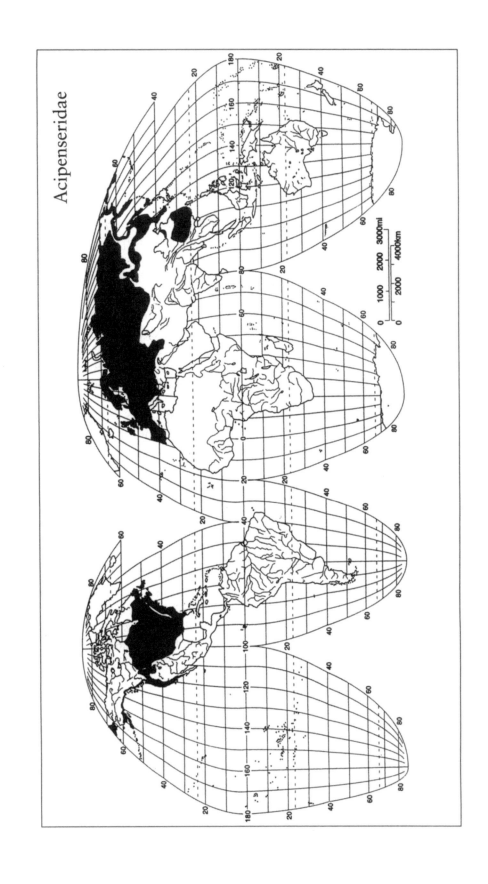

Acipenseridae

The sturgeons of Russia, Ukraine, and Iran are the sources of the most expensive caviar, mainly from *A. gueldenstaedti*. A Caspian-wide fishing ban is thought to be necessary in order to allow the sturgeon population time to recover from tremendous exploitation (Matthews, 1998).

The white sturgeon, *A. transmontanus*, of the Pacific Coast is the largest fish found in North America's rivers. It may reach 6.1 m and weigh 850 kg. Anadromous species are the largest, but not all sturgeons are giants. The shovelnose sturgeon, *Scaphirhynchus platorynchus*, of the Mississippi only grows to 86 cm and *Pseudoscaphirhynchus hermanni* from Russia reaches only 27 cm (Fig. 18).

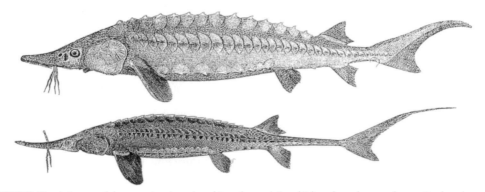

FIGURE 18. *Acipenser fulvescens* (top) as *A. rubicundus* and *Scaphirhynchus platorynchus* as *S. platyrhynchus* (bottom) (Goode, 1884, Plate 244).

The following is a typical life cycle for an anadromous species (Bond, 1996). There is a spawning migration from the feeding area (usually brackish water or the sea, but sometimes a lake) into the breeding grounds, usually a river. Rapidly moving water over gravel is necessary for successful spawning and hatching. The eggs are adhesive and demersal and hatch after about 3–5 days. The 1-cm larvae are swept downstream to rearing areas in the river or sea. Sturgeon development is very slow. It may take 10 or more years for medium-sized species to become sexually mature.

Sturgeons feed on benthic invertebrates and fishes detected with the help of external taste buds and their fleshy barbels. Fishes become a more important item of diet as the sturgeon get larger. *Scaphirhynchus* use their flat snouts to dig up buried invertebrates, hence their name shovelnose sturgeons. Scott and Crossman (1973) provided details of life history and fisheries information for five North American species. Williamson *et al.* (1999) edited the proceedings of a symposium on the harvest, trade, and conservation of North American paddlefish and sturgeons.

Because of overfishing for flesh and caviar, dams that prevent upstream migration, habitat destruction, bycatch mortality, and pollution, sturgeon populations worldwide have suffered. Trautman (1981) described how huge numbers of lake sturgeon, *A. fulvescens*, from Lake Erie were caught before 1850. Their swim bladders were used for isinglass. The carcasses were discarded on the beach, fed to hogs, or stacked like logs and burned. Today this species is very rare in most of Lake Erie. Slow growth, late maturity, and highly prized roe make it difficult for sturgeon populations to recover despite their tremendous fecundity. Atlantic sturgeon, *A. oxyrinchus*, stocks have been reduced 20-fold since the nineteenth century, and despite a profligate spawning strategy that releases 0.4–2.0 million eggs per female, it is estimated that restoration to historical levels will require more than a century (Secor and Waldman, 1999). The cultivation of sturgeons in large hatcheries will probably keep the commercial species from becoming extinct, but the noncommercial species are highly threatened. For example, the U.S. Department of the Interior withdrew a proposal to list the Alabama sturgeon, *S. suttkusi*, as endangered because there was insufficient evidence to show the species was still extant (*Anonymous*, 1994). This is all the more tragic because it was only described as a separate species in 1991 (Williams and Clemer, 1991). Mayden and Kuhajda (1996) substantiated *S. suttkusi* as a valid species and added data from three recently captured Alabama River specimens. *Scaphirhynchus suttkusi* is once again proposed and may be listed soon (D. M. Carlson, personal communication).

The osteology and phylogeny of sturgeons has been studied by Findeis (1997) and Bemis *et al.* (1997). These are 2 of 34 papers on sturgeon biodiversity and conservation assembled in book form by Birstein *et al.* (1997). This book should be the first reference consulted for sturgeon information, including life histories of North American and Eurasian species. Hilton and Bemis (1999) presented a detailed osteological study of 13 individuals of *Acipenser brevirostrum*, the shortnose sturgeon, from a single population in the Connecticut River.

Map references: Bemis and Kynard (1997),* Page and Burr (1991),* Lee *et al.* (1980),* Maitland (1977),* Rostlund (1952),* Scott and Crossman (1973),* Wei *et al.* (1997)*

Class Actinopterygii
Subclass Chondrostei

Order Acipenseriformes; Suborder Acipenseroidei
(1st) Family Polyodontidae—**paddlefishes** (pol'-i-ō-don'-ti-dē)

THERE ARE TWO EXTANT SPECIES IN THIS ODD, PRIMITIVE family with a disjunct distribution: *Polyodon spathula*, the American

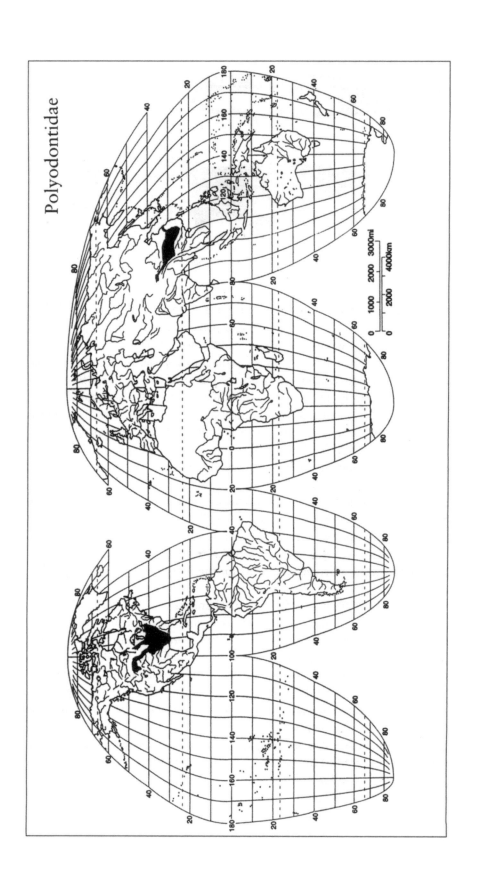

Polyodontidae

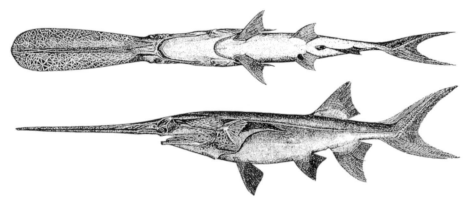

FIGURE 19. *Polyodon spathula* as *P. spatula*, ventral and side view (Goode, 1884, Plate 242).

paddlefish from the Mississippi drainage (Fig. 19), and *Psephurus gladius*, the Chinese paddlefish from the Yangtze River and adjacent areas.

The most distinctive feature of paddlefish is their very long rostrum. Like sturgeons, they have a mostly cartilaginous skeleton. Adult *Polyodon* tails are either homocercal or nearly so (Grande and Bemis, 1991). Two very small barbels are present, and there is an intestinal spiral valve. The gill flaps are elongated posteriorly and cover a spiracle. The skin is naked except for small patches of scales on the caudal peduncle and tail. The eyes are small. Gardiner (1984) gave technical details of the skeletal system that define the family, and many of these details were revised and/or corrected by Grande and Bemis (1991). Grande and Bemis (1991, 1996) and Bemis *et al.* (1997) discussed phylogenetic relationships and provided well-illustrated anatomical details of both extant species and several extinct species. Bemis and Grande (1999) studied development of the median fins of North American paddlefish and found little support for developmental or phylogenetic ideas for the nineteenth-century lateral fin-fold hypothesis.

The function of the paddle is not well understood, but it is covered with many sensory and electroreceptors that aid in detecting zooplankton (Russell *et al.*, 1999). The American species' snout is broad and flat, and it becomes shorter, narrower, and straighter as the fish grows (Hoover *et al.*, 2000), whereas the Chinese species has a sword-like snout. The rostrum may also help the hydrodynamics of the fish swimming with its large mouth open. The North American paddlefish is a continuous swimming, highly aerobic species that depends on ram ventilation (as opposed to buccal pumping) to obtain oxygen (Burggren and Bemis, 1992). Paddlefish that have lost part or all of the rostrum seem to suffer no ill effects (Parr, 1999). The gill rakers of *Polyodon* are long and numerous, and the mouth is nonprotrusible in keeping with its zooplankton and insect larvae diet. It strains water for food as it swims with its mouth wide open. *Psephurus* has fewer and shorter gill rakers and a protrusible mouth. It feeds on fishes.

Polyodon reaches 2 m and 76 kg, whereas *Psephurus* is said to reach 7 m (Nichols, 1943) and 500 kg (Wei *et al.*, 1997), but very little is known about the life history the Chinese species. *Psephurus* is the largest and most endangered freshwater fish in China. See Mims *et al.* (1994) for further conservation information about the Chinese paddlefish.

The Chinese paddlefish is now very rare in the Yangtze River system, and its spawning was severely impaired by the completion of the Gezhouba Dam in 1981. Since 1988 only 3–10 adults have been found below the dam annually. Some spawning takes place above the dam, but the new Three Gorges Dam will further damage paddlefish when it is completed. *Psephurus gladius* has not been bred in captivity, but artificial propagation may be the only hope of saving this species (Wei *et al.*, 1997). In addition to paddlefish, the Yangtze River shares several interesting aquatic animal lineages with eastern North America, including the Chinese sucker (*Myxocyprinus asiaticus*), giant salamanders of the family Cryptobranchidae, soft-shelled turtles (*Trionyx*), and the Chinese alligator (*Alligator sinensis*). The biogeographic connection between China and North America is also apparent in many extinct fish taxa (Grande and Bemis, 1991).

The life history and fisheries of *Polyodon* have been reviewed by Carlson and Bonislawsky (1981) and Dillard *et al.* (1986). Dillard *et al.* provided a cross-indexed bibliography of 555 references to *Polyodon* literature. Graham (1997) reported on the status of *Polyodon* in the various states. Paddlefish spawn over gravel in large rivers during times of increased flow when water temperatures are approximately 15°C. Paddlefish grow rapidly and live a long time—up to 30 years. In Missouri females reach maturity in 7–14 years when they are about 1.4 m and 16 kg. Males are mature in 7–9 years at about 1.3 m and 9 kg. It is thought that females spawn every 2 years. Epifanio *et al.* (1996) found that various North American stocks were sufficiently distinct to merit differential management considerations. Hoover *et al.* (2000) suggested that multiple ecophenotypes existed in populations from the Mississippi delta.

Since paddlefish are filter feeders, they are not caught by the usual angling methods. They are usually snagged with large treble hooks. This is made easier by their habit of congregating in large numbers in spawning areas and, to a lesser extent, in slow-moving waters where they feed. In 1899 a commercial harvest of 1105 metric tons was taken. This was mainly for roe and smoked flesh that served as a substitute for sturgeon. The average harvest from 1965 to 1975 was only 260 metric tons (Carlson and Bonislawsky, 1981). This decline reflects the environmental hazards faced by paddlefish. Dam building and reservoir construction, dredging, and flood control modifications on large rivers prevent access to spawning sites or destroy or permanently flood spawning habitat. Paddlefish are cultured and released by state agencies. References on the harvest, trade, and conservation of North American paddlefish are given in Williamson *et al.* (1999).

Grande and Bemis (1991) determined that fossils of this family dated from the Upper Cretaceous of Montana, and that no fossils were known from outside North America. Since that publication, a well-preserved primitive paddlefish, *Protopsephurus liui*, was described from the Lower Cretaceous (L. Grande, personal communication) of China (Lu, 1994). Grande and Bemis (1991) described the osteology of *Psephurus* and made the interesting observation that reproductively active adult paddlefish lack ossification of the scapulocoroid, vertebral column, and neurocranium. These elements eventually ossify in very large, older individuals. They remark, "In this sense, paddlefish appear to be the axolotls of the fish world, offering one of the best documented cases of paedomorphosis (or neoteny) known to date."

Map references: Burr (1980),* Carlson and Bonislawsky (1981),* Page and Burr (1991),* Rostlund (1952),* Wei *et al.* (1997)*

Class Actinopterygii
Subclass Neopterygii

Order Lepisosteiformes or Semionotiformes
(2nd) Family Lepisosteidae—**gars**
(lep'-i-sos-tē'-i-dē) [lep'-i-sos'-tē-i-dē)

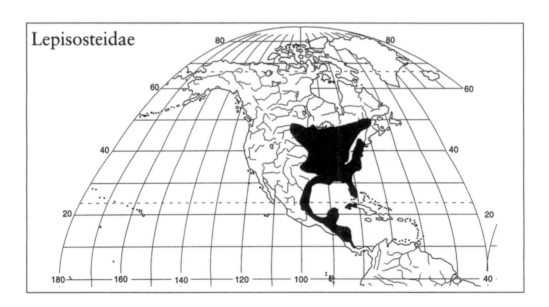

Lepisosteidae

IN THE SUBCLASS NEOPTERYGII, THE NUMBER OF DORSAL AND anal fin rays and the number of their supports are equal. It is thought that neopterygian fishes form a monophyletic group, but there is considerable debate about basal relationships (Nelson, 1994). Following Olsen and McCune (1991), Nelson resurrected the order Semionotiformes, whereas Eschmeyer (1998) used Lepisosteiformes.

Extant gars consist of four species of *Lepisosteus* and three species of *Atractosteus*. They occupy fresh waters in eastern North America, Central America south to Costa Rica, and Cuba. Gars are occasionally taken in brackish and marine waters. The longnose gar, *L. osseus*, is the northern-most species reaching southern Quebec (Scott and Crossman, 1973; Wiley, 1980), whereas the southern limit of the family is reached by the tropical gar, *A. tropicus*, in Costa Rica (Wiley, 1976). This species is the only gar in Pacific coast drainages (from southern Mexico to Honduras). The Cuban gar, *A. tristoechus*, occurs in western Cuba and the Isle of Pines (Wiley, 1976; Burgess, 1983) and enters salt water. Fossil gars date back to the Cretaceous of western North America, South America, Europe, Africa, and India (Wiley, 1976; Wiley and Schultze, 1984). Patterson (1973) and Gardiner *et al.* (1996) maintained that gars are the sister group of a clade containing bowfins plus teleosts (Halecostomi), but the molecular data unite *Lepisosteus* and *Amia* as a clade (Holostei) (Normark *et al.*, 1991). Patterson (1994) considered this problem unresolved.

Gars are primitive, elongate, heavily armored fishes with a long, well-toothed snout. The extended snout represents the elongation of the ethmoid. Their cylindrical body is covered with heavy, diamond-shaped, ganoid scales with some overlap. (The generic name *Lepisosteus* refers to "scale bone.") Their tail is abbreviated heterocercal, and the dorsal and anal fins are situated far posteriorly. Most dorsal and anal fin rays rest on their own bony support. This is an advance over the arrangement found in sturgeons and paddlefishes in which each fin ray support bears many fin rays. The physostomous swim bladder of a gar is highly vascularized and can be used to gulp air at the surface (Graham, 1997). A spiral valve is present. Gar vertebrae are opisthocoelous. This condition is a ball and socket arrangement in which the anterior end of a vertebra is convex and the posterior end is concave. It is common in reptiles but rare in fishes, which usually have biconcave vertebrae (amphicoelous).

Gars are highly predaceous and feed on fishes which they ambush by a sideways slash of their snout. Their large eggs are thought to be toxic to mammals and birds (Scott and Crossman, 1973). The eggs are adhesive and attach to aquatic vegetation. Fertilization is external, and no parental care is given (Suttkus, 1963). The young hatchlings have an adhesive pad on their snout and cling to vegetation. Growth is very rapid.

Preferred habitat of gars includes slow-moving, weedy waters in rivers, larger streams, lakes, bayous, estuaries, and coastal marine waters (Suttkus,

1963). The alligator gar, *A. spatula*, is the largest member of the family at 3 m and 137 kg. It is not known to be dangerous to humans. The shortnose gar, *L. platostomus*, has the shortest, broadest snout (relative to other gars) and is the smallest gar at 83 cm TL (Fig. 20).

FIGURE 20. *Lepisosteus platostomus* as *Lepidosteus platystomus* (Goode, 1884, Plate 241).

Map references: Burgess (1983a),* Lee *et al.* (1980),* Miller (1966),* Page and Burr (1991),* Rostlund (1952),* Scott and Crossman (1973),* Suttkus (1963), Wiley (1976)*

Class Actinopterygii
Subclass Neopterygii

Order Amiiformes
(1st) Family Amiidae—**bowfin**   (a-mī'-i-dē) [ā'-mē-i-dē]

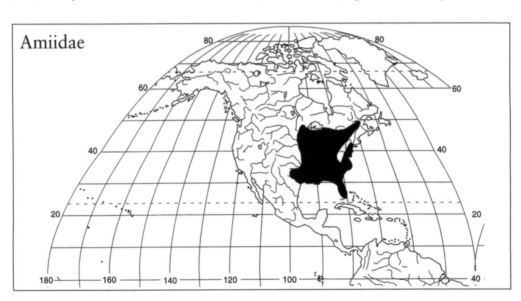

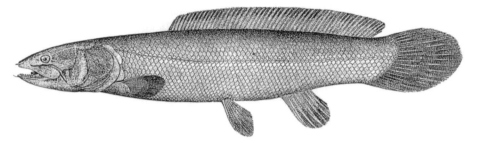

FIGURE 21. *Amia calva* (Goode, 1884, Plate 240).

THIS PRIMITIVE FAMILY IS REPRESENTED BY ONLY ONE LIVING species, the bowfin *Amia calva* (Fig. 21), which is found in lowland fresh waters of the eastern half of North America from the Great Lakes south to the Gulf of Mexico (Burgess and Gilbert, 1980). The fossil record of the family is much more geographically extensive, with specimens from North America, Brazil, Europe, Asia, Saudi Arabia, and Africa (Nelson, 1994; Schultze and Wiley, 1984; Grande and Bemis, 1998). Some of these specimens are from marine sediments. A very large fossil amiid from the Eocene of Mali, west Africa, was described and estimated to reach about 3 m (Patterson and Longbottom, 1989). The oldest fossils date back to the Jurassic. Grande (1996) and Grande and Bemis (1998) discussed how the fossil taxa relate to living *Amia calva*. Grande and Bemis (1998) investigated the comparative osteology, phylogenetic relationships, and historical biogeography of all known taxa of fossil and living amiid fishes. Their monumental work was followed with a paper on the historical biogeography and paleoecology of the Amiidae (Grande and Bemis, 1999).

Patterson (1973) concluded that *Amia* is more closely related to teleosts than to *Lepisosteus* based on a study of jaw structure. A recent parsimony analysis of morphological characters came to the same conclusion; however, molecular data indicate that *Amia* and *Lepisosteus* form a monophyletic holostean clade (Gardiner *et al.*, 1996) (Fig. 1).

Members of the subfamily Amiinae (common name = bowfins) have a long, cylindrical body with an abbreviated heterocercal tail, even though the caudal fin is rounded. Their skeleton is bony rather than cartilaginous. The dorsal fin extends more than half the length of the back. This feature is responsible for the common name "bowfin." Each fin ray articulates with a bony basal element as in gars, unlike the condition in sturgeons and paddlefish. There is a bony gular plate on the underside of the head between the lower jaws. The body scales are cycloid, but with unusual parallel bony ridges. The head is scaleless. The partitioned swim bladder can be used as a lung, and a vestigial spiral valve is present. Nostrils are tubular. Young specimens and adult males usually have a distinct black spot with a yellow-orange halo near the base of the upper caudal fin rays. Adult females lack this spot. Females may reach a meter in length. Males are smaller. The largest reported specimen is 9.75 kg, the current world angling record, from South Carolina [International Game Fish Association (IGFA), 1993].

Bowfins are highly predaceous and usually associated with swampy, weedy, slow-flowing habitats. Their lung-like swim bladder enables them to survive in stagnant waters with little or no oxygen (Graham, 1997). This is especially important in summer when water temperature is high and oxygen content is low. Air breathing begins when water temperatures exceed 10°C. As water temperatures increase, so does the air breathing rate (Horn and Riggs, 1973). A bowfin was found alive in a damp, spherical chamber under hard mud in a dry swamp of the Savannah River (Neil, 1950), although McKenzie and Randall (1990) maintained that *Amia* cannot be induced to estivate.

Life history information is given by Reighard (1940), Breder and Rosen (1966), Scott and Crossman (1973), and Pflieger (1975). Spawning occurs in spring, and the male constructs a nest by removing vegetation from a circular area about 30–60 cm in diameter. The male guards the nest and fans the eggs with his pectoral fins. The eggs are adhesive and hatch in about 8–10 days. Like gars, the hatchlings have an adhesive organ on their snout. The young form compact schools and are aggressively guarded by the male. Larval bowfins resemble tadpoles from about 7 to 10 mm TL. At about 25 mm TL the larvae look remarkably like miniature placoderms (Mansueti and Hardy, 1967). Young bowfin feed on tiny crustaceans and insects until about 10 cm, when they switch to small fish. Adults prey voraciously on fish and crayfish. Sexual maturity is reached in 2 or 3 years. Bowfin live about 10 years in nature and up to 30 years in captivity.

Map references: Page and Burr (1991),* Burgess and Gilbert (1980),* Grande and Bemis (1998),* Rostlund (1952),* Scott and Crossman (1973)*

Class Actinopterygii
Subclass Neopterygii

Order Osteoglossiformes; Suborder Osteoglossoidei
(1st) Family Osteoglossidae—**bonytongues** (os'-tē-ō-glos'-i-dē)

THE OSTEOGLOSSIDS OR BONYTONGUES ARE CIRCUMTROPical in South America, Africa, Southeast Asia, southern New Guinea, and northern Australia. There are four genera and seven species in two

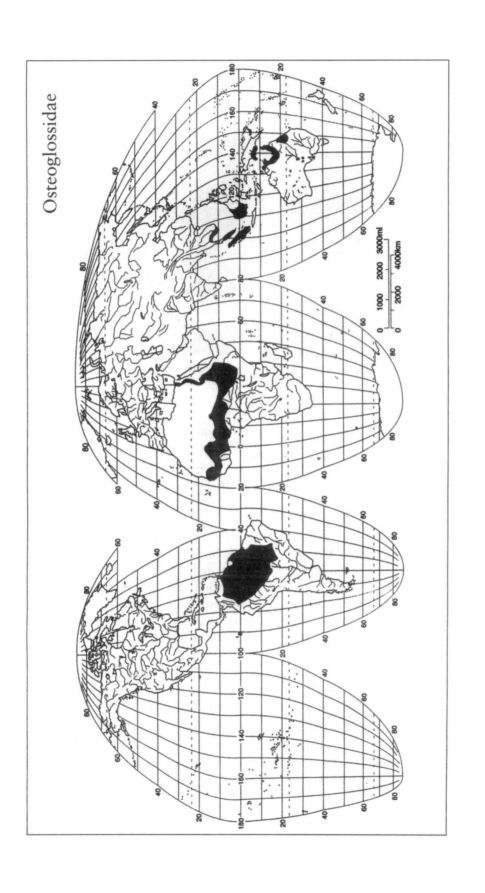

Osteoglossidae

FIGURE 22. *Arapaima gigas* (Lydekker, ca. 1903, p. 479).

subfamilies (Nelson, 1994). The Heterotidinae consists of the South American pirarucú, *Arapaima gigas* (Fig. 22), and the west African *Heterotis niloticus*. The African species occurs in savannah areas in the Corubal, Gambia, Senegal, Niger, Volta, Oueme, and Nile Rivers (Daget, 1984a; Paugy, 1990a). It has been introduced into the Congo River (D. Paugy, personal communication). The Osteoglossinae is composed of the South American arawanas, *Osteoglossum bicirrhosum* and *O. ferreirai*, as well as *Scleropages formosus* from Southeast Asia, *S. jardinii* from southern New Guinea and northern Australia, and *S. leichardti* from the Fitzroy River of Queensland, Australia. The South American osteoglossids are confined to the Amazon drainage, the western Orinoco, and the Rupununi and Essequibo systems of the Guianas (Goulding, 1980). The two species of *Scleropages* are considered to be the only true primary division freshwater fishes that cross the line between the Asian and Australian biogeographical realms. This Southern Hemisphere distribution, similar to the lungfishes, may reflect a previous Gondwanian pattern with subsequent breakup and drifting of the supercontinent. The earliest osteoglossid fossil is from the Early Cretaceous of Brazil (Maisey, 1996). Other fossils date back to freshwater deposits of the Eocene of Wyoming, Europe, Africa, and Australia. Marine dispersal has probably played a role in the distribution pattern of this family (Lundberg, 1993).

Osteoglossids are elongate (60–100 vertebrae), compressed fishes with large, heavy scales and long, posteriorly positioned dorsal and anal fins. The caudal fin is homocercal. The pelvic fins are well behind the pectoral fins. The shearing bite of osteoglossids is between the toothed tongue (basi-hyals and glossohyal) and the toothed bones of the roof of the mouth (parasphenoid) rather than between the lower and upper jaws as in most fishes. Aerial respiration is possible with their lung-like swim bladder.

Bonytongues are primitive teleosts. They are placed in the division Teleostei along with all subsequent fishes discussed in this book (Nelson, 1994). This represents an advance over previously discussed families such as lungfishes, sturgeons, and bowfins. Patterson (1968) defined teleosts (= "end bone") based on the caudal skeleton. Elongate uroneurals, modified from neural arches, support the homocercal caudal fin. The teleosts are the most successful, diverse group of all vertebrates. They comprise about 96% of all extant fish species (Nelson, 1994). De Pinna (1996a) reviewed the literature on the history of teleostean classification and defined the Teleostei as the most inclusive actinopterygian group excluding *Amia* and its relatives (Halecomorphi) and *Lepisosteus* and its relatives (Ginglymodi). De Pinna concluded that teleosts are monophyletic and recent members are composed of osteoglossomorphs, elopomorphs, clupeomorphs, and euteleosts) (Fig. 1). Basal teleosts and teleostean phylogeny have been debated by Patterson (1998) and Arratia (1998), and Arratia (1999) discussed the monophyly of teleosts. Li and Wilson (1996) dealt with the phylogeny of the Osteoglossomorpha.

Arapaima gigas is one of the largest freshwater fishes and may reach 4.5 m and 200 kg. It is an obligatory air breather using a highly vascular-ized and subdivided physostomous swim bladder (Graham, 1997). It is a powerful predator and highly regarded as a food fish by indigenous peoples along the Amazon River who hunt it with harpoons. *Osteoglossum bicir-rhosum*, the silver arawana, is sympatric with the arapaima, and reaches 1 m. It is an important food fish taken by bow-and-arrow hunters. Small specimens are also popular aquarium fish. It feeds on fishes and inverte-brates as well as animals such as frogs that fall into the water from over-hanging rain forest vegetation. It has even been known to snatch birds, snakes (Goulding, 1980), and bats (Lowe-McConnell, 1964) from over-hanging branches by leaping from the water. It has two chin barbels and an upturned mouth which help it feed at the surface. *Scleropages leichardti* is also a surface predator with a strongly upturned mouth and chin barbels. It has been known to leap onto the bank to catch insects (Merrick and Schmida, 1984). Osteoglossids make spectacular aquarium specimens, but leaping is a problem. *Heterotis niloticus* is the only osteoglossid that is not predatory (Fig. 23). This species, which reaches 1 m, has a modi-fied fourth gill arch that functions as a spiral-shaped filtering apparatus (Lowe-McConnell, 1987; Moyle and Cech, 1996). Phytoplankton and detritus become trapped in mucus secreted by the helical organ and are then swallowed.

Arapaima and *Heterotis* are nest builders and brood the young after hatching. Juvenile *Heterotis* have excessively long gill filaments that project

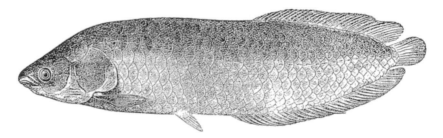

FIGURE 23. *Heterotis niloticus* (Boulenger, 1909, Fig. 121).

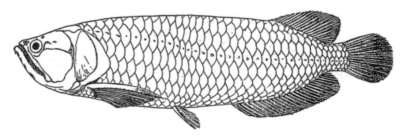

FIGURE 24. *Scleropages leichardti* as *Scleropages leichhardti* (Stead, 1906, Fig. 8).

externally. *Scleropages* and *Osteoglossum* are buccal incubators. *Osteoglossum bicirrhosum* has been found with about 200 eggs in its mouth (Goulding, 1980). *Scleropages leichardti* females carry the eggs and newly hatched young with yolk sacs in their mouth (Lake, 1971; Merrick and Green, 1982) (Fig. 24). Hatching may take 14 days. An incubating female has a conspicuously white chin (Merrick and Schmida, 1984). Young may enter and leave the female's mouth for a few days for brief excursions before they become independent. *Scleropages leichardti* is named for the explorer Ludwig Leichhardt (1811–1848?). The describer, A. Günther, mispelled Leichhardt's name with only one "h." Thus, *S. leichardti* is the accepted spelling of the fish species (Berra, 1989).

Map references: Allen (1989, 1991),* Daget (1984a), Darlington (1957), Norman and Greenwood (1975),*, Paugy (1990a)*, Poll (1973),* Roberts (1989)*

Class Actinopterygii
Subclass Neopterygii

Order Osteoglossiformes; Suborder Osteoglossoidei
(1st) Family Pantodontidae—**butterflyfish**   (pan'-tō-don'-ti-dē)

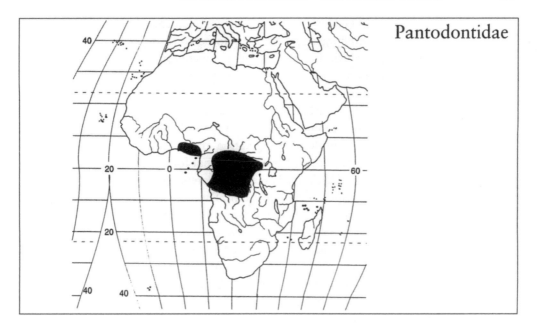

Pantodontidae

THE PANTODONTIDAE IS MONOTYPIC AND RESTRICTED TO the sluggish, weedy backwaters of tropical west Africa, in the lower Niger and the Congo (Zaire) River basins (Gosse, 1984c). The upper Zambezi record mentioned by Gosse (1984c) is doubtful (G. Tuegels, personal communication). *Pantodon buchholzi* may reach 13 cm and can leap a meter or more out of the water, gliding with the help of enlarged pectoral fins (Fig. 25). Large and complicated pectoral muscles are supported on an expanded cleithra. Greenwood and Thompson (1960) discussed its anatomy and flight potential. *Pantodon* is able to move its large pectoral fins during gliding, hence the name butterflyfish. This flapping generates an audible

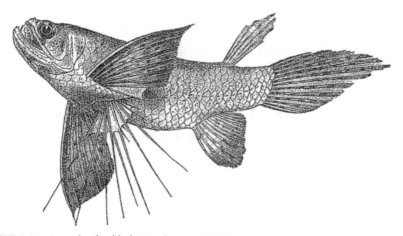

FIGURE 25. *Pantodon buchholzi* (Boulenger, 1909, Fig. 122).

buzzing sound (G. Dingerkus, personal communication). Its gliding or flying has not been well studied and needs to be confirmed. Its pelvic fins have elongate rays which may serve a tactile function in addition to functioning as stilts. Kershaw (1970) discussed the cranial osteology of butterflyfish. Li and Wilson (1996) consider *Pantodon* a member of the Osteoglossidae.

Pantodon is a surface-feeding insect eater and has a large upturned mouth. It is well camouflaged and has the habit of hanging motionless below the surface in floating vegetation and then darting or jumping very fast to capture prey. It lays floating eggs that form a raft at the surface, and the fry float after hatching. There is no parental care of eggs or young (G. Dingerkus, personal communication). The anal fin of males shows some sexual dimorphism, but little is known about the biology of this species. Breder and Rosen (1966) reviewed references to its reproduction. *Pantodon* is capable of breathing air via its large and highly vascularized swim bladder (Greenwood, 1994a). Extensions of the swim bladder penetrate the transverse processes of the vertebrae. Air breathing is the dominant respiratory mode of this species (Graham, 1997). See Teugels (1990) and Gosse (1984c) for additional references.

Map references: Gosse (1984c), Poll (1973),* Teugels (1990)*

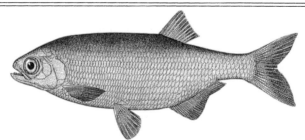

Class Actinopterygii

Subclass Neopterygii

FIGURE 26 *Hiodon tergisus* as *Hyodon tergisus* (Goode, 1884, Plate 219).

Order Osteoglossiformes; Suborder Osteoglossoidei

(1st) Family Hiodontidae—mooneye and goldeye

(hī-ō-don'-ti-dē)

THERE ARE ONLY TWO LIVING SPECIES IN THIS EXCLUSIVELY North American family. The goldeye, *Hiodon alosoides*, is found in the lower Mississippi River basin from Louisiana and Mississippi north to the Ohio River basin and throughout the Great Plains, almost to the mouth of the MacKenzie River into extreme northwestern Canada. There are widely disjunct populations in James Bay tributaries of Ontario and Quebec. It is absent from the Great Lakes drainages. The mooneye, *Hiodon tergisus* (Fig. 26), occurs from St. Lawrence–Great Lakes (except Lake Superior),

Mississippi River, and Hudson Bay basins from Quebec to Alabama south to the Gulf of Mexico and from the Mobile Bay drainage west to the Pearl River (Gilbert, 1980a; Page and Burr, 1991). This is the only osteoglossiform family not found in the tropics (Greenwood, 1994a).

Li and Wilson (1996) and Li *et al.* (1997) treated the Hiodontidae as an order, Hiodontiformes, which they considered a sister group to the Osteoglossiformes. Li and Wilson (1999) presented fossil evidence that suggested that the early divergence and radiation of their Hiodontiformes (*sensu stricto*) occurred in the Lower Cretaceous of northeastern China. Fossil hiodontids from the Eocene occur in North America (Grande, 1979). *Lycoptera*, the world's oldest osteoglossomorph, from the Early Cretaceous (not Late Jurassic as earlier reported) of China, is similar to the mooneye but probably represents a primitive osteoglossomorph unrelated to any living group (Maisey, 1996; Li and Wilson, 1999).

These silvery to golden fishes are strongly compressed with large eyes, and they resemble clupeids. However, unlike shads and herrings, the keel along the belly of *Hiodon* is untoothed and a lateral line is present. The caudal fin is deeply forked, and the anal fin is moderately long. The head is naked, and cycloid scales cover the body. The swim bladder is connected to the skull. An adipose eyelid and an axillary process are present. The keel of *H. tergisus* does not extend anterior to the pelvic fins, and 11 or 12 principal dorsal fin rays are present. Its dorsal fin originates anterior to its anal fin. The keel of *H. alosoides* extends anterior to the pelvic fins and 9 or 10 dorsal fin rays are present. Its dorsal fin originates above the anal fin.

These fishes feed on aquatic insects, other invertebrates, and small fishes. Their eggs are semibouyant and float downstream into quiet waters. This is an uncommon reproductive strategy among North American fishes

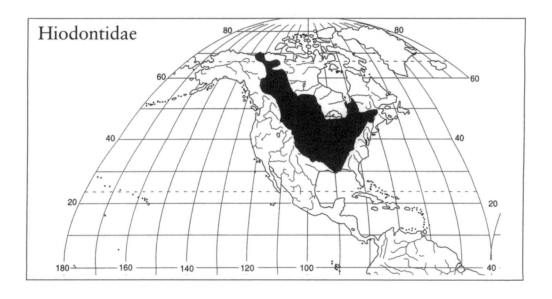

Hiodontidae

(Scott and Crossman, 1973). They prefer deep pools in medium to large rivers and lakes. Goldeye tolerate turbid waters, feed mostly at night, and have only rods but no cones in their retinas. This may be an adaptation to their turbid, low-light habitat. Mooneye are found in clearer rivers and streams. Both species have a golden or gold-silver eye shine due to reflection from the tapetum lucidum. They may reach 51 cm. When smoked, the goldeye is of limited commercial importance in Canada. See Scott and Crossman (1973) and Trautman (1981) for life history information.

Map reference: Gilbert (1980a),* Page and Burr (1991),* Rostlund (1952),* Scott and Crossman (1973)*

Class Actinopterygii
Subclass Neopterygii

Order Osteoglossiformes; Suborder Osteoglossoidei
(1st) Family Notopteridae—featherbacks   (nō-top-ter'-i-dē)

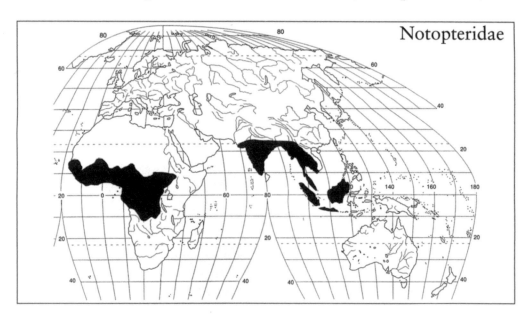

THE FEATHERBACKS, SO CALLED BECAUSE OF THE SINGLE, feather-like dorsal fin, are found in fresh and, occasionally, brackish waters from west and central Africa to India and in Southeast Asia. There are four genera and eight species (Roberts, 1992a). Two species of *Papyrocranus* occur in west Africa mainly from Senegal to Nigeria and the Congo basin, and *Xenomystus nigri* (Fig. 27) is found in the upper Nile, Chad, Niger, and

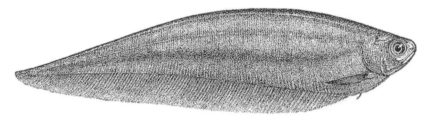

FIGURE 27. *Xenomystus nigri* (Boulenger, 1909, Fig. 120).

Congo basins in tropical Africa (Daget, 1984b). Four species of *Chitala* range from Pakistan and India to Sumatra and Borneo, and *Notopterus notopterus* occurs from India to Sumatra and Java.

Xenomystus lacks a dorsal fin, and all featherbacks have a long flowing anal fin that joins the reduced caudal fin and provides forward and backward propulsion by wave-like rippling movements. Pelvic fins are rudimentary and missing completely in *Papyrocranus*. The scales are very small, with 120–180 in the lateral line series. The body of notopterids is not bent during swimming, probably because the long, extensively branched swim bladder extends from the head to the posterior end of the body cavity. Finger-like projections from the swim bladder extend downward between the internal supports of the anal fin (Greenwood, 1994a). The swim bladder functions as a lung and is also a sound-producing organ (Greenwood, 1963; Graham, 1997). Some featherbacks live in oxygen-poor swampy habitats. *Chitala chitala* (Fig. 28) can reach 1.5 m, whereas the smallest notopterid, *X. nigri*, only reaches 15 cm. *Papyrocranus afer* is intermediate at about 62 cm TL (Fig. 29).

Notopterus is carnivorous and breeds in stagnant or running water in the rainy season. The female deposits eggs in small clumps on submerged vegetation which may be guarded by the males (Talwar and Jhingran, 1992). Little is known of the biology of notopterids. *Chitala chitala* is a valuable food fish and may be transported to market in water-filled rice barges (Smith, 1945). Some notopterids have found their way into the aquarium trade as knifefishes. A small population of *Chitala ornata* was

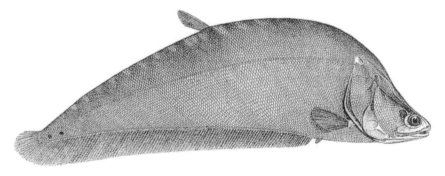

FIGURE 28. *Chitala chitala* as *Notopterus chitala* (Day, 1878b, Plate CLIX, Fig. 5).

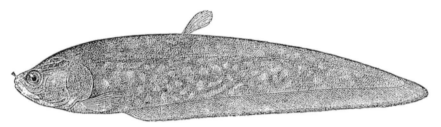

FIGURE 29.   *Papyrocranus afer* as *Notopterus afer* (Boulenger, 1909, Fig. 119).

recently discovered in Palm Beach County, Florida. Its occurrence there is probably due to an aquarium release (Fuller *et al.*, 1999).

Another African family, Gymnarchidae, superficially resembles featherbacks. However, gymnarchids have electric generation and sense organs and utilize their long dorsal fin (versus a long anal fin in notopterids) for propulsion. Li and Wilson (1996) and Li *et al.* (1997) united the Notopteridae, Mormyridae, and Gymnarchidae in the suborder Notopteroidei of the Osteoglossiformes.

Map references: Daget (1984b), Kottelat *et al.* (1993), Nelson (1976),* Poll (1973),* Roberts (1989, 1992a)

Class Actinopterygii
Subclass Neopterygii

Order Osteoglossiformes; Suborder Osteoglossoidei
(1st) Family Mormyridae—elephantfishes
(mor-mir'-i-dē) [mōr-mī'-ri-dē]

THE MORMYRIDS OCCUR IN THE TROPICAL FRESHWATER rivers, swamps, and lakes of Africa including the Nile River and the warmer parts of South Africa. According to Boden *et al.* (1997) there are 18 genera and 195 species. This number included 189 species listed by Gosse (1984d) (erroneously reported as 198 in the checklist). In addition to some taxonomic rearrangements, 8 new species have been described since Gosse's checklist; thus, there are 195 species total (Boden *et al.*, 1997). R. Bigorne (personal communication) considered that taxonomic rearrangements have reduced the total species number to 180. Lavoué *et al.* (2000) found the family to be monophyletic.

All mormyrids have a narrow caudal peduncle and a deeply forked caudal fin, but the family is best known for the diversity of peculiar

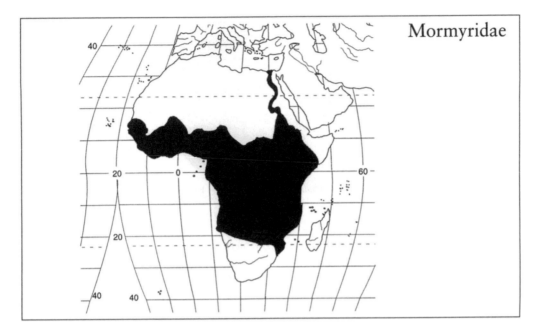

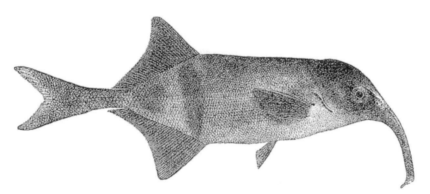

FIGURE 30. *Campylomormyrus numenius* as *Gnathonemus ibis* (Boulenger, 1909, Fig. 102).

snouts among its species that gives rise to the name elephantfishes. *Campylomormyrus* species, such as *C. numenius* (Fig. 30), and many *Mormyrus* have a downcurved, elongated proboscis with a tiny terminal mouth at the end of this snout. *Campylomormyrus tamandua, C. elephas,* and *Mormyrus tapirus* are named after mammals with a trunk-like snout (tamandua, elephant, and tapir). *Mormyrus caballus* has a more horse-like snout as the specific name implies (Fig. 31). The genus *Gnathonemus,* of which *G. petersii* is a prominent example, has a downcurved elongate lower jaw with a terminal mouth above a fleshy "chin" more properly called a submental swelling. The genus *Petrocephalus,* represented by *P. ballayi,* has a bluntly rounded snout with an inferior mouth (Fig. 32). This genus is the sister group of all other mormyrids (Lavoué *et al.,* 2000). In the pet trade these fishes are often called "whales" because they lack the snout of other

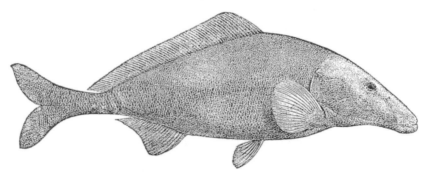

FIGURE 31. *Mormyrus caballus* (Boulenger, 1909, Fig. 110).

elephantfishes. Other genera may have a chin with globular swellings inter-mediate between those of the snouted species and *Petrocephalus*. See Bigorne (1990a) and Skelton (1993) for drawings of the diversity of mormyrid shapes and Gosse (1984d), Poll and Gosse (1995), Moller (1995), and Boden *et al.* (1997) for a listing of species. Body shape may be deep and compressed in some genera or elongate in others such as *Isichthys* and some species of *Mormyrops*, such as *M. boulengeri* (Fig. 33). Dorsal and anal fins are long and well developed in most species, but *Hyperopisus* has a tiny dorsal fin. *Mormyrus* has a very small anal fin. Most mormyrids have small eyes (except *Petrocephalus*) generally covered by skin.

Mormyrids are also known for their ability to produce an electrical current from modified muscle tissue located in the caudal peduncle. The constant, weak pulses produced by the mormyid's electrical system create an electric field around the fish and function like radar in the detection of obstacles, food, and mates in the turbid waters which they inhabit. The electric signals may also be used to communicate with other fish for spatial and social cohesion (Moller and Serrier, 1986; Moller *et al.*, 1989), courtship

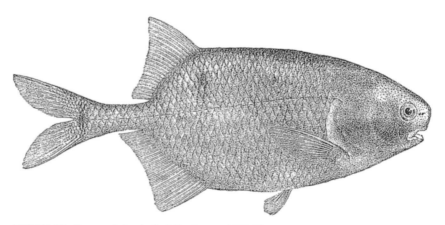

FIGURE 32. *Petrocephalus ballayi* (Boulenger, 1909, Fig. 37).

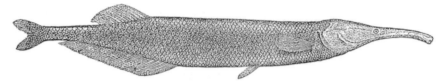

FIGURE 33. *Mormyrops boulengeri* (Boulenger, 1909, Fig. 27).

(Bratton and Kramer, 1989), and aggression (Hopkins, 1974, 1999; Kramer, 1974, 1978; Xu-Friedman and Hopkins, 1999). Electroreceptors on superior and inferior parts of the body, but mostly on the head, alert the fish when its surrounding electric field is interrupted. This electrical field is not an offensive shocking system such as is found in the electric eel and electric catfish. Mormyrids generate an electric field and perceive the resulting current via epidermal electroreceptors. Objects located nearby distort the electric field, thereby altering the current perceived through the electroreceptors. The fish can measure the distance of most objects accurately, independently of size, shape, or material (von der Emde *et al.*, 1998). Heiligenberg (1993), Moller (1995), and Kramer (1996) reviewed the biology and electrical properties of the Mormyridae and Gymnarchidae, and Turner *et al.* (1999) edited a collection of papers reviewing the current knowledge of the neuroethology of electroreception and electrogenesis. Alves-Gomes and Hopkins (1997), Alves-Gomes (1999), and Lavoué *et al.* (2000) discussed the phylogeny and systematic biology of the same two families. The electric signals are often species specific and can be used as taxonomic characters (Hopkins, 1981; Crawford and Hopkins, 1989). See Hopkins (1986) for a discussion of mormyrid behavior.

Mormyrids swim slowly and rather stiffly; presumably this avoids distortion of the electrical field. The presence of the electrical early warning system, the turbidity of the water, and nocturnal behavior probably explain why most mormyrids have small eyes. Mormyrids are absent or uncommon in highly alkaline waters. This may be because highly alkaline lakes also have high conductivity which places special demands on current-generating capacity of electric organs (Hopkins, 1999; Greenwood, 1994a).

The impulse-generating tissue is coupled with an unusually large cerebellum that presumably coordinates the system. Relative to body size, the mormyrid brain is as large as the human brain. Aquarium observations confirm that mormyrids are fast learners and seem to engage in play behavior by pushing small objects around with their heads. Mormyrids also have a large electrosensory lateral line lobe.

Another interesting characteristic of the mormyrids is the presence of a small balloon-like vesicle derived from the swim bladder that is in contact with the saccular otolith of the inner ear within the skull. The inner ear on each side of the head contains a gas-filled ear bladder linked directly to the sacculus. Pressure variations caused by underwater sound are translated to mechanical displacement of hair cells in the sacculus. The semicircular canals, which are concerned with equilibrium, are separated from the portion of the inner ear concerned with hearing. This too is unusual. The role of the swim bladder in sound amplification is discussed by Werns and Howland (1976), and Crawford (1997) reviewed mormyrid hearing and acoustic communication. Sound production by mormyrids consists of an elaborate repertoire of clicks, hoots, grunts, and moans used in courtship behavior (Crawford, 1997).

Chapman and Chapman (1998) reported that *Petrocephalus catostoma* swims inverted at the surface in response to severe hypoxia. This upside-down position places the subterminal mouth at the oxygen-rich air–water interface. Such aquatic surface respiration allows the fish to survive in low-oxygen swamps and wetlands where it is safe from the depredations of the introduced Nile perch, *Lates niloticus*, that requires well-oxygenated waters. Introduced *Lates* have decimated cichlids and mormyrids in Lake Victoria and elsewhere. The ability to utilize oxygen efficiently during hypoxia is thought to be important in protecting the large brain of mormyrids from hypoxial damage (Nilsson, 1996).

Some mormyrids feed on aquatic insects. Those with inferior mouths or downturned snouts feed on benthic invertebrates. The snout may be used to probe the bottom for food. Adult *Mormyrops anguilloides* is one of the few piscivorous species (Joubert, 1975; Blake, 1977; Skelton, 1993). Little is known about mormyrid reproduction. The anal fin of males of some species shows a marked concavity at its anterior end that, when opposed to the female's anal fin, forms a cup which may be used to hold eggs during fertilization (Iles, 1960; Breder and Rosen, 1966). Most mormyrid species are 9–50 cm, but some may reach 1.5 m. See Bigorne (1990a) for 20 references, mostly in French.

Map references: Bigorne (1990a),* Gosse (1984d), Skelton (1993)*

Class **Actinopterygii**
Subclass **Neopterygii**

Order **Osteoglossiformes; Suborder Osteoglossoidei**
(1st) Family **Gymnarchidae—aba**   (jim-nar'-ki-dē)

THIS MONOTYPIC FAMILY, RELATED TO THE MORMYRIDS, IS composed of *Gymnarchus niloticus*, known as the aba. It is distributed in a broad longitudinal band across tropical Africa in the Senegal, Gambia,

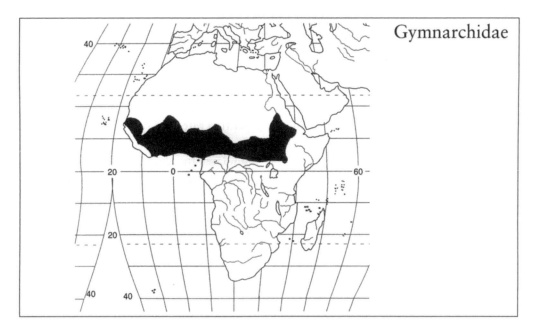

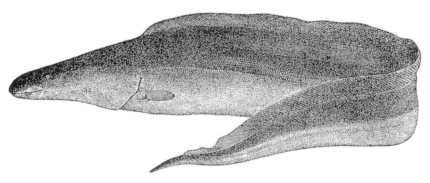

FIGURE 34. *Gymnarchus niloticus* (Boulenger, 1909, Fig. 118).

Volta, Niger, and Nile Rivers, Lake Chad basin, and Lake Rudolph (Bigorne, 1990b; Gosse, 1984e). Its latitudinal range is rather narrow between about 5 and 18°N (Greenwood, 1994a). *Gymnarchus niloticus* (Fig. 34) is the sister group to the family Mormyridae (Alves-Gomes and Hopkins, 1997; Lavoué *et al.*, 2000).

This compressed, elongate species has no anal, caudal or pelvic fins. Gentle undulations of the very long dorsal fin (183–230 rays) propel *Gymnarchus* forward or backward while the body remains rigid. Pectoral fins are present, and the tail ends in a finger-like tip. Despite its classification within the Osteoglossiformes, teeth are absent from the parasphenoid and tongue. Mok (1981) discussed the circulatory anatomy of *Gymnarchus*.

Like its mormyrid relatives, *Gymnarchus* generates a weak electrical field around its body from four elongate electric organs on each side of its tail region. Disturbances of this electrical field help the fish navigate in its turbid water habitat and detect prey, predators, mates, and obstacles. Unlike the mormyrids, however, *Gymnarchus* produces continual, wave-like discharges of between 200 and 500 Hz instead of pulse-type discharges. The duration, shape, frequency, repetition rate, and spectral content of electric organ discharges by weak electrical fishes may be species specific (Hopkins, 1988). *Gymnarchus* modulates the discharge by turning it off and by slight shifts in frequency. It also has a jamming avoidance response. The biology and electrophysiology of *Gymnarchus* are summarized by Moller (1995).

The breeding habits of *Gymnarchus* are better known than are those of the mormyrids. A large floating nest about 60 cm long and 30 cm wide is constructed of plant material in shallow swamps (Breder and Rosen, 1966). Part of the nest projects above the water. About 1000 large eggs, 4 mm in diameter, are deposited. Only the left gonad is developed and functional. The parents guard the nest for at least 3 or 4 days until the eggs hatch. The newly hatched young have long gill filaments and an elongate yolk-sac which anchors them to the bottom. Juveniles feed on insects and small invertebrates while the adults prey on fishes. Air breathing via a lung-like swim bladder begins about 10 days after hatching (Graham, 1997). Maximum size is about 1.6 m.

Map references: Bigorne (1990b),* Gosse (1984e), Roberts (1975)

Class Actinopterygii
Subclass Neopterygii

Order Anguilliformes; Suborder Anguilloidei
(Per) Family Anguillidae—**freshwater eels**   (ang-gwil'-i-dē)

THIS EEL FAMILY, COMPOSED OF THE GENUS *ANGUILLA* with 16–20 species (Smith, 1989a; J. McCosker, personal communication), is widespread but disjunctly distributed on all continents except Antarctica. The distribution of this family is largely due to suitably deep ocean spawning areas of proper temperature and salinity with favorable currents that can disperse the leptocephalus larvae to river systems on distant landmasses. In North America, eels (*A. rostrata*)

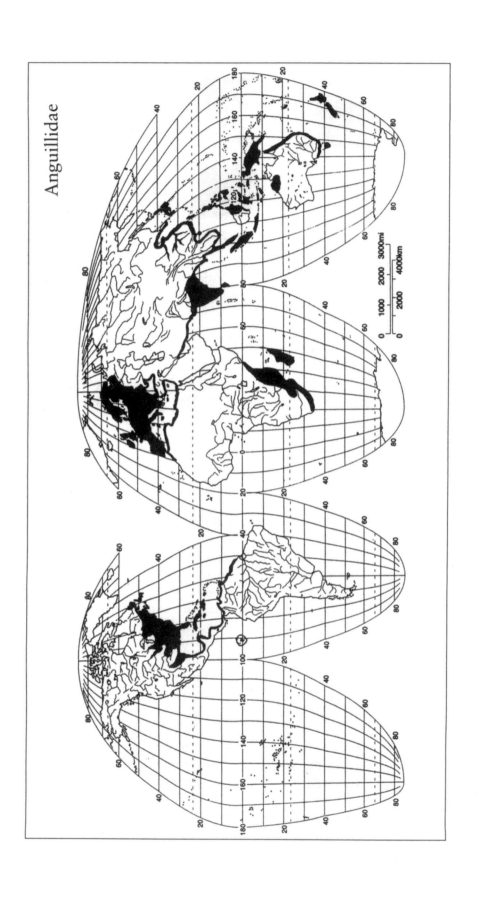

Anguillidae

FIGURE 35. *Anguilla rostrata* as *A. vulgaris* (Goode, 1884, Plate 239).

(Fig. 35) occur from Greenland to the Atlantic coast of Canada and the United States and along the Gulf Coast to the West Indies and Bermuda. They occur throughout the Mississippi drainage and reach as far upstream as Minnesota. The southern limit is not well-known, but American eels have been recorded from Panama, the Bahamas, and most of the West Indies to northern Brazil (Smith, 1989a; J. McCoster, personal communication). The European eel, *A. anguilla*, occurs in the Atlantic drainage from Norway to Morocco and throughout the Mediterranean basin into the Black Sea.

In the Indian Ocean eels occur from Somalia to Cape Agulhas in South Africa, Madagascar, the Mascarenes, Comoros, and Seychelles. Eels are absent from the Red Sea and Persian Gulf. Eels range across India southward through Indonesia, Sulawesi (= Celebes), eastern Borneo, and throughout Southeast Asia. In Australia, *A. australis* inhabits coastal rivers from the South Australian border to the Queensland border, and *A. reinhardtii* occurs in northern and eastern Tasmania and from rivers along the east coast of Cape York in the north to about Melbourne in the south. *Anguilla bicolor* occurs in northern Western Australia and *A. obscura* is found in coastal northeast Queensland. No eels occur in Gulf of Carpentaria drainages. Eels are also found in New Guinea, New Caledonia, and New Zealand. In the North Pacific, eels occur in the Philippines, China, Korea, and Japan. This family is absent from the eastern Pacific and south Atlantic.

Anguillid eels, of course, have an elongate body. Their dorsal and anal fins are very long and confluent with the caudal fin. The pectoral fins are well developed, and the pelvic fins are absent. The body is covered with tiny cycloid scales, but these scales are embedded in thick skin and are easily overlooked. The lateral line is complete. The snout is short with tubular anterior nostrils. The gill opening is crescent shaped. Anguillid eels form a monophyletic group characterized by many synapomorphies of which a high percentage are reductive (Forey *et al.*, 1996).

Anguillid eels have a catadromous life cycle. Eels from North America and Europe migrate from their respective river systems to the Sargasso Sea (between Bermuda and the West Indies). Spawning takes place at an unknown depth, and then the adults die. A leaf-like larvae, called a lepto-cephalus, hatches from the egg and drifts with the currents to a landmass with suitable rivers. In the case of the American eel, this journey takes 1 year. The European leptocephalus must drift for 3 years before reaching European rivers. The larva transforms while at sea into a small transparent "glass eel" about 40–50 mm. Then an elver (small miniature eel with pigment) develops and ascends fresh water. After 2 years the elver reaches about 100 mm and developes a yellowish-brown color. The males usually remain near the river mouth while the females ascend far upstream. This yellow phase may last 4–20 years. The adult male eel may reach 300–400 mm, whereas the females reach 1 m or more. Eventually, the eels stop feed-ing and become sexually mature. This initiates a second metamorphosis. The eels become silvery, and the eyes enlarge. Migration to the Sargasso Sea ensues. Adult eels have rarely been taken at sea. See Smith (1989a) for more details of eel biology.

The discovery of the eel's spawning grounds in the Sargasso Sea is one of the great detective stories in zoology. A Danish ichthyologist, Johannes Schmidt, patiently searched for the spawning grounds from 1904 to 1922 by plotting the distribution of the smallest leptocephali from trawl collec-tions that crisscrossed the Atlantic Ocean. The smallest leptocephali collected were 4 or 5 mm. The spawning grounds of the American and European eels were found to overlap in the Sargasso Sea.

It has been suggested that the American eel and the European eel are really the same species (Tucker, 1959). The differences between them are slight, with American eels having 103–111 vertebrae compared to 110–119 vertebrae in European eels. It was speculated that the longer migration time to Europe resulted in a greater vertebral count. Tucker's single species hypothesis was very clever and helpful as a stimulus to study eel biology. Today, biochemical and chromosomal studies make the case that the two eels are probably distinct species, each with its own spawning grounds and migration route. Williams and Koehn (1984) suggested that the two forms are subspecies that produce intergrades in northern Europe and Iceland. See Smith (1989a) for a complete review of the problem and Avise et al. (1986) for significant mitochondrial DNA differences between the populations.

Studies by Tsukamoto et al. (1998) indicated that Atlantic and Pacific eels taken from the ocean have spent their entire lifetimes in the sea and have never entered fresh water. This was determined by measuring the ratio of strontium to calcium in the otoliths of eels. Fresh water has little stron-tium, and the sea has a high concentration. All river eel samples show a high strontium level at the core of the otolith corresponding to the lepto-cephalus migration from spawning area to rivers. Specimens collected from the sea showed no evidence of having been in fresh water. Tsukamoto et al.

suggested that eels have a facultative catadromy with ocean residents as an ecophenotype. They suspect that genes are maintained between generations only by eels that grow in the sea and that marine eels alone contribute primarily to future recruitment.

Eels are of commercial importance as food fishes in many parts of the world. They are considered a delicacy in Europe and Japan. Small fisheries exist for export to these areas from North America, Australia, and New Zealand. Farm-raised eels are important in Japan.

McCosker (1989) reported that specimens of *A. anguilla*, *A. australis*, and *A. rostrata* have been taken in California. Freshwater eels are not native to the western United States and these records probably represent introductions to satisfy the increasing popularity of live anguillids in California restaurants and markets that cater to the Asian-American communities and young urban professionals. Because of their migratory requirements, it is unlikely that eels would reproduce in California, but they could live for 30 years and would almost certainly cause problems for the native fauna. For this reason, live anguillids are prohibited in California.

McDowall *et al.* (1998) reported transoceanic dispersal of *A. reinhardtii* from Australia to New Zealand and suggested that this is how much of the New Zealand freshwater fish fauna was derived. The same argument is applied to the Galaxiidae (Berra *et al.*, 1996).

Wallets, shoes, belts, and other leather goods labeled as "eel skin" are actually made from hagfishes, not anguillid eels. In addition to Smith's (1989a,b) modern review, Tesch's (1977), Moriarty's (1978), and Bertin's (1956) classic work should be consulted for further information.

Map references: Allen (1989),* Bauchot (1986),* Bertin (1956),* Lee *et al.* (1980),* McDowall (1990),* Maitland (1977).* Moriarty (1978),* Page and Burr (1991),* Scott and Crossman (1973),* Skelton (1993),* Tesch (1977)*

Class Actinopterygii
Subclass Neopterygii

Order Clupeiformes; Suborder Denticipitoidei
(1st) Family Denticipitidae—**denticle herring**
[den'-ti-ci-pit'-i-dē]

THE SUBDIVISION CLUPEOMORPHA CONTAINS THE ORDER OF herring-like fishes, the Clupeiformes. These are relatively primitive fishes

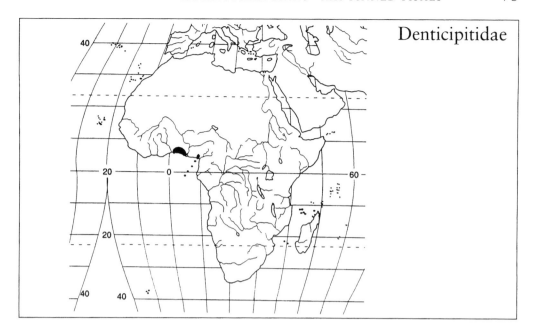

Denticipitidae

(Fig. 1). Ancestral features (plesiomorphies) such as cycloid scales, soft fin rays, and abdominal pelvics are present. Clupeomorphs are physostomous; a pneumatic duct connects the gut and swim bladder. There is a unique connection between paired anterior extensions of the swim bladder and the inner ear. The structure of the caudal fin skeleton helps define the group. Many species have a saw-toothed belly of abdominal scutes. The lepto-cephalus larvae present in eels and other elopomorph fishes is absent in the clupeomorphs. Whitehead (1985a) described this group, and Lecointre and Nelson (1996) reviewed the Clupeomorpha and suggested a relationship between clupeomorphs and ostariophysans. The Denticipitidae is the most primitive living family of this large group of fishes.

This monotypic family, composed of *Denticeps clupeoides* (Fig. 36) which was described in 1959, is known only from west African coastal

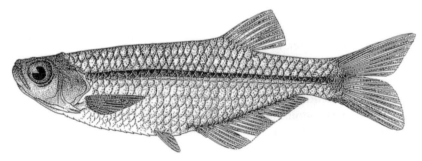

FIGURE 36. *Denticeps clupeoides* (reproduced with permission from Lévêque *et al.*, 1990, Fig. 9.2).

rivers from southeast Benin through Nigeria to southwest Cameroon (Clausen, 1959; Daget, 1984c; Teugels, 1990). *Denticeps* is the most primitive herring-like fish, and it is considered a living fossil (Greenwood, 1984b). Like herrings, it has a sawtooth, keeled belly (ventral scutes). It is distinguished from other herrings by the presence of odontodes (small tooth-like bodies) on the exposed surfaces of the skull roofing bones. Odontodes are also found on the jaw bones, opercular and infraorbital bones. The spacing of these small tooth-like projections imparts a furry appearance to the bone. The paired fins are small, and the anal fin is long. There are 16 principal caudal fin rays and one uroneural and a complete lateral line. Other clupeoids have 19 principal caudal fin rays, three uroneurals, and an incomplete lateral line. Greenwood (1968a,b) described the osteology and soft anatomy of this fish.

Little is known about the biology of *Denticeps*. Its gill rakers are few and widely spaced so it is probably not a plankton strainer like most clupeoids. Adults reach 6 cm in length. An extremely well-preserved fossil species, *Palaeodenticeps* from the Miocene of east Africa (Tanzania), is nearly identical to the living species (Greenwood, 1960). The 3500-km distance separating the fossil and extant species suggests a formerly more widespread distribution of the family (Greenwood, 1984b).

Map references: Daget (1984c), Roberts (1975), Teugels (1990)*

Class Actinopterygii
Subclass Neopterygii

Order Clupeiformes; Suborder Clupeoidei
(Per) Family Engraulidae—**anchovies**   (en-graw'-li-dē)

THE ANCHOVIES ARE A MARINE FAMILY WITH A FEW REPRE-sentatives in fresh water, mostly in South America. There are 16 genera and 139 species, of which about 17 are freshwater inhabitants (Nelson, 1994). Anchovies are usually small, silvery, schooling fishes with a huge mouth. The maxilla extends well behind the eye in most species. They swim through the water, mouth agape, and strain plankton with 90 or more gill rakers on the first gill arch. The snout projects beyond the lower jaw. The prominent snout contains a rostral organ, not visible externally, of unknown sensory function. The lateral line system is absent from the body, but it occurs on the head. Anchovies are delicate fishes, and their cycloid

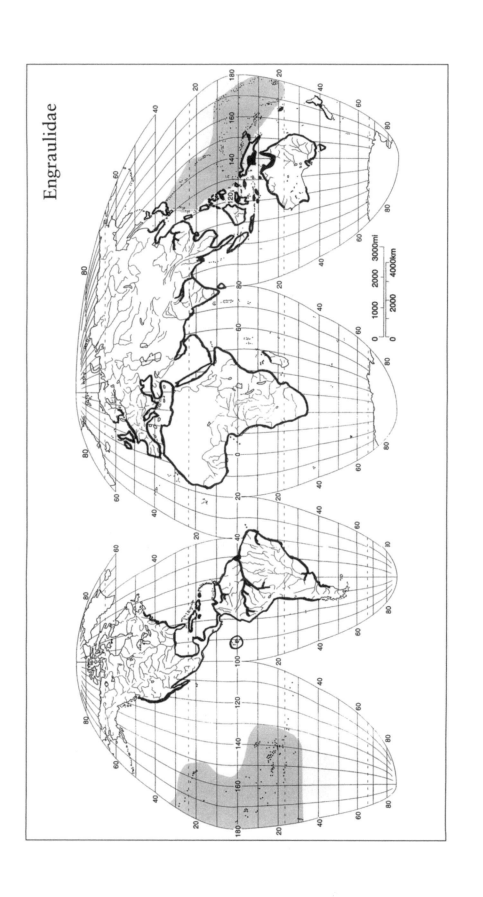

Engraulidae

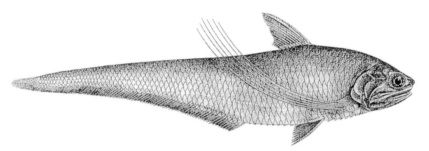

FIGURE 37. *Coilia ramcarati* (Day, 1878b, Plate CLIX, Fig. 2).

scales are easily shed (deciduous). This usually results in death. They are difficult to keep in captivity. There are two subfamilies, Coiliinae and Engraulinae (Nelson, 1994).

The Coiliinae has ventral scutes behind the pelvic fin and a relatively long anal fin. There are five Indo-West Pacific genera found in eastern Africa, Asia, and Australia (Nelson, 1994). Members of the genus *Coilia*, with 13 species, are called rat-tailed anchovies because of their very long tapering body that ends in a pointed tail continuous with the anal fin. *Coilia brachygnathus* and *C. nasus* are important in the commercial catch from the Yangtse River of China (G. Nelson, 1994). *Coilia ramcarati* (Fig. 37) is the largest rat-tailed anchovy of India at 16 cm TL and forms an important element in artisanal fisheries in West Bengal (Talwar and Jhingran, 1992). *Thryssa scratchleyi* is the freshwater anchovy from Fly–Strickland and Lorentz Rivers of New Guinea (Allen, 1991) and the Roper and Alexandria Rivers (Gulf of Carpentaria) of northern Australia (Larson and Martin, 1989; Merrick and Schmida, 1984; Taylor, 1964). It is the largest member of the anchovy family, with a maximum SL of 41 cm, and is carnivorous (Merrick and Schmida, 1984). It has been found as far as 900 km upstream in the Fly River system (Allen, 1991).

The subfamily Engraulinae lacks scutes behind the pelvic fin and has a short anal fin. This subfamily is mostly found in North, Central, and South America, but some species occur in other parts of the world. The bay anchovy, *Anchoa mitchilli* (Fig. 38), occurs in lower fresh waters and estuaries of coastal rivers from Maine to Yucatan (Burgess, 1980a).

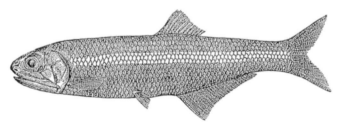

FIGURE 38. *Anchoa mitchilli* as *Engraulis vittatus* (Goode, 1884, Plate 218).

Engraulids of the genera *Anchoviella, Jurengraulis*, and *Lycengraulis* occur in the large rivers of South America such as the Orinoco, Amazon, Parana, and São Francisco. The 19-mm scaleless *Amazonsprattus scintilla* of the Rio Negro of Brazil is the smallest clupeomorph yet discovered. For species descriptions see Whitehead *et al.* (1988).

Because they occur in huge, plankton-feeding schools, anchovies are extremely significant ecologically and commercially. They are vitally important as a link in the food chain of larger fishes. The small Peruvian anchoveta, *Engraulis ringens*, which weighs only 20–30 g, is probably the most exploited fish in history (G. Nelson, 1994). In the 1960s and 1970s this species was harvested at the rate of 10 million tons per year. It was utilized to produce fish meal, oil, and fresh food. Its abundance was due to the Humboldt current that brings upwellings of nutrient-laden waters north along the coast of South America. This enriches the zooplankton that the anchoveta consume. The enormous guano deposits off the South American coast were due to the defecations of large numbers of cormorants and other seabirds that fed on the anchoveta. Even after the inevitable precipitous decline in the mid-1970s due to overfishing, 1 million tons per year of anchoveta were taken. Anchoveta populations, seabirds, and fishers they support are highly susceptible to the vagaries of El Niño, which causes warm waters to replace the rich, cold waters off the coast of South America. This results in population crashes of fishes and seabirds.

Map references: Allen (1991),* Burgess (1980a),* Whitehead (1984a),* Whitehead *et al.* (1988)*

Class Actinopterygii
Subclass Neopterygii

Order Clupeiformes; Suborder Clupeoidei
(Per) Family Pristigasteridae—**pristigasterids**
[prīs'-ti-gas-ter'-i-dē]

THIS IS A COASTAL MARINE FAMILY WITH NINE GENERA AND 37 species. They occur on both sides of the Atlantic and Pacific Oceans and throughout the Indian Ocean. Only 4 South American species have extensive distributions in fresh water and 1 species occurs in the rivers of Burma.

Pristigasterids are moderate to large (20–70 cm) clupeoid fishes with a long anal fin (30–60 or more rays) and a short dorsal fin. They have the usual clupeoid saw-toothed belly. Previously, they have been included with

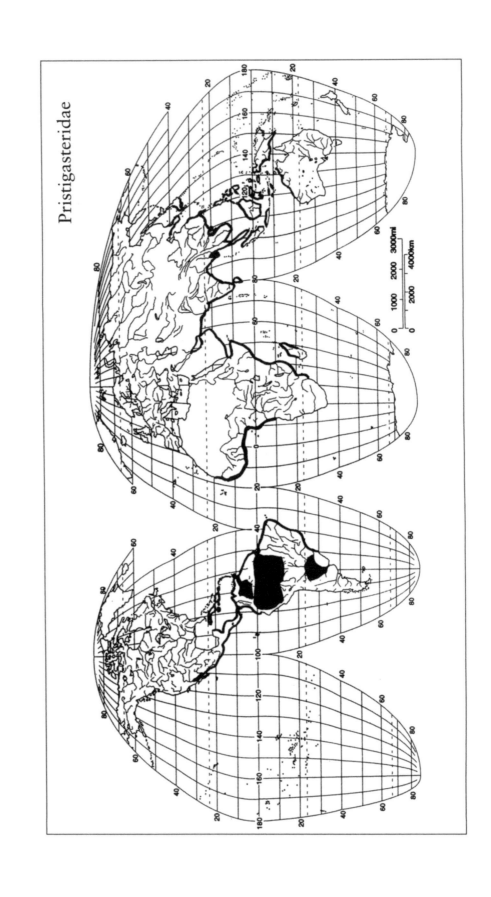

Pristigasteridae

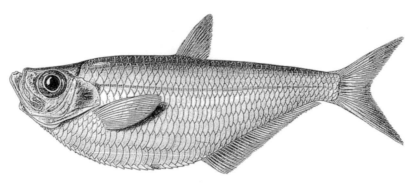

FIGURE 39. *Ilisha africana* (reproduced with permission from Lévêque *et al.*, 1990, Fig. 11.1).

the Clupeidae as a subfamily. Grande (1985) demonstrated that the predorsal bones of this group are oriented vertically or anterodorsally. Nearly all other teleosts have posterodorsally inclined predorsal bones.

Ilisha amazonica and *Pellona castelnaeana* occur in the Amazon River all the way to Iquitos in Peru and *I. novacula* is found in the Irrawaddy and Sittang Rivers of Burma (Whitehead, 1985b). *Ilisha africana* (Fig. 39) enters fresh water along the Atlantic coast of Africa from Senegal to Angola. *Pellona flavipinnis* has an extensive distribution in South American rivers, including the Orinoco, the Amazon from the mouth to Peru, Rio de la Plata, Rio Paraná, and Rio Uruguay. This species is one of the largest pristigasterid at 50 cm SL.

Pristigaster cayana is a small (to 15 cm), extremely deep-bodied and compressed freshwater clupeoid that occurs throughout the Amazon drainage. It is strikingly similar in appearance to the characoid hatchet or flying fishes *Thoracocharax*; however, *Pristigaster* has small pectoral fins and musculature and is not known to be a leaper (Whitehead, 1985b). Since *Pristigaster* and *Thoracocharax* are sympatric and have even been caught in the same net, it has been suggested that their superficial similarity is due to mimicry (Weitzman and Palmer, 1996).

Map reference: Whitehead (1985b)*

Class Actinopterygii
Subclass Neopterygii

Order Clupeiformes; Suborder Clupeoidei
(Per) Family Clupeidae—**herrings** (klōō-pē'-i'-dē)

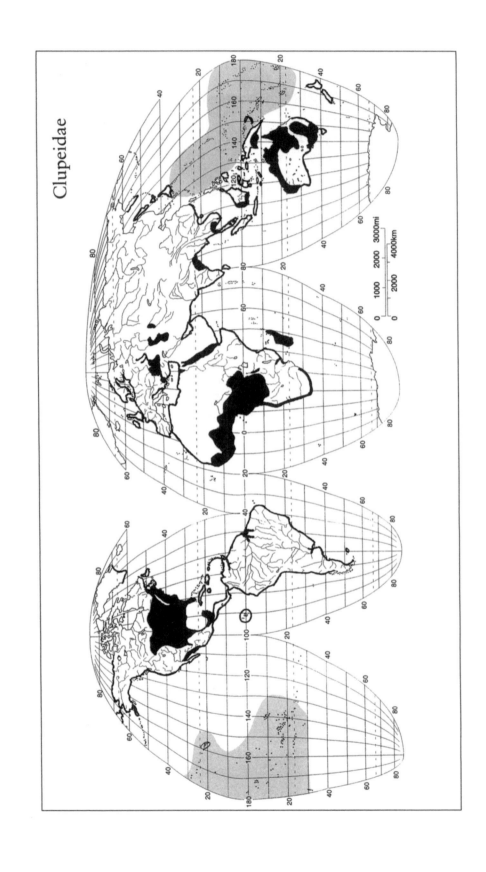

Clupeidae

THE HERRINGS, SARDINES, MENHADEN, SHAD, AND THEIR
relatives are a worldwide, mostly tropical, marine family of about 64
genera and 214 species (G. Nelson, 1994). Only about 50 species are
found in fresh water. Many other species are anadromous. Most clupeids
are silvery coastal fishes that feed on planktonic crustaceans and form
huge schools. As such, they are vitally important in the food web as
converters of minute food items into a size suitable for larger predatory
fishes, birds, and marine mammals. As a group they are unsurpassed in
economic importance as food, fish protein concentrate, fertilizer, or an oil
source.

The body of clupeids is highly variable from compressed to rounded
and is without a lateral line. The mouth is usually terminal, but in
the gizzard shads, Dorosomatinae, the mouth is inferior. Teeth are
either very small or absent. Abdominal scutes and deciduous cycloid
scales are present. Most species are 10–20 cm SL as adults. The hearing
of herrings is especially acute due to a connection of the swim bladder to
the ear.

The family is divided into five subfamilies: Dussumieriinae,
Pellonulinae, Clupeinae, Alosinae, and the Dorosomatinae (Nelson, 1994).
The Dussumieriinae, or round herrings, are marine coastal fishes and are
not present to any significant degree in fresh waters. They have W-shaped
pelvic scutes.

The Pellonulinae are called freshwater herrings. Many of the 44 species
in 23 genera occur in lakes and rivers of west Africa. *Cynothrissa,
Odaxothrissa, Pellonula, Poecilothrissa, Microthrissa, Potamothrissa,
Sierathrissa, Laeviscutella,* and *Congothrissa* are some of the more widespread west African genera (Poll *et al.*, 1984; Gourene and Teugels, 1990).
Other members of this subfamily are found off India, in Southeast Asia, and
in Australia. The small Eocene fossil freshwater herring, *Knightia*, from
North America and China, is placed in this group. *Knightia* is very abundant in the Green River fossil bed of Wyoming (Grande, 1984). More than
20,000 complete specimens have been collected, making it one of the most
common vertebrate fossils in the world (Maisey, 1996). One often finds
beautiful specimens of *Knightia* for sale in rock shops. This fish is said to
be double-armored because it has large scutes along its dorsal and ventral
surface. In Australia *Potamalosa richmondia* (Fig. 40) occurs in the coastal
drainages of New South Wales, especially north of Sydney (Merrick and
Schmida, 1984). It also enters eastern Victoria (Cadwallader and
Backhouse, 1983).

The Clupeinae contains 16 genera and 72 species. Most of these
are marine coastal schooling fishes, but a few species are found in fresh
waters. *Platanichthys platana* and *Ramnogaster melanostoma* are found
in the lower reaches of the Rio de la Plata in Argentina and the Uruguay
River (Whitehead, 1985b). *Rhinosardinia amazonica* occurs in the lower

FIGURE 40. *Potamalosa richmondia* as *Potamalosa novae-hollandiae* (McCulloch, 1917, Plate XXIX).

reaches of the Orinoco and Amazon Rivers. The marine cool-water genera collectively known as "sardines" constitute some of the most valuable fisheries in the world. Sardines are eaten by people throughout the world, but the word has little taxonomic meaning. Sardines from Portugal are likely to be adult pilchards, *Sardinia pilchardus*. Sardines packed in Norway most probably are young sprats, *Sprattus sprattus*, whereas sardines from the Atlantic coast of North America are young herrings, *Clupea harengus* (G. Nelson, 1994).

The shad subfamily, Alosinae, has seven genera and 31 species, half of which are in the genus *Alosa*. They may be marine, freshwater, or anadromous. Shads have a distinct median notch in the upper jaw into which the lower jaw symphysis fits. The mouth is terminal. Shads are large for herring-like fishes, 20–30 cm, with some species reaching 60 cm SL. In North America the blueback herring, *Alosa aestivalis*, and the American shad, *A. sapidissima*, are anadromous and ascend most coastal rivers from Nova Scotia to Florida (Burgess, 1980b). The latter species has been introduced to the west coast. The Alabama shad, *A. alabamae*, ascends Gulf coast rivers from the Suwannae to the Missisippi. The skipjack herring, *A. chrysochloris*, migrates up the Mississippi to Minnesota and South Dakota. The alewife, *A. pseudoharengus*, was introduced from the Atlantic coast into Lake Ontario and has spread throughout the Great Lakes. Other species of *Alosa* are found throughout the Caspian Sea and Black Sea. *Ethmalosa fimbriata* occurs in west African rivers, whereas various species of *Tenualosa* are found in the Ganges River of India, in Malaysia and Indonesia, and in the Mekong River. *Gadusia chapra* occurs in rivers of India and Bangladesh.

The gizzard shads, Dorosomatinae, are made up of six genera with 22 species. Like the shads, their upper jaw is notched, but the mouth is inferior. The gill rakers are fine and numerous. Most species have a long filament on the last dorsal fin ray. The name gizzard shad derives from the muscular gizzard-like stomach of these plankton straining species. The gizzard shad, *Dorosoma cepedianum* (Fig. 41), is distributed throughout

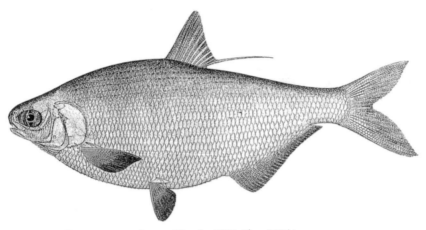

FIGURE 41. *Dorosoma cepedianum* (Goode, 1884, Plate 217A).

most of the eastern half of the United States, whereas the threadfin shad, *D. petenense*, has a more southerly distribution (Megrey, 1980; Burgess, 1980b). Both species are widely introduced as forage species. They may become so abundant that they compete with juvenile game species for zooplankton. This results in decreased growth and survival rates for the game species. These species are very sensitive to temperature changes and massive fish kills of *Dorosoma* may result when stratified lakes "turn over" in spring. High reproductive potential enables the populations to recover rapidly. The bony bream, *Nematalosa erebi*, occupies a hugh swath of northern and central Australia (Allen, 1989). Two species of *Nematalosa* are known from the Fly–Strickland system of New Guinea (Allen, 1991).

Map references: Allen (1989, 1991),* Gourene and Teugels (1990),* Lee *et al.* (1980),* Poll *et al.* (1984), Springer (1982),* Whitehead (1984b, 1985b)*

Class Actinopterygii
Subclass Neopterygii

Order Clupeiformes
(Per) Family Sundasalangidae—**sundasalangids**
[sun'-da-sa-lan'-gi-dē]

UNTIL RECENTLY, THE GENUS *SUNDASALANX* WAS PLACED IN its own family within the Osmeriformes and considered a relative of the

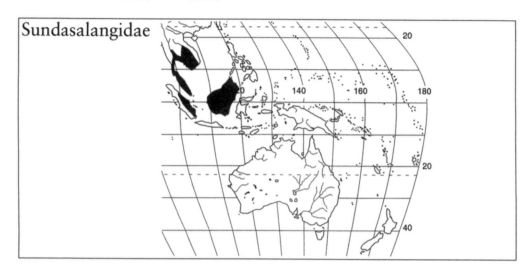

Salangidae (Roberts, 1981; Nelson, 1994). There are seven described species. Siebert (1997) argued that *Sundasalanx* is a highly paedomorphic clupeid, and that the characters used by Roberts (1981) that allied it with the Salangidae were paedomorphic to an extreme. The presence of a prootic bulla and a cavity lateral to the prootic bulla called a *recessus lateralis* define this group as a clupeiform, not a salangoid, according to Siebert (1997). Siebert maintained that *Sundasalanx* development is equivalent to that of the late-stage unmetamorphosed larvae of other clupeiforms, but *Sundasalanx* does not look like any other juvenile or adult clupeid. Siebert postulated that the derived, highly consolidated, caudal skeleton of *Sundasalanx* suggested a sister-group relationship with *Jenkinsia*, a marine Caribbean endemic clupeid. Britz and Kottelat (1999a) considered that conclusion premature and retained Sundasalangidae as a valid family within the Clupeiformes. Its small size and light ossification make the study of *Sundasalanx* difficult.

Roberts described *Sundasalanx* with two species, *S. praecox* and *S. microps* (Fig. 42), from the fresh waters of southern Thailand near Tal Sap, the Malay Peninsula, and the Kapuas River of western Borneo, respectively (Roberts, 1981, 1989a; Kottelat, 1991b). These fishes are minute, with both sexes maturing at less than 15 mm. Maximum size is about 22 mm.

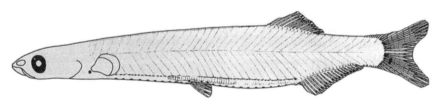

FIGURE 42. *Sundasalanx microps* (reproduced with permission from Roberts, 1981, Fig. 2b)

These are among the smallest of all adult vertebrates. Siebert (1997) added four new species from Borneo that are only slightly larger at 30 mm SL: *S. malleti, S. mesops, S. megalops,* and *S. platyrhynchus.* Britz and Kottelat (1999a) described *S. mekongensis* from the Mekong basin in Laos and Thailand.

Roberts (1984c) presented a detailed account of the skeletal anatomy, and Siebert (1997) reviewed both soft and skeletal anatomy. The skeleton of the Sundasalangidae is largely cartilaginous, and the transparent, scaleless body form resembles a leptocephalus larvae of *Elops.* The dorsal and anal fins are posteriorly positioned. There is no apidose fin, and the pelvic fin has five rays. The anal fin rays are not enlarged. Ribs and supraneurals are lacking. Unusual structures called parapelvic cartilages (Roberts, 1981) are not known from other teleosts. The skeleton of the pectoral girdle consists of a single median element. This is different from all other adult teleosts, but it is similar to some larval teleosts (Roberts, 1984c). *Sundasalanx* is transparent in life except for silvery pigment around the eyeballs. In preserved specimens black pigment is revealed, and males have more vivid pigment patterns than females or juveniles (Siebert, 1997). Gut contents include aquatic insect larvae and segmented worms (Roberts, 1981).

Map references: Roberts (1981, 1984c, 1989a), Siebert (1997)

Class Actinopterygii
Subclass Neopterygii

Order Gonorynchiformes; Suborder Knerioidei
(1st) Family Kneriidae—**kneriids** (ne-rī'-i-dē) [ne'-rē-i-dē]

TELEOSTS ABOVE THE LEVEL OF THE CLUPEOMORPHS ARE classified in the subdivision Euteleostei. This group is the largest of the four major radiations of teleosts, the others being the osteoglossomorphs, elopomorphs, and clupeomorphs (Fig. 1). Within the Euteleostei the superorder Ostariophysi is composed of five orders of fishes (Fink and Fink, 1981, 1996; Nelson, 1994) which release a pheromone from the skin (alarm substance) when injured. This causes a fright reaction in nearby fishes of the same species. Ostariophysans are predominantly freshwater fishes and comprise approximately 75% of the world's freshwater species (Fink and Fink, 1996) and about 27% of all fish species. Their dominance in fresh water may be related to the presence of the

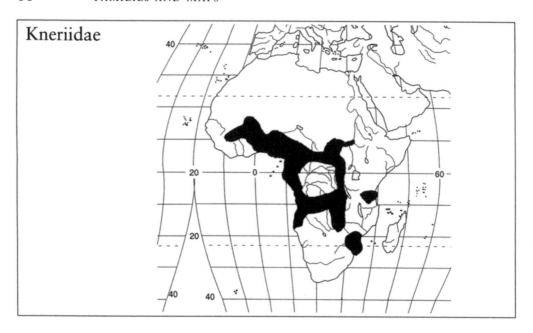

Weberian apparatus, which consists of modified vertebrae that transmit sound waves received by the swim bladder to the inner ear. Ostariophysan fishes are divided into two series, the Anotophysi and the Otophysi. In the Anotophysi (Gonorynchiformes; also spelled Gonorhynchiformes) the first three vertebrae are modified and associated with the cephalic ribs in a primitive Weberian apparatus. The kneriids provide insight into what the early ancestors of the otophysan fishes may have been like.

This African family is made up of four genera and 27 species from the fresh waters of tropical Africa and the Nile (Poll, 1984a,b). The distribution by genus is given by Poll (1973). This family appears to be absent from the middle Congo (Zaire) River system. Its place is taken by the Phractolaemidae (Poll, 1973). Maximum size is about 15 cm. Kneriids are slender loach-like fishes with a protractile upper jaw and an inferior or subterminal mouth. Many species prefer cool, fast-flowing streams at high altitude. Their mouth, sucker-like pectoral fins, and flattened ventral surface are adaptations to this type of habitat. *Kneria* (12 species) and *Parakneria* (13 species) have cycloid scales and a lateral line. Male *Kneria* exhibit sexual dimorphism, with the males possessing a horny rosette of densely packed enlarged scales on the operculum. *Parakneria* males lack this contact organ (Skelton, 1993). See Peters (1967) for an anatomical study of this unique structure. Peters (1973) speculated that this organ may function to hold the female to the male during spawning. *Kneria auriculata*, known as the "shell-ear" because of the cup-shaped structure on the operculum of males, is probably an air-breathing species and is sometimes found in stagnant pools. Apparently, it is also a very good climber and can

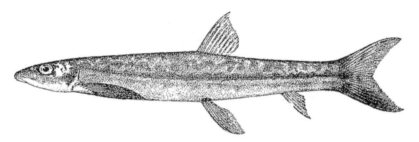

FIGURE 43. *Parakneria cameronensis* as *Kneria cameronensis* (Boulenger, 1909, Fig. 136).

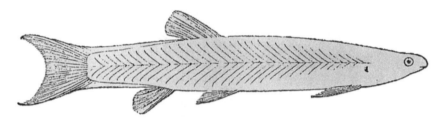

FIGURE 44. *Cromeria nilotica* (Boulenger, 1909, Fig. 137).

negotiate waterfalls 154 m high (Wheeler, 1975). *Parakneria cameronensis* reaches 83 mm TL in the Congo River (Fig. 43). Lenglet (1973) described the internal anatomy of kneriids.

Two monotypic genera, *Cromeria nilotica* (Fig. 44) and *Grasseichthys gabonensis*, are neotenic. They retain juvenile characteristics in that they are small, transparent, and scaleless even after reaching sexual maturity, but they are quite different from each other and from the other two genera and are placed in different families (Cromeriidae and Grassichthyidae) by Poll (1984a; Poll and Gosse, 1995). Paugy (1990b) remarked on the superficial similarity of *Cromeria* to the family Galaxiidae.

Map references: Pool (1973,* 1984a), Paugy (1990a),* Skelton (1993)*

Class Actinopterygii
Subclass Neopterygii

Order Gonorynchiformes; Suborder Knerioidei
(1st) Family Phractolaemidae—**snake mudhead**
[frak-tō-lē'-mi-dē]

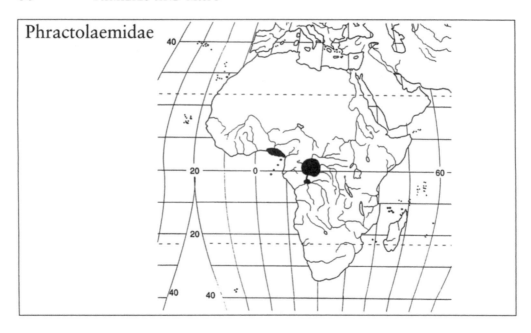

THIS ODD FAMILY IS COMPOSED OF A SINGLE SPECIES, *Phractolaemus ansorgii*, from tropical Africa (Fig. 45). It is found in the Niger Delta, the Malebo Pool (= Stanley Pool), and central Congo (Zaire) system (Thys van den Audenaerde, 1984). This elongate fish (maximum TL 19 cm) has a superior mouth and a protractile upper jaw. It feeds on detritus and small organisms that it extracts from the bottom mud with its tubular mouth. The head is depressed and strongly ossified, and the body is heavily scaled. The anterior nares form a tube that extends in front of the head, and the posterior nares are situated at the base of the nasal tube.

Phractolaemus is sexually dimorphic. The males have four large tubercles that surround the eye on each side of the head. They also develop spines on the caudal peduncle. These features are illustrated by Lévêque

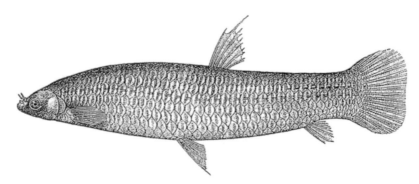

FIGURE 45. *Phractolaemus ansorgii* (Boulenger, 1909, Fig. 133).

(1990b). Thys van den Audenaerde (1961a,b) described the anatomy of this species.

This species lives in small streams and swamps. It prefers muddy bottoms in which it burrows. Its swim bladder is long, richly supplied with alveoli, and functions in air breathing (Graham, 1997). The gills are small, and the circulatory system shows some modifications that suggest air breathing is important to this fish (Thys van den Audenaerde, 1961b). The Phractolaemidae and Kneriidae are sister groups (Poyato-Ariza, 1966).

Map references: Lévêque (1990b),* Poll (1973),* Thys van den Audenaerde (1984)

Class **Actinopterygii**
Subclass **Neopterygii**

Order **Cypriniformes**
(1st) Family **Cyprinidae—minnows, carps**
(si-prin'-i-dē) [sī-prin'-i-dē]

THE CYPRINIFORMES IS ONE OF FOUR ORDERS WITHIN THE series Otophysi. The others are Characiformes, Siluriformes, and Gymnotiformes. Together with the Gonorynchiformes, the five orders make up the superorder Ostariophysi (Fig. 1). The Otophysi ("ear bladder") possess modifications to the anterior four or five vertebrae that allow for sound transmission by connecting the swim bladder and inner ear. The modified vertebral appendages are known as Weberian ossicles and, when combined with muscles, ligaments, other parts of the vertebrae, and the swim bladder, form the Weberian apparatus. This provides these fishes with an acute sense of hearing that is especially useful at night or in turbid waters. The Weberian apparatus may be one of the principal reasons for the dominance of otophysan fishes in shallow freshwater habitats as well as their absence from marine habitats.

Otophysan fishes also possess an alarm substance secreted by damaged skin that warns other members of the species. Primitive teleosts have pharyngeal jaws of limited mobility, but in cypriniforms the pharyngeal jaws are highly mobile (Lauder and Liem, 1983.) This has the advantage of separating the capturing and chewing function of the mouth. The classification of these fishes was devised by Fink and Fink (1981, 1996), and the arrangement presented here follows Nelson (1994).

The Cypriniformes, including the minnows, suckers (Catostomidae), and the loaches (three families), possess protractile mouths without teeth. A variety of shapes of pharyngeal teeth are present. Some species have

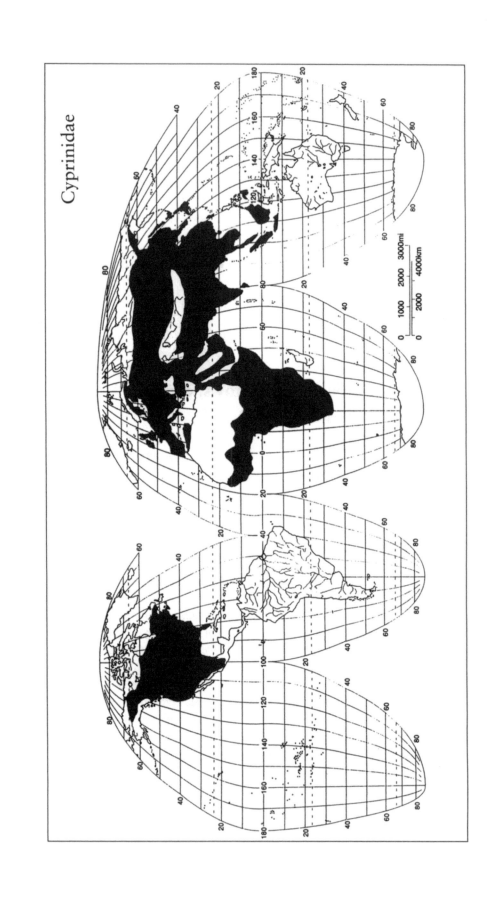

Cyprinidae

molariform pharyngeal teeth used for grinding, others have sharp pharyngeal teeth for piercing prey, whereas others have fine, comb-like teeth used as strainers. The pharyngeal bone is the modified last branchial arch that bears teeth which bite against a tough pad on the basioccipital process at the rear of the skull. Except for some loaches, Cypriniform fishes lack an adipose fin. Most have soft-rayed fins, but a few species, such as carp and goldfish, have hardened rays at the anterior edge of the dorsal and anal fins. Cypriniforms tend to be fully scaled and physostomous.

The Cyprinidae is the largest fish family in the world with 210 genera and more than 2010 species (Nelson, 1994). Its nearest rival in number of species is the predominantly marine Gobiidae with fewer than 1900 species. If 41% of the world's 24,618 fish species occur in fresh waters, then the cyprinids make up about 20% of the world's freshwater fishes and 8% of all fishes (Berra, 1997a). The Cyprinidae is the dominant fish family in the world's fresh waters with an extensive, continuous distribution in Eurasia, Africa, and North America. It is absent from South America and the Australian region, although some species have been introduced. In South America the ecological role of minnows is fulfilled by characiforms, and in Australia the minnow niche is occupied by melanotaeniids.

Eurasia is home to about 1270 species of cyprinids. The greatest center of diversity is China and Southeast Asia. Africa accounts for about 475 species in 23 genera, and North America has about 300 species in 50 genera (Nelson, 1994; M. Coburn, personal communication). The Cyprinidae is the largest family of fishes in North American fresh waters. The oldest fossil cyprinids date to the Eocene of Asia (Jiajian, 1990). Cyprinids appear in the fossil record of North America and Europe in the mid-Oligocene and in Africa in the mid-Eocene (Cavender, 1991). American species are probably of more recent origin than Eurasian cyprinids (Novicek and Marshall, 1976). This is consistent with the view that the Orient is the center of origin of cyprinids. Cyprinids may have crossed the Bering land bridge into North America around 32 million years ago during the low sea level of the Oligocene. The systematics, biogeography, and biology of this family are addressed in Winfield and Nelson (1991).

Cyprinids and osteoglossids are the only primary division freshwater fish families (fish that have little salt tolerance and cannot travel through the sea) to have dispersed east of Wallace's line without human help. *Rasbora* and *Puntius* are found on Lombok, and *Rasbora* also reaches Sumbawa (Darlington, 1957). Cyprinids occur on Borneo but do not cross the Makassar Strait to the Celebes (Sulawesi). In the Philippines about 37 native cyprinids (including *Rasbora* and *Puntius*) are confined to Mindanao, Palawan, Mindoro, and other southern islands (Herre, 1953).

Cyprinid males, in breeding condition, may develop tubercles on various regions of the head, body, and fins. These structures and their significance have been reviewed by Wiley and Collette (1970), Collette (1977), Roberts (1982a), and Chen and Arratia (1996). In some species females also possess tubercles, and they may be permanent in both sexes or seasonal.

Such a huge family has a great diversity of fishes known by a variety of names throughout the world. In North America, cyprinids are thought of as minnows, shiners, dace, and chubs. In Eurasia, carp and goldfish are the archetypical cyprinids. In Africa and south Asia barbs are common cyprinids. Do not let the term "minnow" fool you. Technically, a minnow is any member of the family Cyprinidae, not any small fish. (In England, only *Phoxinus phoxinus* is known as a minnow.) Some cyprinids can be quite large. For example, a tetraploid species from Thailand, *Catlocarpio siamensis*, and *Tor putitora* from India reach nearly 3 m (Smith, 1945; Talwar and Jhingran, 1992). However, specimens of such size are extremely rare today. *Ptychocheilus lucius*, the endangered Colorado squawfish, reaches 1.8 m and 45 kg and is the largest North American minnow (Fig. 46). Aspects of its life history have been reported by Osmundson and

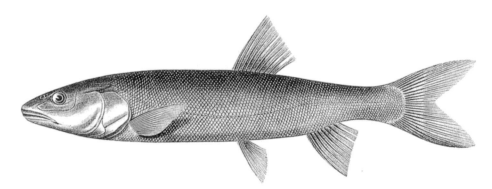

FIGURE 46. *Ptychocheilus lucius* (Girard, 1859, Plate CX).

Burnham (1998) and Osmundson *et al.* (1998). In our age of political correctness, the common name "pikeminnow" has been substituted recently for the traditional "squawfish" (Nelson *et al.*, 1998). The northern pikeminnow, *P. oregonensis*, which reaches 63 cm TL, is an important predator on juvenile samonids in the Pacific Northwest of North America (Petersen and Ward, 1999). Most cyprinids are considerably smaller and many are less than 5 cm. *Danionella translucida* from Burma is a candidate for the world's smallest freshwater fish. The largest adult specimen known is 12 mm (Roberts, 1986a). *Boraras micros* from the swamps of the Mekong basin in northeastern Thailand reaches a maximum known size of 13.3 mm SL (Kottelat and Vidthayanon, 1993).

Carp, *Cyprinus carpio*, originated in Europe in the rivers around the Black Sea and the Aegean basin, especially the Danube. It is widely utilized as food by many people and has been introduced throughout the world. In

North America and Australia it is considered a pest species because it uproots vegetation and muddies the water by probing the bottom. It is angled for in Europe and raised in pond culture in Israel, Europe, and Asia. The goldfish, *Carassius auratus*, has been widely bred for ornamental colors and shapes. Grass carp, *Ctenopharyngodon idella*, is extensively stocked to control aquatic vegetation in lakes. This Asian fish has become established in the Mississippi drainage of North America. Many relatively small cyprinids are utilized in the aquarium trade. Ecologically, minnows are vitally important forage fishes for larger species and cycle nutrients and energy around the food web. Their diversity contributes to the stability of aquatic ecosystems.

The systematics of the Cyprinidae is not settled, and Howes (1991) noted that its monophyly is not established. Mayden (1989) proposed the most extensive nomenclatural overhaul of American minnows in nearly 40 years. This was shortly followed by further, independent reassessment of interrelationships using different characters by Coburn and Cavender (1992). Snelson (1991) remarked on the areas of divergent interpretation between Mayden (1989) and Coburn and Cavender (1992). In their diagnosis of cyprinids, two subfamilies were recognized by Cavender and Coburn (1992), Leuciscinae and Cyprininae, and North American cyprinids were placed in the Leuciscinae, except *Notemigonus*. The eight subfamilies listed here follow Nelson (1994). Simons and Mayden (1999) published a phylogenetic assessment based on molecular data of North American cyprinids, and Broughton and Gold (2000) described the phylogenetic relationships of 30 species of *Cyprinella*.

The Cyprininae includes the barbs (*Barbus*), carp, goldfish, grass carp, and about 600 other species mostly from India, Southeast Asia, and Africa. There are 69 species of *Puntius* (Jayaram, 1981) including *Puntius filamentosus* (Fig. 47). Breeding males of this species develop elongated dorsal fin rays. Juveniles are a brilliant pink with brick-red fins. Some members of the genus *Labeo* are called "sharks" in the aquarium trade. *Labeo calbasu* is an

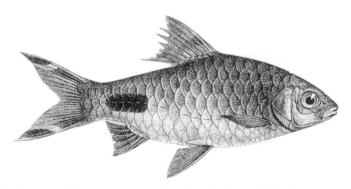

FIGURE 47. *Puntius filamentosus* as *Barbus mahicola* (Day, 1878b, Plate CXL, Fig. 5).

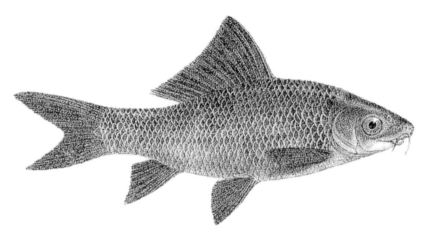

FIGURE 48. *Labeo calbasu* (Day, 1878b, Plate CXXVI, Fig. 4).

important food and game fish in India, where it reaches 90 cm TL (Fig. 48). There are also Eurasian species. *Garra* is found in northeast Africa and southern Asia including the Arabian peninsula (Krupp, 1983). Members of this subfamily usually have two pairs of barbels. Cavender and Coburn (1992) considered grass carp in the Leuciscinae.

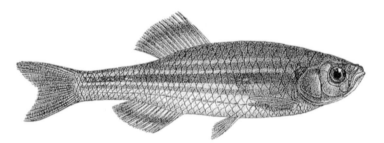

FIGURE 49. *Danio aequipinnatus* (Day, 1878b, Plate CL, Fig. 5).

The Gobioninae is an eastern Asian subfamily, but the gudgeons (*Gobio*) occur from Europe to the Far East. The Rasborinae occurs in Africa and southern Eurasia through Indonesia. Many species of this colorful subfamily are popular in the aquarium trade such as *Rasbora* and *Danio*. *Danio aequipinnatus* (Fig. 49) is a brilliant blue species widely distributed in Indian hillstreams up to 300 m elevation (Talwar and Jhingran, 1992). It is called the giant danio and reaches 15 cm TL. The zebrafish, *Brachydanio rerio*, popular among aquarists, is now the new "white rat" of developmental biologists and genetic researchers. It is valued for its transparent embryos, and its entire genome will likely be sequenced within a few years.

The Acheilognathinae is found in Europe and eastern Asia but not central Asia. Females possess an elongate ovipositor and deposit eggs into the mantle cavity of freshwater mussels. The female bitterling, *Rhodeus sericeus*, has a 5-cm ovipositor that she maneuvers into the mussel while the males hovers nearby waiting to release sperm near the mussel's intake siphon. Fertilization takes place inside the mantle cavity as the mussel draws in the sperm-laden water. Johnston and Page (1992) surveyed eight types of reproductive strategies found in North American cyprinids.

The Leuciscinae is a very large subfamily found in North America and northern Eurasia but not in India and Southeast Asia. All 300 or more species of North American minnows belong to the Leuciscinae. Mayden (1989) extensively revised the North American members of this subfamily. Old World genera include *Abramis* (bream), *Leuciscus* (chub), *Rutilus* (roach), *Scardinius* (rudd), and perhaps *Tinca* (tench). The dace genus *Phoxinus* is found in Europe and North America. Familiar North American genera include *Campostoma* (stonerollers), *Cyprinella* (30 species of shiners; Broughton and Gold, 2000), *Gila* (3 species of endangered western chubs, Minckley and DeMarais 2000), *Luxilus* (shiners), *Notropis* (70 species of shiners), *Pimephales* (minnows), *Ptychocheilus* (pikeminnows), *Rhinichthys* (dace), and *Semotilus* (chubs).

The blue-nose shiner, *Pteronotropis welaka*, from clear, vegetated coastal streams of the Gulf slope, is one of the most beautiful freshwater species in North America. Breeding males develop bright blue snouts, black pigment in their enlarged dorsal fin, and black and yellow pigment in their elongated pelvic, anal, and caudal fins. The life history of this spectacular fish was described by Johnston and Knight (1999) and Fletcher (1999). *Cyprinella* species, including *C. whipplei* (Fig. 50), hide their eggs in the interstices between rocks and may be able to communicate with each other

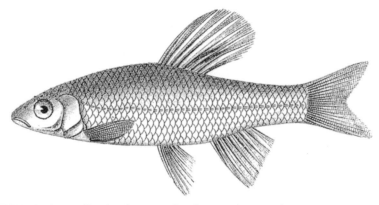

FIGURE 50. *Cyprinella whipplei* as *C. whipplii* (Girard, 1858, Plate LVII).

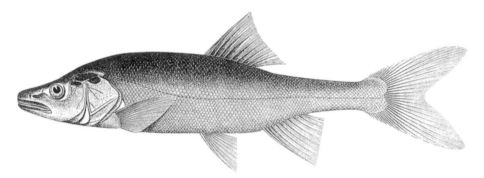

FIGURE 51. *Gila robusta* as *G. grahami* (Girard, 1859, Plate C).

via sound (Page and Burr, 1991). *Gila robusta* (Fig. 51), *G. intermedia*, and *G. nigra* have a deep, compressed body and slender caudal peduncle. They are considered to be primitive minnows (Page and Burr, 1991). The golden shiner, *Notemigonus crysoleucas* (Fig. 52), with its compressed body,

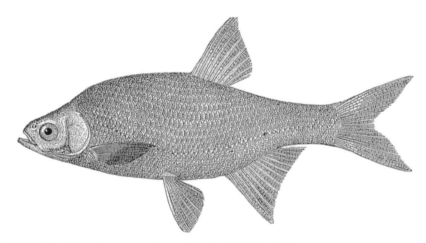

FIGURE 52. *Notemigonus crysoleucas* (Goode, 1884, Plate 227).

depressed lateral line, and scaleless ventral keel, is excluded from a mono-phyletic group of all other North American minnows by Cavender and Coburn (1992). They consider *Notemigonus* to be more closely related to Eurasian than North American species.

The Cultrinae, compressed fish with a keel-like ventral surface, is found in eastern Asia and the Alburninae has a disjunct distribution in Europe and Asia. The Psilorhynchinae or mountain minnow group was formerly placed in its own family. There are five species in two genera, *Psilorhynchoides* and *Psilorhynchus*, and they occur in fast-flowing, high-altitude streams of

Nepal, India, and Burma. Adaptations to the fast-flowing habitat include reduced swim bladder, flattened ventral surface of the head, subterminal mouth with sharp-edged jaws, and broad fleshy lips. This morphology suggests that psilorhynchids cling to rock faces, probably out of the main stream of current.

Map references: Banarescu (1990),* Berg (1948/1949), Darlington (1957), Krupp (1983),* Lagler *et al.* (1977),* Miller (1958, 1966),* Morrow (1980),* Norman and Greenwood (1975),* Roberts (1989a),* Scott and Crossman (1973),* Skelton (1993)*

Class Actinopterygii
Subclass Neopterygii

Order Cypriniformes
(1st) Family Gyrinocheilidae—**algae eaters** [jī'-rī-nō-kī'-li-dē]

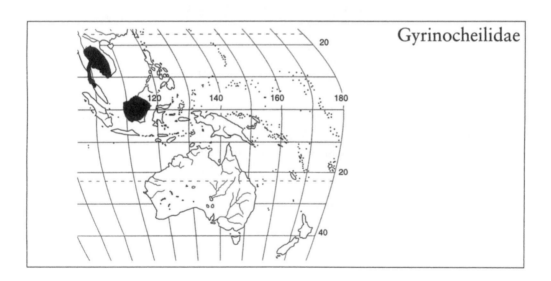

THIS FAMILY CONTAINS A SINGLE GENUS, *GYRINOCHEILUS*, with three highly specialized species from mountain streams of Thailand, Laos, Cambodia, and Borneo (Roberts and Kottelat, 1993). Gyrinocheilids are remarkable for two morphological features. It is the only cyprinoid family to lack pharyngeal teeth, and it possesses both inhalant and exhalant gill openings. Water enters the gill cavity by an upper opening, passes over the gills, and then exits through a lower, external gill opening. This unusual modification allows the mouth to be free from respiratory movements. The mouth, with its well-developed lips and file-like inner surface, adheres to rocks. This enables the fish to hold its position in swift-flowing current and to scrape algae off the rocks. Cool, swift mountain streams are highly oxygenated, but the respiratory rate of *Gyrinocheilus* is very rapid— 240 times per minute for a 12-cm fish. This compensates for the small inhalant opening. A reduced swim bladder is another adaptation to the swift current that prevents the fish from rising off the bottom except by swimming effort.

Gyrinocheilus aymonieri is popular with aquarists because of its ability to clean algae from the walls of aquariums. It occurs in the Chao Phraya, Meklong, lower and middle Mekong basins, and peninsular Thailand (Roberts and Kotelat, 1993). *Gyrinocheilus pustulosus* from western Borneo is the largest species at 30 cm (Fig. 53). It possesses parallel rows of

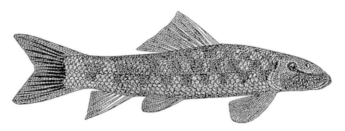

FIGURE 53. *Gyrinocheilus pustulosus* (Weber and DeBeaufort, 1916, Fig. 89).

breeding tubercles on the side of the head (Roberts, 1989a). *Gyrinocheilus pennocki* occurs in the Mekong basin of Thailand, Laos, and Cambodia (Roberts and Kottelat, 1993). See Smith (1945) for further references to life history details and Nelson (1994) and Roberts and Kottelat (1993) for systematic references.

Map references: Roberts (1989a), Roberts and Kottelat (1993), Smith (1945)

Class Actinopterygii
Subclass Neopterygii

Order Cypriniformes
(1st) Family Catostomidae—**suckers** (kat'-ō-stom'-i-dē)

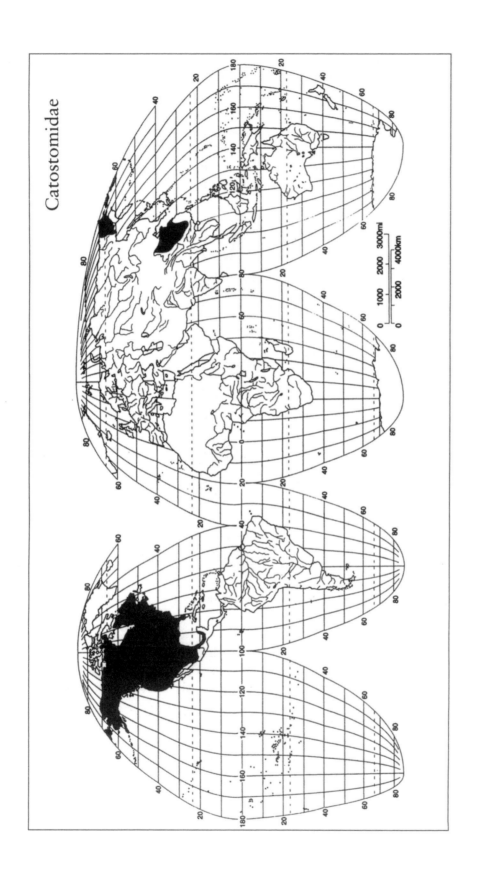

Catostomidae

THE SUCKER FAMILY IS FOUND IN CHINA, NORTHEASTERN Siberia, and North America. There are 14 or 15 genera and about 66 species currently recognized (Nelson, 1994; Mayden *et al.*, 1992); however, at least 12 species remain to be described (R. E. Jenkins, personal communication). About 60% of the species belong to the genera *Catostomus* and *Moxostoma* or related genera. This family is almost exclusively North American, with the exception of *Myxocyprinus asiaticus* (Fig. 54) from the Yangtse and Hwang Ho drainages of eastern China and *Catostomus*

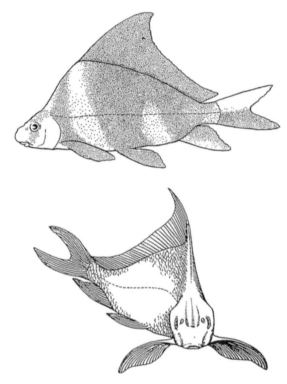

FIGURE 54. *Myxocyprinus asiaticus* (reproduced with permission from Nichols, 1943, Fig. 15).

catostomus, a North American species that invaded Siberia from Alaska during the Pleistocene interglacial period. The distribution of suckers in North America extends from northern Alaska (*C. catostomus*) to Guatemala (*Ictiobus meridionalis*), with the greatest diversity in the southeastern United States and the Mississippi drainage.

The evolution of suckers is generally thought to have proceeded from deep-bodied species that live in large, slow-flowing rivers to slender species that inhabit small to medium streams of moderate flow (Smith, 1992). This may be explained by the fact that larger fish in the larger rivers are stronger swimmers with larger home ranges and are therefore not as isolated as weaker swimming smaller fishes. Thus, gene flow would be more extensive within the larger species, thereby preventing reproductive isolation. Likewise, smaller species in headwater streams are isolated by the barrier of the larger downstream river, and their small populations are more likely to evolve under the influence of the less stable local habitat (Smith, 1992). Selection for smaller body size may be an adaptation to these restricted local conditions that include cooler temperatures and swift current. This, in turn, could have made colonization of more northerly waters possible.

Suckers are small to medium-sized soft-rayed benthic fishes with thick, fleshy lips studded with plicae and papillae. The mouth is protrusible and devoid of teeth. Their vacuum cleaner, bottom-feeding habits account for the name "sucker." Most species feed on aquatic insect larvae, benthic invertebrates, detritus, and algae. A few species (*Chasmistes*) are midwater planktivores and some *Moxostoma* are specialized for mollusk feeding. There is one row of 16 or more pharyngeal teeth as opposed to one to three rows of fewer teeth in the cyprinids. The dorsal fin of most suckers has more than 10 rays (except the *Catostomus* subgenus *Pantosteus*, which may have 8–10), a character that also serves to distinguish catostomids from most cyprinids, which have fewer than 10 rays. No barbels or adipose fin are present, scales are cycloid, and the head is naked. Males develop tubercles during the breeding season, especially on the anal fin and lower caudal lobe (Wiley and Collette, 1970). Many suckers make an upstream spring spawning migration, sometimes extensive. Page and Johnston (1990) reviewed sucker reproductive behavior that often includes a *ménage à trois* involving a female flanked by two males. Most species are smaller than 60 cm. As a member of the Otophysi, a Weberian apparatus is present. Suckers are tetraploid (Uyeno and Smith, 1972) but functionally diploid due to some suppression of duplicate gene function (Buth, 1979). Many western species hybridize extensively. Hubbs *et al.* (1943) conducted a classic study of hybridization in suckers. The presence of suckers, especially the fast-water species, can be an indicator of good water quality. For life histories of various sucker species, see Scott and Crossman (1979), Becker (1983), and Jenkins and Burkhead (1994). Smith (1992) recognized three subfamilies: Cycleptinae, Ictiobinae, and Catostominae.

The subfamily Cycleptinae is the most morphologically primitive group of suckers. Currently, there are two described species in two genera. *Myxocyprinus asiaticus* of eastern China is deep bodied, especially as a juvenile. This transformation is unusual for a large-river species; however, the deep body is triangular in cross section with a flat belly. Water flowing over the back may force the ventral surface of the fish onto the river bottom and help hold it in place. The blue sucker, *Cycleptus elongatus*, is more slender and also inhabits large rivers (Fig. 55). It occurs in the Mississippi, Missouri, and Ohio Rivers, the Rio Grande, and some Gulf slope rivers.

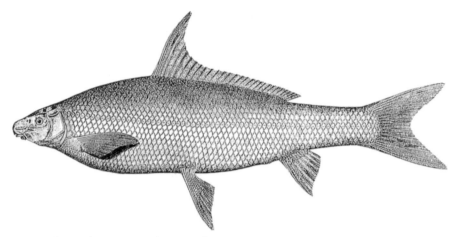

FIGURE 55. *Cycleptus elongatus* (Goode, 1884, Plate 224).

The subfamily Ictiobinae is composed of two, large, deep-bodied, long-finned genera with seven species. This species-poor subfamily is also considered morphologically primitive (Smith, 1992). There are three species of *Carpiodes* (quillback and carpsuckers) and five species of *Ictiobus* from large rivers and lakes in the eastern two-thirds of North America, including *I. labiosus* of Mexico. The pharyngeal pad in this group is very thick and nearly fills the mouth and throat leaving a very narrow passageway for food. These carp-like fishes are sometimes found in fish markets, and they are cultured on a small scale in the southern United States. The quillback, *Carpiodes cyprinis*, has an extremely long first dorsal fin ray (Fig. 56). The

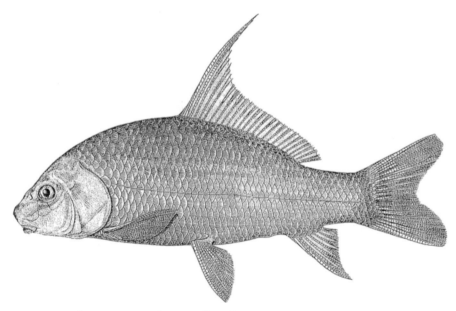

FIGURE 56. *Carpiodes cyprinus* (Goode, 1884, Plate 225).

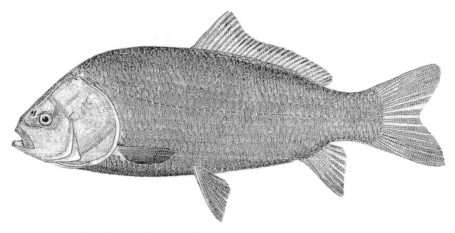

FIGURE 57. *Ictiobus bubalus* (Goode, 1884, Plate 226).

bigmouth buffalo, *I. cyprinellus*, is one of the largest suckers, reaching 1 m and 36 kg (Fig. 57; Harlan and Speaker, 1969). The smallmouth buffalo, *I. bubalus*, grows to 78 mm TL. The Eocene fossil genus *Amyzon* clusters with the ictiobines and is found in North America and Asia (Bruner, 1991; Smith, 1992). The fossil record is consistent with the basal groups (Cavender, 1986; Smith, 1992). The more advanced subfamily, Catostominae, probably originated near the early Miocene.

The Catostominae consists of 10 genera with 57 species. This subfamily is grouped into two tribes. The Catostomini is largely found in western North America, but some species are in eastern North America. The principal genus is *Catostomus*, with 24 species (Mayden *et al.*, 1992). Smith (1966) studied the distribution and evolution of the *Catostomus* subgenus *Pantosteus*. Smith and Hoehn (1971) reported phenetic and cladistic studies of biochemical and morphological characteristics of *Catostomus*. The white sucker, *C. commersoni*, is common in midwestern and northern North American streams and rivers (Fig. 58). The razorback sucker,

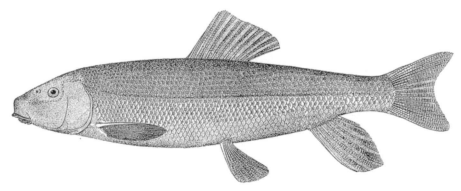

FIGURE 58. *Catostomus commersoni* (Goode, 1884, Plate 223).

Xyrauchen texanus, is so called because of the sharp keel on its nape. It is the only sucker species with such a humpback. As a juvenile, it is shaped like *Catostomus*. It is rare and is only found above the Grand Canyon and in Lakes Mead, Mohave, and Havasu on the lower Colorado River (Page and Burr, 1991). Formerly, it was more widely distributed in the Colorado basin. *Deltistes luxatus*, the Lost River sucker of Oregon and California, is also rare and endangered. Four species of western lake suckers (*Chasmistes*) have a large terminal or subterminal mouth and are midwater planktivores (Miller and Smith, 1981). These species are endangered or extinct.

The other tribe, Moxostomatini, consists of seven genera, most of whose species are found in eastern North America. *Moxostoma*, the redhorses, is the major genus with 10 described species. Another 10 species that were formerly included within *Moxostoma* are now recognized as

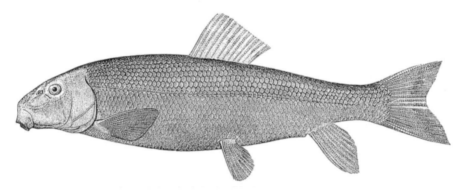

FIGURE 59. *Moxosoma macrolepidotum* (Goode, 1884, Plate 222A).

belonging to the genera *Scartomyzon* (7 species) and *Thoburnia* (3 species) (Smith, 1992; Mayden *et al.*, 1992; Jenkins and Burkhead, 1994). *Moxostoma macrolepidotum* (Fig. 59), the shorthead redhorse, is the most widespread and polytypic member of the genus (Jenkins and Burkhead, 1993). The harelip sucker, *Moxostoma* (= *Lagochila*) *lacerum*, from the middle and lower Ohio basin, the White drainage of the Ozarks, and the Maumee system of Lake Erie, is extinct. This sucker holds the dubious distinction of being the fish species lost from the largest number of American states—eight (Rahel, 2000). Jenkins and Burkhead (1994) summarized what is known about this species. The other genera of this tribe are 3 species of *Erimyzon* (chubsuckers), *Minytrema melanops* (spotted sucker) (Fig. 60), and 3 species of *Hypentelium* (hogsuckers).

Today, suckers show a disjunct distribution in North America and eastern Asia. It is generally assumed that suckers originated in Asia (Darlington, 1957; Gilbert, 1976; Bond, 1996). A primitive sucker, *Myxocyprinus*, is found there today. Likewise, possible sister groups to the catostomids, the

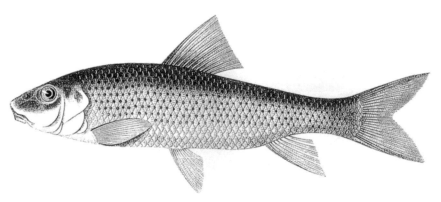

FIGURE 60. *Minytrema melanops* as *Moxostoma victoria* (Girard, 1859, Plate C).

barbs (Cyprinidae) or the loaches (Cobitidae), are Asian (Smith, 1992). However, in the past the holarctic areas were one landmass. Movement could have occurred in either direction. Other primitive suckers are found in North America, and the vast majority of species are North American. The most ancient fossil suckers are found on both continents. The place of origin of the suckers is not certain.

Map references: Lagler *et al.* (1977),* Miller (1958, 1966),* Scott and Crossman (1973),* Smith (1966)*

Class Actinopterygii
Subclass Neopterygii

Order Cypriniformes
(1st) Family Cobitidae—**loaches**   (co-bit'-i-dē) [co-bī'-ti-dē]

LOACHES ARE A FAMILY OF SMALL, SLENDER FISHES FROM Europe, Asia, and North Africa (Morocco). Their greatest diversity occurs in Southeast Asia. There are about 21 genera and 164 species (M. Kottelat, personal communication). Loaches are adapted for a benthic existence in streams, rivers, lakes, and swamps. Most species have elongate, worm-like bodies. Some species are barb-like with fusiform bodies and flat ventral surfaces. The mouth is subterminal and surrounded by three to six pairs of barbels. Scales are reduced or absent. The swim bladder is encased in bone. There usually is a spine below or anterior to the eye that can be erected for defense. Some loaches are brightly colored. Loaches are sexually dimorphic with the pectoral fins of males enlarged and the fin rays modified. The largest loaches, such as *Botia macracanthus* and *Misgurnus fossilis*, reach

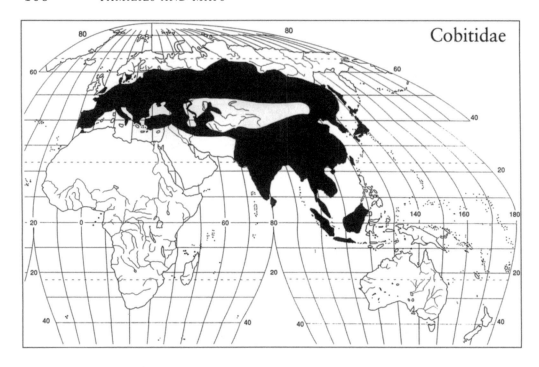

about 30 cm TL (Sterba, 1966). Nelson (1994) recognized two subfamilies: Cobitinae (115 species) and Botiinae (49 species).

Roberts (1984a) reported that *Cobitis taenia* has one of the largest ranges of any freshwater fish in Eurasia. However, Kottelat (1997) indicated that the "species" probably included up to 5 species and acknowledged its range to be most of Europe.

Loaches are popular aquarium fishes (Axelrod and Burgess, 1982) and include such genera as *Cobitis* and *Pangio* (= *Acanthopthalmus*) (coolie loaches), *Botia* (clown loaches), and *Misgurnus* (weatherfish). However, some species, such as *Botia histrionica* (Fig. 61), are very rare (Talwar and Jhingran, 1992). Most loaches feed on worms and insect larvae located by

FIGURE 61. *Botia histrionica* (Day, 1878b, Plate CLIV, Fig. 4).

the sensitive barbels. Some species eat algae. Weatherfish, such as M. *fossilis* from central and eastern Europe, are sensitive to changes in atmospheric pressure and become restless before a thunderstorm (Wheeler, 1975). This may be related to the encasement of the swim bladder in bone. As atmospheric pressure changes the fish would need to expel or take in air to keep the bladder inflated to the normal level. Some species gulp air and intestinal respiration may be possible. Air may be released via the anus.

The oriental weatherfish, M. *anguillicaudatus* (Fig. 62), native to eastern Asia, is established in the headwaters of the Shiawassee River,

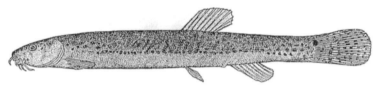

FIGURE 62. *Misgurnus anguillicaudatus* (reproduced with permission from Nichols, 1943, Fig. 106).

Michigan, and in several flood control canals in Orange County, California (Hensley and Courtenay, 1980a). It is thought to have escaped from aquarium supply companies in the 1930s. Populations are also established in Florida, Hawaii, Idaho, Illinois, and Oregon (Fuller *et al.*, 1999) and Germany (Kottelat, 1997).

For species descriptions, see Nichols (1943), Berg (1948/1949), Roberts (1989b), and Kottelat *et al.* (1993). For nomenclatorial corrections, see Kottelat (1987).

Map references: Banarescu (1990),* Darlington (1957), Grzimek (1973),* Maitland (1977),* Masuda *et al.* (1984), Nelson (1976),* Pethiyagoda (1991a),* Roberts (1984a, 1989b*), Talwar and Jhingran (1992)

Class Actinopterygii
Subclass Neopterygii

Order Cypriniformes
(1st) Family Balitoridae—**river or hillstream loaches**
[bal'-i-tor'-i-dē]

THIS FAMILY WAS PREVIOUSLY KNOWN AS THE Homalopteridae. Kottelat (1988) explained the name change. River or hill-

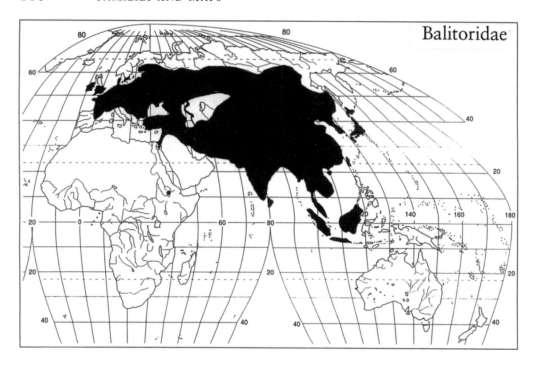

Balitoridae

stream loaches occur in the fresh waters of Eurasia. There are about 58 genera with 485 valid species, and about 70 more species remain to be described (M. Kottelat, personal communication). The Balitoridae, as constituted here, is the second largest Eurasian freshwater fish family (Roberts, 1989b). The Cyprinidae is the largest. These fishes have three or more pairs of barbels near the mouth. They are benthic, and many have a depressed head and body. They occupy slow to fast-flowing, even torrential, streams. Talwar and Jhingran (1992), Kottelat (1990a), and Kottelat *et al.* (1993) provided illustrated keys to many Indian, Indochinese, and Indonesian species, respectively.

Sawada (1982) recognized two subfamilies, Nemacheilinae and Balitorinae. The Nemacheilinae is the larger subfamily with about 340 species. In previous classifications (Nelson, 1984) this subfamily, spelled Noemacheilinae, was assigned to the Cobitidae. They are less compressed than members of the other subfamily. The Nemacheilinae occur in Iran, India, China, Thailand, and Malaysia. *Nemacheilus* is a common genus well-known among aquarists, and more than 450 species have been attributed to it (Kottelat, 1982b). The 8-cm TL, reddish-brown *N. rubidipinnis* (Fig. 63) from Burma is a typical example of the genus, whose species are difficult to distinguish due to remarkable similarity in morphology. *Nemacheilus barbatulus* is found in streams, rivers, and lakes throughout Europe except in the extreme north and south (Maitland, 1977). Kottelat (1997) concluded that *Nemacheilus* should include only species from Southeast Asia and used the name *Barbatula barbatula* for the European

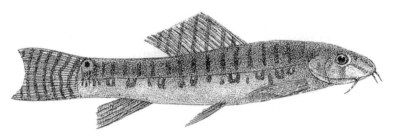

FIGURE. 63. *Nemachilus rubidipinnis* (Day, 1878b, Plate CLII, Fig. 4).

species. A species of the genus *Yunnanilus* (= *Triplophysa*) occurs in hot springs in Tibet at the highest elevation of any known fish species, 5200 m (Kottelat and Chu, 1988). "*Nemacheilus*" *abyssinicus* is known only from the holotype collected from Lake Tsana (Tana), the source of the Blue Nile in Ethiopia (Roberts, 1984a). Kottelat (1990a) reviewed the subfamily from Indochinese waters.

The Balitorinae (hillstream loaches) has about 145 species and is found from India and China through Southeast Asia to Borneo. Jayaram (1974) plotted the discontinuous distribution of this subfamily. Southern China and Borneo have the most species (Roberts, 1989b). In many species the body is flattened, and the pectoral and pelvic fins are expanded and function as suckers that grip stones in rushing water. The pectoral and pelvic fins are often inserted horizontally. The mouth is inferior, and gill openings are reduced and often placed high on the body. Kottelat (1988)

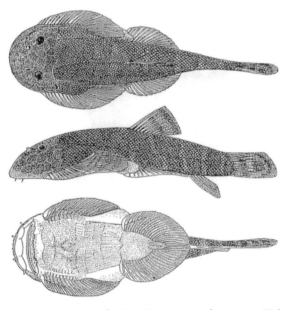

FIGURE 64. *Neogastromyzon nieuwenhuisii* as *Gastromyzon borneensis* (Weber and DeBeaufort, 1916, Fig 1).

reviewed Indian and Indochinese *Balitora*, Roberts (1989b) provided photographs of species from Borneo, and Kottelat *et al.* (1993) illustrated all species from Indonesia. *Balitoria, Gastromyzon*, and *Homaloptera* are common genera. *Gastromyzon borneensis*, which Roberts (1989b) recognized as *Neogastromyzon nieuwenhuisii* (Fig. 64), is the largest *Gastromyzon* at 100 mm SL.

Map references: Banarescu (1990),* Grzimek (1973),* Jayaram (1974),* Maitland (1977),* Masuda *et al.* (1984), Pethiyagoda (1991a),* Roberts (1984a, 1989b),* Talwar and Jhingran 1992

Class   Actinopterygii

Subclass   Neopterygii

Order   Characiformes; Suborder Citharinoidei
(1st) Family   Distichodontidae—**distichodontids**
[di'-sti-kō-don'-ti-dē]

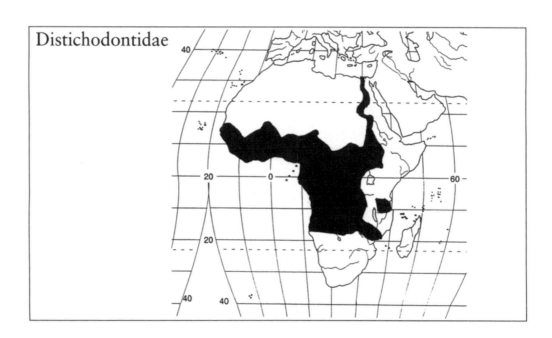

THE CHARACIFORMES ARE OTOPHYSAN FISHES THAT USUALLY have an adipose fin, well-developed jaw teeth, and a scaled body. Barbels are absent, and pharyngeal teeth are usually unspecialized. The caudal fin has 19 principal rays. There is a great diversity of species with many sizes, shapes, and colors that make this group very attractive to aquarists. Characiforms as a group occur in tropical fresh water in Central and South America with 1 species extending into Texas, and others are African. The classification of this large order is unsettled and various authorities recognize from 1 to 18 families. Nelson (1994) listed 10 families with about 237 genera and at least 1342 species, of which 208 are from Africa. Géry (1977) provided keys and descriptions to most of the genera and species.

A recent book on the phylogeny and classification of Neotropical fishes provided a superb tool for characiform studies (Malabarba *et al.*, 1998). Within this edited volume, Vari (1998) gave a historical overview of higher level phylogenetic concepts of the order, and Buckup (1998) provided a preliminary phylogeny and reclassification of major groups. I follow this recent classification which recognized 18 families in two suborders, the Citharinoidei and Characoidei.

Based on Buckup's (1991, 1998) tentative hypothesis of relationships among the Characiformes, there are three African clades: Distichodontidae + Citharinidae, Alestidae, and Hepsetidae. The first clade is the most primitive characiform group, and the Hepsetidae is a monotypic family and presumed sister group to the South American Ctenoluciidae. The implications of this for the biogeography of characiform fishes and the African–South American connection are discussed by Lundberg (1993). Buckup (1998) argued that three African and at least seven Neotropical monophyletic groups originated before the final breakup of Gondwana and the opening of the South Atlantic Ocean 115 million years ago. Weitzman and Weitzman (1982) reviewed the history of South American–African connections in the ichthyological literature.

The Citharinoidei (composed of the Distichodontidae and Citharinidae) is considered to be the primitive sister group to all other characiforms (Fink and Fink, 1981; Buckup, 1991, 1998). This suborder is found only in African fresh waters. Most species have ctenoid scales, unlike the cycloid scales of the vast majority of other characiform fishes. There are 20 genera and about 98 species in the two families which Nelson (1994) and Eschmeyer (1998) treated as subfamilies. Vari (1979) treated these as separate families, as did Buckup (1998). Vari demonstrated the monophyly of each family.

The Distichodontidae occurs in the Gambia, Senegal, Niger, Volta, Nile, Congo, and Zambezi basins and in Lakes Turkana and Chad (Daget and Gosse, 1984). The Congo River system is home to more species than the other rivers. The family contains two groups. One group is very slender and has a movable upper jaw that performs a sissors-like action. These elongate fishes are carnivores in the sense that most have the peculiar habit of feeding on the fins of other fishes (Roberts, 1972, 1990). Their teeth are closely spaced for fin feeding. The name "distichodont" refers to the two rows of teeth these fishes possess (Skelton, 1993). Formerly, these fishes were placed in the family Ichthyboridae. The other lineage of the Distichodontidae consists of herbivores and micropredators with nonprotractile upper jaws whose body shape ranges from slender to deep. There are 17 genera and about 90 species in the family, including *Distichodus, Ichthyborus, Nannocharax,* and *Neolebias,* which are illustrated in Gosse and Coenen (1990). *Ichthyoborus besse* (Fig. 65) from the White Nile, Chad basin, and Benue River reaches 245 mm TL. *Hemigrammocharax*

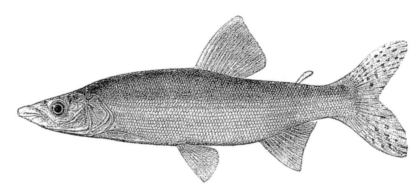

FIGURE 65. *Ichthyoborus besse* (Boulenger, 1909, Fig. 192).

and *Nannocharax* are so-called African darters because of their superficial similarity to North American darters of the family Percidae (Buckup, 1993a). *Distichodus niloticus* is the largest species at 83 cm TL and 6.2 kg (Daget and Gosse, 1984). *Distichodus lusosso* (Fig. 66) from the Congo

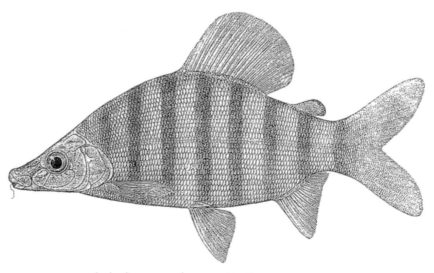

FIGURE 66. *Distichodus lusosso* (Boulenger, 1909, Fig. 212).

basin is more typical at 38 cm TL. Vari (1979) discussed the osteology and soft anatomy of the distichodontids.

Map references: Daget and Gosse (1984), Gosse and Coenen (1990),* Poll (1973),* Roberts (1975), Skelton (1993)*

Class Actinopterygii
Subclass Neopterygii

Order Characiformes; Suborder Citharinoidei
(1st) Family Citharinidae—**citharinids** [sith'-a-rī'-ni-dē]

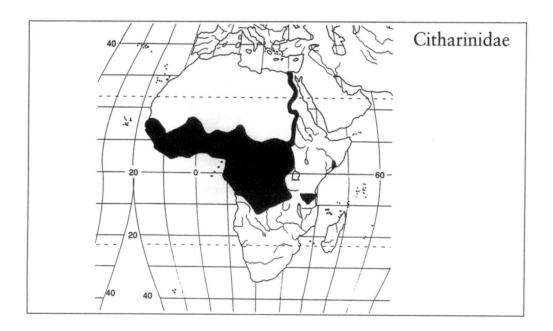

THE SECOND FAMILY WITHIN THE SUBORDER CITHARINOIDEI, Citharinidae, is composed of three genera. *Citharinus* has six species, and *Citharinops* and *Citharidium* are monotypic. These fishes have a deep body and small, toothless maxillae. The mouth is adapted for benthic feeding. There are many gill rakers, and the intestines are very long. In this respect they are similar to the South American curimatids and prochilodontids, which are also detritus eaters. This provides an interesting example of ecological parallelism. Scales are usually cycloid except for the ctenoid scales of *Citharidium* (Géry, 1977).

Citharinids occur in west Africa, the Congo (Zaire) River basin, Nile River, and in Lakes Albert and Tanganyika of East Africa (Daget, 1984d). Species diversity is greatest in the Congo basin. *Citharinops distichodoides* and *Citharinus latus* are the largest species at 84 cm TL and 18 kg (Daget, 1984d). Such large fish are utilized as food by the local people. Smaller species include *Citharinus congicus* (Fig. 67) at 43 cm TL. Vari (1979) examined the anatomy of this family and concluded that it constitutes a monophyletic group.

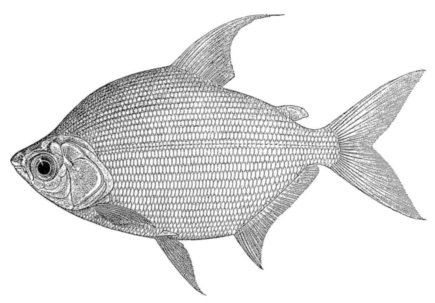

FIGURE 67. *Citharinus congicus* (Boulenger, 1909, Fig. 224).

Map references: Daget (1984d), Gosse (1990b),* Poll (1973),* Roberts (1975)

Class Actinopterygii
Subclass Neopterygii

Order   Characiformes; Suborder Characoidei
(1st) Family   Parodontidae—**parodontids**   [par'-ō-don'-ti-dē]

THIS NEOTROPICAL FAMILY RANGES FROM THE PACIFIC SLOPES of Panama, Colombia, and Ecuador to the Río Orinoco, the Guianas, the

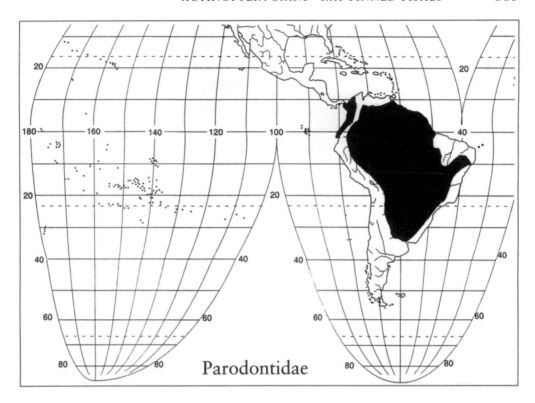

Parodontidae

Amazon, São Francisco, and the Río Paraná basin except in the lowest portions of major basins and the higher Andes (Roberts, 1974a; Starnes and Schindler, 1993). These fishes are benthic in mountain streams from 100 to 1100 m (Roberts, 1974b), but they are absent in the high plateaus between Rio Grande do Sul and Santa Catarina, Brazil, and in the mountains around Rio de Janeiro (P. Buckup, personal communication). The pectoral fins are expanded horizontally which holds the fishes in position in strong water currents. The Parodontidae has large and highly mobile premaxillaries and a ventral mouth. Their upper jaw teeth are modified for scraping periphyton off rocks. A series of from 4 to 30 replacement teeth lie in a trench of the premaxillary bone behind the functional teeth and succeed each other, one at a time, as the functional teeth are shed (Roberts, 1974b). Nuptial tubercles occur in many species (Roberts, 1974b).

Nelson (1994) treated this group as a subfamily of the Hemiodontidae, but Buckup (1998) showed that parodontids are not closely related to hemiodontids. Roberts (1974a), Starnes and Schindler (1993), Buckup (1998), and Langeani (1998) placed the parodontids in their own family, which has been demonstrated to be monophyletic (Vari, 1995).

The three genera *Apareiodon, Parodon,* and *Saccodon* include about 23 species, the largest of which reach 20 cm SL. See Roberts (1974b) for a review of *Saccodon* and its dental morphology and Starnes and Schindler

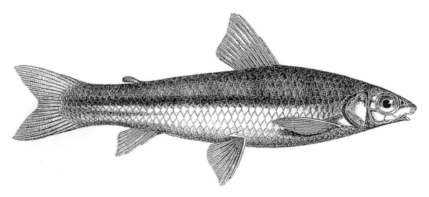

FIGURE 68. *Apareiodon affinis* as *Parodon affinis* (Steindachner, 1879a, Plate III, Fig. 3).

(1993) for a discussion of *Apareiodon*. *Apareiodon affinis* (Fig. 68) is 1 of 14 nominal species assigned to this genus.

Map references: Géry (1977), Miller (1966), Roberts (1974a,b),* Starnes and Schindler (1993)*

Class Actinopterygii
Subclass Neopterygii

Order Characiformes; Suborder Characoidei
(1st) Family Curimatidae—**curimatids** [kū-rē-ma'-ti-dē]

THIS FAMILY OF EIGHT GENERA AND 97 SPECIES OCCURS IN THE fresh waters of southern Central America and South America, including regions of endemism along the western slopes of the Andean Cordillera (Vari, 1988). The Curimatidae is abundant in the Amazon (Goulding, 1980). Vari (1992b) wrote many papers about this family between 1982 and 1992. He considered the groups Curimatidae, Prochilodontidae, Anostomidae, and Chilodontidae to be of family rank (Vari, 1983), as did Buckup (1998). Nelson (1994) and Eschmeyer (1998) treated the first two groups as subfamilies of the Curimatidae and the second two groups as subfamilies of the Anostomidae.

The Curimatidae has a buccopharyngeal complex composed of three longitudinal folds of soft tissue hanging from the roof of the oral cavity (Vari, 1989a). Some species also have secondary folds. Mucous-producing tissue is associated with the folds. Vari (1991) suggested that the buccopharyngeal complex serves as a particle filtering system with detritus, algae, and other fine particles becoming trapped by the mucous on the folds.

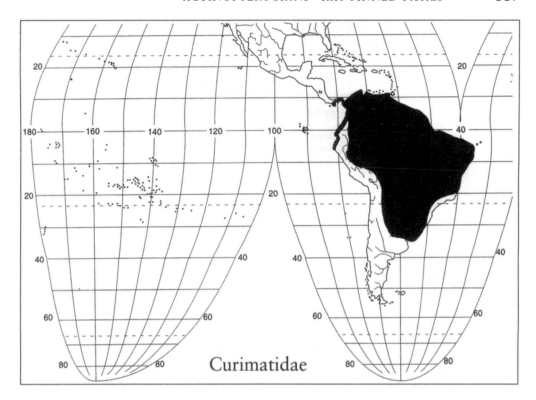

The Curimatidae are toothless detritivors. The biggest of the 97 species can reach 35 cm TL and is eaten by the local people, who exploit the large spawning runs. Vari (1989a) provided a phylogenetic study and keys (Vari, 1992b) to eight genera: *Curimatopsis, Potamorhina, Curimata, Psectrogaster, Curimatella, Steindachnerina, Pseudocurimata*, and *Cyphocharax*. This last named genus contains about one-third of the species in the family. *Steindachnerina* contains 21 species including *S. semiornatus* (Fig. 69; Vari, 1991). Additional species have been described

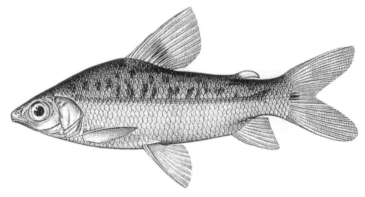

FIGURE 69. *Steindachneria semiornatus* as *Curimatus semiornatus* (Steindachner, 1915, Plate V, Fig. 5).

by Vari and Blackledge (1996) and Vari and Reis (1995). Smaller species, such as those of the genus *Curimatopsis*, exhibit pronounced sexual dimorphism, with highly colored males utilizing their tail fins for display (Weitzman and Vari, 1994). Vari and Weitzman (1990) used the Curimatidae to test alternative hypotheses about the diversity of the South American freshwater fish fauna.

Map references: Miller (1966), Vari (1982a,* 1984,* 1988,* 1989b,*c,*d,* 1991,* 1992a,*b*),* Vari and Barriga (1990),* Vari and Weitzman (1990)*

Class   Actinopterygii
Subclass   Neopterygii

Order   Characiformes; Suborder Characoidei
(1st) Family   Prochilodontidae—**prochilodontids**
(prō-kī'-lō-don'-ti-dē)

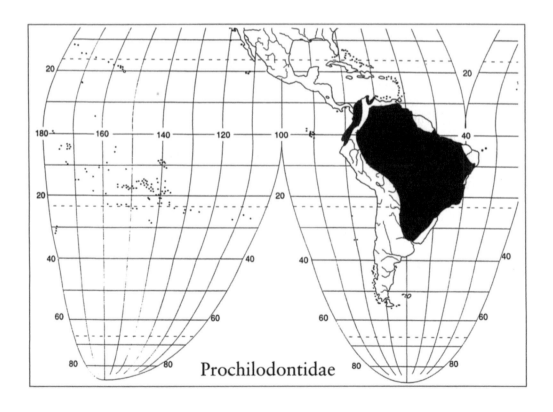

Prochilodontidae

THIS FAMILY OCCURS FROM THE PACIFIC SLOPES OF COLOMBIA
and Ecuador through the Rio Magdalena, Orinoco, Amazon, and São
Francisco basins to the Rio Paraná basin (Roberts, 1973). There are about
21 species in three genera *Ichthyoelephas*, *Prochilodus*, and
Semaprochilodus (R. P. Vari, personal communication) *Prochilodus argenteus* (Fig. 70) is the type of a genus that has little variation, which makes
identification difficult (Géry, 1977).

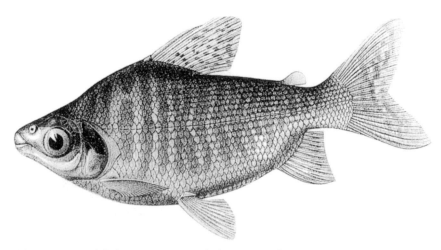

FIGURE 70. *Prochilodus argenteus* (Steindachner, 1915, Plate V, Fig. 4).

The Prochilodontidae has many minute teeth on the jaws and
a protractile mouth with enlarged, fleshy lips often beset with fine papillae,
which earns them the name "flannel-mouths." They are detritivors, as
are their curimatid relatives. The group as a whole is the commercially most important fish family in Neotropical fresh waters, and many
species undertake lengthy upstream spawning migrations during the
wet season. Some species form huge schools in the Rio de la Plata; these
may be the largest schools of any South American freshwater fish
family (Roberts, 1973). A loud, grunting noise emitted during spawning
has been compared to the sound of a motorcycle (Roberts, 1973).
Fecundity may be enormous, with some species producing more
than 600,000 eggs. The largest species reach 60 cm TL. Roberts
(1973) described the osteology of the family and summarized life
history information. Vari (1983) presented anatomical evidence to
separate this family from the Curimatidae and to define it as monophyletic.

Map references: Géry (1977), Roberts (1973), Vari (1983)

Class Actinopterygii

Subclass Neopterygii

Order   Characiformes; Suborder Characoidei
(1st) Family   Anostomidae—anostomids, headstanders
[an-os-tōm'-i-dē]

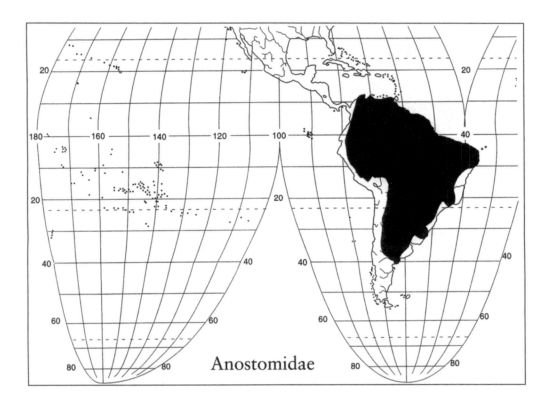

Anostomidae

THE ANOSTOMIDAE IS COMPOSED OF ABOUT 10 GENERA AND about 105 species from the fresh waters of South America. It occurs in the Magdalena, Orinoco, Amazon, São Francisco, Paraguay, Paraná, and Salado River systems (Eigenmann and Eigenmann, 1891; Vari, 1983; Vari and Williams, 1987).

The bodies of most genera are elongate and round in cross section. Anostomids have a short mandible and a single tooth row on each jaw. The premaxilla is enlarged, and the teeth are curved. Many of the smaller species are beautifully colored and are popular with aquarists. Nelson (1994) and Eschmeyer (1998) recognized two subfamilies, Anostominae and Chilodontinae, that are treated as families by Vari

(1983), Vari *et al.* (1995), and Buckup (1998). Géry (1977) provided keys and photographs.

The common name, headstander, derives from the fact that some species orient the long axis of their bodies obliquely about 25–30° to the bottom with the head down. These species feed in this position by picking small organisms off the vertical stems of plants with their small, nonprotractile mouths that open dorsally. Longitudinal stripes camouflage many species among the vegetation in which they feed. Most species are herbivores or detritivores.

Schizodon has 10 species, and *Leporinus* is the largest genus with more than 70 species, some of which have greatly enlarged symphyseal teeth. The name "*Leporinus*" means "a young hare" and may refer to the vaguely hare-lipped appearance of the mouth (Sterba, 1966) or, more likely, the rabbit-like teeth (P. Buckup, personal communication). This odd dentition is taken to the extreme in *Sartor respectus* from the Rio Xingú of Matto Grosso, Brazil, with its awl-like mandibular teeth. The symphysial pair are curved, greatly elongated, and project straight out from the end of the lower jaw. They extend far above the lip of the upper jaw when the mouth is closed (Myers and Carvalho, 1959). The eyes of *S. respectus* are set low on the head and directed upward and backward. It may be able to see backward better than it sees forward.

Winterbottom (1980) discussed the phylogeny of the Anostominae. Some other genera include *Abramites, Anostomus, Gnathodolus, Leporellus,* and *Rhytiodus*. These genera are usually found in the Amazon and Guiana plateau. Most species are less than 15 cm, but *Leporinus arcus* (Fig. 71) can reach 40 cm SL and some species reach 60 cm. These larger species are captured as food fishes during upstream spawning migrations. *Anostomus orinocensis* (Fig. 72) has the typical body shape of members of this family. Goulding (1980) reported on the stomach contents and dentition of four species utilized as food in the Amazon. Vari and Williams

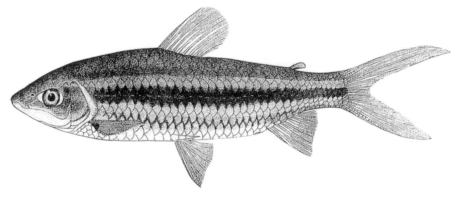

FIGURE 71. *Leporinus arcus* (Eigenmann, 1912, Plate XLII).

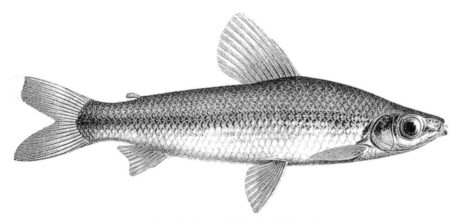

FIGURE 72. *Anostomus orinocensis* (Steindachner, 1879b, Plate II, Fig. 7).

(1987) revised *Abramites* and Vari and Raredon (1991) examined *Schizodon* from northern South America.

Map references: Eigenmann and Eigenmann (1891), Géry (1977), Vari (1983), Vari and Williams (1987),* Winterbottom (1980)*

Class Actinopterygii
Subclass Neopterygii

Order Characiformes; Suborder Characoidei
(1st) Family Chilodontidae—**chilodontids, headstanders**
(kī-lō-don'ti-dē)

THIS FAMILY IS WIDELY DISTRIBUTED IN NORTHERN SOUTH America east of the Andes in the Orinoco and Amazon River basins, the Guianas, the Paraíba basin of northeastern Brazil, and the upper Río Madeira system of southeastern Peru (Vari *et al.*, 1995; Vari and Ortega, 1997). Chilodontids may be distinguished from other characiforms by their very small sixth lateral-line scale.

The Chilodontidae consists of seven species in two genera, *Caenotropus* (three species, represented by *C. labyrinthicus*; Fig. 73), with an inferior mouth, and *Chilodus* (four species), with a terminal or upturned mouth. Their premaxilla is small, but the maxilla is enlarged. Only a few, feeble jaw teeth are present, and they are typically brown-tipped.

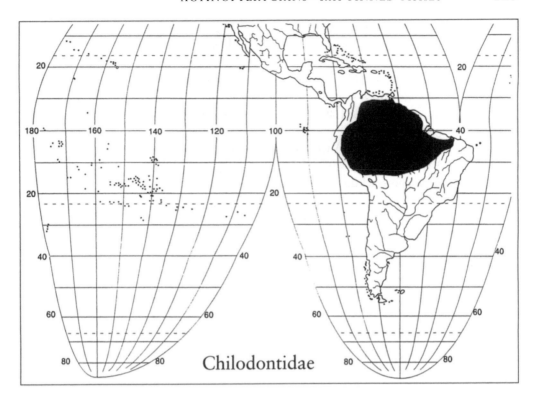

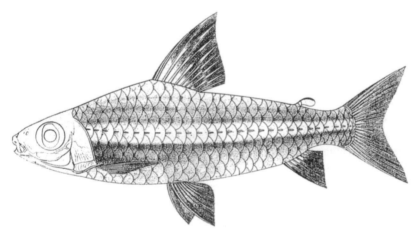

FIGURE 73. *Caenotropus labyrinthicus* as *Microdus labyrinthicus* (Kner, 1859, Plate III, Fig. 5).

Chilodontids have highly modified gill arches and epibranchial organs. The bodies are elongate with elevated backs. They share the peculiar headstanding behavior with their close relatives, the Anostomidae; however, chilodontids orient their head downward at a steeper angle, about 60–70° (S. H. Weitzman, personal communication). They have a series of distinctive modifications of the axial skeleton which may be correlated with their unusual head-down orientation (Vari *et al.*, 1995). Isbrücker and Nijssen (1988) reviewed the genus *Chilodus* and Vari and Ortega (1997) added a new species, *Chilodus fritillus*, whose range extends into southeastern Peru. Vari *et al.* (1995) revised *Caenotropus*, and Vari and Ortega (1997) provided additional distributional records for this genus.

Nelson (1994) treated this group as a subfamily of the Curimatidae. Vari (1983) and Vari *et al.* (1995) recognized the group as a monophyletic unit, and Buckup (1998) considered it as a family.

Map references: Géry (1977), Isbrücker and Nijssen (1988),* Vari (1983), Vari and Ortega (1997),* Vari *et al.* (1995)*

Class Actinopterygii
Subclass Neopterygii

Order Characiformes; Suborder Characoidei
(1st) Family Crenuchidae—crenuchids, South American darters [kren-ooch'-i-dē]

NELSON (1994) LISTED THESE FISHES IN TWO SUBFAMILIES OF the Characidae: the Crenuchinae and the Characidiinae. Buckup (1998) treated the two groups as monophyletic subfamilies of the Crenuchidae with a total of 11 genera and 64 species. Crenuchids are found in most freshwater drainages from the Pacific slope of eastern Panama to Argentina (Miller, 1966; Buckup, 1993a; Buckup and Reis, 1997). They possess unique lateral frontal foramina located behind the orbits on top of the head. These frontal foramina (one on each side) are very large and have a pad of connective tissue in front of them. Opthalmic nerve fibers pass through the foramina. The function of this organ is not known.

The Crenuchinae is from northern South America. There are only three species in two genera. *Crenuchus spilurus* resembles a dwarf cichlid except for the presence of an adipose fin. The two species of *Poecilocharax* are cyprinodont-like and are missing the adipose fin (Géry, 1977).

The Characidiinae includes interesting little darter-like characins. Some live a benthic existence in fast-flowing waters and are convergent with North American darters of the family Percidae, just as the distichodontid characiform genera *Nannocharax* and *Hemigrammocharax*

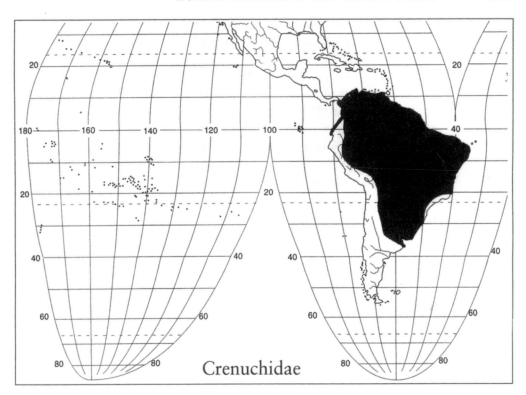

Crenuchidae

are known as African darters (Buckup, 1993a). There are nine genera (*Ammocryptocharax, Characidium, Klausewitzia, Elachocharax,* and others) and about 64 species. *Charadidium* is the most speciose genus with about 42 valid species, including *C. hasemani* (Fig. 74; Buckup and Reis, 1997; Buckup and Hahn, 2000). Buckup (1993b) provided keys to genera and species.

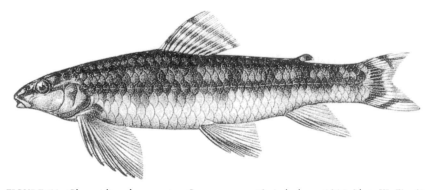

FIGURE 74. *Characidium hasemani* as *C. surumuense* (Steindachner, 1915, Plate III, Fig. 3).

Most characidiins are small, rarely exceeding 100 mm SL (Buckup, 1993a). Many are miniatures (Weitzman and Vari, 1988) with an adult size less than 20 mm SL. Miniaturization is a common phenomenon among Neotropical fishes and often results in the presence of numerous apparently paedomorphic characters and maturation at greatly decreased adult body size (Buckup, 1993c). At least 85 miniature fish species (adult size less than 26 mm SL) have been described from South American fresh waters, and 88% of these are characiform and siluriform fishes (Weitzman and Vari, 1988). Costa and Le Bail (1999) noted that an additional 23 South American miniatures have been described as new species since 1988, and Géry and Römer (1997) described an additional miniature characid not counted by Costa and Le Bail for a total of 109 South American miniatures.

The Characidiinae is also abundant in small, sluggish, lowland rain forest streams and coastal plain marshes (Buckup, 1993b). They are the most abundant characiform fishes in leaf-litter, bank fish communities in central Amazonian blackwater streams. They are usually indicative of healthy streams and associated forests and, like their environments, some species are endangered (P. Buckup, personal communication). Buckup (1993b) discovered a major radiation of characidiin fishes in northern South America. See Weitzman and Kanazawa (1976), Weitzman and Géry (1981), Buckup (1993a,b,c), and Buckup and Reis (1997) for evidence of monophyly, interrelationships, and species descriptions.

Map references: Buckup (1993a,b),* Buckup and Reis (1997),* Géry (1977), Miller (1966), Weitzman and Géry (1981)*

Class Actinopterygii
Subclass Neopterygii

Order Characiformes; Suborder Characoidei
(1st) Family Hemiodontidae—**hemiodontids** (hem-i-ō-don'-ti-dē)

THE HEMIODONTIDAE, AS CONSTITUTED BY LANGEANI (1998), is composed of five genera and about 37 species. The family name, which means "half-toothed," refers to the fact that the lower jaw lacks teeth in most species. The mouth is subterminal. These fish are widespread in tropical fresh water in South America in the Amazon and Orinoco basins, the rivers of Guyana, Suriname, and French Guiana, and south in the Paraná and Paraguay Rivers (Roberts, 1974a; Langeani, 1998). Some species occur at elevations higher than 1000 m in Andean streams (Géry, 1977). Géry provided keys to species and photographs. Langeani

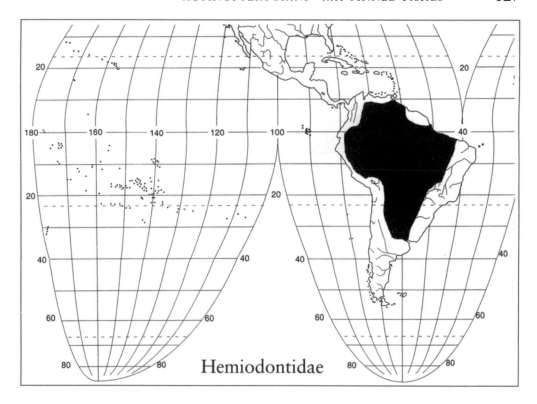

Hemiodontidae

confirmed the monophyly of the Hemiodontidae and excluded the Parodontidae, which some authors had aligned with the Hemiodontidae (Nelson, 1994). This conclusion was also supported by the results of Buckup (1998).

The Hemiodontidae is pelagic rather than benthic and has an adipose eyelid that is absent in the Parodontidae. The largest species reach 30 cm. *Micromischodus sugillatus* is the only hemiodontid with teeth on the lower jaw (Roberts, 1971a). *Bivibranchia* and *Argonectes*, with 7 species, are the only characiform fishes with protrusible upper jaws (Vari and Goulding, 1985). These fishes feed by sucking in bottom sand and extracting the small invertebrates living in the sand. Three species of *Anodus* (= *Eigenmannina*) lack all jaw teeth but have a large number of gill rakers (up to 200 on first arch) and pharyngeal structures with which they strain plankton (Roberts, 1972). *Hemiodus* (= *Hemiodopsis, Pterohemiodus*) (22 species, represented by *H. immaculatus*; Fig. 75) often has a black lateral stripe that extends from the opercular margin into the upper and/or lower lobe of the caudal fin. There is also a lateral spot between the posterior edge of the dorsal and anal fin. The color patterns exhibited by *Hemiodus* are very distinct and conspicuous. The presence of the midlateral body spot is widespread and considered a plesiomorphic state for the family. Langeani (1999) discussed the evolution of color patterns and tooth shapes in *Hemiodus*. Many of the smaller, more colorful species are popular aquarium animals.

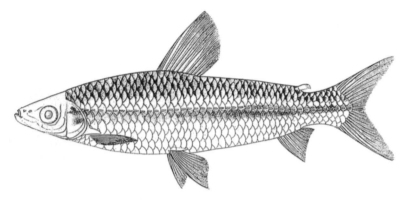

FIGURE 75. *Hemiodus immaculatus* (Kner, 1859, Plate V, Fig. 10).

Map references: Géry (1977), Langeani (1998), Roberts (1971a, 1974a), Vari and Goulding (1985)*

Class Actinopterygii
Subclass Neopterygii

Order Characiformes; Suborder Characoidei
(1st) Family Alestidae—**African characins or tetras** [a-les'-ti-dē]

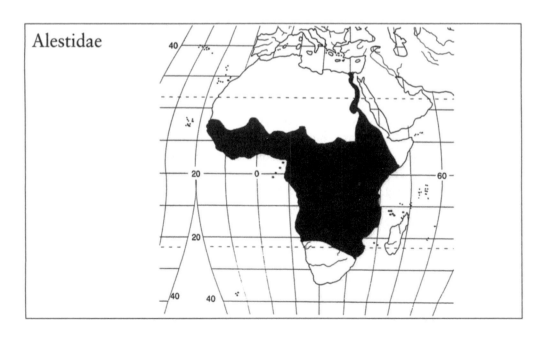

THIS GROUP WAS TREATED AS A SUBFAMILY OF THE Characidae by Nelson (1994). Ortí (1997) demonstrated its monophyly based on molecular data. Buckup (1998) concluded that including the African tetras in the Characidae would make the latter family polyphyletic and treated this group as a family, as did Géry (1977). In references that show a map of the "Characidae," the African distribution belongs to the Alestidae as constituted by Buckup (1998). The distribution of the Alestidae includes the Nile system and sub-Saharan Africa exclusive of the Red Sea coastal regions and extreme southern Africa. African characins, such as

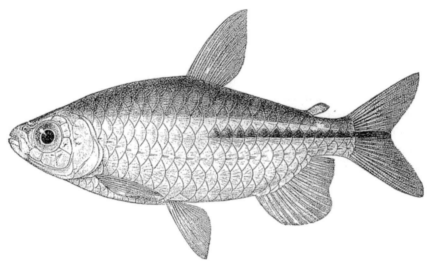

FIGURE 76 *Alestes taeniurus* (Boulenger, 1909, Fig. 161).

Alestes taeniurus (Fig. 76), are most common in the coastal rivers of west Africa and in the Congo (Zaire) basin (Paugy, 1984).

The Alestidae includes about 102 species of African characins in 13 genera such as *Alestes, Hydrocynus, Lepidarchus, Micralestes*, and *Rhabdalestes*. See Paugy (1984) for a list of species throughout Africa and Paugy (1990d) for keys and illustrations of west African species. Fossils assigned to this family date to the Eocene of western Europe (Lundberg, 1993). Géry (1977) divided the family into two subfamilies: The piscivorous Hydrocyninae have piercing teeth in one row on both jaws, and the omnivorous Alestinae have teeth in two or more rows on the upper jaw. This subdivision may be artificial (P. Buckup, personal communication). See Vari (1979) and Brewster (1986) for evidence that the Alestinae is nonmonophyletic. Many alestins are small and popular among tropical fish hobbyists as African tetras.

There are six species of tigerfish, *Hydrocynus*, in Africa (Paugy, 1984). The largest, *H. goliath* (Fig. 77), is a brightly striped, streamlined fish with

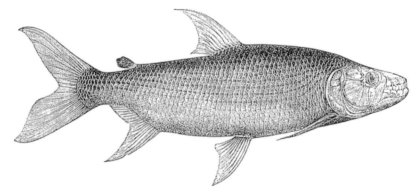

FIGURE 77. *Hydrocynus goliath* as *Hydrocyon goliath* (Boulenger, 1909, Fig. 142).

large, sharp, widely separated, interlocking upper and lower jaw teeth. This specialized predator has a lower jaw hinged at its anterior apex that allows a huge, lateral gape. It may reach 1.5 m and 50 kg in the Congo River system, and it is regarded as one of the finest freshwater game fishes in the world (Skelton, 1993). (Its South American angling equivalent would be *Salminus* of the Characidae.) Kenmuir (1973) reviewed the ecology of *H. vittatus* in Lake Kariba. See Brewster (1986) and Paugy and Guégan (1989) for a review of *Hydrocynus*.

Map references: Géry (1977), Norman and Greenwood (1975),* Paugy (1984, 1990c*), Skelton (1993)*

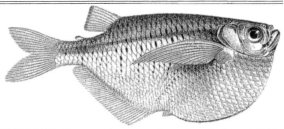

Class Actinopterygii

Subclass Neopterygii

FIGURE 78 *Gasteropelecus maculatus* (Steindachner, 1879b, Plate I, Fig. 4).

Order Characiformes; Suborder Characoidei

(1st) Family Gasteropelecidae—hatchetfishes, flying **characins** (gas'-te-rō-pe-les'-i-dē)

THE HATCHETFISHES ARE FOUND FROM PANAMA TO THE RÍO de la Plata in Argentina (Miller, 1966; Weitzman, 1954; Fraser-Brunner, 1950). There are three genera (*Carnegiella*, *Gasteropelecus*, and *Thoracocharax*) and about nine species. *Gasteropelecus maculatus* (Fig. 78)

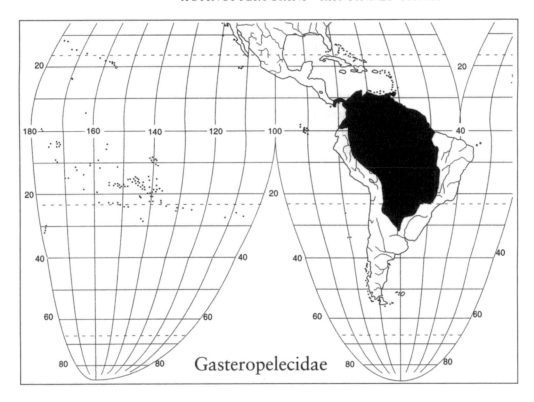

Gasteropelecidae

occurs on the Pacific slopes of Panama and on the Pacific and Atlantic drainages of Colombia to the Maracaibo basin of Venezuela (Miller, 1966; Weitzman, 1996). *Thoracocharax stellatus* occurs in the Amazon system as well as the southward-flowing La Plata system at Asuncion, Paraguay, and Santa Fe, Argentina (Fraser-Brunner, 1950). Several nearly miniature species of *Carnegiella* occur in various parts of the Amazon basin (Weitzman, 1966).

Hatchetfishes have a highly developed pectoral fin musculature attached to the sternum. This keel-like structure is a modification of broad coracoid bones of the pectoral girdle. The chest muscles and sternum may account for 25% of the animal's weight (Fraser-Brunner, 1950). The resulting shape resembles a hatchet, hence the common name hatchetfish. The dorsal profile is nearly a straight line.

These small, extremely compressed, deep-bodied fishes are often cited as the only true flying fishes because they have been reported to flap their greatly enlarged pectoral fins during flight, as opposed to merely gliding on rigid "wings," as do the marine flying fishes of the family Exocoetidae. The fin flapping was said to create an audible buzzing noise (Weitzman, 1954; Weitzman and Palmer, 1996). However, Weitzman's suggestion of propelled flight was based on incidental observation and plausible suppositions and not direct observation (Weitzman and Palmer, 1996). High-speed video and motion analysis demonstrated that the pectoral fins were not utilized while

Carnegiella was airborne (Wiest, 1995). The pectoral fins were used as part of the takeoff mechanism by downward thrusts against the water that, along with tail thrusts, propel the fish from the water. When airborne, the pectoral fins return to the sides of the body for the remainder of the leap. It is the motion of the tail fin in air that generates the buzzing noise heard by Weitzman (Wiest, 1995; Weitzman and Palmer, 1996), but this motion does not appear to contribute to the distance of the leap, which can be 2 m long in nature. In the lab the aerial distance could equal 30 times the length of the fish. Wiest's (1995) observations seem to indicate that hatchetfishes need a rest after several leaps before they can get airborne again. This is thought to be due to muscle fatigue caused by lack of sufficient oxygen and nutrients.

Some hatchetfish have a lateral line that passes downward across the body from behind the head to a point anterior to the front of the anal fin. This arrangement ensures that part of the lateral line remains in the water while the fish is taxiing (Fraser-Brunner, 1950). During takeoff the large keel cleaves the water, and the pectoral fins, positioned high on the body, beat the surface (Fraser-Brunner, 1950). Flight is a mechanism for escaping predation. It is doubtful that leaping is useful for catching airborne insects.

The maximum size is about 10 cm for some members of the genus *Thoracocharax*. Smaller members of the family are approximately 3 cm. *Carnegiella*, the smallest and prettiest genus with a marbeled, spotted or lined body, lacks an adipose fin. Weitzman and Weitzman (1982) studied *Carnegiella* in relation to possible vicariant forest refuge events and speciation. Hatchetfishes tend to prefer forest streams, but *Thoracocharax* can be found in more open waters and even lakes. Hatchetfishes feed on insects at the water's surface. These fishes are popular aquarium animals (with a lid on the tank, of course). Fraser-Brunner (1950) and Weitzman (1954, 1960) reviewed the family. Géry (1977) provided keys and photographs. Buckup (1998) considered the phylogenetic position of the gasteropelecids as uncertain.

Map references: Fraser-Brunner (1950), Miller (1966), Weitzman (1954, 1960), Weitzman and Weitzman (1982)*

Class Actinopterygii
Subclass Neopterygii

Order Characiformes; Suborder Characoidei
(1st) Family Characidae—**characids** (ka-ras'-i-dē)

Characidae

THE CHARACIDAE, AS DEFINED BY BUCKUP (1998), IS THE Characidae of Nelson (1994) minus the Acestrorhynchidae, Alestidae, Crenuchidae, and *Roestes* and *Gilbertolus*. This arrangement, of course, gives a very different family distribution by removing the Characidae from Africa. Currently, the Characidae cannot be diagnosed as a monophyletic group (Weitzman and Fink, 1983; Buckup, 1991, 1998; Weitzman and Malabarba, 1998), and the limits of the family should be considered tentative because some taxa traditionally included in the family are still poorly known (Buckup, 1998). Characid systematics awaits a comprehensive phylogenetic analysis, and its taxonomy does not enjoy universal acceptance. Papers within Malabarba *et al.* (1998) will go a long way toward eventually establishing classification and phylogenetic stability. Géry (1977) provided keys to various groups. The common name "characin," especially in the aquarium trade, is imprecisely used to denote anything from all characiform fishes to members of a specific subfamily of the Characidae. To avoid misunderstanding, "characid" will be used as the common name for members of this family.

The Characidae is a large and diverse family of about 138 genera and more than 700 species that occurs in fresh waters from the Rio Grande in southern Texas (*Astyanax mexicanus*) through Mexico and Central America to about latitude 41°S in Chile and Argentina (*Cheirodon australe*). *Cheirodon galusdae* (Fig. 79) of central Chile occupies the ecological niche

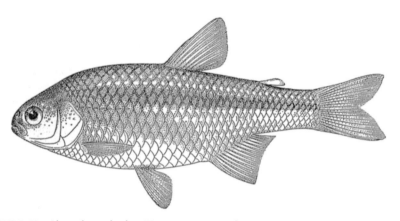

FIGURE 79. *Cheirodon galusdae* (Eigenmann, 1928, Plate IX).

filled by *C. australe* in more southern waters. Characids are absent from most of North America, all of Eurasia, the Orient, and Australia. In these areas, except Australia, cyprinids occupy many niches similar to those filled by characids in South America. Conversely, the cyprinids are absent from South America where characids dominate. These two families can be considered more or less ecological equivalents. Most characids have one or both of the following characteristics: jaw teeth and an adipose fin. These features distinguish characiforms from cyprinids.

With a great deal of morphological diversity in the family, one would expect a wide variety of feeding habits. Some species are herbivorous, some omnivorous, and some carnivorous. A few species have become specialized as fin and scale eaters (Roberts, 1970a; Sazima, 1983). Most lepidophagous species are small, less than 20 cm, and have modified teeth. They usually hunt from ambush and attack with a jarring strike at the prey's flank.

In South America a few characids have evolved into very large, fruit- and seed-eating fishes. This unique niche has no parallel elsewhere. The tambaqui, *Colossoma macropomum* (Fig. 80), is the largest characid in the Amazon basin and reaches 1 m SL and a weight of 30 kg. It is also the most important commercial species in the western Amazon. Its massive molariform and incisive dentition have evolved to crush hard nuts that fall

FIGURE 80. *Colossoma macropomum* as *Myletes nigripinnis* (Steindachner, 1882a, Plate VII, Fig. 1).

from the forest into the water. During the annual floods fishes move into the flooded forest and feed on fruits and seeds. An effort is currently under way to raise this important species in aquaculture (Araujo-Lima and Goulding, 1997). Plant material is an important part of the diet of adult *Colossoma, Mylossoma, Myleus*, and *Brycon*. These fishes play a role in the dispersal of the plants whose fruits they eat. Many of these fishes reach large size and are utilized by the local people as food. If the South American rain forests are destroyed for lumber and cattle pastures, so will be these important food fishes. Goulding (1980), Araujo-Lima and Goulding (1997), and Barthem and Goulding (1997) detailed the relationship between fishes and the flooded forests. Lundberg *et al.* (1986) reported fossils of an extant species, *Colossoma macropomum*, dating back 15 million years to the Miocene—a long time for a species to remain unchanged. This reflects a conservative history for the adaptation of feeding on fallen fruits and seeds and establishes a minimum age for the origin of this ecological specialization.

The subfamily Serrasalminae includes species of all gastric persuasions. The body of serrasalmins is typically compressed and deep. The belly is armed with a variable number of scutes which gave rise to the subfamily name which means "serrated salmons." Some species may have a single spine anterior to the dorsal fin which tends to be larger than in other characids. The scales are small and numerous. Teeth and digestive tract are variable depending on the diet of the species. This group is listed as a family by Géry (1977), whereas Ortí *et al.* (1996) treated it as a subfamily of the Characidae. *Colossoma* and the other herbivores mentioned previously such as the pacus (*Myleus* and *Mylossoma*) and the scale eater, *Catoprion mento*, are among the 13 genera and approximately 60 species in the subfamily Serrasalminae, as are the carnivores discussed next.

Related to the herbivorous *Colossoma* is a group of characids that everyone has heard of—the "piranhas" (pronounced pir-an'-yaa) and their close relatives. The Spanish equivalent of piranhas (Portuguese) is "caribes." Not all members of the subfamily are flesh eaters. For example, the silver dollar fishes, *Metynnis*, make peaceful aquarium animals. On the other hand, the approximately 50 species of *Pygopristis*, *Pygocentrus*, *Pristobrycon*, and *Serrasalmus* (represented by *S. hollandi*; Fig. 81) are to

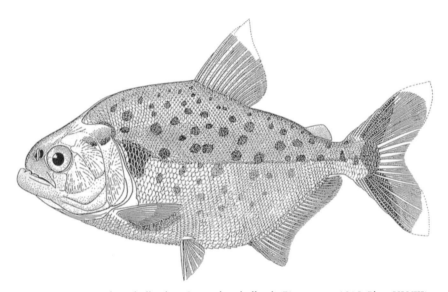

FIGURE 81. *Serrasalmus hollandi* as *Serrasalmo hollandi* (Eigenmann, 1915, Plate XLVIII).

be treated with respect due to their robust, powerful jaws and sharp teeth. A school of frenzied red-bellied piranha, *Pygocentrus nattereri*, may skeletonize a bleeding steer in a matter of minutes (Webster, 1998, p. 21). Conniff (1999), however, presented a more benign picture of piranhas. At quieter times, collectors have often swum with them without incident (Weitzman and Vari, 1994; Conniff, 1999). Some of the horror stories of piranha attacks on humans were the result of scavenging on dead bodies (Sazima and Guimaraes, 1987). See Sazima and Machado (1990) for underwater observations of piranhas. Piranhas and their relatives are found in the Orinoco, Amazon, Paraguay, and São Francisco River basins and the streams of the Guianas.

Fink (1989) studied the changes in body shape and diet during ontogeny. Most piranha species begin as insect feeders and add fruit and flesh to their diet as they grow. In the clade composed of *Pygocentrus, Serrasalmus,* and *Pristobrycon* scales and fins are included in the juvenile diet.

Bennett *et al.* (1997) studied cold tolerance and overwintering of *P. nattereri* in the United States. They found that below 14°C red-bellied piranhas were unable to catch goldfish prey. Below 12°C they would not accept frozen brine shrimp. At 10°C they lost equilibrium. Piranhas most likely could establish permanent populations where January water temperatures do not fall below 14°C. In the United States, this would include Hawaii, regions of southern California, Texas, and Florida. Cyclic recurrence of lethal low temperatures (10°C) make it unlikely that piranha populations could become established in southern Louisiana, Mississippi, Alabama, or Georgia. Fuller *et al.* (1999) reviewed the occurrences of piranhas in 14 states of the United States.

A very diverse group of well-known aquarium animals called "American tetras" includes genera such as *Bryconops, Hemibrycon, Hyphessobrycon, Paracheirodon,* and *Tetragonopterus.* See Weitzman and Fink (1983) and Weitzman and Vari (1988) for discussion of some members of this group. The American tetras are a complex grouping of subfamilies, and there is no agreement among ichthyologists as to the definition of these groups. Some of the more prominent groups are the Tetragonopterinae, Cheirodontinae, and Glandulocaudinae. The exact relationship of *Brycon,* represented by *B. stolzmanni* (Fig. 82), is not clear but probably lies with the groups just mentioned (Weitzman and Malabara, 1998). Nelson (1994) listed 9 subfamilies of the Characidae, but several of these have been elevated to family level and others have been rearranged (Buckup, 1998). Géry (1977) provided keys and illustrations for

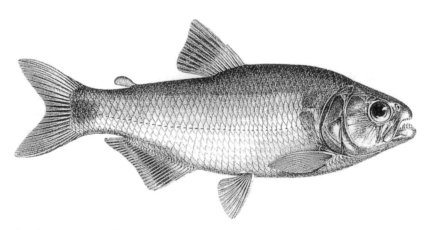

FIGURE 82. *Brycon stolzmanni* (Steindachner, 1879b, Plate II, Fig. 6).

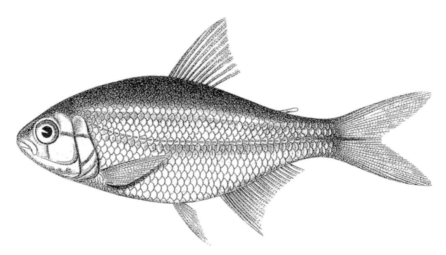

FIGURE 83. *Astyanax mexicanus* as *A. argentatus* (Girard, 1859, Plate LXXXIV).

12 subfamilies of American tetras. Many of the species discussed later belong to this broad assemblage.

The only species of characid to occur naturally in the United States is the Mexican characid, *Astyanax mexicanus*, a tetragonopterin (Fig. 83). This small (up to 12 cm), silvery fish prefers rocky and sandy bottomed pools in creeks and small to large rivers and springs. It originally was restricted to the Nueces, lower Rio Grande, and lower Pecos River drainages, but it is now also found in the streams of the Edwards Plateau in central Texas and throughout the Pecos River system in New Mexico. This range extension has been brought about by use of this species as bait. There are also scattered reports from Oklahoma and Louisiana (Birkhead, 1980). *Astyanax mexicanus* is considered conspecific with troglobitic populations of the blind Mexican cavefish (called *Anoptichthys jordani* in the older literature) (Avise and Selander, 1972; Kirby *et al.*, 1977; Romero, 1985a) and is probably not specifically different from the ubiquitous *A. fasciatus* (A. Romero, personal communication). See Mitchell *et al.* (1977) for a review of Mexican eyeless characids.

Eye development begins normally in cavefish embryos for the first 24 hr, but then programmed cell death (apoptosis) halts development and the eye degenerates. Yamamoto and Jeffery (2000) performed reciprocal transplantation experiments with *A. mexicanus* that showed that an inductive signal from the lens determines the morphological adaptation. When a lens is transplanted into another embryo, it behaves as if it were still in its original embryo. A cavefish lens results in a sunken, degenerate eye when transplanted into a surface embryo, and, conversely, a surface fish lens transplanted into a cavefish embryo results in a normal eye.

There is a second species of hypogean characid called *Stygichthys typhlops*, an extremely unusual fish not only because it lacks eyes and pigmentation but also because of many other aberrant features. Only one specimen has been seen, and it is known from only one locality, a well at Jaiba, Minas Gerais, Brazil (Romero and McLeran, 2000). The single specimen was 23.6 mm SL. The subfamily has not been determined.

There are many differing hypotheses to explain eyelessness in cave dwellers (Romero, 1985b). One view is that the loss of developmentally and energetically expensive tissue as found in an eye would be selected for in a food-poor environment such as a cave. Another view favors the random accumulation of "loss" mutations (Poulson, 1986). However, since there are thousands of hypogean species that are eyeless and depigmented, it is difficult to imagine neutral, nonselectionist pressure that could bring this about. It seems more plausible that convergent evolution is selecting for similar traits in hypogean species, perhaps by turning off regulatory genes.

Astyanax occurs further north than any other genus of the family, and *Cheirodon australe* (subfamily Cheirodontinae) occurs at the southern limit of the family in the Americas at about 41°30'S in Chile (Eigenmann, 1928). A rival as the most southerly characid is *Gymnocharacinus bergi*, a rare, nearly scaleless fish of the monotypic subfamily Gymnocharacininae that occurs in an isolated thermal stream in northeastern Patagonia at 40°50'S (Menni and Gómez, 1995; Ortubay *et al.*, 1997). Malabarba (1998) proposed a new classification of *Cheirodon* and some other genera in a monophyletic Cheirodontinae. Weitzman and Malabarba (1999) discussed phylogenetic relationships of the cheirodontins. The members of one subgroup, the Compsurini, are internally fertilizing (Burns *et al.*, 1997; Malabarba, 1998).

Characids figure prominently in the aquarium trade because of their diversity of form, brilliant color, interesting behavior, and often small size. A case in point is the gaudy neon tetra, *Paracheirodon* (= *Hyphessobrycon*) *innesi*, one of the most popular aquarium fishes in the world with its electric blues and reds. Many species in this family are less than 3 cm, whereas the smallest species may reach adulthood at only 13 mm (Weitzman and Vari, 1988).

The Glandulocaudinae is a South American subfamily of about 50 species in which adult males have modified caudal fin scales associated with well-developed glandular tissue from which a pheromone is pumped during courtship (Weitzman and Fink, 1985). Internal insemination and sperm storage are typical even though males lack an obvious intromittent organ (Burns *et al.*, 1995, 1998). They do, however, possess hooks on the anal fin which seem to be involved in male to female contact and thus in sperm transfer. Some of the 19 genera are *Corynopoma, Diapoma,*

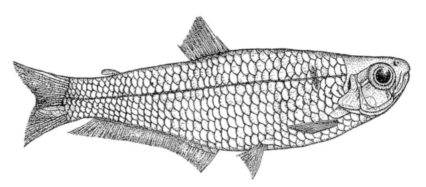

FIGURE 84. *Diapoma speculiferum* (Eigenmann, 1921, Plate 61).

Gephyrocharax, Glandulocauda, Mimagoniates, Tyttocharax, and *Xenurobrycon* (Weitzman and Menezes, 1998). The opercle of *Diapoma speculiferum* (Fig. 84) lacks almost all of its upper half, and the lower half forms a point. Its lateral line scales are present anteriorly and posteriorly but missing from the middle of the body. Weitzman *et al.* (1988) discussed the biogeography of this subfamily. See Menezes and Weitzman (1990) for a key to glandulocaudin fishes, Weitzman and Menezes (1998) for relationships within the subfamily, and Burns *et al.* (1998) for an analysis of sperm ultrastructure.

The Iguanodectinae is a South American subfamily with about six species in *Iguanodectes* and *Piabucus* (Vari, 1977). The Rhoadsiinae is a small group of three genera—*Carlana, Rhoadsia,* and *Parastremma*—found from Costa Rica to western Ecuador (Weitzman *et al.*, 1986).

Map references: Arratia (1981),* Eigenmann (1909a,* 1928), Géry (1977), Harold and Vari (1994),* Miller (1966),* Norman and Greenwood (1975),* Paugy (1984, 1990c*), Skelton (1993),* Vari and Harold (1998),* Weitzman *et al.* (1988)*

<hr>

Class Actinopterygii
Subclass Neopterygii

Order Characiformes; Suborder Characoidei
(1st) Family Acestrorhynchidae—**acestrorhynchids**
(a-ses'-trō-ring'-ki-dē)

THIS SMALL GROUP OF ONE GENUS (*ACESTRORHYNCHUS*) AND the 15 species recognized by Menezes and Géry (1983) was previously included in the Characidae (Nelson, 1994). Lucena and Menezes (1998)

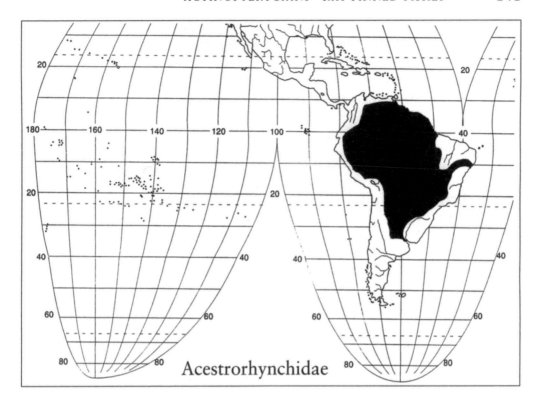

Acestrorhynchidae

demonstrated its monophyly and proposed that it is the sister group to the Cynodontidae. Buckup (1998), in his classification of the Characiformes, followed Lucena and Menezes and treated it as a family.

The Acestrorhynchidae occurs in the Amazon, Orinoco, and Paraná River basins as well as in the rivers of the Guianas and the Rio São Francisco (Gèry, 1977). Some species live in rivers and others prefer bays and lagoons near river banks. *Acestrorhynchus microlepis* (Fig. 85) is a

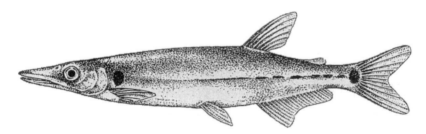

FIGURE 85. *Acestrorhynchus microlepis* (reproduced with permission from Sterba, 1966, Fig. 80).

typical representative of these elongate, pike-like piscivores. The largest species reaches about 40 cm TL (Menezes, 1969; Mago-Leccia, 1970). They have sharp, conical teeth and lack the multicuspid teeth found in many other characiform fishes. They were once united in a group with *Oligosarcus*, a characid, because of the similar pike-like appearance, but a study of their osteology revealed that this superficial similarity is the result of convergence (Lucena and Menezes, 1998).

Map references: Banarescu (1990),* Géry (1977)

Class   Actinopterygii
Subclass   Neopterygii

Order   Characiformes; Suborder Characoidei
(1st) Family   Cynodontidae—**cynodontids**
(sin-ō-don'-ti-dē) [sī-nō-don'-ti-dē]

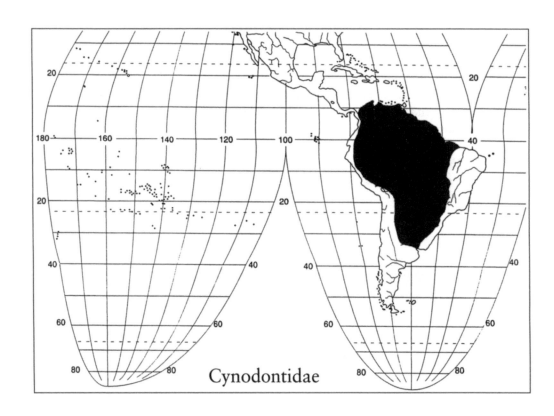

Cynodontidae

THIS SMALL FAMILY WAS FORMED BY REMOVING *CYNODON*, *Rhaphiodon*, *Hydrolycus*, *Gilbertolus*, and *Roestes* from the Characidae as recognized by Nelson (1994) and placing them in the Cynodontidae, which is now considered a monoplyletic group (Lucena and Menezes, 1998) and sister to the Acestrorhynchidae. Howes (1976) studied the cranial musculature and taxonomy of some members of this group.

The family includes 16 species in five genera that occur in the Magdelena, Maracaibo, Orinoco, Amazon, and the Paraná River basins and the Guianas (Géry, 1977). More species remain to be described. Lucena and Menezes (1998) recognized two subfamilies: the Cynodontinae and Roestinae. The Cynodontidae is composed of *Cynodon, Rhaphiodon*, and *Hydrolycus*. The Roestinae is made up of *Roestes* and *Gilbertolus*.

The cynodontines comprise about 10 species and are characterized by long canines on the anterior region of the lower jaw, oblique mouth, ventral keel, well-developed pectoral fins, and reduced pelvic fins reminiscent of the Gasteropelecidae, but cynodontids are not known to "fly" (Howes, 1976; Géry, 1997). The body is covered with minute scales. The third tooth may be very enlarged and perforate the cranial roof (Lucena and Menezes, 1998). *Raphiodon vulpinus* from the Rio Uruguay may reach 80 cm TL (Sverliji *et al.*, 1998). *Raphiodon* is the only genus of the subfamily occurring in the Paraguai and Paraná River basins. The other genera are distributed in the Rio Amazonas and Rio Orinoco basins and Atlantic slope rivers of the Guianas (Géry, 1977; Toledo-Piza *et al.*, 1999). *Hydrolycus armatus* (Fig. 86) and 3 other species were reviewed by Toledo-Piza *et al.* (1999).

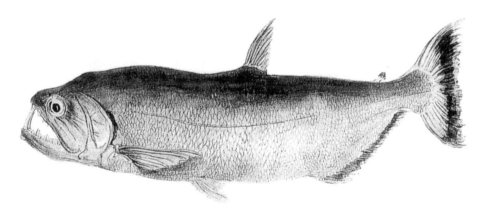

FIGURE 86. *Hydrolycus armatus* (reproduced with permission from Toledo-Piza *et al.*, 1999, Fig. 12).

The roestines include six species of medium-sized fishes that may reach a maximum size of 20 cm. They possess an oblique mouth, as do the cynodontines, but the canines of roestines are smaller. *Gilbertolus* is the only genus with a ventral keel. This genus has a transandean distribution: Lago Maracaibo basin and Rio Magdalena and Rio Atrato basins. *Roestes* only occurs east of the Andes in the Rio Orinoca, Rio Amazonas, and Rio Tocantins basins (Menezes and Lucena, 1998).

Map references: Géry (1977), Menezes and Lucena (1998),* Toledo-Piza *et al.* (1999)*

Class Actinopterygii
Subclass Neopterygii

Order Characiformes; Suborder Characoidei
(1st) Family Erythrinidae—**trahiras** [e-ri-thrī'-ni-dē]

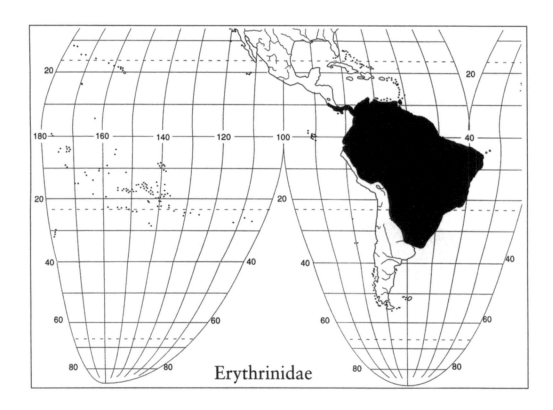

Erythrinidae

THIS SMALL FAMILY OF CYLINDRICAL PREDATORS IS DISTRIButed from the Pacific coast of Costa Rica to southern Ecuador in the west and to Buenos Aires, Argentina, in the east (Miller, 1966; Géry, 1977; Weitzman and Vari, 1994). It is the only characiform family in which the adipose fin is always absent. (Some members of the Crenuchidae,

Lebiasinidae, and Characidae also lack the adipose fin.) There are at least 10 species in three genera, *Erythrinus*, *Hoplerythrinus*, and *Hoplias* (Nelson, 1994). Erythrinids are common in shallow waters of lakes, streams, and flooded forests (Goulding, 1980).

The gape of these fish eaters is very long, extending past the anterior eye margin, and the caudal fin is rounded. Superficially, erythrinids resemble the North American bowfin, *Amia calva*. Although specialized for predation, the trahiras, as the Brazilians refer to them, were thought to be primitive (Géry, 1977). Vari (1995) and Buckup (1998), however, showed that erythrinids are actually advanced characiforms despite possessing features such as maxillary teeth and an unusual pelvic girdle that some precladistic researchers considered archaic. Teeth are also present on the palate.

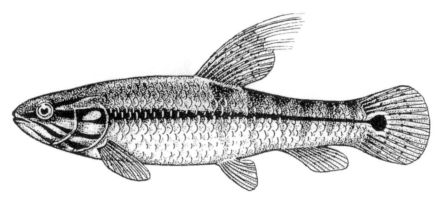

FIGURE 87. *Erythrinus erythrinus* (reproduced with permission from Sterba, 1966, Fig. 91).

Erythrinus erythrinus (Fig. 87) and *Hoplerythrinus unitaeniatus* have been shown to be continuous, facultative air breathers that use a modified portion of the posterior swim bladder as an accessory respiratory organ (Graham, 1997). *Hoplias* is a non-air breather that can tolerate hypoxic waters. At least one species is capable of moving overland (Saul, 1975; Géry, 1977). *Hoplias macrophthalmus* may reach 1.0 m and is utilized as food. Weitzman (1964) discussed the osteology of members of this family. Vari (1995) demonstrated the monophyly of the family and considered the Erythrinidae to be the sister group of the Ctenoluciidae. Buckup (1998) had a slightly different interpretation of sister-group relationships in that part of the characiform phylogeny and considered the Erythrinidae to be the sister group of the Lebiasinidae and the clade of Ctenoluciidae + Hepsetidae.

Hoplias malabaricus was found in the Little Manatee River of south-eastern Florida in 1974 and 1975, but the species presumably died out during the extremely cold temperatures of January 1977 (Fuller *et al.*, 1999).

Map references: Géry (1977), Miller (1966), Vari (1995)

Class Actinopterygii
Subclass Neopterygii

Order Characiformes; Suborder Characoidei
(1st) Family Lebiasinidae—**lebiasinids** [lē-bē-a-sīn'-i-dē]

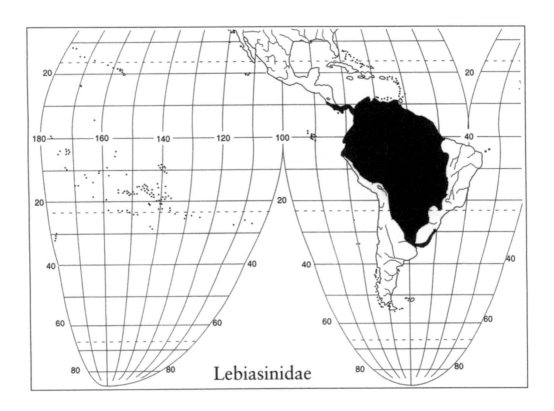

Lebiasinidae

THIS FAMILY CONSISTS OF ABOUT 51 SPECIES IN SIX GENERA distributed from the Pacific slope of Costa Rica through the Amazon River basin and south to La Plata, Argentina. The adipose fin may be irregularly present. The mouth gape is short. There are two subfamilies, Lebiasininae and Pyrrhulininae.

The Lebiasininae is found in streams and ponds from Costa Rica and Panama through Andean areas in Peru, Ecuador, Colombia, and into the Guiana upland region of Venezuela and the Guianas (Miller, 1966; Weitzman and Vari, 1994). There are two genera, *Lebiasina* and *Piabucina*, with about 11 species. These fishes are cylindrical with well-toothed mouths. They resemble small erythrinids with a smaller gape and smaller teeth. They seem to tolerate stagnant waters and some species are able to

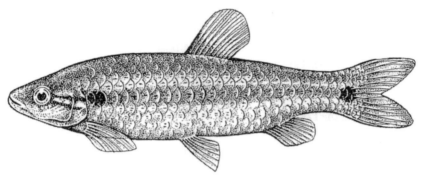

FIGURE 88. *Lebiasina bimaculata* (reproduced with permission from Sterba, 1966, Fig. 229).

breathe air. The presence of air breathers and non-air breathers within the same family is an unusual feature of the Lebiasinidae (Graham, 1997). Mosquito larvae and other insects make up most of the diet. *Lebiasina bimaculata* (Fig. 88) has been introduced to other countries for mosquito control (Géry, 1977). Some species may reach 18 cm.

The Pyrrhulininae are very colorful and are utilized extensively by the aquarium trade. About 25 species of *Copeina*, *Copella*, and *Pyrrhulina* occur in lowland areas of the Orinoco, Amazon, and Paraguay basins and coastal streams of the Guianas (Weitzman and Vari, 1994). Species of *Pyrrhulina* and *Copella* have an elongated upper caudal fin lobe. The larger species reach about 10 cm. They feed on a wide variety of insects, insect larvae, and plant matter (Knöppel, 1970).

The spraying characin from the lower Amazon and the Guianas, often cited as *Copeina arnoldi*, is a species of *Copella* (Myers, 1956). It lays its eggs out of the water on emergent vegetation in a gelatinous mass. The male and female leap simultaneously out of the water to deposit a few eggs on the undersurface of a leaf. This is repeated until a clump of 60 or more eggs has been deposited. The male splashes the eggs with water with its tail every 20–30 minutes to prevent desiccation. The male does not splash the eggs when rain is falling. The young drop into the water when they hatch after about 3 days (Krekorian, 1976). This unusual mode of reproduction may be an adaptation to the low oxygen content of the swampy backwaters inhabited by this species.

Another lineage within this subfamily is the genus *Nannostomus* (= *Poecilobrycon*) with about 15 species in the group. They live in the midwaters, usually near shore in forest streams and major rivers. Weitzman and Weitzman (1982) discussed forest refuges as a possible isolating factor in speciation of *Nannostomus*. These small (2–4.5 cm) colorful fishes with cylindrical bodies and tiny mouths usually have at least one lateral dark stripe. Two species, *N. eques* and *N. unifasciatus*, often stand obliquely vertical on their tails in the water column. The common name, pencilfishes, is derived from the slender body of these fishes. Their peculiar behavior and attractive coloration make them highly valued aquarium subjects. The presence of the adipose fin is a very labile character, and even among specimens from the same nest of *N. marginatus* the adipose fin may be present or absent (Hoedeman, 1974). Consult Weitzman (1964, 1966, 1978), Weitzman and Cobb (1975), Weitzman and Weitzman (1982), Géry (1977), and Fernandez and Weitzman (1987) for a review of the taxonomy of this family. Vari (1995) treated the Lebiasinidae as the sister group to a clade composed of the Hepsetidae, Erythrinidae, and Ctenoluciidae. Buckup (1998) provided a slightly different view of this part of the unsettled characiform cladogram.

> Map references: Géry (1977), Miller (1966), Weitzman (1964, 1966, 1978), Weitzman and Cobb (1975), Weitzman and Vari (1994), Weitzman and Weitzman (1988)*

Class **Actinopterygii**
Subclass **Neopterygii**

Order **Characiformes; Suborder Characoidei**
(1st) Family **Ctenoluciidae—pike characins**
(tē'-nō-lū-sī'-i-dē) [tē'-nō-lū-sē'-i-dē]

THESE DISTINCTIVE FISHES ARE FOUND IN THE FRESH WATERS of Panama, northern South America, and the Amazon basin. The pike characins are well named with their elongated *Esox*-like body, posteriorly positioned dorsal and anal fins, and piscivorous feeding habits. They have numerous, short, conical, recurved teeth in a beak-like jaw, and the snout or chin may possess a fleshy flap. A small adipose fin is present, and the caudal fin is deeply forked.

There are two genera: *Ctenolucius* (two species), with strongly ctenoid scales, and *Boulengerella* (five species), with finely denticulated scales (Géry, 1977; Vari, 1995). *Ctenolucius hujeta* (Fig. 89) and *C. beani* are

FIGURE 89. *Ctenolucius hujeta* as *Luciocharax insculptus* (Steindachner, 1879a, Plate XIII, Fig. 2).

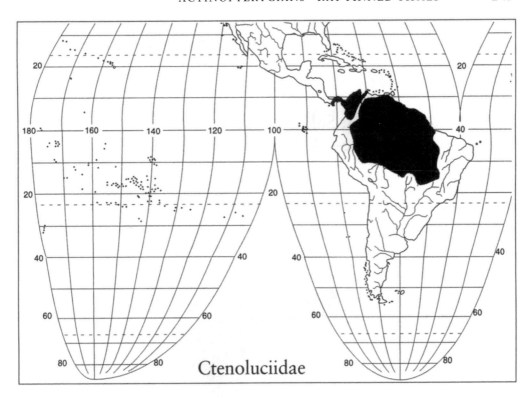

Ctenoluciidae

found from the Pacific rivers of Panama through northwestern and northern Colombia, the Río Magdalena, to the Lake Maracaibo basin of Venezuela. The five species of *Boulengerella* occur in the Orinoco, Amazon and Tocantins River basins as well as the shorter coastal rivers of Guyana and French Guiana (Vari, 1995). Pike characins are among the larger New World characiforms. The largest species, *B. cuvieri*, may reach 68 cm SL. In addition to their major ecological role as a high-level predator, members of this family are important as food fishes and in the aquarium trade.

Roberts (1969) studied their osteology and suggested that ctenoluciids most resemble the African Hepsetidae. However, it was not clear at that time if this similarity was due to convergence or an evolutionary relationship. Vari (1995) concluded that the Ctenoluciidae and Erythrinidae are sister groups. He further considered the Hepsetidae to be the sister group to the Ctenoluciidae plus Erythrinidae and the Lebiasinidae to be the sister group to the clade formed by the other three families. Buckup (1989) considered the Ctenoluciidae and Hepsetidae as sister groups. He treated the clade of Ctenoluciidae + Hepsetidae as sister to the Lebiasinidae. These three families were treated as sister to the Erythrinidae.

Map references: Géry (1977), Miller (1966), Vari (1995)*

Class Actinopterygii
Subclass Neopterygii

Order Characiformes; Suborder Characoidei
(1st) Family Hepsetidae—**African pike characins** [hep-sē'-ti-dē]

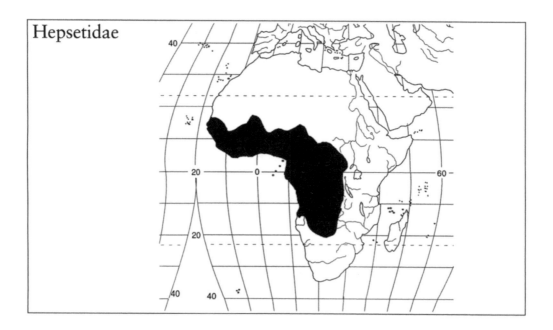

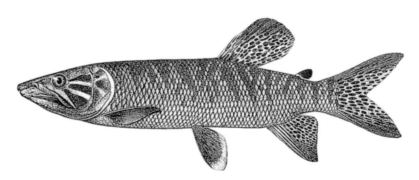

FIGURE 90. *Hepsetus odoe* (reproduced with permission from Lévêque *et al.*, 1990, Fig. 19.1).

THIS MONOTYPIC FAMILY CONSISTS OF *HEPSETUS ODOE* (Fig. 90), which is widespread in west and central tropical Africa. It inhabits large and small rivers, lakes, and swamps, including coastal rivers from

Senegal to Angola such as the Niger, Volta, Chad, Ogowe, Congo (Zaire), and Upper Zambezi, but it is absent from the Nile basin (Roberts, 1984b). This species is pike-like, with an elongate body, posteriorly positioned dorsal and anal fins, long snout, a depressed head, and large mouth with large canine teeth and many smaller conical teeth. *Hepsetus* is a voracious fish predator that stalks and ambushes its prey with a swift lunge (Skelton, 1993). An adipose fin is present, and the caudal fin is deeply forked. Scales are cycloid. Membranous flaps project from the lateral margins of the jaws. It reaches about 70 cm and 4 kg (Paugy, 1990b) and is considered a sport fish. *Hepsetus* constructs a bubble nest that is guarded by the parents (Géry, 1977; Merron *et al.*, 1990).

Roberts (1969) described the osteology and relationships of this species. He considered *Hepsetus* to be a very primitive characiform and unrelated to any other African group of characiforms. He suggested that *Hepsetus* may be related to the South American pike characins, such as *Ctenolucius*. Vari (1995) confirmed that the similarities were due to common ancestry rather than convergence and concluded that a vicariance event predating or associated with the final separation of Africa and South America about 85 MYA separated the Hepsetidae from the clade composed of the Ctenoluciidae plus Erythrinidae. Buckup (1998) independently supported the sister-group relationship between the Ctenoluciidae and Hepsetidae.

Map references: Paugy (1990b),* Roberts (1984a), Skelton (1993)*

Class Actinopterygii
Subclass Neopterygii

Order Siluriformes
(1st) Family Diplomystidae—**diplomystids** [dip-lō-mis'-ti-dē]

THE CATFISHES (SILURIFORMES) REPRESENT THE FOURTH group of the Otophysi (formerly called Ostariophysi). The four C's (cyprinids, catostomids, characoids, and catfishes and their relatives) have distinctive modifications of four or five anterior vertebrae that connect the swim bladder to the inner ear. Siluroids have five vertebrae in the Weberian apparatus, whereas nonsiluroids have four. These ossicles confer an acute sense of hearing upon this group and may account for the dominance of the Otophysi in the fresh waters of the world. See Alexander (1964) and Chardon (1968) for a description of catfish Weberian apparatus. The swim bladder acts as a resonator and amplifier so that gas volume

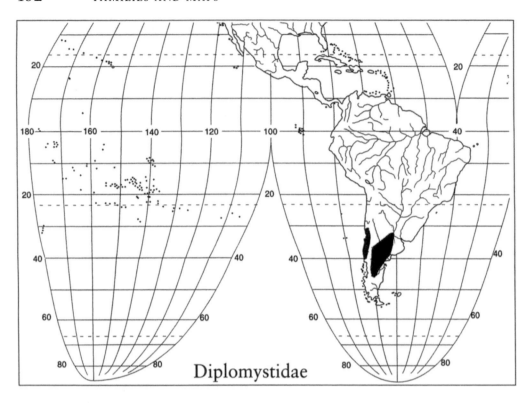

Diplomystidae

changes due to sound waves in the water are transmitted to the ear by the ossicles.

Catfishes usually have spine-like rays at the anterior edge of the dorsal and pectoral fins, up to four pairs of barbels around the mouth, an adipose fin, and a scaleless or armored body. The maxilla is usually toothless and small. A great deal of information about catfishes is summarized by Burgess (1989) and updated by Burgess and Finley (1996). Teugels (1996) provided an overview of catfish taxonomy, phylogeny, and biogeography.

Catfish relationships are not agreed on by all workers. Nelson (1994) accepted 34 families with about 412 genera and 2400 species. De Pinna (1998) reviewed the phylogenetic relationships of Neotropical siluriforms and his work will go a long way toward establishing relationships among the major groups of catfishes, although multifamily catfish relationships are unclear in some regions of the cladogram (Mo, 1991; de Pinna, 1998). The gymnotiforms (knifefishes) are considered the sister group to the catfishes within the Otophysi (Fink and Fink, 1981). The sequence in which I have listed the catfishes follows de Pinna (1998) and has phylogenetic meaning, but I try to keep taxa from the same continent adjacent as much as possible.

The Diplomystidae is the most primitive catfish family (Fink and Fink, 1981; Grande, 1987; Arratia, 1992; Mo, 1991; de Pinna, 1998). It has a limited distribution in central and southern Chile and Argentina. It is the only living catfish family with well-developed teeth on the maxilla. The North American fossil catfish family Hypsidoridae also has maxillary teeth, but all other recent catfishes have edentulous maxillae (Grande, 1987; Grande and de Pinna, 1998). A common interpretation is that teeth were present on the maxillae of catfish ancestors but have been lost in lineages beyond the Diplomystidae and Hypsidoridae. However, most ostario-physians lack maxillary teeth (Gymnotiformes, Cypriniformes, and Gonorynchiformes). The presence of maxillary teeth in characiforms and siluriforms could be considered derived rather than primitive (de Pinna, 1998). Diplomystids also have 18 principal caudal fin rays, whereas most other catfish families have 17. The maxillary barbels are the only barbels present. The dorsal and pectoral fins have a strong spine.

The family is currently considered to include six species in one genus, *Diplomystes* (Azpelicueta, 1994a). Three species are found on the western slope of the Andes in the Chilean and Valdivian provinces of central and southern Chile: *D. chilensis* (Fig. 91), *D. camposensis*, and *D. nahuelbu-taensis* (Arratia, 1987). Three east Andean species occur in different river basins such as the Desaguadero–Salado system, the Colorado River, and the Negro River from Mendoza (central Argentina) to southern Patagonia: *Diplomystes* (= *Olivaichthys*) *viedmensis*, *D. cuyanus*, and *D. mesembrinus* (Arratia, 1987; Azpelicueta, 1994a).

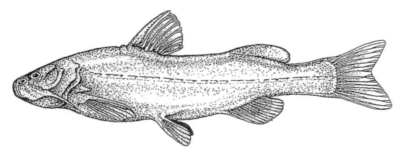

FIGURE 91. *Diplomystes chilensis* as *D. papillosus* (reproduced with permission from Burgess, 1989, p. 24).

Very little is known of the biology of *Diplomystes*. Arratia and Huaquin (1995) examined the lateralis systems, neuromast lines, and skin of benthic siluroids including *Diplomystes*. Karyotypes of *D. camposensis* and *D. nahuelbutaensis* revealed a diploid number of 56 chromosomes (Campos *et al.*, 1997). Specimens of *D. nahuelbutaensis* have been taken at moderate elevation (370–520 m) in fast-flowing tributaries of the Rio Biobio and their diet was composed of aquatic insect larvae, especially chironomids, and a decapod crustacean (Ruiz and Berra, 1994). Diplomystids in general are carnivorous, consuming annelids, mollusks, and arthropods (Arratia, 1987). Reproduction is probably in summer since that is when females are taken with eggs. *Diplomystes viedmensis* has been taken from sea level to about 1900 m. Maximum size of *Diplomystes* is about 32 cm SL (Azpelicueta, 1994a).

Map references: Arratia (1981*, 1987), Arratia *et al.* (1985),* Azpelicueta (1994a, 1994b*)

Class Actinopterygii
Subclass Neopterygii

Order Siluriformes
(1st) Family Cetopsidae—**whale-like catfishes**   [sē-top'-si-dē]

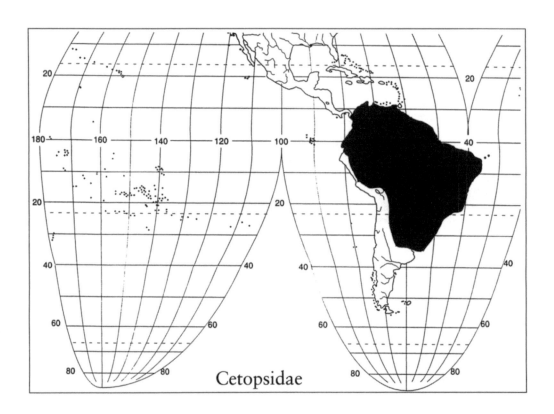

Cetopsidae

THE CETOPSIDAE IS A SMALL FAMILY OF SOUTH AMERICAN catfishes found from Colombia to Paraguay. Their habitat includes small rain forest streams to large rivers. They occur in the Amazon, Essequibo, Orinoco, São Francisco, and Paraná–Paraguay basins, the rivers of the

Pacific slope of Ecuador and Colombia, and rivers draining from the Caribbean slope of Colombia and northwestern Venezuela (de Pinna and Vari, 1995). De Pinna and Vari united the Helogenidae with the Cetopsidae into an expanded Cetopsidae and described juvenile specimens of *Helogenes* for the first time. They consider the enlarged Cetopsidae to be the sister group of all other nondiplomystid, nonhypsidorid siluriforms. Mo (1991) and de Pinna (1998) agree that this group occupies a basal position within the siluriforms. Nelson (1994) treated the cetopsids and helogenids as separate families, and de Pinna and Vari (1995) and de Pinna (1998) considered them as subfamilies (Cetopsinae and Helogeninae) within the Cetopsidae. There are seven genera (including *Helogenes*) and about 21 species (Schultz, 1944; Burgess, 1989; Ferraris and Brown, 1991; de Pinna and Vari, 1995; Ferraris, 1996).

The Cetopsinae are referred to as whale-like catfishes because of the cylindrical, streamlined, naked body that is more or less oval in cross section and the blunt snout. A small dorsal fin, which may have a weak spine, is present, but there is no adipose fin in adults, although juveniles have one (de Pinna and Vari, 1995). The anal fin is moderately long. Conical or incisor-like teeth are present in the jaws, and the vomer is toothed. The eyes are very small and nearly concealed in the skin. Three pairs of barbels are present. Opercular odontodes ("spines" that are actually teeth outside the mouth) are absent, unlike in the Trichomycteridae. The swim bladder is often greatly reduced and may be encased in a bony sheath in some species.

The six genera of the Cetopsinae are *Hemicetopsis* (seven species), *Pseudocetopsis* (four species), *Cetopsogiton* (= *Paracetopsis*) (one species), *Cetopsis* (two species), *Bathycetopsis* (one species), and *Denticetopsis* (two species). *Cetopsogiton occidentalis* occurs in rivers near Guayaquil on the Pacific slope of Ecuador. *Cetopsis* is from Brazil and Venezuela; *Pseudocetopsis* (represented by *P. macilentus*; Fig. 92) occurs in Brazil, Peru, Venezuela, and eastern Ecuador; and *Hemicetopsis* inhabits

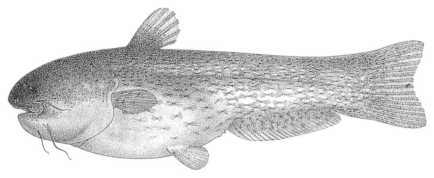

FIGURE 92. *Pseudocetopsis macilentus* as *Hemicetopsis macilentus* (Eigenmann, 1912, Plate XXIII).

Venezuela, the Guianas, Ecuador, and the Amazon. Lundberg and Rapp Py-Daniel (1994) described a small, blind, and depigmented *Bathycetopsis oliveirai* from a deep water channel of the Brazilian Amazon. The similarity of this species to cave-dwelling fishes is discussed by Lundberg and Rapp Py-Daniel (1994), and they speculate that light levels in silt-laden deep water channels may exert selective pressures similar to those of a cave. Poulson (1986) reviewed evolutionary reduction by neutral mutations in troglophiles and troglobites. *Denticetopsis* occurs in southern Venezuela and has a cluster of prominent caniniform teeth at the symphysis of the dentary (Ferraris, 1996).

Cetopsis and *Hemicetopsis* can be a nuisance for fishers who are seeking the large catfishes such as pimelodids. The much smaller cetopsids (15–25 cm) rip flesh from hooked or netted large fish. They may even enter the wound they create, eating the larger fish from the inside out and creating a bloody mess. This mode of eating may be an adaptation to feeding on carrion (Goulding, 1980). *Hemicetopsis candiru* is named after the infamous tricomycterid catfish that enters the urethra of mammals, including humans who urinate while under the water. The burrowing and biting activity shared by cetopsids and some trichomycterids is due to convergent evolution since the two families are not closely related (de Pinna and Vari, 1995). Cetopsids are not important food fishes.

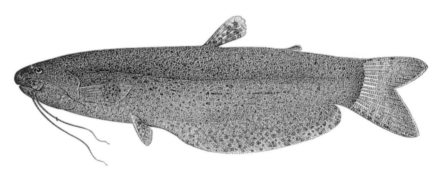

FIGURE 93. *Helogenes marmoratus* (Eigenmann, 1912, Plate XXII).

The Helogeninae is a small subfamily of one genus and four species that occurs in clear or blackwater streams in rain or gallery forests of tropical South America. *Helogenes marmoratus* (Fig. 93) is found throughout the Atlantic drainages of the Guianas and the Upper Río Orinoco of Venezuela and in the Amazon basin of Venezuela, Brazil, Peru, and Ecuador. *Helogenes castaneus* occurs in eastern Colombia in tributaries of the Río Orinoco. *Helogenes uruyensis* is endemic to an Orinoco tributary in southeastern Venezuela around Auyantepui, and *H. gouldingi* is known from the Rio Madeira in Brazil (Vari and Ortega, 1986). *Leyvaichthys* is a synonym of *Helogenes*.

These are small fishes with a maximum size of 73 mm SL. The body is naked, and there is a groove extending from the posterior end of the head to the origin of the short, spineless dorsal fin. The dorsal fin is placed near the center of the body. The pectoral fins also lack spines. An adipose fin may be present or absent; when present, it is very small. The anal fin is elongate with 32–49 rays. The caudal fin is forked. The eyes are small and dorsally situated, and the fishes are probably nocturnal. Six barbels are present. The maxillary barbels fit into a groove under the eye. The outer row of teeth on the dentary bone are enlarged and widely spaced. Helogenins are generalized predators of terrestrial insects (Vari and Ortega, 1986).

Chardon (1968) described the anatomy of the five Weberian vertebrae. Lundberg and Baskin (1969) reported on caudal skeleton morphology, and Lundberg (1975b) discussed cranial anatomy.

Map references: Burgess (1989), Eigenmann (1910a, 1912), Ferraris (1996),* Lundberg and Rapp Py-Daniel (1994),* de Pinna and Vari (1995)

Class Actinopterygii
Subclass Neopterygii

Order Siluriformes
(1st) Family Ictaluridae—**North American or bullhead catfishes**
(ik-ta-lur'-i-dē)

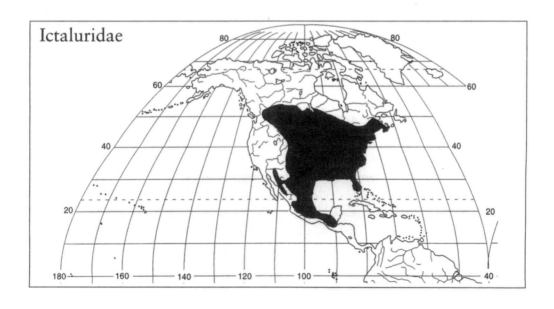

THE ICTALURIDS ARE DISTRIBUTED IN THE FRESH WATERS OF North America from the Rocky Mountains eastward and from southern Canada south into drainages of western Mexico and Guatemala. It is the largest family of freshwater fishes endemic to North America. There are seven extant genera and about 45 species (Nelson, 1994). They have four pairs of barbels and a spine (fused rays) in the dorsal and pectoral fins. In some species the spines are serrated. An adipose fin is present. The name "catfish" is related to the presence of barbels (= whiskers). The barbels and parts of the body may be covered with cells similar to taste buds. This enables catfishes to locate food by following a chemical gradient in the water. This sensory system also functions in social behavior (Todd, 1971). The chemical senses are particularly important for catfishes since most species are usually nocturnal and have very small eyes. All species for which data are available are cavity nesters or nest builders and provide some degree of parental protection (Taylor, 1969). In fact, no other North American fish family provides as much parental care to eggs and young as do the ictalurids. The oldest fossil ictalurids date back to the Paleocene (Lundberg, 1975a; Cavender, 1986). The family is considered to be monophyletic (Lundberg, 1992).

Ictalurids and many other catfishes are capable of social communication via sound which is used in both courtship and agonistic behavior. The sounds are produced either by specialized extrinsic muscles that insert on the swim bladder or by stridulation of the pectoral spine within the pectoral girdle (Fine *et al.*, 1997a). Moreover, individual catfish may have a preference for sound production with the right pectoral fin rather than the left fin (Fine, 1997). The pectoral spine may be locked in a fully abducted (erect) position. This defensive posture increases the size of the catfish and makes it difficult for predators to swallow it. This also makes handling catfish hazardous for humans, as any angler knows.

Ictalurus includes channel catfishes with about nine large species. These fish have forked caudal fins and a small adipose fin base. The channel catfish, *I. punctatus* (Fig. 94), is a commercially important species and is

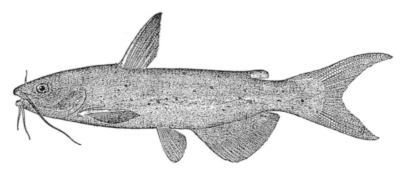

FIGURE 94. *Ictalurus punctatus* (Jordan and Evermann, 1905, p. 21).

raised by fish farmers in warm areas of the United States. It can reach about 1.3 m and 26.3 kg (IGFA 1993). It has been widely stocked throughout the United States. The huge amount of literature on the biology of this species has been admirably summarized by Becker (1983). The blue catfish, *I. furcatus*, is the largest species in the family. It can grow to 1.7 m and an astounding 143 kg (Cross, 1967). The blue catfish occurs from the Mississippi river basin, south into Belize and Guatemala (Greenfield and Thomerson, 1997). The larger species may live 15 or more years.

Ameiurus is the genus of bullheads, with seven medium-sized species. These fish usually have a rounded or strait-edged caudal fin (except *A. catus*) and a small-based adipose fin. The bullheads were formerly classified with *Ictalurus*. These game fishes are often caught by anglers. Bullheads range in size from about 28 to 62 cm. The biology of the black, brown, and yellow bullheads [*A. melas* (Fig. 95), *A. nebulosus*, and *A. natalis*] is summarized by Becker (1983).

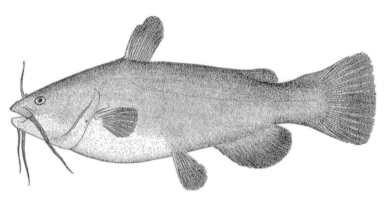

FIGURE 95. *Ameiurus melas* as *Amiurus melas* (Goode, 1844, Plate 233).

Pylodictis olivaris is the flathead catfish, so called because of its wide, flat head. This large catfish can reach 1.6 m and 42.6 kg (Robison and Buchanan, 1989). Fossils of this species date to the Miocene (11–15 MYA) (Lundberg, 1975a, 1982). Flatheads prefer deep holes in large rivers, streams, and lakes, and they tolerate heavy turbidity. They occur in the large rivers of the Mississippi, Missouri, and Ohio basins south into Mexico. Because of its large size and good flavor, this species is popular among anglers. Local people in the southern United States may capture flatheads by noodling. This exciting sport involves feeling under logs, banks, or other cover and grabbing the fish by its lower jaw or operculum. Much splashing follows. Noodling (also called tickling or hogging) requires a great deal of strength and an adventurous spirit.

Noturus is the genus of the madtoms. There are about 25 species of these cryptic catfishes. Most species are small, less than 10 cm, but

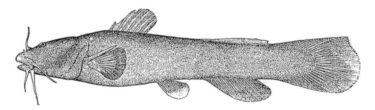

FIGURE 96. *Noturus flavus* (Jordan and Evermann, 1905, p. 34).

N. flavus (Fig. 96) can reach 31 cm and 0.5 kg (Trautman, 1981). These species possess a venom gland at the base of the pectoral spines and a toxic integumentary sheath around the spines that can inflict a painful sting (Birkhead, 1972). The adipose fin is long and low and often joined to the caudal fin. Some species of *Noturus* are rare and endangered and restricted to very limited habitats. For example, *N. trautmani* is known only from 18 specimens from Big Darby Creek in south-central Ohio, the last of which was collected in 1957 (Trautman, 1981). Phylogenetic relationships within *Noturus* were reviewed by Grady and LeGrande (1992).

There are three unrelated genera of blind catfishes. *Satan eurystomus* and *Trogloglanis pattersoni* are eyeless species known from artesian wells 305–582 m deep in the San Antonio Pool of the Edwards Aquifer near San Antonio (Bexar County), Texas (Cooper and Longley, 1980). The water temperature in the wells is 27°C. This habitat is the sole source of drinking water for the city of San Antonio. These odd fishes have a whitish or pink body and a maximum size of about 10–14 cm. No swim bladder is present. This adaptation allows the troglobitic species to withstand the great hydrostatic pressure present in the deep wells. A generous accumulation of adipose tissue provides buoyancy. They feed on shrimp, amphipods, and isopods (Longley and Karnei, 1979). *Satan* may prey on *Trogloglanis*, although they only occur together at two or three of the five known artesian wells. *Satan* may be related to *Pylodictis* and *Trogloglanis* may be allied with *Ameiurus*. *Trogloglanis* is the only ictalurid to lack jaw teeth. A third blind catfish, *Prietella phreatophila*, is known from a well in northeastern Mexico, and a recently described species, *Prietella lundbergi*, is known only from a thermal spring in Tamaulipas, Mexico (Walsh and Gilbert, 1995). *Prietella* is thought to be related to *Noturus*.

Poulson (1986) argued against the idea that loss of eyes and other structures are selected for by a food-poor environment. He compared cave dwellers from habitats of varying food resources and showed no correlation between resource availability and degree of eye loss. For more information on the blind catfishes, see Hubbs and Bailey (1947), Suttkus (1961), Longley and Karnei (1979), Lundberg (1982), and Walsh and Gilbert (1995). For details of catfish relationships, see Taylor (1969),

Grady and LeGrande (1992), and Lundberg (1992). LeGrande (1981) discussed chromosome evolution in the Ictaluridae. For keys, illustrations, and life history information consult Trautman (1981), Becker (1983), Page and Burr (1991), Etnier and Starnes (1993), and Jenkins and Burkhead (1994).

Map references: Lee *et al.* (1980),* Miller (1958, 1966),* Page and Burr (1991),* Scott and Crossman (1973)*

Class Actinopterygii

Subclass Neopterygii

Order Siluriformes
(1st) Family Claroteidae—**claroteid catfishes** (klar-ō-tē'-i-dē)

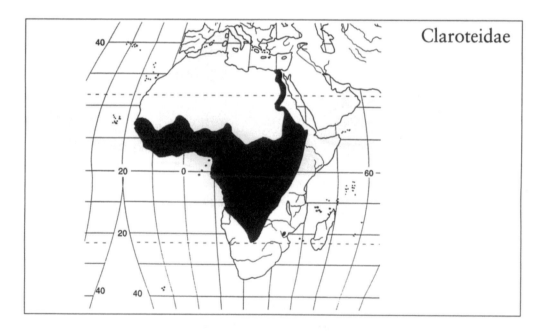

THIS FAMILY WAS CARVED OUT OF THE TRADITIONAL BAGRIDAE by Mo (1991) to reflect a monophyletic group of African catfishes. Although this work was discussed by de Pinna and Ferraris (1992), who found fault with some aspects of it, it was considered to be an improvement over previous classifications. Skelton (1993), Teugels (1996), and de Pinna (1998) follow Mo's classification.

Mo (1991) recognized two subfamilies, Claroteinae and Auchenoglanidinae, which together contain more than 90 species in 13 genera (Skeleton, 1993). Mo presented cladograms for *Clarotes*-like and *Auchenoglanis*-like groups and systematic keys to genera. Claroteids have a large mouth with three or four barbels. The short-rayed dorsal fin and the pectoral fins are armed with a spine. The anal fin is short-based, and there is an adipose fin.

The distribution of the Claroteidae includes the Nile River basin and most of west and central Africa south to the Tropic of Capricorn including the east African lakes (Risch, 1986; Skelton, 1993). Teugels *et al.* (1991) reviewed the genera *Auchenoglanis* and *Parauchenoglanis*. Generic-level classification is still problematic, but *Chrysichthys* is the largest genus with 41 species listed by Risch (1986). Other prominent genera include *Auchenoglanis* (18 species) and *Leptoglanis* (10 species). *Chrysichthys cranchii* can reach over 1 m and weigh 135 kg, but most species are much smaller, such as *C. longibarbis* (Fig. 97) and *A. ngamensis* (Fig. 98), which grow to about 25 cm SL (Risch, 1986).

Map references: Risch (1986, 1992*), Skelton (1993)*

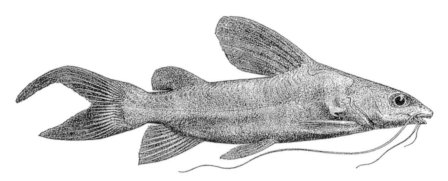

FIGURE 97. *Chrysichthys longibarbis* (Boulenger, 1911, Fig. 258).

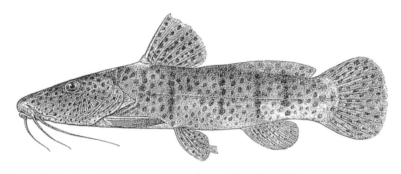

FIGURE 98. *Auchenoglanis ngamensis* (Boulenger, 1911, Fig. 287).

Class Actinopterygii
Subclass Neopterygii

Order Siluriformes
(1st) Family Austroglanididae—**austroglanidid catfishes**
(aus'-trō-glan-id'-i-dē)

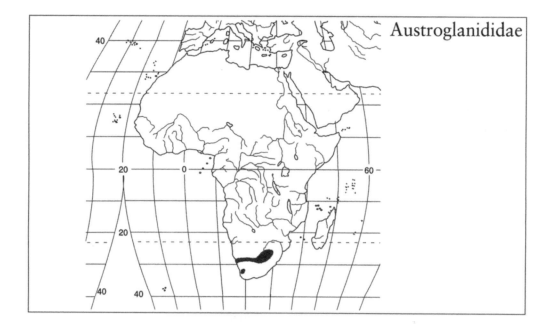

MO (1991) ERECTED THIS FAMILY TO ACCOMMODATE THE
southern African genus *Austroglanis* and its three species. Previously, this
group was included within the Bagridae (Nelson, 1994). *Austroglanis*
barnardi (Fig. 99) and *A. gilli* inhabit the Clanwilliam–Olifants River

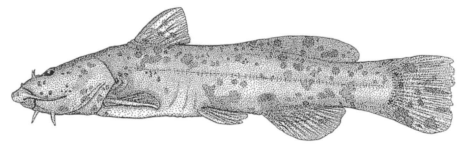

FIGURE 99. *Austroglanis barnardi* by Paul Skelton (used with permission of the director, J. L. B. Smith
Institute of Ichthyology, who holds the copyright).

system in southern Africa, and *A. sclateri* lives in the Orange River basin (Risch, 1986). Mo (1991) considered *Austroglanis* to be more closely related to the Bagridae than to the Claroteidae.

These small (7.5–30 cm SL) riverine catfishes have moderately small eyes, and the nasal barbels are shorter or equal to the posterior nares (Skelton, 1993). The mandibular barbels are on the ventral side of the head. Dorsal and pectoral spines are present, and the adipose fin is large. They feed on aquatic insects, benthic invertebrates, and small fishes. They prefer clear, flowing streams, and all three species are threatened by habitat alteration (Skelton, 1993).

Map references: Risch (1986), Skelton (1993)*

Class Actinopterygii
Subclass Neopterygii

Order Siluriformes
(1st) Family Bagridae—**bagrid catfishes**   (bah'-gri'-dē)

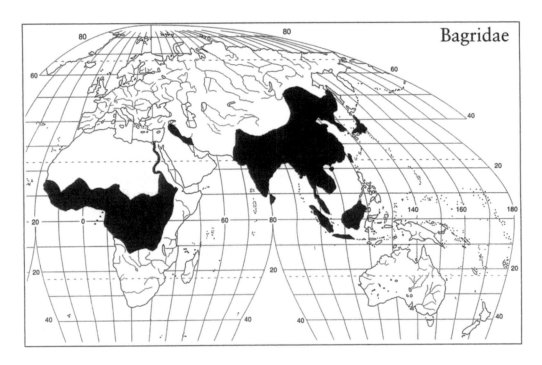

BAGRIDS ARE FOUND IN FRESH WATER IN SUBSAHARAN AFRICA and in the Tigris–Euphrates system of western Asia throughout southern and eastern Asia as far north as Japan and south to the East Indies. There are about 16 genera and about 120 species, all of which are Asian except for the 10 African species of *Bagrus*.

Bagrids often have a very large adipose fin. Four pairs of barbels are usual but many species lack nasal barbels. The dorsal fin has a stout spine, and the pectoral spines are serrated. The caudal fin is forked or deeply emarginate. Some bagrids resemble North American ictalurids and South American pimelodids.

Jayaram (1976) recognized five subfamilies and Bailey and Stewart (1984) corrected the subfamily names. Mo (1991) reorganized the Bagridae by dividing it into three families, Claroteidae, Austroglanididae, and Bagridae *sensu stricto* leaving *Bagrus* as the only African representative of the new Bagridae which he divided into two subfamilies, Bagrinae and Ritinae (including *Nanobagrus*). A cladogram and keys to genera are given by Mo (1991). *Bagrus docmak* can grow to 1.1 m and 22 kg, whereas *B. ubangensis* (Fig. 100) is more typical at 30 cm TL (Risch, 1986).

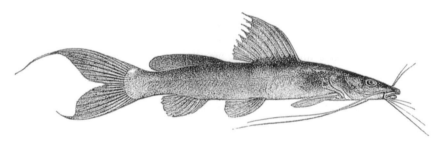

FIGURE 100. *Bagrus ubangensis* (Boulenger, 1911, Fig. 249).

Young of large species and many smaller, colorful species are popular in the aquarium trade, such as *Leiocassis siamensis* (now included in *Pseudomystus*) from Thailand and *Mystus vittatus* (Fig. 101) from

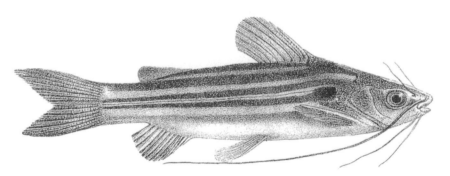

FIGURE 101. *Mystus vittatus* as *Macrones vittatus* (Day, 1878b, Plate XCIV, Fig. 4).

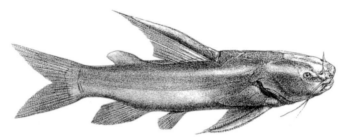

FIGURE 102. *Rita rita* as *R. buchanani* (Day, 1878b, Plate CIII, Fig. 1).

Southeast Asia. *Bagrichthys hypelopterus*, from Sumatra and Borneo, has an extremely long dorsal spine that may reach beyond the caudal fin base (Roberts, 1989). Some other genera in this family are *Pelteobagrus* and *Rita*. *Rita rita* (Fig. 102) reaches 1.5 m and is a common food fish in India (Talwar and Jhingran, 1992). The genus *Mystus* accounts for the western Asian distribution (Khalaf, 1961). Mo (1991) also included *Neotropius*, which was formerly placed in the Schilbeidae. *Hyalobagrus* is a genus of small (up to 40 mm SL) catfishes from Southeast Asia (Ng and Kottelat, 1998a).

Roberts (1989b) and Mo (1991) grouped the monotypic Olyridae with the Bagridae. This small group of hillstream fishes is endemic to eastern India (base of Darjeeling Himalaya, Meghalaya, and Assam), Burma, and western Thailand (Talwar and Jhingran, 1992). There is one genus, *Olyra*, with four species. The body is elongate, and the caudal fin is long, lanceolated, or forked. The eyes are small, and the four pairs of barbels are well developed. The short dorsal fin is without a spine. The adipose fin is long and low. The gill openings are very wide. The pectoral spine is serrated on both edges. The head is depressed and the body flattened anterior to the pelvic fins. These fish reach about 11 cm SL. Some olyrids are loach-like, probably in response to the selective pressures of fast-flowing waters. Hora (1936) reviewed the genus. Jayaram (1981) provided keys to *O. burmanica* (Fig. 103), *O. harai*, *O. longicaudata*, and *O. kempi*. Talwar and Jhingram (1992) provided additional taxonomic data.

FIGURE 103. *Olyra burmanica* (Day, 1878b, Plate CXI, Fig. 5).

See Risch (1986) for a list of African *Bagrus* species. Jayaram (1981) and Talwar and Jhingran (1992) described the bagrid fauna from India and adjacent countries. Smith (1945) provided keys, illustrations, and some life history information for bagrids of Thailand. Kottelat *et al.* (1993) gave descriptions and color photos of Indonesian species. Roberts (1989b) discussed the bagrids of Borneo.

Map references: Berg (1948/1949), Darlington (1958), Masuda *et al.* (1984), Nelson (1976),* Ng and Kottelat (1998a),* Risch (1986), Roberts (1989b),* Talwar and Jhingran (1992)

Class Actinopterygii
Subclass Neopterygii

Order Siluriformes
(1st) Family Pimelodidae—**long-whiskered catfishes**
(pim-ē-lō'-di-dē)

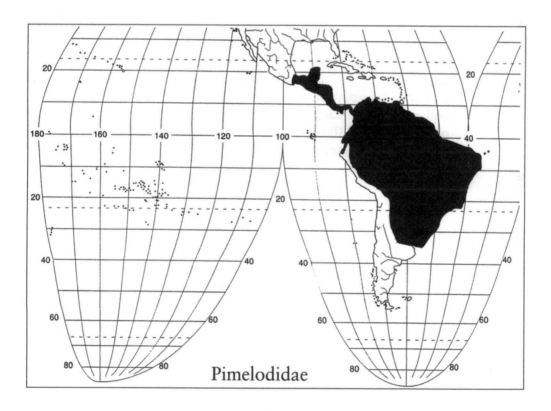

Pimelodidae

THE PIMELODIDAE AS CURRENTLY RECOGNIZED IS CONSID-
ered a heterogeneous assemblage of groups that do not form a mono-

phyletic clade (Mo, 1991; de Pinna, 1998). The group is under study, and the family Pimelodidae is recognized more or less by default until the relationships of the various pimelodid subgroups to other catfishes can be worked out (de Pinna, 1998). The Pimelodidae (*sensu lato*) is composed of three monophyletic groups currently considered to be subfamilies: Pimelodinae, Heptapterinae, and Pseudopimelodinae. Each of these groups is more closely related to other catfishes than to each other, so they eventually will be recognized as families in a phylogenetic classification (de Pinna, 1998). Lundberg *et al.* (1991a,b) recognized three subfamilies: Rhamdiinae (the Heptapterinae of de Pinna, 1998), Pimelodinae, and Pseudopimelodinae. See Stewart (1986a), Lundberg and McDade (1986), Ferraris (1988b), Silfvergrip (1992, 1996), and de Pinna (1998) and the references cited therein for taxonomic information. De Pinna (1998) listed many of the genera in each of the three subfamilies and included the Hypophthalmidae (*Hypophthalmus*) within the Pimelodinae as recommended by Howes (1983a) and Lundberg *et al.* (1991a,b).

Pimelodid catfishes are found from southern Mexico through Central and South America to Río de la Plata. The Pimelodidae (*sensu lato*) is the second largest family of South American catfishes (after the Loricariidae) with about 56 genera and 300 species (Nelson, 1994). Until recently, the genus *Rhamdia* (Heptapterinae) consisted of about 60 species that extended throughout the range of the family, even reaching the Andes and Lake Titicaca (Burgess, 1989). There is a cave-dwelling, depigmented, eyeless form of *Rhamdia quelen* (Fig. 104) as well as a whole range of intermediates to fully eyed and normally pigmented forms (Burgess, 1989). Silfvergrip (1996) revised *Rhamdia* and placed many species in synonymy, thus reducing this genus to about 11 species.

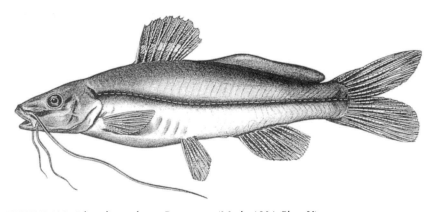

FIGURE 104. *Rhamdia quelen* as *R. oaxacae* (Meek, 1904, Plate V).

Pimelodids are naked, and the head is greatly depressed and shovel-like in some species. Three pairs of barbels are present, and they may be exceptionally long, extending beyond the length of the fish. The mouth is usually terminal to ventral and teeth are present on the jaws. There is a well-developed dorsal fin. Dorsal and pectoral spines may be present or absent, strong or weak, and serrated or smooth. A large adipose fin is present, and the anal fin is short in most genera.

Pimelodids come in a wide variety of sizes. *Brachyplatystoma filamentosum* (Pimelodinae), from the Amazon and Orinoco Rivers, is one of the largest catfishes in South America. It exceeds 2.8 m and 140 kg and is an important commercial species (Barthem and Goulding, 1997). On the other hand, *Microglanis parahybae* (Pseudopimelodinae) reaches only about 8 cm, and *Horiomyzon retropinnatus* (Heptapterinae) is even smaller at 3 cm TL (Stewart, 1986b). *Pseudopimelodus albomarginatus* (Fig. 105)

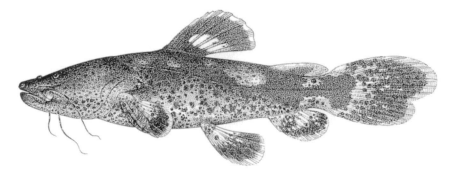

FIGURE 105. *Pseudopimelodus albomarginatus* (Eigenmann, 1912, Plate XI).

is a more typical representative of the Pseudopimelodinae. Other large pimelodids include *Paulicea lutkeni*, at 1.5 m and 100 kg, and *Pseudoplatystoma fasciatum* at 90 cm but only 12 kg (Goulding, 1980). *Pimelodus* is a genus of about 30 species (D. J. Stewart, personal communication), most of which are omnivorous and feed on fruits, fishes, and insects. *Brachyplatystoma* and *Pseudoplatystoma* are fish predators and swallow their prey whole. *Calophysus macropterus* has incisor-like teeth instead of the villiform teeth rows of most pimelodids. This 50-cm voracious piscivore is able to rip out pieces of flesh from its prey. *Piramutana piramuta* (Fig. 106) has flattened, band-like maxillary barbels that extend beyond the base of the pelvic fins. The previous genera are members of the Pimelodinae.

The genus *Hypophthalmus* with four species occurs in the tropical fresh waters of the Guianas, the Amazon, and the Rio Paraná. Nelson (1994) considered it a family, Hypopthalmidae, but de Pinna (1998) grouped it with the Pimelodinae. *Hypophthalmus edentatus* from the rivers

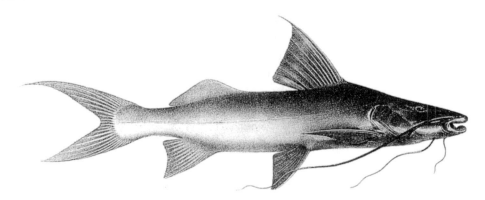

FIGURE 106. *Piramutana piramuta* (Steindachner, 1882b, Plate IV).

of equatorial Brazil is the best known species. The generic name refers to the fact that the eyes are small and positioned ventrolaterally behind and somewhat below the angle of the mouth. This gives rise to the common name of lookdown catfishes. The specific epithet denotes that the jaws and palate are toothless. All members of this family lack teeth but have many long, fine gill rakers that form a sieve for filter feeding on zooplankton (Roberts, 1972). Other species include *H. marginatus, H. oremaculatus,* and *H. perporosus* (Howes, 1983a).

These catfish have a depressed head and compressed body. The dorsal and adipose fins are small, and the dorsal and pectoral fin spines are weak. The anal fin is very long with 63–68 rays. The lateral line is well developed and branched to form a network along each side of the body. Three pairs of barbels are present.

Hypophthalmids are abundant in the expanded mouths of large clearwater rivers, such as the Tocantins, Xingu, and Tapajós, in which zooplankton occur (Goulding, 1980). They may exceed 50 cm in length. Howes (1983a) considered the hypophthalmids to be the sister group of the pimelodids. De Pinna (1998) wrote that the puzzle of their taxonomic placement was an artifact of prephylogenetic methodology that mistook divergence as an indication of a higher level taxon.

The dourada *Brachyplatysoma flavicans* (approximately 2 m TL) and the piramutaba *B. vaillantii* (approximately 1 m TL) are very important commercial species in the Amazon. They make the longest migrations of any freshwater fish species, about 3300 km from the estuaries to the Upper Amazon where spawning takes place (Barthem and Goulding, 1997). They feed on various characiform and siluriform fishes that migrate from the floodplains to the river channels on a seasonal basis. Other important food sources include the pacu fishes (*Mylossoma* sp., Characidae) that feed on fruits, seeds, and invertebrates that fall into the river. The dourada and the piramutaba unite the terrestrial and aquatic habitats and present an important conservation message: To save the fisheries, the Amazon rain forest must be conserved and vice versa (Barthem and Goulding, 1997).

Phractocephalus hemioliopterus (Pimelodinae) is a colorful, omnivorous, large species that reaches 1.3 m and 80 kg. Lundberg *et al.* (1988) reported fossil remains from a 6-million-year-old Miocene formation in Venezuela that appear to be the same species.

The maxillary barbels of some species can be quite fantastic. For example, in *Duopalatinus goeldii* (Pimelodinae), the barbels extend well beyond the tip of the caudal fin, and they may be ossified for about half of their length. Some species with stiff barbels hold them anteriorly like antennae. An eyeless form, *Pimelodella* (= *Typhlobagrus) kronei* (Heptapterinae), is known from a cave area near São Paulo, Brazil. The barbels of the blind fish are moderate in length.

Map references: Burgess (1989), Bussing (1998),* Eigenmann (1909a,* 1910, 1912*), Eigenmann and Allen (1942), Miller (1966),* Stewart (1986)*

Class Actinopterygii
Subclass Neopterygii

Order Siluriformes
(1st) Family Cranoglanididae—**armorhead catfishes**
[crā'-nō-glan-id'-i-dē]

THIS FAMILY CONSISTS OF ONE GENUS, *CRANOGLANIS*, AND three species from large rivers in east Asia (Ng and Kottelat, 2000). *Carnoglanis* is abundant in the West River and its tributary, the Fu River at Wuchow (Nichols, 1943), and in the Xijiang River basin of southern China, the Red River basin in northern Vietnam, and on Hainan Island (Banarescu, 1990; Ng and Kottelat, 2000).

These fishes have a rough bony plate on the top of the head, hence the common name armorhead catfish. They are medium-sized catfishes (20–30 cm TL) with compressed bodies. The caudal fin is deeply forked and the eyes are large. The dorsal and pectoral fins have a large spine. Four pairs of barbels are present. The jaw teeth are fine, and teeth are absent from the palate.

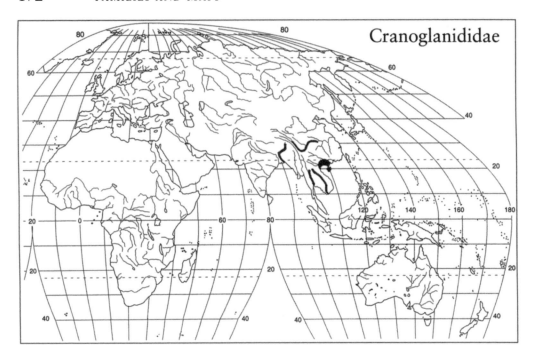

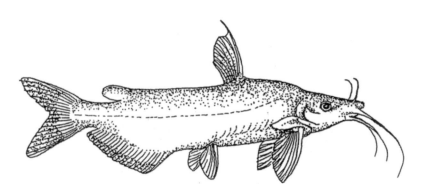

FIGURE 107. *Cranoglanis bouderius* (reproduced with permission from Burgess, 1989, p. 72).

Myers (1931) established the family name. *Cranoglanis bouderius (C. sinensis* is a junior synonym) (Fig. 107) and *C. multiradiatus* are mentioned by Nichols (1943) and Burgess (1989) from China. Ng and Kottelat (2000) determined that *Anopleutropius* is a synonym of *Cranoglanis* and recognized *C. henrici* from the Red River drainage in northern Vietnam. See Jayaram (1956), Lundberg and Baskin (1969), and Ng and Kottelat (2000) for taxonomic information.

Map references: Banarescu (1990),* Burgess (1989), Ng and Kottelat (2000), Nichols (1943)

Class Actinopterygii
Subclass Neopterygii

Order Siluriformes
(1st) Family Siluridae—**silurid catfishes** (si-lū'-ri-dē)

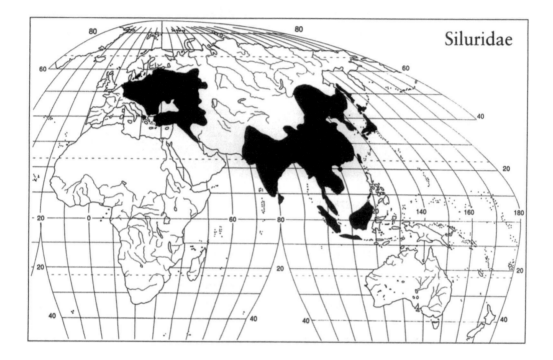

THE SILURIDAE IS AN EURASIAN FAMILY AND THE ONLY CAT-
fish family native to Europe. It ranges from the Rhine River eastward to
eastern Siberia and the Amur River, through Japan, Korea, and China south
into the Malay Archipelago and Palawan in the Philippines and west
throughout India. The family is absent from much of the central Asian
plateau, and it does not occur in Africa (Haig, 1950). Only 2 species of
about 100 in the family occur in European waters, and the rest are Asian.
There are about 12 genera (Nelson, 1994). Haig reviewed the family, and
Bornbusch (1991a) determined that the Siluridae is monophyletic.

The compressed silurid catfishes are distinctive. The dorsal fin lacks a
spine and is very small, having only 1–7 rays. In some species it is missing.
An adipose fin is absent. The pectoral fins have a spine, and the pelvic fins
are small or absent. The anal fin is very long, with 41–110 rays. There are
two or three pairs of barbels, but nasal barbels are not present.

One of the European species is known as the wels, *Silurus glanis*, from central and eastern Europe and across the southern regions of Russia, including the basins of the Black, Caspian, and Aral Seas. Wels may occur in brackish waters. They can grow to gigantic size. A specimen captured in the Dneiper River was 5 m long and weighed 306 kg (Wheeler, 1975). The species commonly reaches 3 m. Young wels are good eating and are of considerable commercial importance in eastern Europe. They have been introduced into England. The other European catfish is *S. aristotelis* from southwestern Greece (Maitland, 1977). *Silurus afghana* (Fig. 108) from Afghanistan and Pakistan is a relatively rare species whose taxonomic status has only recently been recognized (Talwar and Jhingran, 1992).

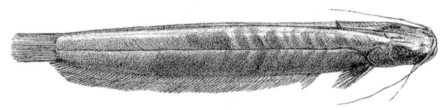

FIGURE 108. *Silurus afghana* (Day, 1878b, Plate CXII, Fig. 1).

Kryptopterus is known from the Malay Peninsula, Thailand, and Indonesia. *Kryptopterus bicirrhis* is translucent and known as the glass catfish. They are popular aquarium animals. The dorsal fin is reduced to only one or two rays, and there is usually one pair of very long barbels. Young specimens of *Ompok* are also nearly transparent and are kept as aquarium specimens. *Ompok bimaculatus* (Fig. 109), which reaches a SL of 45 cm, is a highly priced food fish in India (Talwar and Jhingran, 1992). The taxonomy of *Ompok* was reviewed by Tan and Ng (1996). Other genera include *Ceratoglanis, Belodontichthys, Pterocryptis,* and *Wallago*

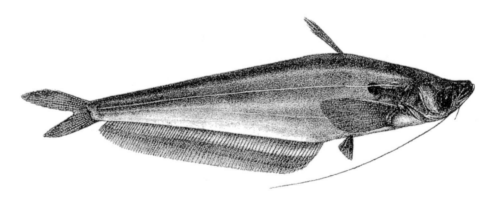

FIGURE 109. *Ompok bimaculatus* as *Callichrous macropthalmus* (Day, 1878b, Plate CX, Fig. 2).

(Ng, 1999a; Kottelat and Ng, 1999). Ng and Kottelat (1998b) described an unpigmented, cave-dwelling *Pterocryptis buccata* from western Thailand. Roberts (1982d) revised the systematics of *Wallago*, a large predatory genus from the Indian subcontinent and Southeast Asia. Ng (1992) clarified the taxonomy of the giant Malayan catfish, *W. leerii*, and reviewed reports that it can reach 1.7 m TL and 46.4 kg.

The family is richest in the Indian subcontinent, Mekong basin, and Borneo. For species descriptions see Jayaram (1981), Talwar and Jhingran (1992), Smith (1945), Kottelat *et al.* (1993), and Roberts (1989b). For systematic discussions see Haig (1950), Bornbusch (1991a,b), and Howes and Fumihito (1991).

Map references: Banarescu (1990),* Berg (1948/1949), Darlington (1957), Haig (1950), Jayaram (1974),* Maitland (1977),* Ng (1999),* Roberts (1982d),* Smith (1945), Talwar and Jhingran (1992)

Class **Actinopterygii**
Subclass **Neopterygii**

Order **Siluriformes**
(1st) Family **Schilbeidae—schilbeid catfishes** [shil-bē'-i-dē]

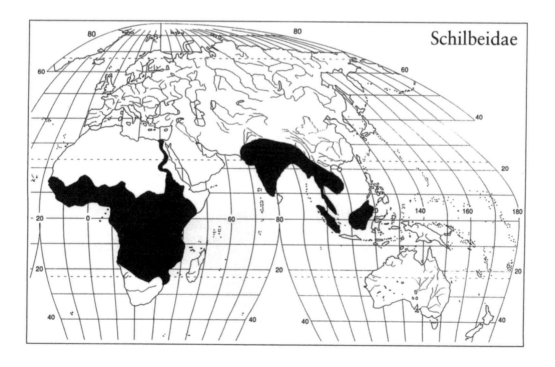

Schilbeidae

THE SCHILBEID CATFISHES ARE FOUND IN FRESH WATER throughout sub-Saharan Africa, Pakistan, India (not Sri Lanka), Bangladesh, Burma, Thailand, Malaysia, Sumatra, and Borneo. There are about 18 genera and 45 species, with the majority of the species in Africa (Nelson, 1994).

A short-based dorsal fin with a spine is usually present, as is an adipose fin. However, either may be lacking in some genera. Pectoral fins have a strong, usually serrated spine. The anal fin is long but not confluent with the forked caudal fin. There may be two to four pairs of very long barbels. The body is elongate and compressed. Eyes are lateral and large.

Thirty-four African species are listed by De Vos (1986). They are classified in five genera: *Eutropiellus*, *Irvineia*, *Parailia* (this genus lacks a dorsal fin, represented by *P. congica*; Fig. 110), *Schilbe*, and *Siluranodon*. *Schilbe marmoratus* (Fig. 111), at 22 cm SL, is representative of the

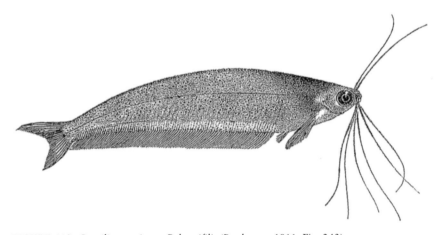

FIGURE 110. *Parailia congica* as *P. longifilis* (Boulenger, 1911, Fig. 243).

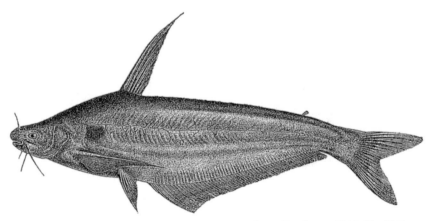

FIGURE 111. *Schilbe marmoratus* as *Eutropius congolensis* (Boulenger, 1911, Fig. 230).

figure 112. *Eutropiichthys murius* as *Pseudotropius murius* (Day, 1878b, Plate CVIII, Fig. 6).

African schilbeids. Asian genera include *Ailia* (lacking a dorsal fin), *Silonia* (vestigial or absent barbels), *Eutropiichthys, Proeutropiichthys, Pseudeutropius,* and *Clupisoma* (Talwar and Jhingrann, 1992). *Eutropiichthys murius* (Fig. 112) is a commercially important Indian species relished for its rich oil content. The Asian genus *Laides* is also assigned to this family (Roberts and Vidthayanon, 1991; Ng, 1999b). Most of the Asian species are confined to the Indian region. Only *Pseudeutropius* occurs in Indonesia (Roberts, 1989b). Previously, the two genera of the Pangasiidae, *Pangasius* and *Helicophagus*, were placed in this family (Roberts and Vidthayanon, 1991).

Larger species are utilized for food, and smaller species are found in the aquarium trade. For example, *Parailia* (= *Physailia) pellucida*, known as the African glass catfish, is nearly colorless and transparent and its vertebral column, swim bladder, and blood vessels are easily seen. A school of these long-barbeled, lively fishes makes a very attractive aquarium display. They reach 15 cm SL and come from the Upper Nile basin.

De Vos (1986, 1992) provided a checklist, drawings, keys, and references to African schildeids. See Jayaram (1981), Talwar and Jhingran (1992), Smith (1945), Kottelat *et al.* (1993), and Roberts (1989b) for descriptions of Asian species.

Map references: De Vos (1986, 1992*), Ng (1999b),* Roberts (1989b), Skelton (1993),* Smith (1945), Talwar and Jhingran (1992)

Class Actinopterygii
Subclass Neopterygii

Order Siluriformes
(1st) Family Pangasiidae—**pangasiid catfishes**
(pan-gā-sī'-i-dē) [pan-gā'-si-i-dē]

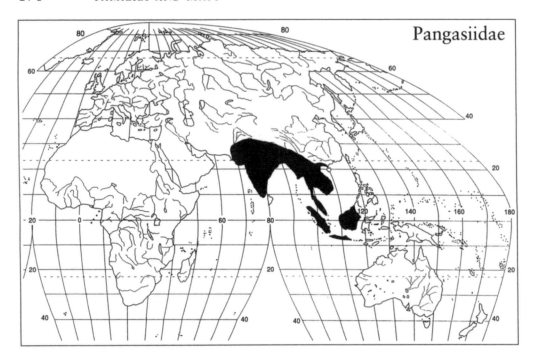

THE PANGASIIDS ARE FOUND IN FRESH WATERS IN SOUTHERN
Asia from Pakistan, India (but not Sri Lanka), Bangladesh, Myanmar
(Burma), Thailand and Indo-China through the East Indies to Borneo. This
family was formerly classified with the Schilbeidae. There are two genera
with 22 species: *Helicophagus* (3 species) and *Pangasius* (19 species)
(Roberts and Vidthayanon, 1991).

These catfishes are large with elongated, moderately compressed
bodies. Only two pairs of barbels (maxillary and mental) are present. Nasal
and a second pair of mental barbels are absent. The dorsal and pectoral fins
have well-developed serrated spines. An adipose fin, an elongate anal fin,
and a deeply forked caudal fin are present.

The genus *Helicophagus* consists of three species from Thailand to
Sumatra. As the generic name implies, this group feeds on snails. Some
Pangasius species have an unsually developed axial gland which secretes a
mucus-like substance through a pore above the origin of the innermost
pectoral fin rays. There may be additional pores over the gland. The local
people believe that this secretion is used to nourish the catfish's young
(Smith, 1945; Roberts, 1989b). Very little is known about this gland, which
is found, to some degree, in other catfish families.

The Indian region contains only one species, *P. pangasius* (Fig. 113). It
inhabits large rivers in India, Pakistan, and Bangladesh (Roberts and
Vidthayanon, 1991). It can reach 1.5 m and is a food fish, although its flesh

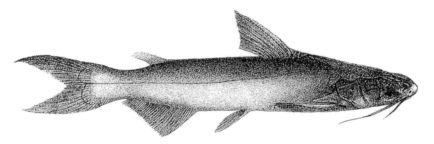

FIGURE 113. *Pangasius pangasius* as *P. buchanani* (Day, 1878b, Plate CVIII, Fig. 5).

is usually not relished due to its scavenging feeding habits (Talwar and Jhingran, 1992).

Pangasianodon (= *Pangasius*) *gigas* of the main channel of the Mekong River of Thailand, Laos, Cambodia, and Vietnam reaches enormous size. Roberts and Vidthayanon (1991) reported specimens as large as 3 m and 300 kg. *Pangasius gigas* is one of the largest freshwater fishes in the world. Smith (1945) included photographs of this remarkable animal. No juvenile specimens had been seen until recently when the species was bred in captivity (Fumihito, 1989). This giant catfish is unusual in that adults lack teeth and feed only on plant matter. Juveniles possess teeth in the jaws and on the palate (Fumihito, 1989). This fish has what may be the most rapid growth of any fish species. It can reach 200 kg in its first 3 years. The digestive tract, including the anus, is capable of remarkable expansion. This adaptation accommodates the gluttonous predatory feeding behavior in juveniles that supports the exceptional growth rate (Roberts and Vidthayanon, 1991). When flood waters subside *P. gigas* makes a spawning migration of several thousand kilometers up river. Stored fat is used during the migration and for gonadal development. The flesh is better tasting after the migration is completed. It was heavily fished during migration (Smith, 1945) but is now very rare. Two pair of barbels are present in this species (maxillary and mandibular), contrary to the original generic diagnosis that specified only one pair.

Pangasius sanitwongsei from Thailand is another gigantic species that rivals *P. gigas* at approximately 3 m (Smith, 1945). It has elongate filaments on the dorsal, pectoral, and pelvic fins and teeth on the roof of the mouth. These fin filaments make identification easy. According to Smith, since this species has a fondness for dogs, whose floating carcasses are common throughout the river, some fishermen remove the fins before sending the fish to market, thereby hiding the identity of this scavenger. Its enormous pectoral spines may be 60 cm long. Smith reported that a fisherman received a lethal stab wound from a pectoral spine while trying to remove a fish from a net.

Map references: Roberts and Vidthayanon (1991), Talwar and Jhingran (1992)

Class Actinopterygii
Subclass Neopterygii

Order Siluriformes
(1st) Family Chacidae—squarehead, angler, or frogmouth
catfishes [kas'-i-dē]

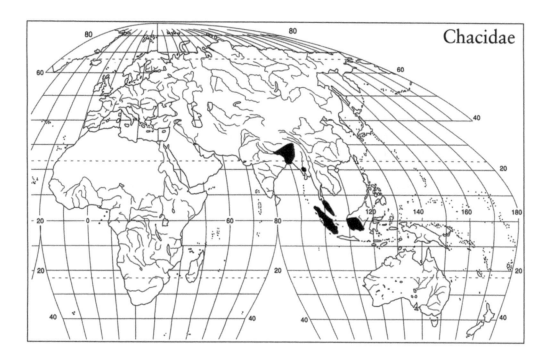

THIS BIG-HEADED FAMILY OF THREE SPECIES OF *CHACA* IS found in eastern India, Bangladesh, Nepal, Burma, the Malay Peninsula, Indonesia, and Borneo (Jayaram, 1981; Talwar and Jhingran, 1992; Roberts, 1982b, 1989a; Brown and Ferraris, 1988).

The head of these fishes is broad, depressed, very large, and nearly square when viewed from above. The body is strongly compressed behind the anal fin. The eyes are minute and dorsally situated. There are three or four pairs of barbels. The dorsal fin has a very strong spine and four rays. The pectoral fin spine is strongly serrated. The lateral line is complete and is marked by a prominent papillated ridge from operculum to caudal fin. The anal fin is short with only 8–10 rays. The head is covered with cutaneous flaps or cirri on its dorsolateral surface and the skin is granulated. The mouth is terminal, very wide, and fringed by barbellike appendages. The adipose fin is a low ridge confluent with the rounded caudal fin which extends along the dorsal and ventral surfaces. These fishes reach a maximum length of about 24 cm (Roberts, 1982b).

FIGURE 114. *Chaca chaca* as *C. lophioides* (Day, 1878b, Plate CXII, Fig. 2).

Chaca chaca (Fig. 114), from the Indian subcontinent, occurs in the Ganges–Brahmaptura drainage. Brown and Ferraris (1988) described a Burmese species, *C. burmensis*, from the Irrawaddy basin. *Chaca banka-nensis* is distributed from the Malay Peninsula to Sumatra and Borneo, where they inhabit lowland streams and lakes (Roberts, 1982b, 1989a). They are ambush fish predators. Their disruptive coloration and cutaneous cirri allow them to sit unnoticed like an algae-covered rock on the bottom. They are said to wriggle their barbels to simulate the movement of a small worm (Roberts, 1982b), hence the name angler catfish. Prey species such as cyprinids and anabantoids are inhaled into the huge mouth as they investi-gate the lure. Very little is known about the biology of *Chaca*. Jayaram and Majumdar (1964) and Roberts (1982b) reviewed the genus. Brown and Ferraris (1988) described the comparative osteology of the family.

Map references: Roberts (1982b,* 1989a)

Class Actinopterygii
Subclass Neopterygii

Order Siluriformes
(Per) Family Plotosidae—**eeltail catfishes** [plō-tō'-si-dē]

THE EELTAIL CATFISHES ARE A MARINE GROUP WIDELY DISTRIB-uted in the Indo-West Pacific region and from Japan to Fiji and Australia

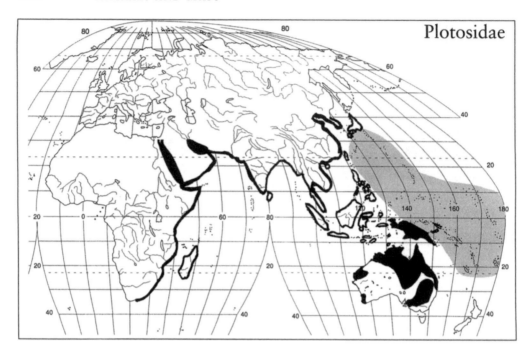

Plotosidae

(Springer, 1982). The 10 genera and 33 species inhabit coastal seas and estuaries (Allen and Feinberg, 1998; Allen, 1998a). More than half of the species are found in fresh waters of Australia and New Guinea. The Plotosidae and Ariidae are the only families of otophysan fishes in Australia and New Guinea.

The family takes its common name from the elongated, eel-like body and from an extension of the caudal fin that forms a caudodorsal fin that lacks interneurals and is confluent with the long anal fin. This wraparound fin usually ends in a point. There is a short-based dorsal fin armed with a strong spine immediately behind the head. No adipose fin is present. The pectoral fins also have strong spines and can inflict a very painful, venomous sting. Herre (1949) provided a scary, firsthand account of his own envenomation, the effects of which lasted 6 months. There are four pairs of barbels. Some marine species have dendritic organs between the anus and the anal fin. This highly vascularized structure may function in osmoregulation by secreting excess salt.

There are 13 species in five genera (*Anodontiglanis, Neosilurus, Neosiluroides, Tandanus,* and *Porochilus*) in Australian fresh waters (Allen, 1989; Allen and Feinberg, 1998). There are 13 species in four genera (*Neosilurus, Oloplotosus, Plotosus,* and *Porochilus*) in the rivers of New Guinea (Allen, 1991). Only *N. ater, N. brevidorsalis,* and *Porochilus obbesi* are shared between northern Australia and southern New Guinea.

Tandanus tandanus (Fig. 115) is widespread throughout the Murray–Darling River system and in coastal drainages of eastern Australia.

FIGURE 115. *Tandanus tandanus* (Waite, 1923, p. 67).

It is a bottom dweller in slow-moving streams. It is one of the best known plotosids. Its life history has been detailed in a series of papers by Davis (1977a,b,c,d). *Tandanus tandanus* spawns in the spring and large females (53 cm TL) may produce 20,000 or more eggs. The male excavates a circular nest in the gravel from 0.6 to 2.0 m in diameter. There is an elaborate courtship. The eggs hatch in about 7 days, but the male usually guards the nest and young for an additional 2 weeks after hatching. They are not mouth brooders like the ariid catfishes. Young fish feed on zooplankton, eventually switching to larger benthic invertebrates and fish. They can reach 90 cm and 6.8 kg, but most adults are less than 2 kg. They are good eating and a popular sport fish.

Neosilurus is one of the largest freshwater plotosid genera. These are small fishes, mostly less than 30 cm, and appear to be an ecological equivalent of the North American madtoms, *Noturus*. Their caudodorsal does not extend as far anteriorly as in *Tandanus*. *Neosilurus hrytlii* (Fig. 116) is a common fish in the Northern Territory and is found in the Timor, Gulf, and Lake Eyre drainages in a variety of habitats from clear creeks to muddy billabongs. It can reach 40 cm TL and 2 kg but is usually much smaller. The dorsal and pectoral spines are very sharp, and their

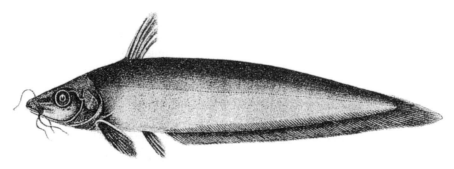

FIGURE 116. *Neosilurus hyrtlii* (Waite, 1923, p. 68).

venom is quite painful (Larson and Martin, 1989). *Neosilurus equinus* occurs in torrential streams with small waterfalls in central New Guinea. It may be the only fish in some headwater streams at 1500 m (Allen, 1991). *Neosilurus gloveri*, one of the smallest plotosids at 85 mm SL, is endemic to Dalhousie Springs, a warm artesian bore in the central desert. It has been collected in waters as warm as 40°C (Allen and Feinberg, 1998). *Neosiluroides cooperensis* is confined to the Cooper Creek system of the internal drainage, northeast of Lake Eyre. Its peculiarities include a thick epidermal covering around the posterior nostril which forms a separate outer chamber, and it has a dense covering of minute papillae over its body surface (Allen and Feinberg, 1998). *Oloplotosus luteus* has been taken as far as 1000 km from the sea in southern New Guinea. *Plotosus papuensis* from the Lorentz and Fly Rivers is one of the few freshwater plotosids that retains a dendritic organ.

For life histories and color illustrations of freshwater plotosids, see Merrick and Schmida (1984), Allen (1989, 1991), and Pollard *et al.* (1996).

Map references: Allen (1989, 1991),* Pollard *et al.* (1996),* Springer (1982)*

Class Actinopterygii
Subclass Neopterygii

Order Siluriformes
(1st) Family Clariidae—**air-breathing or walking catfishes**
(cla-rī'-i-dē) [cla'-rē-i-dē]

THE AIR-BREATHING CATFISHES ARE WIDELY DISTRIBUTED IN Africa to Asia Minor and from southern and southeastern Asia to the Philippines and Java. There are about 14 genera and approximately 90 species, with the greatest diversity in Africa where 12 genera and 74 species have been recorded (Teugels, 1986, 1996).

The dorsal and anal fins of air-breathing catfishes are very long and spineless. These fins may or may not be joined to the rounded caudal fin. Most members of this family lack an adipose fin, but an adipose fin is present in *Heterobranchus*, *Bathyclarias*, *Dinotopterus*, *Encheloclarias*, and one species of *Clarias*. Some species have very stout pectoral fin spines, but others may lack pectoral and pelvic fins altogether. The four pairs of barbels are usually very long. The head is depressed and may be covered with bony plates, and the eyes are small. The body is tubular and may be extremely elongated in some species.

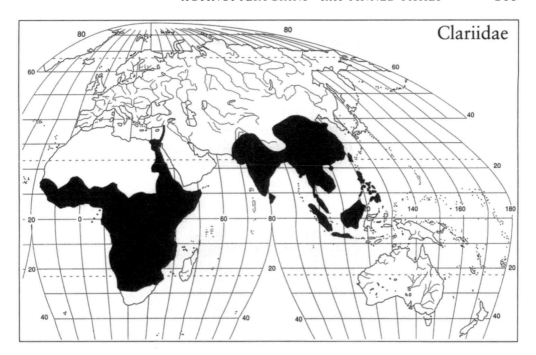

Clariidae

The common name of the air-breathing catfishes derives from the presence of a unique accessory respiratory organ that occupies the upper part of each branchial cavity (Fig. 117). These suprabranchial arborescent organs (sometimes called respiratory trees) arise from the gill arches and

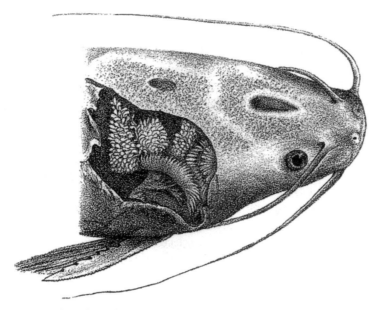

FIGURE 117. *Clarias batrachus* as *C. magur* (Day, 1878b, Plate CXII, Fig. 5a).

enable the fish to utilize atmospheric oxygen by providing a highly vascularized and extensive gill epithelium-covered surface area for absorption
(Graham, 1997). The gills are quite small, and it is reported that some
species must gulp air to survive (Smith, 1945). Graham reviewed various
studies that suggest that some species are not obligatory air breathers. Air
breathing enables these catfishes to live in some very foul habitats. Air
breathing has been established for at least four genera: *Clarias,
Bathyclarias, Heterobranchus*, and *Dinotopterus* (Graham, 1997).
Greenwood (1961) described the anatomy of the suprabranchial organs in
the Clariidae.

FIGURE 118.   *Clarias batrachus* as *C. magur* (Day, 1878b, Plate CXII, Fig. 5).

The walking catfish, *Clarias batrachus* (Fig. 118), from Asia has well-
developed pectoral spines and is able to wriggle overland by tail movements. This fish can remain out of the water for extended periods and is
often sold alive in baskets in native markets in Asia. Its hardiness has
enabled humans to transport it and introduce it to other areas. Its presence
in the Celebes (Sulawesi), across Wallace's line, is probably the result of
human transport (Darlington, 1957; Lever, 1996). This species is a very
important food fish in Asia and is cultured in ponds. It can walk overland
at night in search of prey or new habitats. It was accidently introduced into
southern Florida as an aquarium culture escapee, where it has become
established in canals and streams (Courtenay *et al.*, 1984; Fuller *et al.*,
1999). The fish that escaped either from the culture ponds or transport
trucks were albinos originally from Thailand. Natural selection has since
selected for the normal brown-gray color (Courtenay and Stauffer, 1990).
This took only 2 or 3 years. Within 10 years walking catfish had colonized
about 18% of Florida. The occasional unusually cold winter has killed
many walking catfish but has also selected for cold-tolerant forms that
eventually may be able to colonize northern Florida and adjacent states
(Courtenay, 1978). It can reach 55 cm, and it preys on fishes. This represents an ecological danger for Florida's native fish fauna (Taylor *et al.*,
1984). It has also been introduced into Guam (Lever, 1996) and New

Guinea (Allen, 1991). Much of what was stated previously about *C. batra-chus* is applicable to many other species of *Clarias*. The widespread *C. batrachus* does not occur naturally in Africa, but it is found from India to the Philippines. *Clarias batrachus* is actually a species complex of several undescribed species (G. G. Teugels, personal communication).

Clarias is the largest genus in the family, with 32 species in Africa alone. Another 10 or so species occur in southern Asia. The range of this genus more or less approximates the range of the family. *Clarias gariepinus* (Fig. 119) has a nearly pan-African distribution. It also occurs in Asia

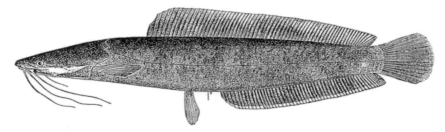

FIGURE. 119. *Clarias gariepinus* (Boulenger, 1911, Fig. 193).

Minor (Jordan, Israel, Lebanon, Syria, and South Turkey), where it can reach 1.5 m TL (Teugels, 1986) and 60 kg (Burgess, 1989). It is capable of weak electric discharge activity (Moller, 1995). It has been introduced for fish culture purposes throughout the world, including Southeast Asia and South America (G. G. Teugels, personal communication). Observations of *C. gariepinus* in mud of porridge-like consistency revealed that they take air in through the mouth and release bubbles from the opercular openings. They are able to survive in drying mud pools by suprabranchial respiration as long as they have access to atmospheric air (Donnelly, 1973).

Clariallabes and *Dolichallabes* are African genera with 13 species each. Other African genera include *Channallabes, Dinotopterus, Gymnallabes, Heterobranchus, Platyallabes, Platyclarias, Tanganikallabes, Uegitglanis*, and *Xenoclarias*. *Dinotopterus cunningtoni* (Fig. 120) is a very large species

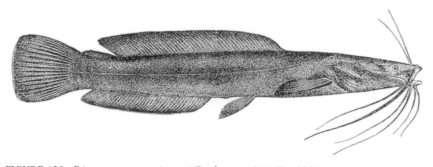

FIGURE 120. *Dinotopterus cunningtoni* (Boulenger, 1911, Fig. 228).

from Lake Tanganyika and can grow to 1.75 m SL. *Encheloclarias* is the only Asian genus with an adipose fin (Roberts, 1989a). Ng and Lim (1993) provided keys to the five species of secretive *Encheloclarias* from peat swamps and backwaters. *Horaglanis krishnai* from wells in India lacks eyes (Talwar and Jhingran, 1992), as does *Uegitglanis zammaranoi* from the caves of Somalia. The pectoral fins of *Horaglanis* are vestigial. Some members of *Channallabes, Dolichallabes micropthalmus*, and *Gymnallabes* have reduced or absent pectoral and pelvic fins and tiny eyes in association with burrowing. *Dolichallabes micropthalmus* is very elongate and eel-like. *Heterobranchus longifilis*, from Africa, has a high fish culture potential (Legendre *et al.*, 1992).

See Teugels (1986) for an African species list and Teugels (1992a) for keys and drawings of west African species and a list of references. See Smith (1945), Jayaram (1981), Roberts (1989a), and Talwar and Jhingran (1992) for information on Asian clariids. Burgess (1989) summarized the family and provided a key to the genera.

Map references: Darlington (1957), Herre (1953), Nelson (1976),* Ng and Lim (1993),* Skelton (1993),* Smith (1945), Talwar and Jhingran (1992)

Class Actinopterygii
Subclass Neopterygii

Order Siluriformes
(1st) Family Heteropneustidae—**air sac or stinging catfishes**
[het'-er-op-nū'-sti-dē]

THIS SMALL FAMILY OF ONE GENUS AND TWO SPECIES IS FOUND in Pakistan, India, Sri Lanka, Nepal, Bangladesh, Burma, Thailand, and Vietnam. Jayaram (1981) reported that it is absent from Malaya and Indonesia. *Heteropneustes* has an elongated body that is compressed posterior to the pelvic fins. The head is depressed and covered with bony plates. Four pairs of long barbels are present. The anal fin is very long. *Heteropneustes* is similar to clariid catfishes and is probably related to that family. However, the dorsal fin of *Heteropneustes* is short with six to eight rays, not long as in the clariids. There is no dorsal spine, nor is there an adipose fin. The mouth is small and terminal. The pectoral fins bear a moderately strong spine associated with a venom gland. These fish can deliver a very painful sting, hence their common name of stinging catfish. Smith (1995) maintained that human death has resulted from envenomations.

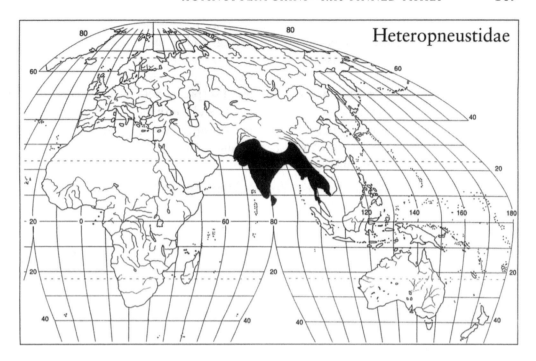

Their other common name, air sac catfish, is due to the presence of a pair of elongated, tubular cavities that extend posteriorly from the gill chamber through the back musculature almost to the tail. These air sacs serve as "lungs" enabling the fish to utilize atmospheric oxygen. Their natural habitat includes stagnant pools and swamps that are often deficient in oxygen. During the dry season they can burrow into damp mud. *Heteropneustes* can live out of water for many hours and can move overland. They are sold alive in local markets (Smith, 1945; Talwar and Jhingran, 1992) and are considered very good eating. They are, however, nonobligatory air breathers (Graham, 1997).

The two species are *H. fossilis* and *H. microps. Heteropneustes fossilis* (Fig. 121) is better known. It spawns in flooded rice paddies and is said to

FIGURE 121. *Heteropneustes fossilis* as Saccobranchus fossilis (Day, 1878b, Plate CXIV, Fig. 1).

be aggressive. Males and females excavate nest depressions in the rice paddies, and both parents guard the eggs and young. Workers fear stinging at this time. Likewise, fishers handle them with respect. First-aid measures include immediate immersion of the affected limb in very hot water which denatures the proteinaceous venom. This species can reach 30 cm.

Hora (1936) reviewed the family and Graham (1997) summarized what is known of their air-breathing anatomy and physiology and provided numerous references. Smith (1945) and Talwar and Jhingran (1992) gave life history information.

Map references: Jayaram (1981), Smith (1945), Talwar and Jhingram (1992)

Class Actinopterygii
Subclass Neopterygii

Order Siluriformes
(1st) Family Malapteruridae—electric catfishes
(mal-ap'-tē-rōō'-ri-dē)

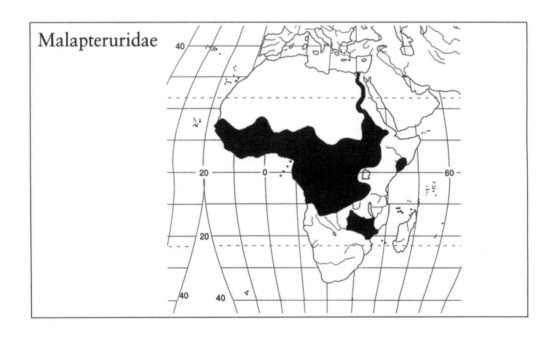

THE ELECTRIC CATFISHES, AS CURRENTLY RECOGNIZED, consist of three African species. *Malapterurus electricus* (Fig. 122) has a

FIGURE. 122. *Malapterurus electricus* (Boulenger, 1911, Fig. 382).

Nilo-Sudanic distribution including the Nile system (but excluding Lake Victoria), Lake Chad system, and all of sub-Saharan west Africa except "Upper Guinea" (Liberia, Sierra Leone, and Guinea) (Gosse, 1986a; Skelton, 1993). *Malapterurus microstoma* occurs in the Congo (Zaire) basin (Poll and Gosse, 1969). Sagua (1987) described a third species, *M. minjiriya*, from the Niger River of West Africa, and this species is listed by Teugels (1992b). There is one record from Lower Gebba River, Somalia (S. M. Norris, personal communication). The taxonomic diversity of this group has been severely underestimated, and there may be approximately 20 species in the family (S. M. Norris, personal communication).

These heavy-bodied fishes are covered with a thick, smooth skin. There is no dorsal fin. The adipose fin is situated posteriorly near the rounded caudal fin, and the anal fin is similarly located opposite the adipose fin. The pectoral fins lack spines. Spines and body armor common among other catfishes are probably unnecessary for defense of a catfish with such a powerful electric organ discharge (Alexander, 1965). There are three pairs of long barbels (nasal barbels absent). The eyes are small, and the mouth is terminal.

The electric organs form a mantle of tissue just under the skin that extends the length of the body and is derived from modified muscle tissue (Bennett, 1971a). The polarity of *Malapterurus*, negative at the anterior end and positive at the posterior end, is the reverse of the electric eel (Electrophoridae). The electric organs are composed of numerous elements arranged in series like the plates of a battery. Thus, the longer the fish, the greater the potential difference between head and tail and the greater the shock that can be delivered. Specimens of *M. electricus* as large as 1.2 m and 27 kg have been collected (Sagua, 1979). They can deliver a shock between 100 and 350 V, enough to render a human unconscious. The unique electrical ability of this catfish caused it to be recognized early in historical representation. A pictogram of an electric catfish is known from a slate palette of the predynastic Egyptian ruler Na'rmer around 3100 BC (Howes, 1985). The ability to discharge appears voluntary, and even small specimens can deliver a shock.

This stunning ability appears to be used to capture prey since electric catfish are very sluggish swimmers. Details of the electrical abilities of these fishes can be found in Grundfest (1960), Bennett (1971a,b), and Belbenoit *et al.* (1979). Moller (1995) provided a review of electric catfish biology. Sagua (1979) described the food and feeding habits of *Malapterurus* and reported that it is a voracious predator of cichlids, clupeids, and schilbeids. *Malapterurus* emits a powerful high-frequency electric discharge in proximity to a school of fishes, and thereby paralyzes many fishes simultaneously.

Howes (1985) suggested that the Malapteruridae may be related to the Siluridae and the Schilbeidae, but Bornbush (1991a) was critical of this study. Poll and Gosse (1969) examined the family and described *M. microstoma*.

Map references: Gosse (1986a), Skelton (1993),* Teugels (1992b)*

Class Actinopterygii
Subclass Neopterygii

Order Siluriformes
(Per) Family Ariidae—sea catfishes   (a-rī'-i-dē) [a-rē'-i-dē]

THE ARIIDAE IS A COASTAL MARINE FAMILY WITH A WORLD-wide distribution in tropical, subtropical, and warm temperate seas. Many species enter estuaries and ascend rivers a short distance. A few species may spend their entire life cycle in fresh water. There are about 14 genera and 120 species (Nelson, 1994).

Ariids are robust, medium to large catfishes with a small adipose fin, forked caudal fin, and stout spines in the dorsal and pectoral fins. The head is covered by a bony granular shield that may be visible under the skin or covered by thick skin. Two or three pairs of barbels are present. The ventral surface of the skull resembles a crucifix, so sea catfishes are sometimes called "crucifix fishes" by people who find a skull washed up on a beach.

Ariids are mouthbrooders. Spawning of freshwater species occurs near the beginning of the wet season. Claspers, a hook-like thickening on the dorsal surface of the pelvic fins of females, enlarge as the breeding season approaches. These may be used to hold the egg mass. The large eggs are carried in the mouth of the male until hatching. It is not known how the eggs get into the male's mouth. Hatching occurs 4 or 5 weeks after fertilization and may take 5–7 days. The fry, until they absorb their yolk sacs, may retreat into the male's buccal cavity for some time after hatching. Normally, the male fasts while incubating the eggs and young.

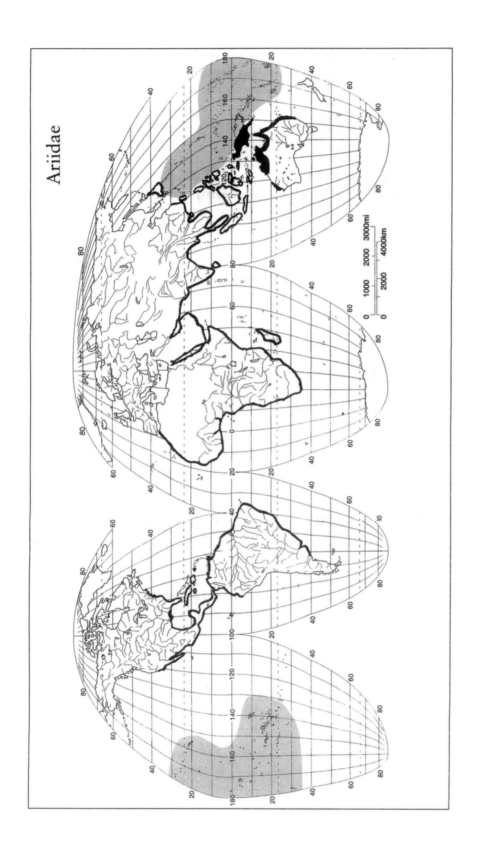

Ariidae

Lake and Midgley (1970) reported a 55-cm TL male *Arius* (= *Hexanematichthys) leptaspis* with 123 eggs of 14 mm diameter in its mouth from the Dawson River, a tributary of the Fitzroy River, Queensland. A photograph of eggs in a catfish's mouth is included in Lake (1971), in which the fish is identified as *Arius* (= *Hexanematichthys) australis*. Today, this species is recognized as *A. graeffei* (Kailola, 1989). A series of papers by Rimmer (1985a,b,c) described reproduction, early development, and growth of this species.

Arius is a large genus with approximately 80 species worldwide. About 17 of these occur in Australia and New Guinea. *Arius graeffei* is widely distributed in coastal rivers of northern Australia from Western Australia around the northern and eastern coasts to the Hunter River in New South Wales (Kailola, 1989). Three other species of *Arius* and *Cinetodus froggatti*

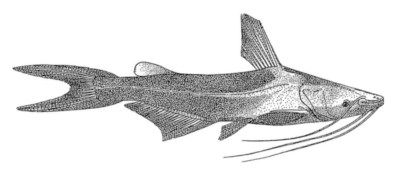

FIGURE 123. *Doiichthys novaeguineae* (Weber, 1913, Fig. 12).

are included within that range. Most of these species also occur in southern New Guinea (Allen, 1991). *Doiichthys novaeguineae* (Fig. 123) of southern New Guinea was formerly referred to its own family Doiichthyidae. Other New Guinea genera include *Cochlefelis* (2 species), *Nedystoma* (1 species), and *Tetranesodon* (1 species) (Allen, 1991). Many species of *Arius* were assigned to *Trachysurus*, but this genus and its family Trachysuridae are no longer recognized (Wheeler and Boddokway, 1981).

Talwar and Jhingran (1992) listed 11 species of *Arius* from the Indian region as well as five other genera. *Batrachocephalus mino* from the Indian subcontinent, Thailand, and Indonesia has a frog-like head with large eyes, and the barbels are reduced to a tiny pair on the chin. It reaches 25 cm. Smith (1945) considered 18 species of Ariidae (= Tachysuridae) from Thailand. Herre (1953) reported 15 species from the Philippines.

Taylor (1986) listed four genera [*Ancharius* (2 species), *Arius* (8 species), *Galeichthys* (2 species), and *Netuma* (1 species)] and 13 species from the coastal waters of Africa. Mo (1991) considered *Ancharius* as a member of the Mochokidae. Daget (1992a) reported that *Arius gigas* is common in the Niger and Volta basins of west Africa and is taken only in fresh water, not the sea. Stiassny and Raminosoa (1994) listed 5 species of ariids from the fresh waters of Madagascar, including *Ancharius* from the eastern coast and *Arius madagascariensis* from the western coast.

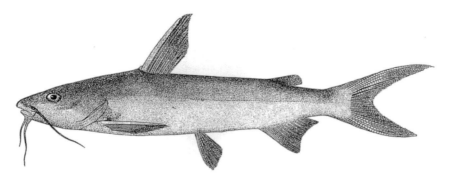

FIGURE 124. *Arius felis* (Goode, 1884, Plate 236).

Gosline (1945) listed the ariids that occur in fresh water in South and Central America. Miller (1966) recorded 19 species of ariids from Central America. In North America *Arius felis* (Fig. 124), the hard-head catfish, is found along the Atlantic and Gulf slopes from Cape Cod to Mexico (Platania and Ross, 1980). This species prefers shallow waters with sand or mud bottoms. It enters rivers in the southern part of its range. On the Pacific coast of North America *Bagre panamensis* is rare north of southern Baja (Eschmeyer and Herald, 1983).

Map references: Allen (1989, 1991),* Eschmeyer and Herald (1983), Gosline (1945), Kottelat *et al.* (1993), Masuda *et al.* (1984), Miller (1966), Taylor (1986), Talwar and Jhingran (1992)

Class Actinopterygii
Subclass Neopterygii

Order Siluriformes
(1st) Family Mochokidae—**upside-down catfishes, squeakers** [mō-kō'-kid-dē]

THE MOCHOKIDAE IS FOUND THROUGHOUT AFRICA (BUT NOT Madagascar) except the desert regions. There are 10 genera and about 175 species, of which about 66% are in the genus *Synodontis* (Gosse, 1986b).

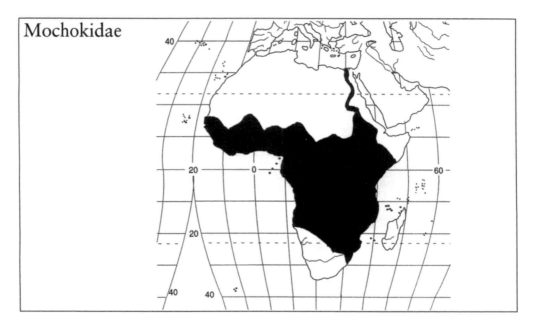

Mochokidae

There is debate among ichthyologists whether 2 genera (*Brachysynodontis* and *Hemisynodontis*) are valid. They may be synonyms of *Synodontis* (D. Paugy, personal communication). Mo (1991) considered *Ancharius*, with its 2 species, to be a mochokid, but Nelson (1994) treated that genus under the Ariidae. Mochokids have a bony head and nape shield and three pairs of barbels (nasals absent). In many species the barbels are subdivided into a feathery appearance. They have a stocky body, and the adipose fin may be very large. The 2 species in the genus *Mochokus* have a rayed adipose fin. The dorsal and pectoral fin spines are very strong and lockable. The pectoral fin, when moved in its socket, emits a low grunting sound which is why some species of *Synodontis* are called "squeakers."

As the common name indicates, some species, such as *Brachysynodontis batensoda* (Fig. 125) and *Synodontis nigriventris*, actually swim with their ventral surface up. This position is related to feeding on algae growing on the undersurface of aquatic plants and other objects. They may also evade predators by hiding upside down under a log (M. de Pinna, personal communication). Some species that swim in this unusual position may exhibit reversal of countershading with the dorsal surface light and the ventral surface dark, as the specific name *nigriventris* denotes. *Synodontis nigriventris* occurs in the central Congo (Zaire) basin and reaches 96 mm TL.

Synodontis, with 116 species (Teugels, 1996), is the second largest otophysan genus in Africa. Only the cyprinid genus *Barbus* is larger. Because the genus *Synodontis* is so large, these fishes are often considered synonymous with the family Mochokidae, especially by aquarists. Many

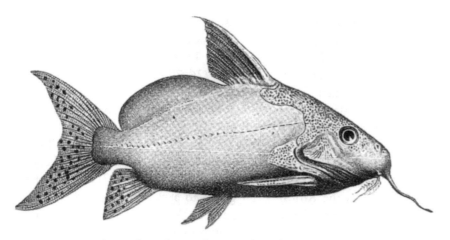

FIGURE 125. *Brachysynodontis batensoda* as *Synodontis batensoda* (Cunningham, 1912, Plate XXXI).

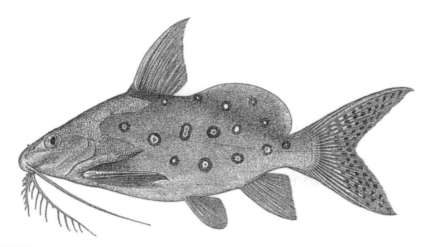

FIGURE 126. *Synodontis ocellifer* (Boulenger, 1911, Fig. 308).

Synodontis are attractively colored and patterned, such as *S. ocellifer* (Fig. 126), and are highly prized in the aquarium trade. *Synodontis alberti* is an aquarium favorite with exceptionally long barbels. The maxillary barbels may reach beyond the caudal fin. *Synodontis xiphias* from the Niger basin is the largest species in the family at 80 cm TL. *Synodontis multipunctatus* occupies the niche of an "aquatic cuckoo" in that this species allows its eggs to be injested and hatched in the mouths of mouth-brooding cichlids (Sato, 1986). The catfish eggs hatch earlier than the cichlid's own eggs, and the catfish fry feed on the fry of the host while still in the mouth. This species is apparently an obligate brood parasite of mouth brooders and represents the only example of true brood parasitism recorded among fishes (Sato, 1986). Poll (1971) revised the genus. Russell (1987) explained that the lizardfishes have priority on the usage of the family name Synodontidae. A few species of *Synodontis* can emit weak electric potentials (Moller, 1995).

The genus *Chiloglanis* (38 species; Teugels, 1996) is quite different from *Synodontis*. These small catfishes (26–70 mm TL) inhabit fast-flowing waters, and their lips are modified into a sucking disc. They are called suckermouth catfishes and are very efficient algae scrapers due to their mandibular teeth. Roberts (1989b) revised the genus.

See Gosse (1986b) for a checklist of species and Paugy and Roberts (1992) for keys and illustrations. Burgess (1989) provided a general discussion of the group and keys to the genera. Lundberg (1993) considered the Mochokidae as the African sister group to a clade formed by the South American Doradidae, Ageneiosidae, and the Auchenipteridae and discussed the biogeographical implications of this arrangement.

Map references: Gosse (1986b), Paugy and Roberts (1992),* Poll (1973),* Skelton (1993)*

Class Actinopterygii
Subclass Neopterygii

Order Siluriformes
(1st) Family Doradidae—**thorny or talking catfishes**
(dō-rad'-i-dē)

THIS FAMILY IS FOUND IN THE ATLANTIC SLOPE FRESH WATERS of South America from the Río Magdalena in Colombia through the Río Paraguay in Argentina. Doradids occur from lowland areas of the São Francisco to elevations of 1000 m in Bolivia and Peru (Burgess, 1989). They are especially common in Peru, Brazil, and the Guianas. There are about 35 genera and 90 species (Nelson, 1994).

Eigenmann (1925) reviewed the family. Lundberg (1993) presented the Doradidae as the sister group of the Auchenipteridae plus Ageneiosidae. De Pinna (1998) discussed family relationships of "doradoids," including Doradidae, Auchenipteridae, and the African Mochokidae. He noted that some catfish specialists also include the African Malapteruridae, the Asiatic Pangasiidae, and the worldwide Ariidae in this group. De Pinna (1998) provided a cladogram of generic relationships within the Doradidae based on Higuchi (1992). See Burgess (1989) for an overview of this group.

Doradids are called thorny catfish because the robust body has a row of lateral bony plates, each armed with a posteriorly directed hook. The

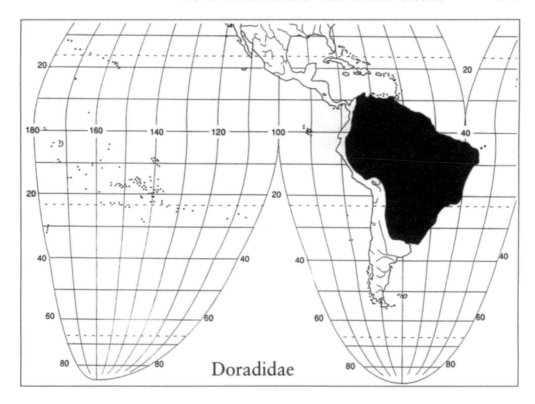

Doradidae

dorsal fin has a strong spine that may or may not be serrated on one or more edges. The large pectoral spines are serrated on both edges. The pectoral and dorsal fins can be locked in the erect position. The other common name, talking catfishes, derives from the groaning, chattering, or purring sounds the fishes make especially when removed from the water or when handled. These sounds emanate from rotation of the pectoral spines in their sockets with amplification by the swim bladder. A swim bladder–Weberian complex called the elastic spring apparatus is also a source of sound generation. The skull extends as a bony plate to the base of the dorsal fin, and the skull sutures are distinct. An adipose fin is present, and there are three pairs of barbels. In some species the barbels are subdivided. A heavy-toothed spine (humeral process) extends posteriorly from the shoulder region. The various spines can inflict a painful wound. Doradids are nocturnal or crepuscular bottom dwellers. They feed on small benthic invertebrates and detritus.

The genus *Opsodoras* is the largest, with about 12 species including *O. boulengeri* (Fig. 127). The genus *Hassar* has about 9 species and is one of the few doradid genera whose name does not end in "doras." Most of the other genera have only 1 to a few species and are restricted to a limited drainage area. *Centrochirs crocidili* is found in the Magdalena and Amazon basins. *Doraops zuloagai* occurs only in the Lake Maracaibo basin of Venezuela. *Franciscodoras marmoratus* is confined to the Río das Velhas,

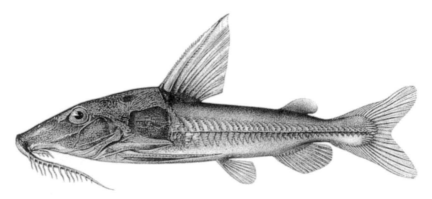

FIGURE 127. *Opsodoras boulengeri* as *Hemidoras boulengeri* (Steindachner, 1915, Plate VIII, Fig. 1).

and *Rhinodoras dorbignyi* is only known from the Río de la Plata. The Amazon and Orinoco basins each have 8–10 endemic genera.

Megalodoras irwini and *Pterodoras granulosus* are large species that can reach at least 60 cm TL. Both feed on snails and fruits. *Lithodoras dorsalis* reaches about 1 m TL and 12 kg. It is the only doradid whose belly is covered with bony scutes (Goulding, 1980). It is a fruit eater. *Oxydoras niger* is the largest South American toothless catfish. It can reach 1.2 m and 20 kg (Goulding, 1980). It lives in rivers of the flooded forest and feeds on detritus with its suctorial mouth. At the other end of the size spectrum is *Physopyxis lyra*, which probably does not reach 50 mm TL.

Acanthodoras spinosissimus and *A. cataphractus* are well-known in the aquarium trade. The sound these fishes make is loud enough to be heard outside an aquarium. Ladich (1997) reviewed sound production in fishes. *Platydoras costatus* is an attractive aquarium fish. It has a bright white band along the side of the body. *Hemidoras microstomus* (Fig. 128)

FIGURE 128. *Hemidoras microstomus* (Eigenmann, 1912, Plate XVIII, Fig. 2).

from Guyana has strongly serrated pectoral spines and fringed barbels. *Wertheimeria maculata* is the most primitive member of the family (de Pinna, 1998). It was previously included in the Auchenipteridae. Its adipose fin is unusual in that its base is as long as the anal fin base.

Map references: Burgess (1989), Eigenmann (1910, 1912),* Eigenmann and Allen (1942), Eigenmann and Eigenmann (1891), Géry (1969)

Class Actinopterygii
Subclass Neopterygii

Order Siluriformes
(1st) Family Auchenipteridae—driftwood catfishes, slopehead catfishes [ow-ken-ip-ter'-i-dē]

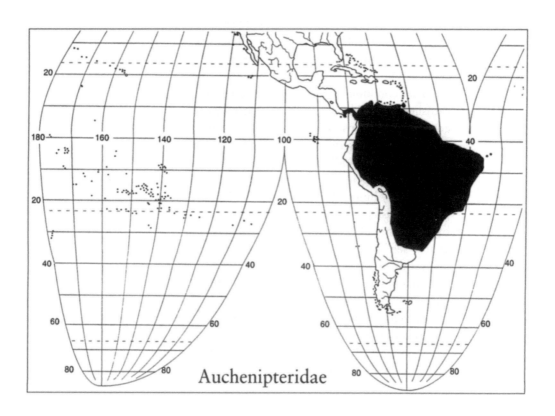

Auchenipteridae

THE AUCHENIPTERIDAE RANGES FROM PANAMA TO THE RÍO de la Plata in Argentina. There are about 20 genera and 70 species (Curran, 1989; de Pinna, 1998). This group has previously been classified with the Doradidae and Pimelodidae. Curran considered the Doradidae to be the sister group to the Auchenipteridae. Nelson (1994) treated the Ageneiosidae and Auchenipteridae as separate families. De Pinna reviewed the confusing classification history of the Auchenipteridae and included the Ageneiosidae within this family, as did Ferraris (1988a), who showed that ageneiosids were deeply internested within the auchenipterids (reflected in the cladogram in de Pinna, 1998). I use this view of the expanded Auchenipteridae.

Auchenipterids are naked, lacking the armor of the doradids. Three pairs of barbels are usually present, except in ageneiosids. The short-based dorsal fin usually has a stout spine and is located close to the head. The dorsal spine is partially surrounded by processes from the nuchal bones. The pectoral fins have a strong spine. Soares-Porto *et al.* (1999) described a miniature, pedomorphic species of *Gelanoglanis* that lacks ossified dorsal and pectoral spines. The adipose fin is small or absent. The lateral line often has a zigzag arrangement with short branches at each angle.

The male's genital papilla is supported by the anterior rays of the anal fin and functions as an intromittent organ. Auchenipterids are the only siluriforms, other than *Scoloplax*, with internal fertilization (de Pinna, 1998). Sperm storage is possible and fertilization occurs at spawning time, independently of the male. The posterior portion of the testes produces gelatinous material that forms a plug in the female's oviduct after copulation.

The common name of some members of this family, driftwood catfishes, probably derives from their association with woody debris. In aquaria, specimens often hide around driftwood decorations during the day. Some species reach a length of 30 cm and feed on fruits as well as invertebrates (Goulding, 1980). *Tatia* and *Parauchenipterus* are two of the most speciose genera. Males of *Parauchenipterus fisheri* (Fig. 129) have a dramatically humped back at the dorsal fin and a concave head profile. The jaguar catfish, *Liosomadoras oncinus*, is a beautiful, spotted catfish prized by aquarists. Some species of *Pseudauchenipterus* can tolerate full salt water. See Burgess (1989) for photographs and aquarium information about this family.

Soares-Porto (1998) reviewed 30 species in three genera of auchenipterids. Ferraris and Vari (1999) reviewed *Auchenipterus* and recognized 11 species. *Auchenipterus nuchalus* (Fig. 130) has a zigzag lateral line with many dorsal and ventral branches. Males of this species have thick maxillary barbels that are ossified almost to their tips, and the first few anal fin rays are fused. They considered this genus monophyletic

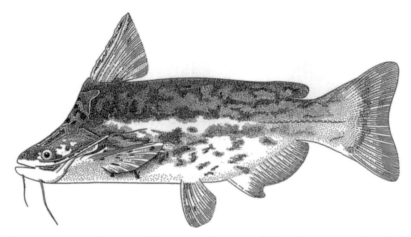

FIGURE 129. *Parauchenipterus fisheri* as *Trachycorystes fisheri* (Eigenmann, 1922, Plate V, Fig. 1).

FIGURE 130. *Auchenipterus nuchalus*, female, as *A. demerare* (Eigenmann, 1912, Plate XXI, Fig. 1).

based on the shared presence of grooves in the ventral surface of the head that accommodate the bases of fully adducted mental barbels. Similar grooves are absent in all species of *Pseudepapterus* and *Epapterus*. Vari and Ferraris (1988) discussed *Epapterus* and considered it to be the sister group to *Pseudepapterus*.

The ageniosids are known as slopehead or barbelless catfish among aquarists. The head of "slopehead" catfishes is flattened anteriorly and deep posteriorly. This is very obvious in *Ageneiosus polystictus* (Fig. 131), but less so in some other species such as *A. marmoratus* (Fig. 132). The mouth gape is large. The body is naked and streamlined, in keeping with its pelagic and piscivorous habits. The dorsal fin is located immediately behind the head and is armed with a moderately to weakly pungent spine. In breeding males the dorsal spine becomes elongated and serrated along the anterodorsal margin. The adipose fin is very small, and the anal fin is long. The pectoral fins have weak to moderate spines. The swim bladder is

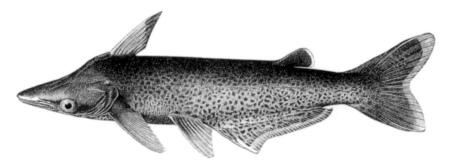

FIGURE 131. *Ageneiosus polystictus* (Steindachner, 1915, Plate VII, Fig. 1).

FIGURE 132. *Ageneiosus marmoratus* (Eigenmann, 1912, Plate XXII, Fig. 1).

reduced and is encased in bone in one group. There is only one pair of small barbels, which are easily overlooked; hence the name "barbelless" catfishes. However, the maxillary barbels of males elongate during the breeding season.

There are two genera and about 14 species of ageneiosids. The genus *Ageneiosus*, with 13 species, has short maxillary barbels, and *Tetranematichthys quadrifilis* has short mandibular barbels. *Ageneiosus pardalis* occurs on the Pacific slope of Panama as well as in the Río Magdalena and Lake Maracaibo basin (Miller, 1966). *Ageneiosus brevifilis* has the broadest distribution of any ageneiosid from the Orinoco River and Guyana through the Amazon basin into Peru, Uruguay, and Argentina. Its feeding habits include gulping whole fish. It reaches 55 cm TL and 2 kg and is utilized as a food fish. Walsh (1990) produced a systematic revision of the ageneiosids.

Map references: Burgess (1989), Eigenmann (1909a, 1910, 1912*), Eigenmann and Allen (1942), Eigenmann and Eigenmann (1891), Ferraris and Vari (1999),* Géry (1969), Miller (1966),* Soares-Porto *et al.* (1999)*

Class Actinopterygii
Subclass Neopterygii

Order Siluriformes
(1st) Family Amblycipitidae—**torrent catfishes**
[am-bli-si-pit'-i-dē]

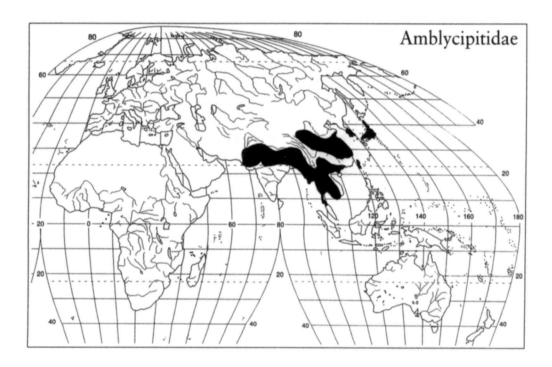

THESE DIMINUTIVE (100 MM) LOACH-LIKE CATFISHES ARE found in swift water along the foothills of the Himalayas from Pakistan, India, Nepal, and Bangladesh through southern China, Japan, Burma, Thailand, Laos, and Malaysia (Jayaram, 1981; Talwar and Jhingran, 1992; Chen and Lundberg, 1995). There are three genera and about 18 species: *Amblyceps* (7 species), *Liobagrus* (10 species), and *Xiurenbagrus xiurenensis* (Chen and Lundberg, 1995; M. Kottelat, personal communication). The generic distributions are nonoverlapping. *Amblyceps* is widespread from Pakistan and northern India to the Malay Peninsula. *Liobagrus* occurs in Korea, Japan, and central China, and *Xiurenbagrus* is distributed in southern China (Chen and Lundberg, 1995).

De Pinna (1996b) erroneously included Sumatra, Java, and Borneo as part of their range, but amblycipitids are not listed by Roberts (1989a) or Kottelat *et al.* (1993) for any part of Indonesia, and they are not shown in Indonesia on the distribution map of Chen and Lundberg (1995). The Amblycipitidae is considered to be monophyletic and the sister taxon to Akysidae plus Parakysidae by Chen (1994). De Pinna also considered the Amblycipitidae to be monophyletic, but hypothesized it as the sister group to the clade formed by Akysidae, Erethistidae, and Aspredinidae.

The amblycipitid body is depressed anteriorly. The dorsal fin is covered with thick skin and has a spine capable of inflicting a painful wound. An adipose fin is present and may or may not be confluent with the caudal fin. The anal fin is short-based. Four pairs of barbels are present. The paired fins are also enveloped in thick skin, and a pectoral spine is present. The lateral line is absent or incomplete. The swim bladder is reduced and encapsulated in bone. The most unusual feature of this family is the presence of a cup-like fold of skin in front of the pectoral fin. The gill membranes rest on this fold when the gill openings are closed. Hora (1933) considered this a respiratory adaptation to life in fast currents. The amblycipitid respiratory pattern is similar to that of other hillstream fishes. The fishes make a series of rapid inhalations followed by a quiescent period of 1–4 minutes. Graham (1997) disputed Hora's (1935) claim that *Amblyceps* is a facultative air breather.

Amblyceps from the Indian subcontinent possess the skin fold mentioned previously, as do the 10 species of genus *Liobagrus* from China and Japan. *Xiurenbagrus* from southern China does not. Chen and Lundberg (1995) consider this feature to be a synapomorphy for the sister-group relationship between *Amblyceps* and *Liobagrus* and that *Xiurenbagrus* is the sister taxon to the other two genera.

FIGURE 133. *Amblyceps mangois* (Day, 1878b, Plate CXVII, Fig. 1).

Amblyceps mangois (Fig. 133) is said to be able to live out of water for a long time and to be able to deliver a vicious bite to careless fishers (Smith, 1945). It only reaches 13 cm. When its streams dry out seasonally, *A. mangois* buries itself in the bottom to seek groundwater, or it can wriggle over dry stream beds into deeper pools of water.

Map reference: Chen and Lundberg (1995)*

Class Actinopterygii
Subclass Neopterygii

Order Siluriformes
(1st) Family Akysidae—**stream catfishes** [a-kī'-si-dē]

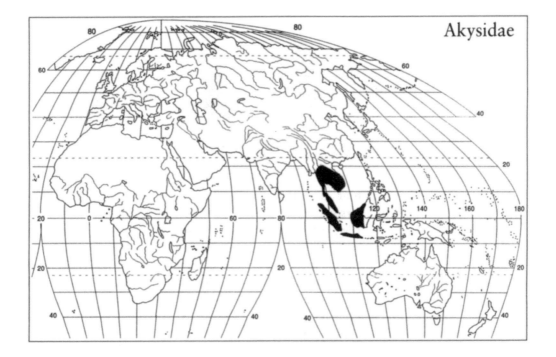

THE AKYSID CATFISHES ARE KNOWN FROM FAST-FLOWING
waters in Myanmar (Burma), Thailand, Malayasia, Java, Sumatra,
and Borneo (Jayaram, 1981). De Pinna (1996b) hypothesized that this
group is the sister taxon to the Sisoridae and Erethistidae plus the
Aspredinidae. Roberts (1989b) erected a separate family, Parakysidae,
for two species of *Parakysis* from the Malay Peninsula and Indonesia;
however, de Pinna (1998) does not recognize this family. He considered
Parakysis to be deeply nested within the Akysidae and that recognition of
a separate Parakysidae would result in a paraphyletic Akysidae (de Pinna,
1996.). Ng and Kottelat (1998c) reviewed *Akysis*, and Ng and Lim (1995)
revised *Parakysis*. The Parakysidae is now recognized as a subfamily,
Parakysinae.

The short-based dorsal fin has a strong spine, and the adipose fin is
usually long. The pectoral spine is well developed and serrated on most

species. There are four pairs of barbels, and the eyes are small. The swim bladder is slightly reduced. The entire surface of the head, body, and fins is covered with tubercles similar to those of the Sisoridae (Roberts, 1982a) but normally more pronounced. Some of these tubercles are enlarged and arranged in longitudinal rows along the body. Akysids shed their skin periodically. The only other catfish family known to molt is the Aspredinidae, which also has integumentary tubercles. Most akysids are small to minute, cryptically colored, and secretive.

FIGURE 134. *Akysis leucorhynchus* (reproduced with permission from Fowler, 1934, Fig. 44).

There are four genera (including *Parakysis*) and about 30 species (M. Kottelat, personal communication). Some species, such as *Akysis leucorhynchus* (Fig. 134), are miniatures and are sexually mature at 3 cm (Smith, 1945). *Breitensteinia insignis* can reach 19 cm (Roberts, 1989a). This elongate species is lacking an adipose fin but has two dorsal fin spines found in most siluriforms. Ng and Siebert (1998) revised *Breitensteinia* and provided keys to 3 species. Six nominal species of *Acrochordonichthys* are listed by Burgess (1989). *Akysis* is the largest genus with about 19 species.

Parakysis verrucosa from the Malay Peninsula, Sumatra, and Sarawak and *P. anomalopteryx* from the Kapuas basin of Western Borneo were placed in the Parakysidae erected by Roberts (1989a). Two other species have been described (Ng and Lim, 1995). These fishes have an exceptionally stout dorsal spine and only four soft rays. The dorsal and pectoral spines are not serrated. There are four pairs of barbels, but the mandibular barbels may have small, accessory barbels. Gill rakers are absent. The lateral line is reduced, and the eyes are tiny. The adipose fin is represented by a long, low keel along the caudal peduncle. The skin of *Parakysis* is covered with rounded tubercles of uniform size evenly distributed over the head and body, unlike the tubercles of other akysids, which are arranged in longitudinal rows. These fish reach about 60 mm in length. See Herre (1940) and Roberts (1989a) for further details and Kottelat *et al.* (1993) for color photos.

Map references: Jayaram (1981), Kottelat *et al.* (1993), Ng and Lim (1995),* Ng and Siebert (1998),* Roberts (1989a)

Class Actinopterygii
Subclass Neopterygii

Order Siluriformes
(1st) Family Sisoridae—**sisorid catfishes** [si-sō'-ri-dē]

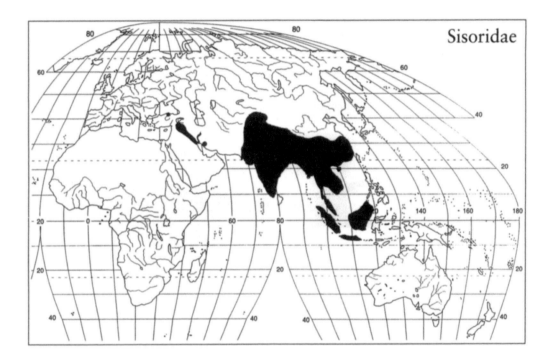

THIS WIDESPREAD ASIAN CATFISH FAMILY IS FOUND THROUGH-
out south and Southeast Asia, Turkey and Iran in the west (Coad, 1981a;
Coad and Delmastro, 1985), and Afghanistan, Pakistan, India (but not Sri
Lanka), Bangladesh, China, Korea, Burma, Thailand, Vietnam, Malaysia,
Indonesia, and Borneo (de Pinna, 1996b; Talwar and Jhingran, 1992). It is
also represented north of the Himalayas in central Asia west to the Aral
region (Berg, 1948/1949). The majority of species prefer fast-flowing water,
but some species occur only in slow or still waters. There are about 25
genera and 120 species (de Pinna, 1996b).

Sisorids have a long, low adipose fin and a short-based dorsal fin, with
or without a spine. In many species the skin is leathery and thickened with
horny epidermal projections (Roberts, 1982a). Four pairs of barbels are the
rule except in *Sisor rhabdophorus* (Fig. 135), which has six pairs (one
maxillary and five mandibular pairs). Some species (in the Glyptosterninae)

FIGURE 135. *Sisor rhabdophorus* (Day, 1878b, Plate CXV, Fig. 1).

have a thoracic adhesive disc formed by radiating ridges or transverse skin folds. The only other catfish family with such a structure is the Erethistidae. Other species have flattened lips and/or modified pectoral and pelvic fins that serve as a sucking organ. These organs help the fish hold station in torrential streams. The swim bladder is small and partly enclosed in bone. Some species are cryptically colored.

FIGURE 136. *Bagarius yarrellii* (Day, 1878b, Plate CXV, Fig. 3).

The Indian region has the most species, and these are identified and illustrated in Jayaram (1981) and Talwar and Jhingran (1992). The wide-ranging, predatory *Bagarius yarrelli* (Fig. 136) reaches more than 2 m and 113 kg and is one of the largest catfishes in Asia (Roberts, 1989a), whereas other members of the family such as *Glyptothorax stocki* reach only 55 mm (Talwar and Jhingran, 1992). Some torrent-living sissorids include *Pseudecheneis sulcatus* (Fig. 137) and *Glyptothorax*. The latter genus is the

FIGURE 137. *Pseudecheneis sulcatus* (Day, 1878b, Plate CXV, Fig. 1).

largest in the family with about 35 species (Burgess, 1989). *Glyptosternon* occurs from central Asia through China.

De Pinna (1996b) published a phylogenetic analysis of the family and determined that it was paraphyletic as currently constituted. He removed *Conta, Laguvia, Pseudolaguvia, Erethistoides, Hara*, and *Erethistes* from the Sisoridae and placed them in the family Erethistidae. The Sisoridae was then considered to be monophyletic and sister group to a clade composed of Aspredinidae + Erethistidae. De Pinna divided the Sisoridae (*sensu stricto*) into two subfamilies, the Sisorinae (*Bagarius, Sisor, Gagata*, and *Nangra*) and the Glyptosterninae (*Glyptothorax, Pseudecheneis*, and "glyptosternoids"). The map presented here is the same for the Sisoridae considered *sensu lato* or *sensu stricto*. Ng and Kottelat (1999) reviewed the "glyptosternoid" genus *Oreoglanis* and its three species.

For revisions and species descriptions see Hora and Silas (1952), Coad (1981a), Coad and Delmastro (1985), and de Pinna (1996b). For fauna and life histories see Jayaram (1981), Talwar and Jhingran (1992), Smith (1945), Kottelat *et al.* (1993), and Roberts (1989a).

Map references: Berg (1948/1949), Coad (1981a), Coad and Delmastro (1985), Jayaram (1974),* Nichols (1943), de Pinna (1996b), Smith (1945), Talwar and Jhingran (1992)

Class **Actinopterygii**
Subclass **Neopterygii**

Order **Siluriformes**
(1st) Family **Erethistidae—erethistid catfishes**   (er'-ē-thiz'-ti-dē)

DE PINNA (1996B) RESURRECTED THE FAMILY ERETHISTIDAE, originally created by Bleeker, to accommodate six genera of Asian catfishes from the Sisoridae: *Conta, Laguvia, Pseudolaguvia, Erethistoides, Hara*, and *Erethistes*. The first three named genera constitute the subfamily Continae. These fishes possess a long, narrow adhesive apparatus formed by integumentary ridges and grooves on the thorax or abdomen. This structure is unique within the Siluriformes and is found only in the Continae and in some members of the sisorid subfamily Glyptosterninae. The adhesive apparatus helps a fish maintain its position in a strong current. The latter three genera make up the subfamily Erethistinae. This group lacks an adhesive apparatus (de Pinna, 1996b).

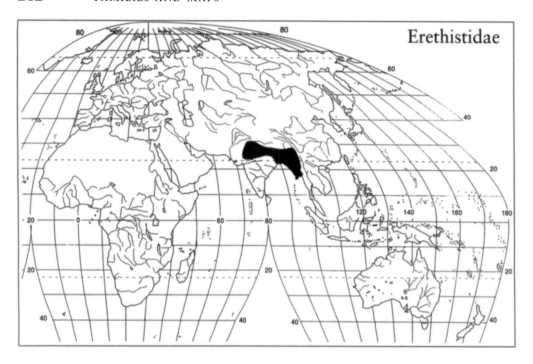

Members of this family generally resemble sisorids, but erethistids do not reach nearly the same size as the much larger sisorids. Some erethistids are miniature fishes. *Erethistes maesotensis*, for example, attains a maximum size of 21.7 mm SL (Kottelat, 1983). See Weitzman and Vari (1988) and Kottelat and Vidthayanon (1993) for lists of miniaturized species from South America and Asia, respectively. *Hara hara* (Fig. 138) only reaches about 25 mm SL (Talwar and Jhingran, 1992). *Hara jerdoni* and all species in *Laguvia* and *Pseudolaguvia* are also tiny but may on occasion exceed the 26-mm SL limit adopted as an arbitrary maximum for miniaturized species (de Pinna, 1996b).

FIGURE 138. *Hara hara* as *Erethistes hara* (Day, 1878b, Plate CII, Fig. 1).

About 12 species of erethistids occur in India, Nepal, Bangladesh, Burma (Myanmar), and China. They are absent from the Malay Peninsula and the islands of Indonesia. *Conta* is monotypic and endemic to India. *Conta conta* inhabits rocky streams at the bases of hills (Talwar and Jhingran, 1992). *Laguvia* includes four small species that inhabit fast waters at the base of mountains. All are Indian, and *L. asperus* reaches China. *Pseudolaguvia tuberculatus* inhabits hillstreams in Upper Burma and has a rough, tuberculated skin (Talwar and Jhingran, 1992).

Two species of *Erethistes* are found in India, with *E. maesotensis* extending from Burma to the Salween River on the Thai–Burmese border. This is the easternmost record for the family. *Erethistoides montana* is the only species in this genus of torrent-inhabiting catfishes from the eastern Himalayas, Assam, Uttar Pradesh, and West Bengal (Talwar and Jhingran, 1992). *Hara*, with three species closely allied to *Erethistes*, is endemic to south Asia from northern and eastern India, Nepal, and Burma.

Map reference: Talwar and Jhingran (1992)

Class Actinopterygii
Subclass Neopterygii

Order Siluriformes
(1st) Family Aspredinidae—**banjo catfishes** (as'-prē-din'-i-dē)

THIS FAMILY IS FOUND IN THE TROPICAL WATERS OF SOUTH America east of the Andes from the Madgalena basin in Colombia to La Plata. They are not known from any Pacific slope drainage (Stewart, 1985). Some species may occur in brackish waters along the coasts of the Guianas and northeastern Brazil. There are 12 genera and about 34 species (de Pinna, 1998).

Howes (1983b) determined that the Aspredinidae is not as closely related to the Loricariidae as previously believed. De Pinna (1996b, 1998) considered the Aspredinidae to be the most aberrant member of the Neotropical catfish fauna and placed them within an internested set of Asiatic sisoroids.

These fishes are referred to as banjo or frying pan catfishes because the head and anterior body is flattened in some cases almost like a disc, and the tail is long and slender. The head is protected by thickened skull bones visible through the skin. The body may be studded with integumentary tubercles and plates arranged in longitudinal rows. The dorsal fin is short

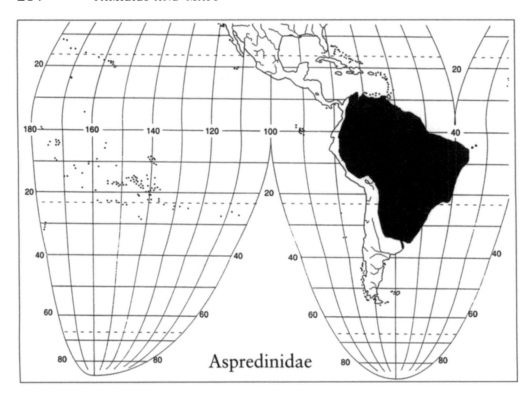

Asolpredinidae

and spineless, or a weak spine may be present. There is no adipose fin. The pectoral fins have strongly serrated spines. The opercular opening is a very small slit. The eyes are tiny. Three pairs of barbels are typically present. Nasal barbels are absent. The underside of the head and belly may be covered with many small barbels. The mouth is terminal, subterminal, or inferior. Villiform teeth are arranged in rows on the jaws in most taxa. The anal fin ranges from short to very long. The swim bladder is well developed in most species. Aspredinids may shed their skin in what looks like a peel of the fish or in large strap-like pieces (Ferraris, 1991a). The new skin is brighter than the shed skin. The only other catfishes that molt are the akysids.

For most of their taxonomic history, two subfamilies have been distinguished by the size of the anal fin. The Bunocephalinae has a small anal fin with up to 18 rays. This group is found only in fresh water and occurs from Colombia to Argentina. There are eight genera and about 28 species in the subfamily. These fish may move with a jerky sort of jet propulsion by rapidly forcing water from the gill chamber out of the small opercular opening (Gradwell, 1971). *Agmus lyriformis* (Fig. 139) has a rugose body and resembles a piece of wood when resting quietly on the bottom. *Amaralia hypsiura* from the central and upper Amazon has a dorsal fin with only 2 rays. Four species of *Hoplomyzon*, less than 25 mm in length, are found in fast-flowing waters in Venezuela and Ecuador. Other genera of this

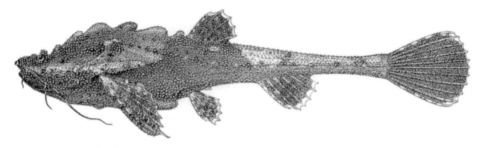

FIGURE 139. *Agmus lyriformis* (Eigenmann, 1912, Plate III).

subfamily include *Dupouyichthys* (1 species), *Ernstichthys* (3 species), *Bunocephalus* (1 species), *Dysichthys* (13 species), *Petacara* (1 species), and *Xyliphius* (5 species). There has been a great deal of confusion regarding the type species of *Bunocephalus*. This genus was formerly the most speciose in the family. Today, all save 1 of the species formerly assigned to *Bunocephalus* are placed in *Dysichthys* (Ferraris, 1991b). *Dysichthys coracoideus* (Fig. 140), known in some areas of South America as "guitarrita" (little guitar), is one of the most common members of this family in the aquarium trade (Burgess, 1989). For other papers on the taxonomy of this subfamily see Myers (1960), Stewart (1985), Mees (1988, 1989), and Friel (1994). De Pinna (1998), following Friel, separated the Bunocephalinae (*Bunocephalus* and *Amaralia*) from the recently erected Hoplomyzontinae (*Xyliphius, Hoplomyzon, Dupouyichthys,* and *Ernstichthys*).

FIGURE 140. *Dysichthys coracoideus* as *Bunocephalus amaurus* (Eigenmann, 1912, Plate II).

The second (or third) subfamily is the Aspredininae. This group has a very low, long anal fin with 50–60 rays. These elongate fishes inhabit the mangrove-lined east coast of South America from the Orinoco delta to the Brazilian state of Maranhão about 2400 km south (Burgess, 1989). These are euryhaline, lowland fishes of river mouths and estuaries. They are not usually found inland beyond the zone of tidal influence. There are four genera, including *Aspredo* and *Aspredinichthys*, each with two species (Mees, 1987; de Pinna, 1998). *Aspredo aspredo* is the largest member of the family at about 42 cm. The skin of the ventral surface of *Aspredo* females has cup-like depressions seasonally. These structures disappear after the reproductive period. It is speculated that as the female rests on eggs deposited in a spawning site the eggs attach to these cups (Burgess, 1989; de Pinna, 1989). Apparently, the young are also ferried about by the female. *Aspredinichthys tibicen*, which reaches 20 cm in length, has about 10 pairs of small mental or postmental barbels. The females carry the eggs on their belly, on which the eggs are attached to spongy tentacles that develop during the breeding season. Other genera include *Pterobunocephalus* and *Platystacus*.

Several members of this family are popular aquarium animals, but little is known of their biology. Most species are nocturnal and relatively inactive. They spend much of the day hiding in aquatic vegetation.

Map references: Burgess (1989), Eigenmann (1910, 1912*), Eigenmann and Allen (1942), Géry (1969), Myers (1960), Schultz (1944)

Class Actinopterygii
Subclass Neopterygii

Order Siluriformes
(1st) Family Amphiliidae—loach catfishes
(am-phil-ī'-i-dē) [am-phil-ē'-i-dē]

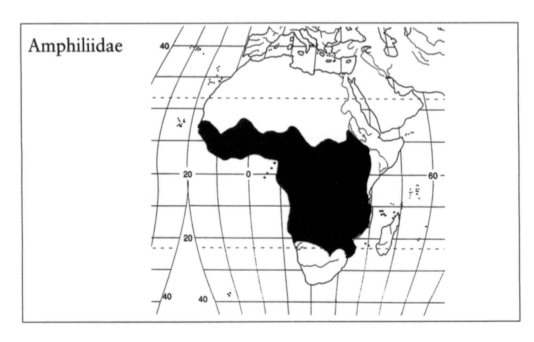

Amphiliidae

THIS AFRICAN FAMILY CONSISTS OF SEVEN GENERA WITH
about 50 species that are found mainly in fast-flowing water of clear
streams and rivers at high elevations. They are especially abundant in the
Congo (Zaire) basin. The pectoral and pelvic fins in many species are broad
and form a weak sucking disc in conjunction with the ventral surface of the
body. This helps the catfish cling to rocks in swift streams. The swim bladder is reduced. The dorsal fin base is short, and the dorsal spine is absent.
Pectoral spines are somewhat flexible. The long, low adipose fin sometimes
has a modified scute at its leading edge. There are three pairs of barbels
(nasal barbels are absent).

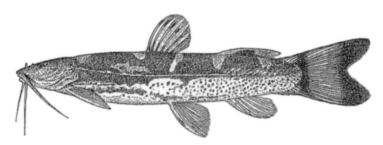

FIGURE 141. *Amphilius atesuensis* (Boulenger, 1911, Fig. 284).

The family is divided into two subfamilies. The Amphiliinae have short
bodies. There are two genera, *Amphilius* and *Paramphilius*, with 26 species.
Amphilius atesuensis (Fig. 141) is a typical subfamily representative that
grows to about 63 mm TL. The Doumeinae have elongate bodies, often
with an attenuated caudal peduncle and specialized bony armor. They
resemble South American loricariid catfishes. There are 24 species in the
following genera: *Doumea, Andersonia, Phractura, Trachyglanis*, and
Belonoglanis. Bailey and Stewart (1984) and Mo (1991) consider
Leptoglanis (9 species) and *Zaireichthys zonatus* as amphiliids instead of
bagrids.

Very little is known about the biology of the species in this family.
Amphilius platychir, a hillstream species from 1800 m in the Zambezi
River system, is imported in the aquarium trade. It feeds on insect
larvae and lays eggs on the undersides of stones. The young resemble
tadpoles (Wheeler, 1975). *Belonoglanis tenuis* from the Congo system
reaches 17 cm SL, but most species are 3–10 cm (Skelton and Teugels,
1989).

Burgess (1989) and Skelton (1992, 1993) provided keys and illustrations. Harry (1953) and Skelton (1984, 1989) provided systematic
information.

Map references: Skelton (1992, 1993),* Skelton and Teugels (1989)

Class   Actinopterygii
Subclass   Neopterygii

Order   Siluriformes
(1st) Family   Nematogenyidae—**nematogenyid catfishes**
[nē'-mat-ō-gen-ē'-i-dē]

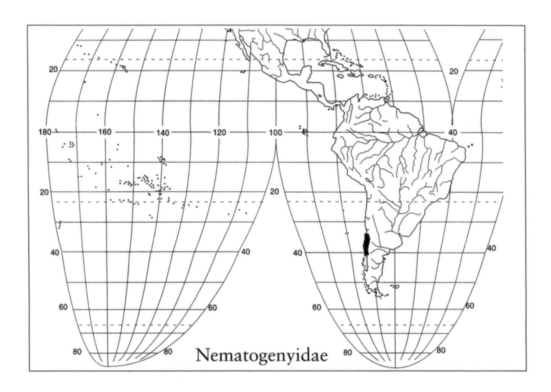

THIS MONOTYPIC, RELICT FAMILY CONSISTS OF *NEMATOGENYS inermis* (Fig. 142), which is endemic to central and southern Chile from

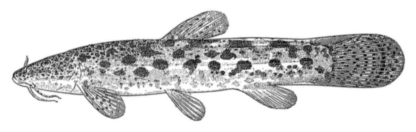

FIGURE 142 *Nematogenys inermis* (Eigenmann, 1928, Plate VII, Fig. 3).

about the latitude of Valparaiso to Osorno (Arratia, 1981). Eigenmann (1918) placed this species in a subfamily of the Trichomycteridae but later (Eigenmann, 1928) elevated *Nematogenys* to a family of its own. Baskin (1973), Howes (1983b), de Pinna (1989), and Schaefer (1990) accepted family status for this species. De Pinna (1992a, 1998) suggested that *Nematogenys* may be the sister group of the trichomycterids.

The mottled body of *Nematogenys* is elongate and naked, and an adipose fin is lacking. The spineless dorsal fin is located more or less in the middle of the back, above the pelvic fins. Pectoral fin spines are serrated on the posterior margin. There are three pair of barbels and, unlike the trichomycterids, there are no opercular odontodes. The barbel patterns of Chilean catfishes from the Biobio River are illustrated by Ruiz and Berra (1994). Very little is known about the biology of *Nematogenys*. Arratia and Hauquin (1996) examined the skin, lateralis system, and neuromast lines of *Nematogenys* and other benthic siluroids. Eigenmann (1928) reported that the species is good eating, and that specimens as large as 41 cm appeared in the Santiago fish market.

The study of loricarioid catfishes (Nematogenyidae, Trichomycteridae, Callichthyidae, Scoloplacidae, Astroblepidae, and Loricariidae) has contributed to evolutionary theory. Schaefer and Lauder (1986) demonstrated that these catfishes form a clade showing a pattern of progressive increase in the mechanical complexity of feeding structures. They used this observation to test a hypothesis that explains the increased potential for structural/functional diversification in descendent taxa (Schaefer and Lauder, 1996).

Map reference: Arratia (1981)*

Class   Actinopterygii
Subclass Neopterygii

Order   Siluriformes
(1st) Family   Trichomycteridae—**parasitic or pencil catfishes**
[tri-kō-mik-ter'-i-dē]

THIS FAMILY, FORMERLY KNOWN AS THE PYGIDIIDAE (Eigenmann, 1918), ranges over most of South America from Costa Rica to Patagonia, on both sides of the Andes, and through a wide range of habitats (de Pinna, 1992a).

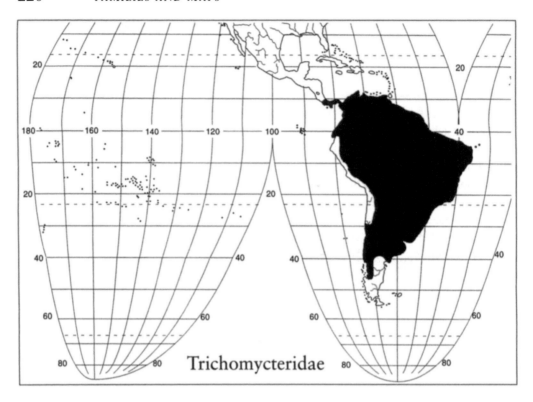

Trichomycteridae

Trichomycterids have a slender, elongated body. The head is depressed, and the eyes are usually small. An adipose fin is absent, except in one subfamily. Nasal barbels and two pairs of "maxillary" barbels (called rictal barbels) are usually present, whereas the mental (chin) barbels are absent. The opercule and interopercle usually have "spines" more properly called odontodes which are actually teeth outside the oral cavity. Pelvic fins may be absent in some groups. Incisiform and conical teeth are present in bands within a subterminal or inferior mouth. The size range of species within this family varies from about 23 to 300 mm. There are approximately 36 genera with 155 species in eight subfamilies according to Weitzman and Vari (1988), de Pinna (1989a, 1992a), and Arratia (1990).

The most primitive of eight subfamilies is the Copionodontinae (de Pinna, 1992a). It is considered the sister group to the rest of the trichomycterids. There are two genera, *Copionodon* and *Glaphyropoma*, with three species from northeastern Brazil. These fishes are the only trichomycterids with a well-developed adipose fin. The dorsal fin originates in the anterior half of the body. The jaw teeth are strongly spatulate in shape (de Pinna, 1992a).

The Trichogeninae is a monotypic subfamily from southeastern Brazil that consists of *Trichogenes longipinnis*. The specific epithet refers to the long anal fin with more than 30 rays.

The Trichomycterinae is the largest subfamily with six genera and more than 100 species. The subfamily is wide ranging from sea level to 4500 m. It is a very difficult group to work with taxonomically due to the few derived features that characterize the group (Arratia, 1990). De Pinna (1989a) discussed the phyletic status of this subfamily. Arratia and Huaquin (1995) studied the microscopic anatomy of the skin and associated structures of members of this subfamily. The widespread genus *Trichomycterus* has about 100 nominal species, such as *Trichomycterus chapmani* (Fig. 143), and many undescribed forms (de Pinna, 1992b). Members of

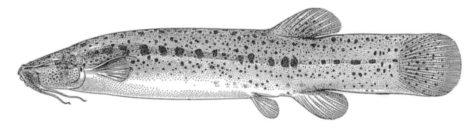

FIGURE 143. *Trichomycterus chapmani* as *Pygidium chapmani* (Eigenmann, 1918, Plate LIII, Fig. 3).

this genus generally have a deep and strongly compressed caudal peduncle. Some *Trichomycterus* species inhabit torrential streams and are said to be able to ascend waterfalls by gripping rocks with their opercular odontodes and inching their way up (Burgess, 1989). Small individuals seem to prefer vegetated shallow waters at the river's edge, whereas larger specimens dwell among stony substrate in river sections of moderate gradient. The odontode patches on the gill covers may be used to resist strong current and ascend rapidly flowing streams by an "elbowing" action (de Pinna, 1998). Other genera include *Bullockia, Eremophilus, Hatcheria, Rhizosomichthys*, and *Scleronema*. Several of these are monotypic. *Bullockia maldonadoi* has a limited distribution in gravel- and sand-bottomed rivers near Concepcion, Chile (Arratia *et al.*, 1978). *Hatcheria macraei*, which reaches a total length of about 21 cm, occurs in both Argentina and Chile between 29 and 45°30'S latitude (Arratia and Menu-Marque, 1981), which is further south than the range of any other primary division freshwater fish (Darlington, 1957).

The Stegophilinae is a subfamily of about 10 genera and 30 species that is widely distributed throughout many of the major basins of South America, including the Amazon, Orinoco, São Francisco, Paraná–Paraguay, and those of southeastern Brazil (de Pinna and Britski, 1991). This group feeds on mucus and scales of other fishes and may also rasp pieces of flesh from prey fishes. Some of the genera include *Acanthopoma, Apomatoceros, Henonemus, Homodiaetus, Megalocentor, Ochmacanthus, Pareiodon*, and *Stegophilus*. Many of these fishes are no more than 50 mm long, but they can be voracious predators (parasites) of scales and mucus, with at least 1 genus (*Pareiodon*) eating chunks of flesh and entering the body cavities of larger fishes. Goulding (1980) accuses *Pareiodon* and *Pseudostegophilus* of attacking large catfish trapped in gill nets or hooked on long lines and of devouring the catch before it can be landed.

FIGURE 144. *Vandellia plazai* (Eigenmann, 1918, Plate LIII, Fig. 1).

The Vandelliinae is a subfamily of about five genera and 18 species including *Vandellia plazai* (Fig. 144). This subfamily, along with the Stegophilinae, is the reason why the family is called "parasitic catfishes," despite the fact that most family members are free living. These fishes are also known as "candiru," an indigenous name that became infamous among readers due to the lurid descriptions in Gudger's (1930) book, *The Candiru*. The feeding strategy of the vandelliines is exclusively hematophagus—that is, they feed only on blood of other fishes. The only other exclusively hematophagous jawed vertebrates are some vampire bats. Vandelliine species enter the branchial chamber of larger fishes and nip the blood-rich gills with their specialized teeth.

An even more extraordinary, but probably accidental, mode of living is attributed to the candiru, a collective name applied to several species such as *Vandellia cirrhosa*. This slender, 6-cm fish has been reported to enter the urethra of men and women who urinate while bathing in the Amazon. It has also been reported to enter cattle that wade in the river. Once a candiru has entered the urethra, excruciating pain is generated, and the fish must be removed surgically before it reaches the bladder. The posteriorly directed opercular and interopercular odontodes which dig into the flesh prevent the removal of the candiru by urination or external pressure. Amazon natives have developed various types of penis sheaths and G-strings to prevent the candiru's admission. Since the water is turbid, it probably locates an appropriate entrance by orienting to the current (positively rheotaxic), whether it be a stream of urine or the exhaled water from a larger fish's gill cavity. Gudger's (1930) book is a fascinating collector's item about this only known vertebrate parasite of humans. This parasitism is almost certainly accidental since it results in the death of the fish with no adaptive value to the individual fish or its genes. Further information on candirus can be found in Kelley and Atz (1964), Masters (1968), and Machado and Sazima (1983). Vinton and Stickler (1941) described several eyewitness accounts and reported that the unripe fruit of a native tree, *Gunipa americana*, is used to purge the fish. Other genera in the subfamily include *Branchioica*, *Paracanthopoma*, *Paravandellia*, and *Pletrochilus*.

The Tridentinae is a small subfamily of four genera and six species with large eyes and relatively long anal fins that contain more than 15 fin rays. The genera are *Miuroglanis*, *Tridens*, *Tridensimilis*, and *Tridentopsis*.

The Glanapteryginae is a subfamily of weird, often transparent, sand- or litter-dwelling, elongate fishes characterized by the absence or reduction of various fins, eyes, pigmentation, and sensory canals (Nico and de Pinna, 1996). The pectoral fin has only three or fewer rays, principal caudal fin rays number 11 or fewer, and the pelvic fin and skeleton are absent in most species and may be present or absent in *Glanapteryx anguilla* (de Pinna, 1989b). This species also lacks an anal fin. The dorsal fin is absent in most genera and present only in the two species of *Listrura*. There are four genera, including *Pygidianops* and *Typhlobelus*, and about seven species (Nico and de Pinna, 1996). Three genera occur in the Orinoco and Amazon basins. Only the two species of *Listrura* are found in the coastal drainages of southeastern Brazil (de Pinna, 1988). More species remain to be described.

The Sarcoglanidinae is known from only a few miniature specimens from the Amazon basin (de Pinna, 1989a). There are four monotypic genera: *Malacoglanis*, *Sarcoglanis*, *Stauroglanis*, and *Stenolicmus* (de Pinna, 1989a; de Pinna and Starnes, 1990).

Map references: Arratia (1981),* Arratia and Menu-Marque (1981),* Arratia *et al.* (1983),* Darlington (1957), Eigenmann (1909a,* 1912, 1918,* 1928), Eigenman and Allen (1942), Eigenmann and Eigenmann (1891), Miller (1966),* Nico and de Pinna (1996),* de Pinna (1989a, 1992a), de Pinna and Britski (1991)*

Class Actinopterygii
Subclass Neopterygii

Order Siluriformes
(1st) Family Callichthyidae—**armored catfishes**
[kal-lik-thē'-i-dē]

THE CALLICHTHYID CATFISHES ARE FOUND FROM SOME Pacific slope drainages in southern Panama to the lower Río Paraná in northern Argentina, but they are absent from the Pacific slope of the Andes (Reis, 1998a). Several genera occur on the island of Trinidad, including *Callichthys* (currently monotypic but certainly contains more species; W. E. Burgess, personal communication) and the widespread *Corydoras*.

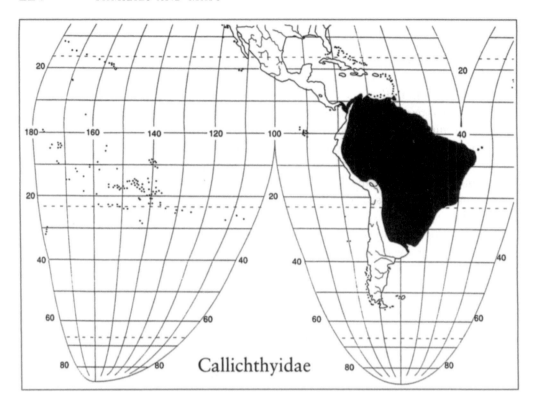

Callichthyidae

Hoplosternum littorale also occurs on Trinidad, whereas *H. punctatum* reaches the Pacific slope of Panama but not the Pacific slopes of Colombia and Ecuador. The Callichthyidae currently contains about eight genera and about 174 species and is divided into two subfamilies: Callichthyinae and Corydoradinae (Reis, 1997, 1998a,b, personal communication). Gosline (1940) provided an early revision of the family, and Nijssen (1971) and Nijssen and Isbrücker (1970, 1979, 1980, 1983, 1986) added many new species to it. Reis (1997, 1998a,b) reviewed the systematics, biogeography, anatomy, and fossil history of the family.

These fishes are called "armored catfishes" due to the presence of two rows of overlapping bony plates on each side of the body. The dorsal and pectoral fins each have a strong spine. An adipose fin is present, and it also is preceded by a spine. The small mouth is inferiorly located and surrounded by two pairs of maxillary barbels. Some genera may have one or two pairs of fleshy mental barbels on the lower lip. The eyes are comparatively large and mobile. A two-chambered swim bladder is enclosed in bone.

Graham (1997) wrote that some species of *Callichthys* and *Hoplosternum* are capable of extracting oxygen from atmospheric air swallowed into their highly vascularized gut. This probably applies to all callichthyids (R. E. Reis, personal communication). These fishes can tolerate swampy, muddy conditions of low dissolved oxygen (Burgess, 1989).

During the rainy season, many species can crawl overland in search of different habitats (Beebe, 1945). This family of small to medium-sized, peaceful species is probably the most utilized of any catfish group by aquarists. *Callichthys* may reach 25 cm and is utilized for food, as is *Hoplosternum* (Goulding, 1980). Burgess (1989) reported that the larger, abundant schooling species are roasted in their armor and cracked open much as one might crack open a peanut.

The subfamily Callichthyinae consists of five genera—*Callichthys, Lepthoplosternum, Megalechis, Dianema*, and *Hoplosternum*—with a total of 12 species pending a revision of *Callichthys* (Reis, 1997, personal communication). These fishes have a depressed head and long maxillary barbels reaching or surpassing the pelvic fins. Spawning behavior in this subfamily involves construction of a bubble nest under a floating object, usually plant material such as a leaf. The 200–300 eggs are guarded by the male. Burgess (1989) provided photographs of courtship and spawning behavior.

The Corydoradinae has a compressed head and short maxillary barbels not surpassing the gill openings ventrally. There are three genera: *Aspidoras* (approximately 16 species), *Brochis* (3 species), and *Corydoras* (approximately 143 species). *Corydoras* is the largest catfish genus and contains about 80% of the members of this family (Reis, 1998a). This is the catfish genus that is most common in home aquariums throughout the world. They are small fishes ranging from 2.5 to 12 cm. They tend to prefer slow-moving streams with sand or mud bottoms. Many species occur in large schools numbering thousands of individuals. Some of these scavenging bottom dwellers have even been observed to feed upside down at the surface (Burgess, 1989). One of the most common species in the aquarium trade, *C. paleatus* (Fig. 145), was first collected by Charles Darwin during the voyage of H. M. S. *Beagle*.

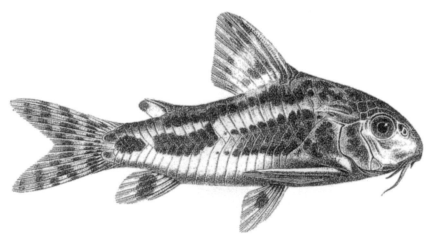

FIGURE 145. *Corydoras paleatus* as *C. marmoratus* (Steindachner, 1879a, Plate V, Fig. 1).

Courtship of another popular aquarium species, *C. aeneus*, is illustrated by Burgess (1989). As in most *Corydoras* species, the male forms a "T" with the female by positioning himself in front of the female and holding her barbels with his pectoral spine. While locked in this embrace the partners quiver and release the sperm and eggs. The female can hold her eggs in a pouch formed by the pelvic fins until she finds a suitable surface, such as a leaf, on which to deposit them.

Nonindigenous populations of callichthyids have been established in Hawaii and Florida (Fuller *et al.*, 1999).

Map references: Burgess (1989), Eigenmann (1912),* Eigenmann and Allen (1942), Hoedeman (1974),* Miller (1966),* Reis (1997, 1998a,b)*

Class Actinopterygii
Subclass Neopterygii

Order Siluriformes
(1st) Family Scoloplacidae—spiny dwarf catfishes
[skol'-ō-plac'-i-dē]

THIS FAMILY OF TINY CATFISHES IS FOUND IN THE FRESH waters of Peru, Bolivia, Brazil, and Paraguay. Previously, they were considered a subfamily of the Loricariidae (Bailey and Baskin, 1976). Isbrücker (1980) recognized them as a distinct family. Howes (1983b) accepted family status for the Scoloplacidae and suggested that this family is the sister group to the Astroblepidae plus Loricariidae. De Pinna (1998) reviewed the taxonomic history of this family and other siluriforms.

There is only one genus, *Scoloplax*, with four species (Schaefer *et al.*, 1989; Schaefer, 1990). *Scoloplax dicra* (Fig. 146) is found in the

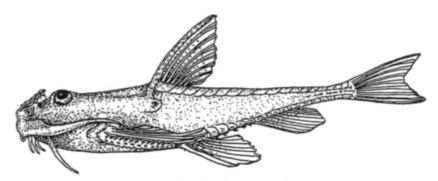

FIGURE 146. *Scoloplax dicra* (reproduced with permission from Burgess, 1989, p. 451).

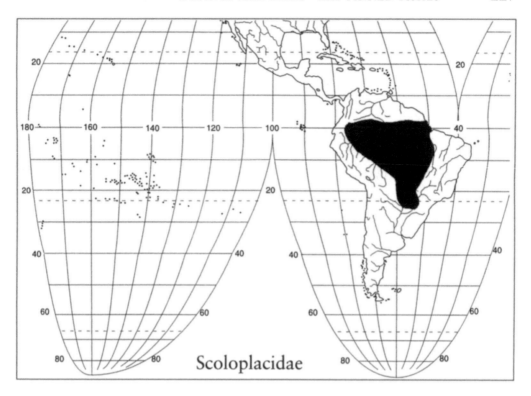

Scoloplacidae

upper to mid-Amazon in Bolivia, Peru, and Brazil. *Scoloplax dolicholophia* is reported only from the mid to lower Rio Negro of Brazil. *Scoloplax distolothrix* occurs in the Xingu, Tocantins, and Paraguay drainages of Brazil and Paraguay, and *S. empousa* inhabits the Paraná, Paraguay, and Guaporé drainages of Brazil and Paraguay (Schaefer, 1990; Schaefer *et al.*, 1989).

The body of scoloplacids is partially covered by five rows of odontode-bearing plates. There are two rows of dorsolateral plates posterior to the dorsal fin and two rows of ventrolateral plates from the anal fin to the caudal fin base. There is a mid-ventral series of plates between the anus and the anal fin. Odontodes (tooth plates) may be present on other parts of the body, including the dorsal surface of the snout. The function of this rostral plate is not known. The dorsal and pectoral fins each have a stout, odontode-bearing spine. There is no adipose fin. The head is depressed, and the mouth is subterminal. There are three pairs of barbels. Nasal barbels are absent. The maximum size of these fishes is about 20 mm, making scoloplacids among the smallest Neotropical fresh-water vertebrates. Weitzman and Vari (1988) reviewed miniature Neotropical fishes.

Scoloplacids live in the leaf litter of lakes, backwater pools, and heavily vegetated streams in which low oxygen levels are common. Armbruster

(1998a) suggested that a modified stomach of *S. dicra* is an adaptation for air breathing.

Map references: Schaefer (1990),* Schaefer *et al.* (1989)*

Class   Actinopterygii
Subclass   Neopterygii

Order   Siluriformes
(1st) Family   Loricariidae—**suckermouth armored catfishes**
(lor'-i-ka-rī'-i-dē) [lor'-i-ka-rē'-i-dē]

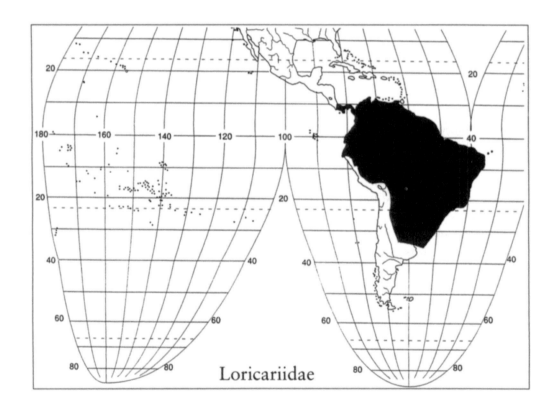

Loricariidae

THE LORICARIIDS RANGE FROM COSTA RICA TO RÍO DE LA PLATA and from sea level to torrential Andean streams at 3000 m. It is the largest catfish family with about 650 species in about 87 genera and includes more than one-fourth of the world's catfishes (Isbrücker, 1980;

Reis and Schaefer, 1998). Montoya-Burgos *et al.* (1998) reviewed the phylogenetic relationships of the Loricariidae based on mitochondrial gene sequences, and de Pinna (1998) discussed the relationships among the six subfamilies.

The suckermouth armored catfishes are well named. The inferior mouth and lips in many species form a circular sucking disc that is used for attachment to the substrate in swift waters. The suctorial mouth combined with rasping teeth allow these catfishes to be effective grazers on algae and detritus-covered stones. Many loricariids consume wood, but if and how they are able to digest it remains a mystery (de Pinna, 1998). Barbels may be present or absent around the mouth. There are no mental or nasal barbels. The depressed body is armored with thick bony scutes (plates) usually arranged in five rows but sometimes reduced (Reis and Pereira, 1999). Each plate may be studded with odontodes (small tooth-like structures). The dorsal fin is well developed. An adipose fin may be present or absent; when present, it possesses a spine. The dorsal, pectoral, and pelvic fins have a nonpungent but often stout spine derived from the first fin ray. These spines are usually armored with odontodes. Armbruster (1998a) discussed several modifications of the loricariid digestive tract that function as accessory respiratory organs or hydrostatic organs. Loricariids are mostly facultative air breathers (Graham, 1997).

Gosline (1947) provided a foundation for current studies. Five subfamilies were recognized as determined by Isbrücker (1980), Isbrücker and Nijssen (1982, 1989, 1991, 1992), and Nijssen and Isbrücker (1987). De Pinna (1998), utilizing Schaefer's (1987, 1990, 1991) papers, provided a cladogram showing the relationships of six loricariid subfamilies. Montoya-Burgos *et al.* (1998) confirmed the monophyly of the Loricariinae but found no support for the monophyly of the other subfamilies.

Lithogenes villosus, known only from the holotype at its type locality in British Guiana (S. A. Schaefer, personal communication), is the only member of the Lithogeninae and is considered to be the most basal loricariid group (de Pinna, 1998; Montoya-Burgos *et al.*, 1998). Nijssen and Isbrücker (1987) considered this poorly scuted species to be an astroblepid.

The Neoplecostominae is made up of six species in the genus *Neoplecostomus* from Brazil. The Hypoptopomatinae consists of 15 genera and about 80 species, most of which are in the genera *Hypoptopoma*, *Otocinclus*, and *Parotocinclus* (Reis and Schaefer, 1998; Schaefer, 1996, 1997, 1998). The greatest taxonomic diversity of this subfamily is in the Atlantic coastal drainages from Guyana to Uruguay (Reis and Schaefer, 1992). Some genera are widespread throughout the families' South American range. The dozen or so species of *Hypoptopoma* have a long snout and a depressed head. There are 13 species of *Otocinclus* whose distal esophageal wall is modified as an accessory blind pouch that probably functions as an accessory respiratory organ for aerial respiration. Schaefer

(1997) reviewed the systematics and biogeography of *Otocinclus* and provided keys to the genera of the Hypoptopomatinae. Schaefer (1998) discussed generic-level phylogeny and classification and reevaluated his previous phylogeny (Schaefer, 1991). Reis and Schaefer (1998) included two new genera in the subfamily.

The Loricariinae is the largest subfamily with 35 genera and about 185 species. The caudal peduncle is strongly depressed in this group, and various body parts may be elongated, such as the snout, the caudal peduncle, and the upper and lower caudal fin rays. The genus *Farlowella* is elongate with prominent caudal filaments and a bony snout. They are sometimes called "stick catfishes" because of their slender shape and woody appearance. *Rineloricaria* is a genus of about 40 species of small fishes, including

FIGURE 147. *Rineloricaria lima* as *Loricaria lima* (Steindachner, 1882b, Plate I, Fig. 1).

R. lima (Fig. 147), that do not exceed 23 cm. The lower lip of brooding males of the genus *Loricariichthys* is enlarged posteriorly as a sheet-like membrane that may extend beyond the pectoral fin base. The eggs are carried on this structure (Burgess, 1989). *Lamontichthys filamentosus* from the Brazilian Amazon has enormous filamentous extensions of the dorsal, pectoral, and caudal fins. Other genera include *Harttia*, *Loricaria*, *Pseudohemiodon*, and *Sturisoma*.

The Ancistrinae contains about 18 genera and 175 species (Burgess, 1989). Scutes in the interopercular region are fused and bear hooks. *Ancistrus* (= *Xenocara*) is the largest genus with about 50 species. Many species in this genus, such as *A. triradiatus* (Fig. 148), have bifurcated bristles on their snouts. These structures can be impressive, especially in males. The group is known as bristle-nosed catfishes among aquarists. Other genera include *Chaetostoma*, *Hemiancistrus*, *Hypancistrus*, *Lasiancistrus*, *Lithoxus*, *Panaque*, *Peckoltia*, and *Pseudacistrus*. Many of these fishes have very long pectoral spines that extend beyond the origin of the pelvic fin. Species of *Peckoltia* are colorful and referred to as clown plecos by hobbyists. *Peckoltia vittata* (Fig. 149) is the most commonly imported clown pleco in the aquarium trade (Burgess, 1989). *Panaque suttoni* has blue eyes and is called the blue-eyed pleco.

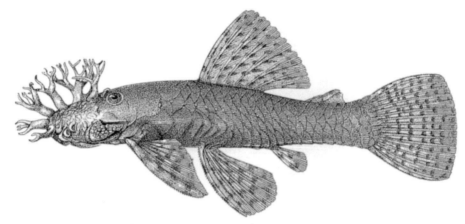

FIGURE 148. *Ancistrus triradiatus* (Eigenmann, 1922, Plate V, Fig. 3).

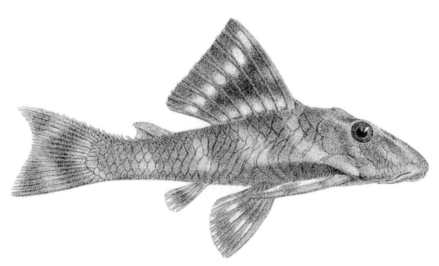

FIGURE 149. *Peckolita vittata* as *Chaetostomus vittatus* (Steindachner, 1882a, Plate II, Fig. 5).

The Hypostominae includes 18 genera and about 175 species (Burgess, 1989), 130 of which are in the genus *Hypostomus* (= *Plecostomus*) (Montoya-Burgos *et al.*, 1998). Many aquarists refer to several loricariid genera as "plecos" after the genus *Plecostomus*, which is now known as *Hypostomus*. See Burgess (1989) for a history of plecostomus as a common name. To further confuse the issue, there is a species named *Hypostomus plecostomus* (Schaefer, 1987). Species in this genus range in size from about 14 to more than 50 cm. Many *Hypostomus*, such as *H. hemiurus* (Fig. 150) have beautiful sail-like dorsal fins. Some of the larger species are prized as food fishes in South America. At least 3 species of *Hypostomus* have been established in the United States due to releases from home aquariums and, in one case, for algae control (Hensley and Courtenay, 1980b).

FIGURE 150. *Hypostomus hemiurus* as *Plecostomus hemiurus* (Eigenmann, 1912, Plate XXV, Fig. 1).

Reproducing populations are known from Nevada, Texas, and Florida. In Florida there is also a population of *Pterygoplichthys multiradiatus* in weedy, mud-bottomed canals (Page and Burr, 1991). Other genera include *Cochliodon* (a prodigious wood eater), *Delturus*, *Hemipsilichthys*, and *Rhinelepis*. Armbruster (1998b) discussed the phylogenetic relationships of the *Rhinelepis* group.

Several loricariid species and members of four other Neotropical catfish families have been taken in United States waters, probably as escapees from fish farms and aquarium releases (Fuller *et al.*, 1999).

Map references: Burgess (1989), Eigenmann (1909a, 1912),* Miller (1966),* Reis and Schaefer (1998),* Schaefer (1997)

Class Actinopterygii
Subclass Neopterygii

Order Siluriformes
(1st) Family Astroblepidae—**climbing** catfishes   [as-trō-blep'-i-dē]

THE ASTROBLEPIDAE CONSISTS OF THE MONOTYPIC GENUS *Astroblepus*, with about 37 species from Panama and at high elevations in fast-flowing Andean streams from Venezuela, Columbia, Ecuador, and Peru including Lake Titicaca.

The astroblepids resemble loricariids except that the astroblepids have a naked body; armor is lacking. The inferior mouth is suctorial, and there

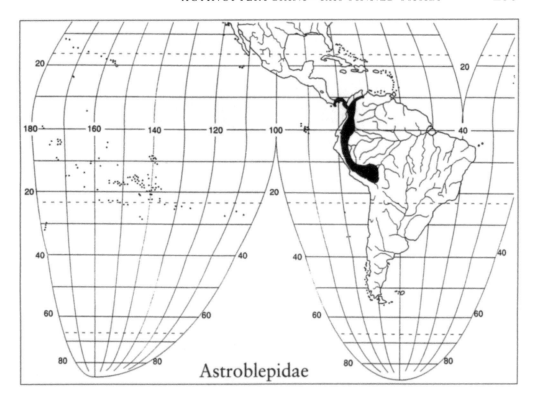

Astroblepidae

are inhalant and exhalent gill openings. Maxillary and nasal barbels are present. The head and body are depressed. There is a stout spine in the short dorsal fin. An adipose fin may be present or absent. When present, it may be long, low, and armed with a spine. The interopercular region is spineless. The swim bladder is reduced. The urogenital papilla of males functions as a copulatory organ.

The common name, climbing catfishes, is earned by virtue of the fishes' ability to ascend steep, algae-covered rock walls (Johnson, 1912). Fishes of this family can climb better than any other fishes (Burgess, 1989). They are well adapted to do this. The sucker-like mouth (as exhibited by *A. chapmani* and *A. pirrense*; Figs. 151 and 152) and the flattened ventral surface are assisted by broad, flattened pelvic fin rays whose surfaces are richly studded with posteriorly pointed denticles. Alternate attachment of the mouth and pelvic apparatus provides a grip as the fish inches up a slippery slope. When the mouth is engaged as a suction cup, the fish can continue to respire due to the incurrent gill openings. Exhaled water is expelled via the excurrent opening. Therefore, the mouth does not have to loosen its grip for respiration to occur.

The largest species is *Astroblepus grixalvii*. It reaches 30 cm and as been reported from torrential streams at elevations higher than 4000 m (Burgess, 1989). It is utilized as a food fish. *Astroblepus longifilis*

FIGURE 151. *Astroblepus chapmani* (Eigenmann, 1922, Plate V, Fig. 2).

FIGURE 152. *Astroblepus pirrense* as Cyclopium pirrense (Meek and Hildebrand, 1916, Plate XVI).

is recorded from the Pacific slope of eastern Panama to Peru (Miller, 1966). Ortega and Vari (1986) reported 15 species of *Astroblepus* from Peru.

Nijssen and Isbrücker (1987) considered *Lithogenes villosus* to be an astroblepid rather than a loricariid, and Nelson (1994) followed their lead. Schaefer (1987) placed this genus in the Loricariidae, a placement accepted by Eschmeyer (1998). De Pinna (1998) and Montoya-Burgos *et al.* (1998) considered *Lithogenes* as the most primitive group of the Loricariidae. Some authors consider the astroblepids as a subfamily of the Loricariidae (Gosline, 1947).

Map references: Burgess (1989), Eigenmann (1909a), Miller (1966)

Class   Actinopterygii
Subclass   Neopterygii

Order   Gymnotiformes; Suborder Sternopygoidei
(1st) Family   Sternopygidae—**glass knifefishes**   [stur'-nō-pij'-i-dē]

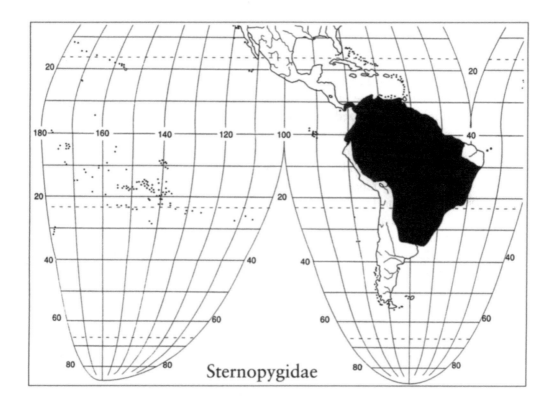

Sternopygidae

THE NEXT SIX FAMILIES UNDER DISCUSSION BELONG TO THE
order Gymnotiformes. They occur only in Central and South America.
Some species inhabit deep channels of large rivers. This order is considered
to be the most advanced otophysan group, and these fishes have electro-
generating abilities. Fink and Fink (1981) determined that the
Gymnotiformes are related to the catfishes (Siluriformes) rather than to the
characoids as suggested by Mago-Leccia and Zaret (1978). Campos-da-Paz
(1996) presented alternative explanations of the evolution of some of Fink
and Fink's homoplastic characters. Mago-Leccia (1994), Alves-Gomes *et al.*
(1995), Alves-Gomes (1999), Campos-da-Paz and Albert (1998), Albert

and Campos-da-Paz (1998), and Albert (2000) described the classification history, taxonomy, and phylogenetic systematics of the knifefishes. Ellis (1913) provided an important early review of the order. Moller (1995) reviewed the history and behavior of these fishes.

As a group the Gymnotiformes are referred to as electric knifefishes because of the electric field they generate and their blade-like shape. They resemble the nonelectric African featherbacks, Notopteridae. The body of most species is elongated and compressed, except for the electric eel (Electrophoridae) and *Gymnotus*, which have cylindrical bodies. There are no dorsal, pelvic, or caudal fins, except for the Apteronotidae which has a tiny caudal fin. A dorsal filament may remain on some species. The pectoral fins are small, and the pelvic girdle is absent. The anus and urogenital pore are located below or anterior to the pectoral fins. The eyes are generally small, in keeping with the nocturnal activity pattern, turbid habitat, and electric sense of this group. Their head morphology reflects an adaptation for specialized feeding (Roberts, 1972). The internal organs are crowded anteriorly. This allows more room for the electric-generating muscle tissue. Gymnotiform fishes are able to regenerate caudal parts of their bodies (Moller, 1995). The gills and gill openings are small. Air breathing is reported in three families—Hypopomidae, Gymnotidae, and Electrophoridae (Graham, 1997).

The anal fin is exceedingly long with at least 140 rays, and it extends from about the origin of the pectoral fin to the posterior region of the body. Undulating movements of this flowing anal fin gently propel kinfefishes forward or backward. Such graceful movements prevent major disruption of the surrounding electric field generated by the fish. The anatomy of the anal fin's basal pterygiophores allows each fin ray to move in a circular path, thus facilitating the anal fin's wave-like motion.

Knifefishes are capable of generating and receiving electrical impulses which they use as an additional sense to locate food, navigate through their turbid habitats, and communicate with conspecifics. The discharges knifefishes produce are measured in millivolts. For this reason, they are called "weak" electrical fishes. The only "strong" member of this group is the electric eel. The electric organs are derived from muscle tissue in all the knifefishes except the Apteronotidae. Other weak electrical fishes include the osteoglossiform mormyrids and gymnarchids. However, most gymnotiform weak electric fishes generate a continuous, high-frequency output versus the pulsed, low-frequency output of the mormyrid electric fishes. The ancestral waveform of the electric organ discharge (EOD) was an intermittent monophasic pulse (Stoddard, 1999). This primitive discharge type has largely been replaced by continuous wave trains or multiphasic pulsed waveforms in gymnotiform fishes. Stoddard (1999) presented experimental evidence that predation pressure seems to have selected for greater signal complexity. Biphasic signals are less detectable by the predatory electric eel (*Electrophorus electricus*) than the primitive monophasic signals. An additional wave phase added to the monophasic EOD is thought to shift the emitted signal above the major sensitivity range of electroreceptive predators. Sexual selection has further elaborated the EOD for signal crypsis (Stoddard, 1999). See Bennett (1971a,b), Bullock and Heiligenberg (1986), Moller (1995), and Turner *et al.* (1999) for discussion of electrical properties of these and other fishes. Hypotheses of neural evolution in gymnotiform fishes were tested by Albert *et al.* (1998).

Gymnotiforms are known from the Atlantic and Pacific slopes of Middle America, the Pacific slopes of Colombia and Ecuador, the Atrato, Magdalena, and Maracaibo basins of Venezuela, Trinidad, the Atlantic drainages of the Guyanas, the Orinoco–Amazon basin east of the Andes, the Rio Parnaiba, the drainages of southeast Brazil, the La Plata basin, and the endorheic Río Dulce–Sali of northeastern Argentina (Albert, 1999). There are 27 recognized gymnotiform genera and 105 species (Albert and Campos-da-Paz, 1998; Albert, 2000).

There are two suborders within the Gymnotiformes: the compressed-bodied Sternopygoidei (Sternopygidae, Rhamphichthyidae, Hypopomidae, and Apteronotidae) and the cylindrical-bodied Gymnotoidei (Gymnotidae and Electrophoridae) (Mago-Leccia, 1978). The Sternopygidae and the Apteronotidae are considered sister groups by Albert and Campos-da-Paz (1989), as are the Rhamphichthyidae and the Hypopomidae.

The Sternopygidae, or glass knifefishes, consists of five extant genera with about 23 species distributed from Panama through northern South America to La Plata (Albert, 2000). Species of *Sternopygus* and *Eigenmannia* occur on the Pacific slope of Panama (Miller, 1966). The common name denotes that they are transparent, although *Sternopygus* is not. The spine and internal organs can be clearly seen through the body. The muscles may appear red where blood vessels pass through, but the muscle tissue is clear. Such a fish is nearly invisible in the water. The upper and lower jaws contain villiform teeth and the snout of sternopygids is relatively short. Sternopygids and apteronotids generate continual wave-type discharges, whereas the other four families in this order produce pulse-type discharges (Heiligenberg, 1993). Alves-Gomes (1998) reported on the phylogenetic position of genera within this family. *Eigenmannia macrops* (Fig. 153) represents the typical family morphology.

FIGURE 153. *Eigenmannia macrops* (Ellis, 1913, Plate XXII, Fig. 1).

Sternopygus macrurus can reach 91 cm and is eaten throughout its range in northeastern South America. It is an insect eater but will also consume crustaceans and fishes. *Sternopygus astrabes*, with its suite of plesiomorphic features, most closely resembles the ancestral gymnotiform condition (Albert and Campos-da-Paz, 1998). *Eigenmannia virescens*, which occurs from Panama to Buenos Aires, grows to about 45 cm and is translucent except for a yellowish head and geeenish tail. Specimens from clear waters may be boldly striped (Wheeler, 1975). Refer to Heiligenberg (1993) for a diagram of the structure of electroreceptors in the skin of *Eigenmannia*. Other genera include *Archolaemus*, *Distocyclus*, and *Rhabdolichops*. The latter genus is often found in deep channels of rivers in the Orinoco and Amazon basins. See Mago-Leccia (1978) and Lundberg and Mago-Leccia (1986) for taxonomic details.

Ellisella kirschbaumi is the only known gymnotiform fossil species. It is from the Upper Miocene of Bolivia, about 10 MYA, and is included in the Sternopygidae (Albert and Campos-da-Paz, 1998).

Map references: Albert (2000), Albert and Campos-da-Paz (1998),* Eigenmann (1909a),* Miller (1966)

Class Actinopterygii
Subclass Neopterygii

Order Gymnotiformes; Suborder Sternopygoidei
(1st) Family Rhamphichthyidae—**sand knifefishes**
[ramf-ik-thē'-i-dē]

THE SAND KNIFEFISHES CONSIST OF SEVEN SPECIES OF *Rhamphichthys*, including *R. rostratus*, found in northeastern South America from the Guianas to Rio de la Plata, in addition to five species of sand-burrowing *Gymnorhamphichthys* (Schwassmann, 1989) and *Iracema caiana* (Triques, 1996).

Sand knifefishes have a long, tubular snout and toothless jaws. The body is similar to those of other gymnotiforms. The dorsal, pelvic, and caudal fins are missing, and the anal fin is very long, extending from under the throat to near the tip of the tail. The finger-like tail end may be used tactilely while the fish is backing up into cover. All rhamphichthyids produce a pulse-type electric organ discharge (Moller, 1995).

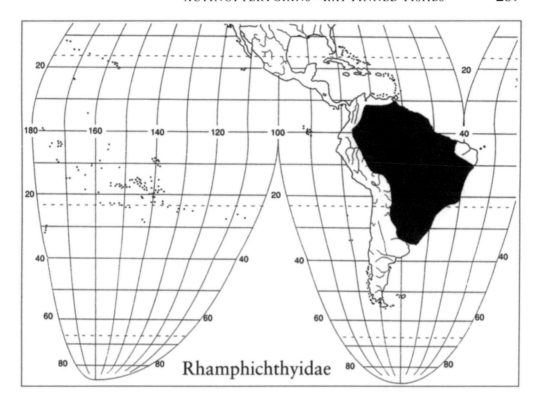

Rhamphichthyidae

Rhamphichthys rostratus is an impressive fish, with its long, trunk-like snout ending in a small mouth and its 1.4 m length. It lives in small, open streams, but it is occasionally found in large rivers. It spends the daylight hours hidden in vegetation and emerges at night to feed on insect larvae and other invertebrates that it locates with its electrical sense organs. It is prized as a food fish. There may be other undescribed species in this genus.

Gymnorhamphichthys is active at night and burrows into the sand during daylight hours. This sand-dwelling behavior is unique among the gymnotoids (Schwassmann, 1989). During nighttime activity its electric discharge frequency is high (70–100 Hz). This frequency decreases to 10–15 Hz during the daytime inactivity. The ecology of *G. hypostomus* (Fig. 154) was studied by Schwassmann (1976), who found that it feeds on small trichopteran and dipteran larvae.

FIGURE 154. *Gymnorhamphichthys hypostomus* (Ellis, 1913, Plate XXIII, Fig. 2).

In older literature sternopygids such as *Eigenmannia* and *Sternopygus* were included in this family.

Map references: Albert (2000), Albert and Campos-da-Paz (1998),* Schwassmann (1976,* 1989)

Class Actinopterygii
Subclass Neopterygii

Order Gymnotiformes; Suborder Sternopygoidei
(1st) Family Hypopomidae—**hypopomid knifefishes**
[hī-pō-pō'-mi-dē]

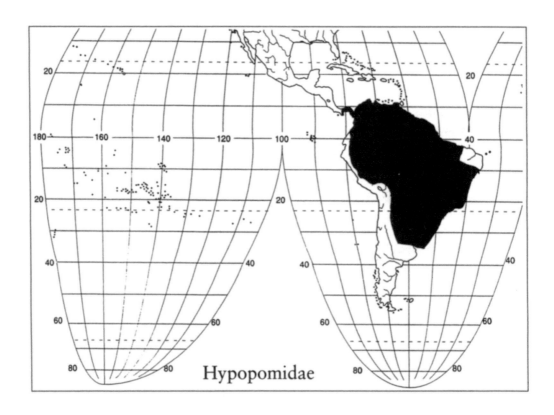

THE DISTRIBUTION OF THE HYPOPOMIDAE IS WIDESPREAD throughout South America in a pattern similar to that of the Sternopygidae.

Hypopomus occidentalis occurs on both slopes of Panama and Colombia to the Maracaibo basin of Venezuela (Miller, 1966). Like most gymnotiform fishes, members of this family lack dorsal, pelvic, and caudal fins and have a very long anal fin. The hypopomids have a rather short snout and toothless jaws. There are six genera—*Brachyhypopomus, Hypopomus, Hypopygus, Microsternarchus, Racenisia,* and *Steatogenys*—with about 16 species (Schwassmann, 1984; Hopkins, 1991; Albert and Campos-da-Paz, 1998; Albert, 2000). This family is the largest gymnotiform family to generate pulse, rather than wave, electric organ discharges. *Steatogenys elegans* is a typical family representative (Fig. 155).

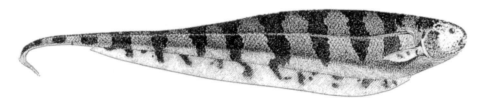

FIGURE 155. *Steatogenys elegans* (Eigenmann and Ward, 1905, Plate IX, Fig. 11).

The caudal filament of male *Brachyhypopomus pinnicaudatus* from the coastal swamps of French Guiana is long, compressed, and feather-like, whereas females and juveniles have shorter cylindrical tails. The electric organ extends into the caudal filament of both sexes. The functional significance of this sexual dimorphism is discussed by Hopkins *et al.* (1990). There are sex differences in the waveform of the electric organ discharge (EOD). The EOD is longer in males than in females; surprisingly, however, the peak-to-peak amplitude of the male's EOD is weaker than that of a same-sized female. Hopkins *et al.* suggested that the male's reduction in EOD amplitude is a consequence of the increase in EOD duration, and that female choice plays a role in the selection of extended EODs among males. Hopkins *et al.* further speculated that the long tails of males compensate for the loss in active space that would otherwise accompany a weaker EOD.

Brachyhypopomus pinnicaudatus is able to gulp air at the surface in low-oxygen streams (Hopkins, 1991). Beebe (1945) observed that *Hypopomus* could travel overland through the rain forest a considerable distance from water, and Saul (1975) reported that it would burrow in soft mud. *Hypopomus artedi* is one of the largest members of this group at 42 cm.

Map references: Albert (2000), Albert and Campos-da-Paz (1998),* Bussing (1998),* Costa and Campos-da-Paz (1992),* Miller (1966)

Class Actinopterygii
Subclass Neopterygii

Order Gymnotiformes; Suborder Sternopygoidei
(1st) Family Apteronotidae—**ghost knifefishes**
[ap'-te-rō-nō'-ti-dē]

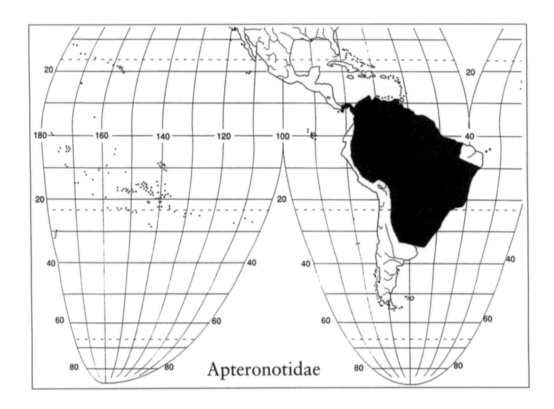

Apteronotidae

THE DISTRIBUTION OF THE GHOST KNIFEFISHES EXTENDS
from the Pacific slope of eastern Panama (*Apteronotus*) through South
America to the Parana–Paraguay basin. Apteronotids have the general
gymotiform body plan of an elongate anal fin, compressed body, and no
pelvic fins. They are the only gymnotiforms with a caudal fin. The tiny
tail fin is not connected to the anal fin. There may be a long dorsal
filament in the middle of the back that is depressed into a groove. Most
have blunt snouts, but *Sternarchorhynchys oxyrhynchus* and *S. mormyrys*
(Fig. 156) have a long, curved trunk-like snout reminiscent of African
elephant-nosed mormyrids. This is an interesting example of convergent
evolution in that both mormyrids and apteronotids are weak electrical

FIGURE 156. *Sternarchorhynchus mormyrus* (Eigenmann and Ward, 1905, Plate VIII, Fig. 8).

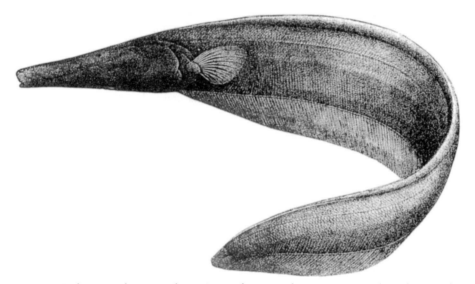

FIGURE 157. *Orthosternarchus tamandua* as *Sternarchus tamandua* (Eigenmann and Ward, 1905, Plate VIII, Fig. 6).

fishes but phylogenetically unrelated. *Orthosternarchus tamandua* has a long, tubular, straight snout (Fig. 157).

The apteronotids are unusual among gymnotiform fishes in that as adults their electric organs are formed from nerve tissue instead of muscle tissue. They produce a wave-type discharge (Moller, 1995). Various aspects of apteronotid electrosensory biology are reviewed in references cited in Turner *et al.* (1999). There are 10 genera with approximately 41 species (Mago-Leccia *et al.*, 1985; Lundberg *et al.*, 1996; Albert and Campos-da-Paz, 1998; Albert, 2000), including *Adontosternarchus,* *Magosternarchus, Porotergus,* and *Sternarchella. Adontosternarchus sachsi* (Fig. 158) and *Apteronotus leptorhynchus* reflect typical apteronotid body

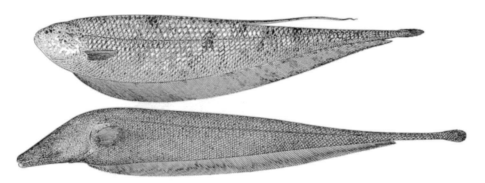

FIGURE 158. *Adontosternarchus sachsi* as *Adenosternarchus sachsi* (top) and *Apteronotus leptorhynchus* as *Sternarchus leptorhynchus* (bottom) (Ellis, 1913, Plate XXII, Figs. 3 and 4).

shapes. Campos-da-Paz (1995, 1999) reviewed the taxonomic history of *Sternarchorhamphus* and *Megadontognathus*, respectively. Fernandes (1998) showed that male *Apteronotus hasemani* have relatively elongated snouts compared with females, whereas *A. bonapartii* showed no sexual dimorphism.

Magosternarchus has greatly enlarged jaws and teeth and preys on the tails of other gymnotiform fishes. This genus has been taken by benthic trawl sampling in large, white- and blackwater river channels of the Amazon basin in Brazil (Lundberg *et al.*, 1996).

Map references: Albert (2000), Albert and Campos-da-Paz (1998),* Campos-da-Paz (1995),* Lundberg *et al.* (1996),* Miller (1966)

Class Actinopterygii
Subclass Neopterygii

Order Gymnotiformes; Suborder Gymnotoidei
(1st) Family Gymnotidae—naked-back knifefishes
(jim-not'-i-dē) [jim-nō'-ti-dē]

THE GYMNOTIDS OCCUR FURTHER NORTH INTO MIDDLE America than other gymnotiforms (Miller and Carr, 1974). The family is the most widely distributed of the Gymnotoidei, ranging from the Río San Nicolás of southeastern Chiapas, Mexico, south to Río de la Plata (Albert and Miller, 1995). *Gymnotus carapo* is widely distributed on both slopes of Costa Rica and Nicaragua and the Atlantic slope of western Panama. Its northern limit is not known with certainty; however, *G. cylindricus* reaches

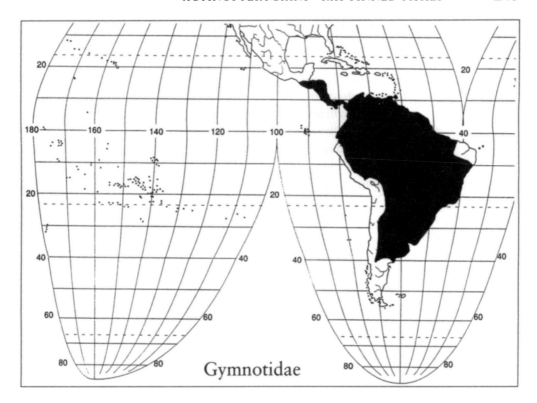

Gymnotidae

Guatemala and extends south to the Atlantic slopes of Costa Rica (Miller, 1966; Campos-da-Paz, 1996). *Gymnotus carapo* occurs south to the Río de la Plata in Argentina and from the Atlantic coast to the Andes. This species has the most extensive geographical range of any gymnotiform; however, much diversity is found within "*G. carapo*" and this taxon may include previously unrecognized species (Albert *et al.*, 1999).

FIGURE 159. *Gymnotus cataniapo* (original drawing by Pedro Nass provided by James Albert; reproduced with permission).

Gymnotus, represented by *G. cataniapo* (Fig. 159), is the only genus in the family, and it includes 12 species if the electric eel, *Electrophorus*, is placed in its own family (Campos-Da-Paz and Costa, 1996; Albert and Campos-da-Paz, 1998; Albert, 2000). Unlike the compressed bodies of other gymnotiforms, *Gymnotus* has a rounder, more cylindrical body covered with small scales. The body is eel-like, and the anal fin is long and flowing and terminates in a fine point at the tip of the tail.

Damage to the cadual area is not fatal, and regeneration may take place. Dorsal, pelvic, and caudal fins are absent, and the pectoral fins are relatively large for a gymnotiform. The internal organs are positioned anteriorly, and the anus is located under the throat. Gymnotids use their weak, pulse-type electrical discharge to locate prey and navigate through their environment. Like other gymnotiform fishes, they glide through the water by undulations of their long anal fin. *Gymnotus carapo* can reach 60 cm and is a local food fish. It feeds on fishes and crustaceans as an adult.

Gymnotus carapo is a facultative air breather when aquatic oxygen levels are low. It utilizes atmospheric air by way of an enlarged, highly vascularized swim bladder. Ventilation is carried out by a unique esophageal force pump (Liem *et al.*, 1984).

Map references: Albert (2000), Albert and Campos-da-Paz (1998),* Albert and Miller (1995),* Bussing (1998),* Miller (1966)*

Class Actinopterygii
Subclass Neopterygii

Order Gymnotiformes; Suborder Gymnotoidei
(1st) Family Electrophoridae—electric eel or electric knifefish
[ē'-lĕc-trō-phō'-ri-dē]

THE ELECTRIC EEL IS MORE ACCURATELY CALLED THE ELEC-tric knifefish, but the common name is so entrenched that the knifefish appellation is unlikely to catch on. The family consists of one well-known species, *Electrophorus electricus* (Fig. 160), from northeast South America including the Guyanas, the Orinoco, and the middle and lower Amazon basin. There may be a second species in Peru (Nelson, 1994), but there is no evidence for this.

The electric eel, of course, is eel-like, with an elongate body that is circular in cross section. It can grow to 2.5 m and 20 kg and has about 240 vertebrae. The very long anal fin contains more than 300 rays and extends to the tip of the tail (Albert and Campos-da-Paz, 1998). Scales are absent, as are dorsal, pelvic, and caudal fins. The viscera are located anteriorly, and the anus is anterior to the pectoral fins. The snout is broad with a wide mouth and equal jaws. The eyes are small, in keeping with the turbid water in which it resides. It inhabits pools and deeply shaded streams.

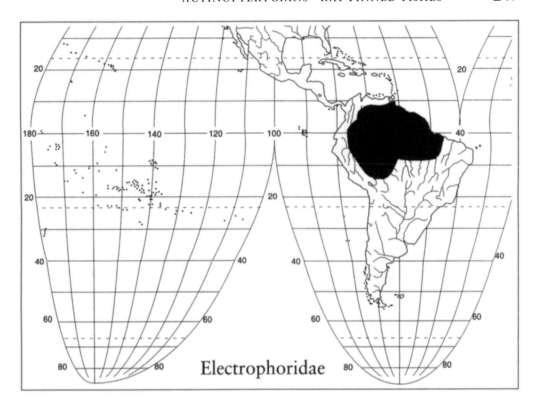

FIGURE 160 *Electrophorus electricus* as *Gymnotus electricus* (Lydekker, 1903, p. 451).

The electric eel is an obligatory air breather that uses the vascularized epithelium of its buccopharyngeal cavity as an accessory air-breathing organ (Graham, 1997). Since *Electrophorus* holds air in its mouth, it is one of the few air-breathing fishes that cannot ventilate the gills while air breathing. The gills are small. Farber and Rahn (1970) demonstrated that electric eels obtain about 78% of their oxygen from the air and release about 81% of their respiratory carbon dioxide through their skin.

This fish can deliver an electrical discharge of 350–650 V of direct current at about 1 amp. Larger fish produce larger discharges because of the greater potential difference between the head and tail. The electric eel has three electric organs, two strong (for offense and defense) and one weak (for orientation and communication). Both weak and strong electric organ discharges are pulsed. Newly hatched electric eel larvae begin to discharge when they are only 15 mm long (Moller, 1995). *Electrophorus* is linearly polar with a positive head and negative tail. The shock is strong enough to stun fishes in the vicinity, which are then consumed. Electric eels may prey on other knifefishes which they detect when the prey utilize their own weak electric discharge (Westby, 1988). *Electrophorus* may ambush its prey by turning off its own electric field in order to remain undetected by the prey. Larger organisms, including humans, may also be injured by the shock of an electric eel, which may be pulsed at 0.002- to 0.0005-sec intervals. The electric organs are derived from muscle tissue arranged in columns of wafer-like electroplates on both sides of the spinal column. These organs make up most of the posterior five-sixths of the animal. Weak electrical signals are also sent out to locate prey and detect intruders. See Bennet (1971a,b) and Grundfest (1960) for more details of the electrical properties of this animal. Wu (1984) reviewed the discovery of animal electricity. Moller (1995) provided a comprehensive review of the biology of electric eel and other electric fishes.

Some authorities group *Electrophorus* with its closest relative, *Gymnotus*, in the family Gymnotidae (Albert, 2000; Albert and Campos-da-Paz, 1998).

Map references: Albert (2000), Albert and Campos-da-Paz (1998),* Ellis (1913)

Class Actinopterygii
Subclass Neopterygii

Order Esociformes
(1st) Family Esocidae—**pikes**   (ē-sos'-i-dē) [ē-sō'-si-dē]

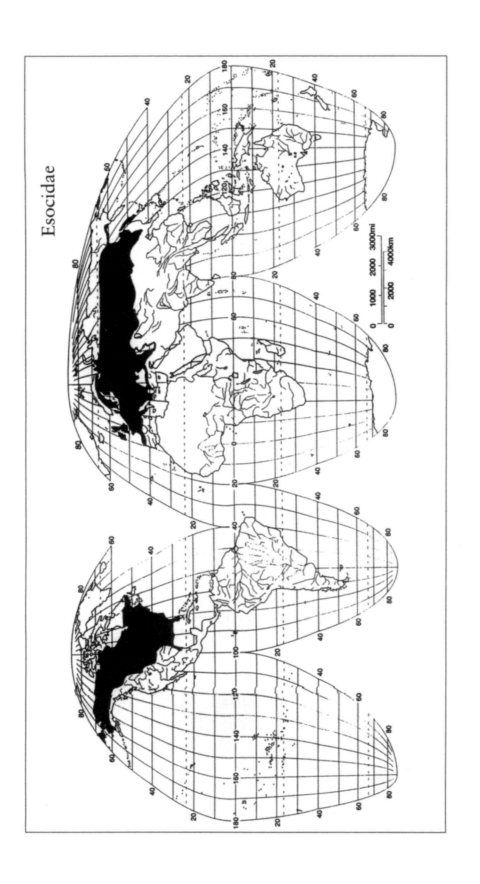

Esocidae

THE ORDERS ESOCIFORMES, OSMERIFORMES, AND Salmoniformes are grouped together in the superorder Protacanthopterygii (Fig. 1). This taxon has undergone many rearrangements over the years and is not completely satisfying to all ichthyologists. Its checkered history is summarized by Nelson (1994), and I follow his usage. The name "protacanthopterygii" means "before spiny fin fishes" and reflects the idea that this group is ancestral to the spiny rayed fishes; however, there is no consensus on what that ancestor may be.

Nelson (1994) considered the esociforms to be the primitive sister group to the osmeriforms + salmoniforms. Fink and Weitzman (1982), Lauder and Liem (1983), and Fink (1984) viewed the esociforms as the primitive sister group of the euteleosts. Williams (1987), on the other hand, considered the osmeroids and argentinoids as the primitive sister group of the neoteleosts. Wiliams also suggested that salmonids and esocids might be nearest relatives and more primitive than osmeroids + higher forms.

Esociform fishes have equally sized and posteriorly located dorsal and anal fins. The dorsal fin is situated more or less directly above the anal fin and no adipose fin is present. The fins are soft-rayed. Teeth are absent from the maxillary, and there are no pyloric cecae. They are physostomes, meaning that the swim bladder is connected to the gut by a pneumatic duct.

The pike family, Esocidae, includes one genus, *Esox*, and five species. These duck-billed, elongate, lie-in-wait predators are distributed holarctically in northern and eastern North America and Eurasia.

Esox lucius, the northern pike, has a circumpolar distribution in North America and Eurasia. It reaches 1.3 m and 25 kg, is highly valued by anglers, and has been widely introduced. It is an important food fish in Europe. *Esox reicherti*, the Amur pike, occurs only in the Amur River drainage system in Siberia and China (Berg, 1948/1949) and is the only pike not indigenous to North America. It has been stocked in Pennsylvania (Fuller *et al.*, 1999). It may reach 1.1 m and 10 kg. *Esox masquinongy* (Fig. 161), the muskellunge or muskie, is the largest pike at about 1.8 m and 45 kg (Scott and Crossman, 1973), but 16–23 kg is a respectable-sized

FIGURE 161. *Esox masquinongy* as *E. nobilior* (Goode, 1884, Plate 184).

muskie that would please any angler. It is native in the upper Mississippi and Great Lakes drainages. It is highly prized by anglers. *Esox niger*, the chain pickerel, so called because of the chain-like markings on its side, is primarily distributed along the Atlantic coastal plain and Gulf coast drainages as far north as Missouri. It reaches about 80 cm and 2.2 kg. *Esox americanus*, the grass pickerel as it is known on the coastal plain, or redfin pickerel (Mississippi drainage), is the smallest pike at approximately 38 cm. See Crossman (1980) for distribution maps of each North American species.

As adults, pikes are voracious piscivores that ambush their prey, usually in vegetation in clear water. They live in slow-flowing streams and rivers, ponds, and lakes. A quick flick of their forked caudal fin swiftly propels their cylindrical body as they lunge toward their prey. The jaw teeth are large and teeth are also present on the vomer, palatines, and tongue. The lateral line is complete. All *Esox* have 50 chromosomes (Beamish *et al.*, 1971), and the position of the chromosomal nucleolar organizer region in the American *Esox* taxa and in European *E. lucius* is the same. This suggests that karyotypes of extant species of *Esox* retained ancestral esocoid characters (Ráb and Crossman, 1994). Spawning usually occurs in early spring on flooded vegetation. The rich and diverse fossil record of this family ranges from the Late Cretaceous to the late Pleistocene (Wilson *et al.*, 1992; Grande, 1999). The fossil species *E. tiemani*, from Paleocene formations in Alberta, is very similar to *E. lucius* even though the two forms are separated by 62 million years (Wilson, 1984).

See Scott and Crossman (1973) for life history details, Crossman (1978) for taxonomic information, and Crossman and Casselman (1987) for an annotated bibliography of *E. lucius*. Raat (1988) provided a synopsis of the biological data on this species. Hall (1986) dealt with the biology of muskies, and Craig (1996) reviewed the biology of pike.

Map references: Berg (1948/1949), Crossman (1966, 1978, 1980),* Darlington (1957),* Grande (1999),* Lagler *et al.* (1977),* Maitland (1977),* Page and Burr (1991),* Rostlund (1952),* Scott and Crossman (1973)*

Class Actinopterygii
Subclass Neopterygii

Order Esociformes
(1st) Family Umbridae—**mudminnows** (um'-bri-dē)

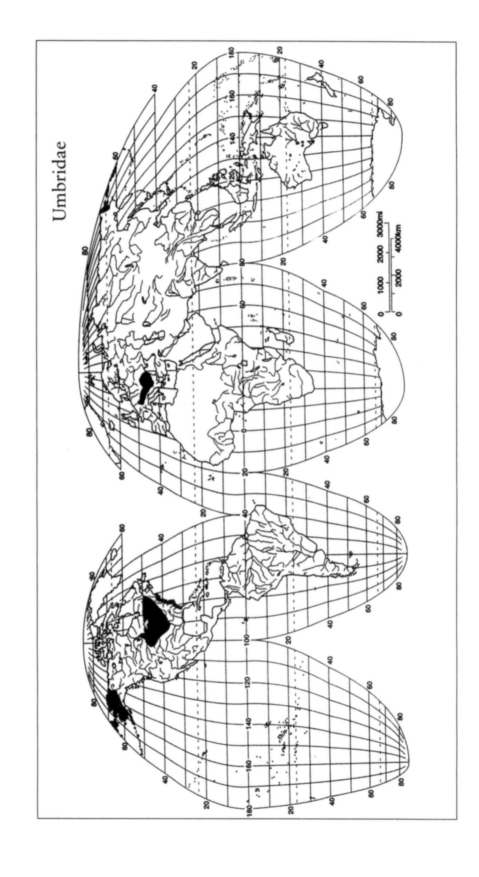

Umbridae

THE MUDMINNOWS ARE FRESHWATER FISHES WITH A HIGHLY
disjunct distribution that includes arctic areas of Siberia and Alaska
(*Dallia*), the Olympic peninsula of Washington (*Novumbra*), upper
Mississippi River drainages, Atlantic coast drainages, and Europe (*Umbra*).
There are three genera and at least five species (Nelson, 1994).

Umbrids are small fishes (7–33 cm) with elongate bodies and posteriorly positioned dorsal and anal fins. Unlike the pikes, mudminnows have a
short, blunt snout; their caudal fin is rounded; and the lateral line is not
prominent and may be absent. Their chromosome number varies widely
from 22 to 78 (Beamish *et al.*, 1971).

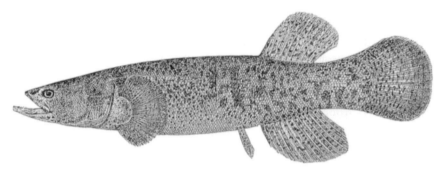

FIGURE 162 *Dallia pectoralis* (Goode, 1884, Plate 185).

Dallia pectoralis (Fig. 162), the Alaska blackfish, occurs in arctic
and subarctic fresh waters of Alaska. It is also known from northeastern
China, northeastern Siberia, and islands in the Bering Sea. *Dallia* has
large, fan-like pectoral fins with 31–36 fin rays. It has tiny pelvic fins with
2 or 3 rays, whereas other members of the Umbridae have 6 or 7 pelvic
rays. *Dallia* has 78 chromosomes (Beamish *et al.*, 1971) and several unusual
osteological features. Crossman and Ráb (1996) suggested that *Dallia* is
closer to *Novumbra* than either is to *Umbra* based on chromosome studies.
Some ichthyologists consider *Dallia* in a separate family, Dallidae (Bond,
1996).

Dallia is the largest umbrid at 33 cm. It prefers vegetated areas in
medium to large rivers and lakes. Spawning takes place soon after ice
breakup in May and continues through July. A short upstream movement
coincides with increase in water temperature. Fertilized eggs are demersal
and adhesive. Hatching occurs in about 10 days. The fish can live up to
8 years. Sexual maturity occurs about age 2 or 3, and they spawn several
times during their life. Preferred food is small invertebrates (Morrow,
1980).

Dallia is well-known for its tolerance to extremely cold conditions.
They can survive exposure to –20°C for up to 40 min and can survive freezing of parts of their body (Morrow, 1980). However, the legendary story of
frozen blackfish fed to sled dogs being thawed by the heat of the dogs'

stomach and then regurgitated alive (Turner, 1886) is doubtful. Complete freezing that allows ice crystals to tear cells would result in death. Blackfish can withstand anoxic environments for up to 24 hr at 0°C (Morrow, 1980). Air breathing has been established for this species and a capillary-dense section of its esophagus serves as an accessory respiratory organ (Crawford, 1974).

The Olympic mudminnow, *Novumbra hubbsi*, occurs in the coastal lowlands of the western and southern parts of Olympic peninsula of Washington. It prefers quiet water, usually with dense vegetation, over mud and detritus bottoms. *Novumbra* has 18–23 pectoral fin rays. Adults reach 70 mm and feed on ostracods, isopods, oligochaetes, mysids, mollusks, and dipterans. Fry have a sticky gland on the top of their head with which they attach to vegetation (Meldrim, 1980). Wydoski and Whitney (1979) summarized life history information for this species.

The three species of *Umbra* have smaller pectoral fins than the other two genera with 11–16 fin rays. The central mudminnow, *Umbra limi*, occurs in southern Canada, the Nelson River drainage, and the central United States from Quebec west through the Great Lakes basin and south in the Mississippi basin. *Umbra limi* reaches 13 cm, but most specimens are much smaller. Sluggish waters with a soft bottom are preferred, and the fish may burrow tail first into the bottom muck to hide or estivate (Trautman, 1981). Gee (1980, 1981) demonstrated that *U. limi* is a continuous, facultative air breather subject to external controling factors such as the presence of predators. Air gulping may also be used for buoyancy control. Mudminnows may survive hypoxia under ice by air breathing in pockets of trapped gas with low concentrations of O_2 (Klinger *et al.*, 1982; Magnuson *et al.*, 1983, 1985). See Graham (1997) for further details of aerial respiration. Because it has only 22 large chromosomes, *U. limi* is an ideal animal for experimental karyotype studies (Mong and Berra, 1979).

The eastern mudminnow, *U. pygmaea* (Fig. 163), occurs along the Atlantic and Gulf slope from New York to Georgia (Gilbert, 1980b). It

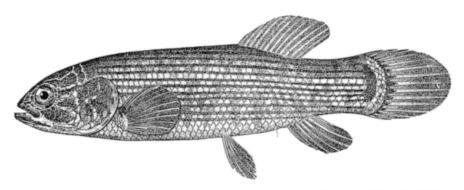

FIGURE 163. *Umbra pygmaea* as *U. pymaea* (Jordan and Evermann, 1900, Plate XCIX, Fig. 268).

reaches about 11 cm in mud-bottomed, vegetated, slow water. It has been introduced into Belgium, France, and Holland. *Umbra krameri* is known from the Danube and Dniester Rivers of central Europe. It reaches a maximum size of 12 cm.

The relationships between *Dallia, Novumbra*, and *Umbra* are not settled. Cavender (1969) linked *Dallia* and *Novumbra*. Wilson and Veilleux (1982) stated that *Dallia* is closer to *Umbra* than to *Novumbra*, and that the latter genus is the most primitive. They also wrote that *U. krameri* is the primitive sister group of the two American species of *Umbra*. Reist (1987) suggested that *Umbra* and *Novumbra* are closely related, and that *Dallia* was closer to *Esox*. Chromosome studies suggest a close relationship among all species of *Umbra* and a possible closer relationship between *Umbra* and *Esox* than between *Esox* and *Novumbra* + *Dallia* (Crossman and Ráb, 1996). Based on molecular studies, Waters *et al.* (2000) suggested that the esocoids may represent a paraphyletic group with the position of *Umbra* unresolved. Fossils date to the Eocene of Europe if one accepts *Palaeoesox* as an umbrid (Cavender, 1986) or to the Oligocene of Siberia (*Proumbra*) and Oregon (*Novumbra*).

Map references: Berg (1948/1949), Lagler *et al.* (1977),* Lee *et al.* (1980),* Maitland (1977),* Miller (1958),* Page and Burr (1991),* Rostlund (1952),* Scott and Crossman (1973)*

Class Actinopterygii
Subclass Neopterygii

Order Osmeriformes; Suborder Osmeroidei
(Per) Family Osmeridae—**smelts** (os-mer'-i-dē)

THE OSMERIFORMES IS COMPOSED OF TWO SUBORDERS: Argentinoidei (deep-sea smelts) and Osmeroidei (true smelts and Southern Hemisphere smelts). The order contains 13 families and about 236 species (Nelson, 1994) and is highly diverse with some very strange deep-sea forms. I consider only the species that spend time in fresh water. Osmeriform fishes tend to be elongate and predaceous. An adipose fin may be present or absent. The scales are cycloid and without radii. Begle (1991, 1992) and Wilson and Williams (1991) discussed the phylogeny of osmeriform fishes.

Within the Osmeroidei are six families and about 72 species. The Osmeridae contains seven genera and about 15 species (Nelson, 1994;

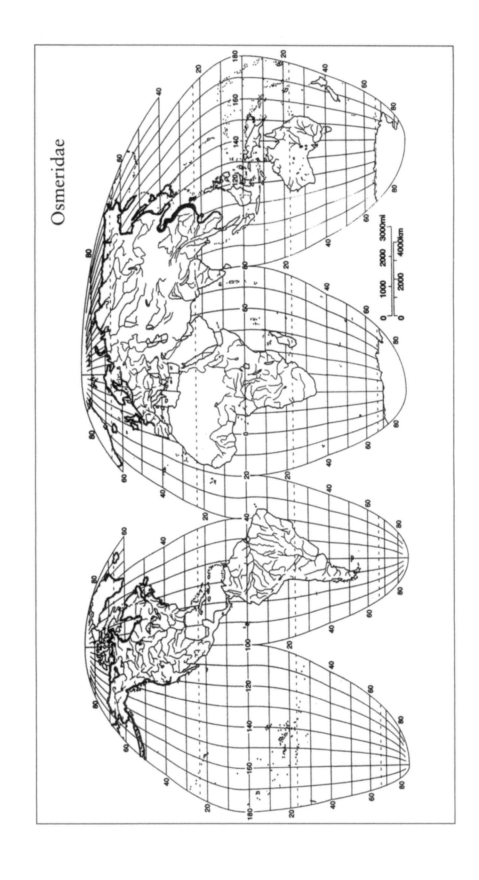

Osmeridae

D. E. McAllister, Personal communication). The smelts are marine, anadromous, and coastal freshwater fishes of the Northern Hemisphere. They have an adipose fin, and they lack a pelvic axillary process. Most are small, silvery, schooling fishes that are less than 20 cm TL. They can be enormously abundant and commercially valuable. Despite their small size most have well-developed teeth and are highly predaceous feeding on zooplankton and small fishes. Species of *Hypomesus* and *Mallotus*, with 6 of the 14 currently recognized species, have fine teeth.

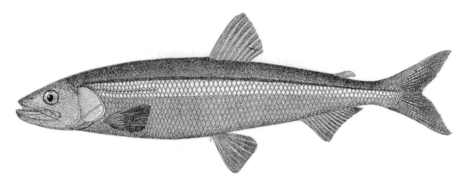

FIGURE 164. *Osmerus mordax* (Goode, 1884, Plate 199).

The rainbow smelt, *Osmerus mordax* (Fig. 164), occurs on the Atlantic coast of North America from Newfoundland to Pennsylvania and on the Pacific coast from Vancouver Island northward around Alaska. Coastal populations ascend rivers to spawn (anadromous). They have been introduced into the Great Lakes. Smelt have been recorded annually since 1936 in Lake Erie (Trautman, 1981). Smelt prefer clear, cool water, and they spawn over gravel. They are an important forage species for game fishes, and O. *mordax* is fished for its delicate flavor. The largest specimens can reach 36 cm and 227 g, but most are about half this size.

The odor of cucumbers emanates from the skin of some smelts. This odor has been demonstrated to come from the same molecule present in cucumbers. This substance, *trans-2-cis-6-nonadienal*, was first identified from *Prototroctes maraena* (Retropinnidae) (Berra *et al.*, 1982) and was later confirmed to be present in European smelt, O. *eperlans*, and Canadian surf smelt, *Hypomesus pretiosus*, as well as *Retropinna* and *Stokellia* (Retropinnidae) (McDowall *et al.*, 1993).

The eulachon, *Thaleichthys pacificus*, occurs along the west coast of North America from Bristol Bay, Alaska, to the Klamath River in California. It spawns a short distance up coastal rivers. It reaches about 20 cm and is fished commercially. The oil of the eulachon has been prized for its nutritional qualities and value as a trade item. Its oiliness meant that it could be burned, which gave rise to one of its common names, candlefish (D. E. McAllister, personal communication). The capelin, *Mallotus villosus*, has a marine, holarctic distribution and is fished commercially in the North Atlantic. This species has teeth so small that they are easily overlooked.

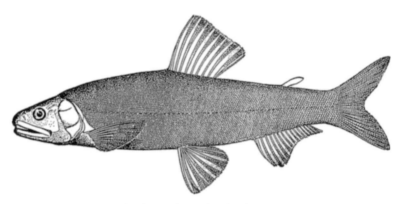

FIGURE 165. *Plecoglossus altivelis* (Jordan and Richardson, 1909, p. 167, Fig. 3).

The ayu, *Plecoglossus altivelis* (Fig. 165), occurs in Japan, Korea, Taiwan, and the Pacific coast of China, where it lives in fresh water or is anadromous. Formerly, this species was placed in its own family, Plecoglossidae. Howes and Sanford (1987), Begle (1991), and Wilson and Williams (1991) discussed the relationship of *Plecoglossus* to other osmerid genera.

Unlike other osmerids that have 0–11 pyloric caeca, ayus have an astonishing 300 or more caeca. This unusual species feeds on phytoplankton and reaches a length of about 30 cm. Spawning takes place in the lower reaches of rivers in fall and early winter, and the young drift down to the sea. They return to fresh water in late winter or early spring. Adults live only 1 year and die after spawning. The young grow rapidly and support fisheries by early summer (Bond, 1996). *Plecoglossus* is the object of the peculiar cormorant fisheries, whereby a trained cormorant, with a ring around its neck to prevent swallowing, scoops up ayus on their spawning run. The bird's owner then turns the cormorant upside down and shakes out its catch. The long-suffering bird is rewarded with a piece of fish. This commercially important species is now raised in aquaculture.

McAllister (1963) revised the family. Consult Scott and Crossman (1973) for details of the biology of some species. The oldest fossil species of osmerid dates to Paleocene freshwater deposits of Alberta and resembles *Plecoglossus* (Wilson and Williams, 1991).

Map references: Berg (1948/1949), Lee *et al.* (1980), McAllister (1984),* Maitland (1977),* Nelson (1976),* Page and Burr (1991),* Rosen (1974),* Scott and Crossman (1973)*

Class   Actinopterygii
Subclass   Neopterygii

Order   Osmeriformes; Suborder Osmeroidei
(Per) Family   Salangidae—icefishes or noodlefishes   [sa-lan'-gi-dē]

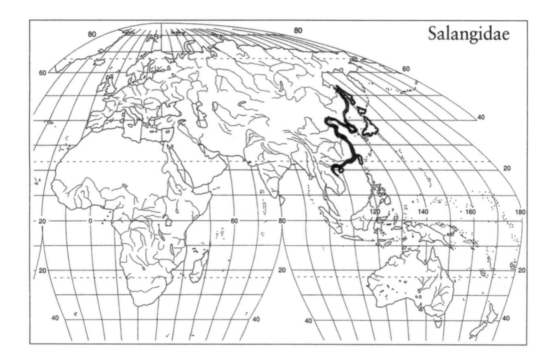

THE NOODLEFISHES OCCUR FROM SAKHALIN AND Vladivostok through Korea, Japan, and China to the Gulf of Tonkin in northern Vietnam in the sea, coastal rivers, and lakes. China and Korea have the greatest species richness. There are four genera (*Neosalanx, Protosalanx, Salangichthys,* and *Salanx*) with 11 species (Roberts, 1984c). Some species are anadromous, and others are confined to fresh waters.

Noodlefishes are delicate, elongate, and transparent or translucent with prominent eyes. *Protosalanx hyalocranius* (Fig. 166) represents typical family morphology. Relatively few melanophores are scattered around the body. There may be a row of melanophores at the junction of the ventral myotomic musculature and the ventral abdominal wall (Roberts, 1984c). These fishes are probably neotenic and superficially resemble leptocephali of elopoid fishes. The cranium is strongly to moderately depressed. Their

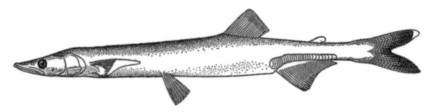

FIGURE 166. *Protosalanx hyalocranius* as *Salanx hyalocranius* (Jordan and Metz, 1913, Fig. 11).

skeleton is largely cartilaginous. An apidose fin is present, and the pelvic fins usually have seven rays.

Salangids are sexually dimorphic. The males usually have larger pectoral, pelvic, and anal fins than the females. The anterior rays of the male's anal fin are greatly enlarged. Mature males also have a row of large scales above the anal fin base. There may also be a few scales near the vent; otherwise, salangids are scaleless (Roberts, 1984c). Some species have breeding tubercles on the head, abdominal keel, and fins. Salangids presumably die after spawning.

Their flesh is a delicacy, and they are the object of valuable local fisheries. Noodlefishes are predaceous and feed on small fishes and crustaceans. Freshwater species consume insects as well. Adults range from about 35 to 163 mm SL.

Roberts (1984c) examined the skeletal anatomy and divided the family into three subfamilies: Protosalanginae, Salanginae, and Salangichthyinae. He provided a key to species and summarized their natural history. This group is called noodlefishes because of their slender, soft-bodied form or icefishes because of their transparent or translucent body.

Map references: Berg (1948/1949), Nichols (1943), Roberts (1984c), Rosen (1974)*

Class Actinopterygii
Subclass Neopterygii

Order Osmeriformes; Suborder Osmeroidei
(Per) Family Retropinnidae—Southern Hemisphere smelts and graylings [re-trō-pin'-ni-dē]

THE SOUTHERN HEMISPHERE SMELTS AND GRAYLINGS ARE found in fresh and brackish waters of southeastern Australia including

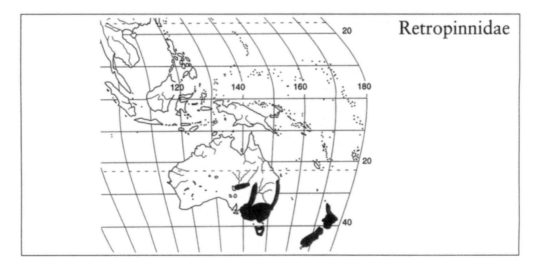

Tasmania, New Zealand, and the Chatham Islands near New Zealand. There are three genera and six species.

Retropinnids possess an adipose fin and a small, midventral horny keel along the abdomen anterior to the anus. There is no lateral line on the body, and the scales are cycloid and easily shed. Only a left gonad is present. A cucumber odor is present in the skin of fresh specimens of most species. The family is divided into two subfamilies, the Prototroctinae (southern graylings) and Retropinninae (southern smelts).

The Southern Hemisphere graylings, *Prototroctes*, were formerly considered in their own family, Prototroctidae (McDowall, 1976a). Their dorsal fin is positioned above the pelvic fins. Graylings can reach at least 300 mm SL, but most recent specimens rarely exceed 250 mm SL (Berra and Cadwallader, 1983). They comprise two species. The New Zealand grayling, *P. oxyrhynchus*, has not been collected since the 1930s and is presumed to be extinct (McDowall, 1990). The Australian grayling, *P. maraena* (Fig. 167), has been extensively studied (Berra, 1987). It occurs in

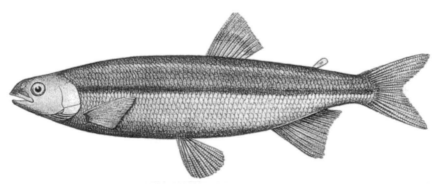

FIGURE 167. *Prototroctes maraena* (Waite, 1902, Plate XLI).

coastal rivers of southeastern Australia from the Grose River near Sydney around the Victoria coast to the Hopkins River and throughout coastal systems of Tasmania (Bell *et al.*, 1980).

The life cycle of the Australian grayling is amphidromous. This is a category of diadromous fishes which is not anadromous or catadromous. That is, their "migration from fresh water to the sea or vice versa is not for the purpose of breeding but occurs regularly at some other stage of the life cycle" (McDowall, 1988). Spawning takes place in the freshwater midreaches of coastal rivers in fall. The eggs are demersal and nonadhesive. The newly hatched fry are swept downstream to brackish water or the sea, where they remain for about 6 months. In spring juveniles ascend to the midreaches of the rivers, where they spend the rest of their lives. Most graylings die before forming a third annulus, but some live to 5 years of age (Berra and Cadwallader, 1983). Males develop breeding tubercles on scales and fin rays (Berra, 1982, 1984). The cucumber odor emanating from their skin has been identified as the same molecule present in cucumbers (Berra *et al.*, 1982). The reproductive anatomy of *P. maraena* was discussed by Berra (1984). Approximately 47,000 eggs of 0.9-mm diameter are produced from the single left ovary. Aquatic insects, algae, and a diatom/organic matrix were found in grayling stomachs. They have a black peritoneum, and this may be related to the presence of plant matter in their diet (Berra *et al.*, 1987). The comb-like jaw teeth of graylings may help remove insects from filamentous algae.

The second subfamily, Retropinninae, is so named because the dorsal fin is situated posterior to the pelvic fins. These elongate, silvery fishes occur in fresh and coastal waters and range in size from about 50 to 150 mm TL. This group was formerly recognized as a family excluding *Prototroctes* (McDowall, 1979).

Retropinna retropinna is widespread in New Zealand, and seagoing stocks have reached the Chatham Islands. *Retropinna semoni* (Fig. 168) occurs around the southeastern coast of Australia from the Fitzroy River in southern Queensland to eastern South Australia. It is also present in

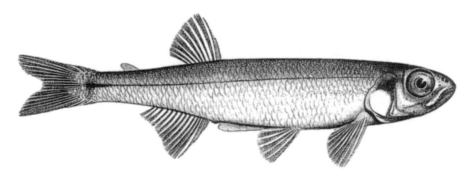

FIGURE 168. *Retropinna semoni* (McCulloch, 1920, Plate XI).

tributaries of Australia's largest river system, the Murray–Darling. It is even represented in Cooper Creek drainages that flow into Lake Eyre (McDowall, 1996). *Retropinna tasmanica* is widespread in coastal streams around Tasmania. The life history of *R. retropinna* has been well studied and is described by McDowall (1990). The fins of both sexes become enlarged near spawning time and breeding tubercles develop. They are anadromous, migrating upstream from the sea to spawn. The larval fish are carried out to sea on the river currents. The life history of the Australian species of *Retropinna* is similar to that of the New Zealand species, but there are many landlocked populations in Australia (Merrick and Schmida, 1984; McDowall, 1996).

Stokellia anisodon occurs in coastal rivers of the Canterbury area of central-eastern South Island, New Zealand. It is distinguished from *Retropinna* by having about 100 very small scales along the side, whereas *Retropinna* has about 60 larger scales. *Stokellia* is the most marine of the retropinnids. It spends very little time in fresh water (McDowall, 1990). In late spring vast numbers of adults congregate in river mouths. They mature in estuaries and the lower parts of rivers. They spawn on sandy shoals just above estuaries. Larval fish drift downstream to the sea. *Stokellia* and *Retropinna* have been shown to possess the same cucumber molecule as *Prototroctes* (McDowall *et al.*, 1993).

Map references: Allen (1989),* Bell *et al.* (1980),* McDowall (1990, 1996)*

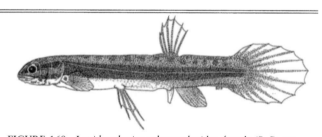

Class Actinopterygii
Subclass Neopterygii

FIGURE 169. *Lepidogalaxias salamandroides*, female (B. Pusey, from Berra and Pusey, 1997; reproduced with permission).

Order Osmeriformes; Suborder Osmeroidei
(?) Family Lepidogalaxiidae—**salamanderfish**
(lep'-i-dō-gal'-ak-sī'-i-dē) [le'-pi-dō-ga-lax'-ē-i-dē]

THIS MONOTYPIC FAMILY IS COMPOSED OF THE BIZARRE salamanderfish, *Lepidogalaxias salamandroides* (Fig. 169), whose scientific name is longer than the fish. *Lepidogalaxias* is found only in the southwest

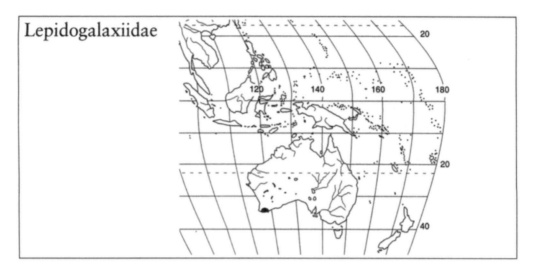

Lepidogalaxiidae

corner of Western Australia in pools and streams on heathland peat flats between the Blackwood and Kent Rivers, a distance of about 180 km centered on the village of Northcliffe (Allen and Berra, 1989).

This species has an elongate, slender, scaled body and a blunt head. The dorsal fin is tall and usually held erect with five to seven dorsal rays. The pelvic fins are abdominal with elongate rays. The anal fin of males possesses modified rays and becomes noticeably sexually dimorphic at 25 mm TL. The anal fin is covered with a scaly sheath. Females develop a genital papilla. The caudal fin is lanceolate.

The phylogenetic relationships of this family are not certain. Recent studies have arrived at different conclusions. *Lepidogalaxias* was originally described as a galaxiid despite the fact that it has scales and all other galaxiids are scaleless (Mees, 1961). Rosen (1974) considered it the only Southern Hemisphere esocoid, a view not supported today (Fink and Weitzman, 1982). Fink (1984) placed *Lepidogalaxias* in an unresolved trichotomy with the Salmonidae as the sister group of the Neoteleostei. Williams (1987) considered *Lepidogalaxias* to be a member of the Southern Hemisphere assemblage of galaxioid fishes, as did Begle (1991). Based on a study of intermuscular bones, Patterson and Johnson (1995) supported Begle's placement of *Lepidogalaxias* among galaxioid osmeroids as the sister group of salangids + galaxiids. Johnson and Patterson (1996) and Patterson and Johnson (1997) placed *Lepidogalaxias* in a subfamily of the Galaxiidae and considered *Lovettia sealii* its closest relative. They severely criticized Begle's (1991) work. Williams (1997), based on bones and muscles of the suspensorium, argued that *Lepidogalaxias* is sister to galaxiids + aplochitonids and that Southern Hemisphere galaxioids are monophyletic. Using both cytochrome b and 16SrRNA data sets, Waters (1996) and Waters *et al.* (2000) reported that *Lepidogalaxias* is the most divergent member of the Galaxioidea and may not belong within the galaxioids.

Waters *et al.* placed salamanderfish basal to the Galaxioidea, a position consistent with Fink (1984) but at odds with Williams (1987, 1997), Begle (1991), and Johnson and Patterson (1996). Therefore, the relationships of this very strange little fish remain enigmatic. It does not neatly fit into a primary, secondary, or peripheral division category. This situation may become clearer when we know its ancestry. It may be a very ancient lineage.

Much has been learned in approximately the past decade about the natural history of salamanderfish. This information has been summarized by Berra (1997b) and Berra and Pusey (1997). Salamanderfish reach a maximum TL of 67 mm. They live in ephemeral pools and streams that dry up during the heat of summer. They survive the desiccation of their habitat by burrowing into the damp substrate and remaining moist as the ground-water retreats from the surface. They reemerge when the autumn rains fill their pools (Berra and Allen, 1989a). Salamanderfish do not feed while underground and apparently enter a state of estivation. While buried, they exchange oxygen and carbon dioxide through their moist skin (Martin *et al.*, 1993) They have no accessory respiratory structures (Berra *et al.*, 1989), nor can they tolerate hypoxic waters (Berra and Allen, 1995). While buried, salamanderfish metabolize stored fat (Pusey, 1990). Urea is stored during estivation and excreted upon arousal (Pusey, 1986).

The fish can emerge within minutes following a rain that partially fills their dry pools. An experiment in which 2700 liters of water was released from a fire truck into a dry pool produced fish in 10 min (Berra and Allen, 1989a). It was truly instant fish, just add water! Attempts to artificially induce estivation by lowering the water level in concrete troughs containing sandy soil and water resulted in the survival of only 3 of 29 fish (Pusey, 1989).

The tea-colored habitat of salamanderfish pools and streams is very acidic, with a pH of about 4.0 (Berra and Allen, 1989a). Salamanderfish feed on aquatic insects and spawn during the winter when their pools are full of water (Berra and Allen, 1991). The modified anal fin of males, illustrated by Rosen (1974), functions in mating and insemination is internal (Pusey and Stewart, 1989). The young grow and mature rapidly (Berra and Allen, 1991; Gill and Morgan, 1999).

Lepidogalaxias has the odd ability to bend its neck at right angles while tracking live food (Berra and Allen, 1989a). This is possible because of the wide space between the skull and first vertebra. There is also a substantial gap between each vertebra. In addition, the abdominal ribs are reduced. All of these anatomical features gives a great deal of flexibility to the spine and enable the salamander-like wriggling locomotion. This may be useful in burrowing into the damp sand. The skull is robust, which is probably an adaptation to supporting the weight of the sand.

In their redescription of *Lepidogalaxias*, McDowall and Pusey (1983) reported that eye muscles are absent. They concluded that because the eye lacks a circumorbital sulcus it cannot be moved and therefore lacks eye muscles. Collin and Collin (1996), however, showed the presence of six extraocular muscles that allow the eye to move freely beneath the spectacle.

Lepidogalaxias is locally abundant within its limited range. Dozens of pools occur in the Northcliffe area and each pool may harbor 120–150 fish (Allen and Berra, 1989a). Christensen (1982) and Pusey and Edward (1990) described the biota associated with these acidic heathland peat flats which have recently been incorporated into D'Entrecasteaux National Park. The future of *Lepidogalaxias* seems secure for now.

Map references: Allen (1989),* Allen and Berra (1989),* Christensen (1982),* McDowall and Pusey (1983)*

Class Actinopterygii
Subclass Neopterygii

Order Osmeriformes; Suborder Osmeroidei
(Per) Family Galaxiidae—**galaxiids**
(gal'-ak-sī'-i-dē) [ga-lax'-ē-i-dē]

THE GALAXIIDS ARE FRESHWATER FISHES, SOME OF WHICH ARE diadromous, that occur in cool temperate waters of the Southern Hemisphere in Australia, Lord Howe Island, New Zealand, the Chatham Islands, Auckland and Campbell Islands, New Caledonia, southern South America, the Falkland Islands, and the southern tip of South Africa (McDowall, 1990). There are eight genera and about 51 species.

Galaxiids are elongate, scaleless fishes that lack an adipose fin. Some species are stocky and tubular. They lack the horny keel and cucumber odor of the retropinnids. Most species have 7 pelvic and 16 caudal fin rays (McDowall, 1990). However, there is wide variation of other elements of the caudal skeleton within and among species (McDowall, 1999). A lateral line is present. Some species may have an accessory lateral line composed of small, widely spaced neuromasts along the dorsolateral trunk from the occiput to the dorsal fin (McDowall, 1997c). This accessory lateral line may function in sensing food or predators at the water's surface. The dorsal fin is situated posteriorly and more or less above the anal fin.

Formerly, the genera *Lovettia* and *Aplochiton* were placed in the family Aplochitonidae. Begle (1991) considered them part of the Galaxiidae. Johnson and Patterson (1996) considered *Aplochiton* to be the sister group

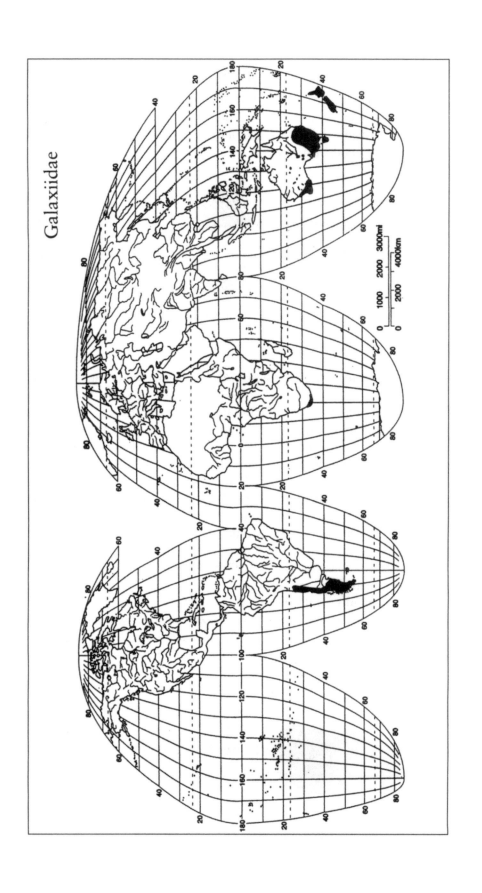

Galaxiidae

of the remainder of the Galaxiidae, and they treated *Lovettia* as the sister group of *Lepidogalaxias*. Williams (1996) studied jaw muscles and suspensoria and concluded that *Aplochiton* and *Lovettia* are each others closest relatives and that they form a monophyletic group. I follow Nelson (1994) and divide the Galaxiidae into three subfamilies: Lovettiinae, Aplochitoninae, and Galaxiinae.

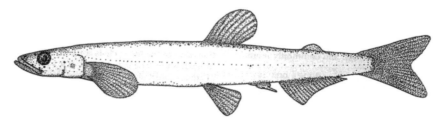

FIGURE 170. *Lovettia sealii* (reproduced with permission from McDowall, 1996, Fig. 11.1).

Lovettia sealii (Fig. 170), known as Tasmanian whitebait, is a small, slender species known only from coastal Tasmania, especially in the north. Its small, low adipose fin is attached along the entire base of the fin. The snout is sharply pointed, and the mouth is large. *Lovettia* is anadromous and spends most of its life at sea. One-year-old fish spawn after migrating into rivers in spring in large schools. The adults die, and the young are carried downstream to the sea. In 1947, commercial catches of whitebait, including young of *Galaxias*, reached 480,000 kg, but today the fisheries are closed. *Lovettia* can reach 77 mm, but most rarely exceed 60 mm. When they leave the sea, whitebait are more or less transparent. Prespawning adults darken from gray to nearly black. The fins of males are larger than those of females. See McDowall (1971a, 1996) for older references and further information.

The subfamily Aplochitoninae is composed of two species of *Aplochiton* from southern South America. *Aplochiton zebra* (Fig. 171) and *A. taeniatus* are both widespread along the western side of the Andes from about 39°S to Tierra del Fuego at about 56°S (McDowall and Nakaya,

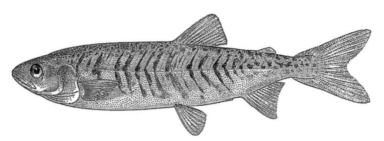

FIGURE 171. *Aplochiton zebra* (Eigenmann, 1928, Plate XI, Fig. 4).

1988). In Chile these species are found in scattered localities from about Concepcion (37°S) to Punta Arenas (53°S) (Arratia, 1981). *Aplochiton zebra* occurs in the Falkland Islands. *Aplochiton marinus* is considered a synonym of *A. taeniatus* (McDowall, 1971a).

The dorsal fin of *Aplochiton* is anteriorly positioned over the pelvic fins. An adipose fin is present, and the caudal fin is forked. *Aplochiton taeniatus* is the larger species, reaching a maximum SL of 36 cm (McDowall and Nakaya, 1988). It has a very large mouth and greatly enlarged teeth which suggest it is more of a predator than the insectivorous *A. zebra*. *Aplochiton* is probably amphidromous, with the larvae being carried to sea after hatching in freshwater streams. McDowall (1984a) reported a postlarval *Aplochiton* taken at sea in southern Chile.

The Galaxiinae includes 5 genera and about 48 species. This sub-family is most diverse in Australia, especially Tasmania, and in New Zealand. However, representatives also occur in South America, and one species, *Galaxias zebratus*, occurs in Cape coastal streams at the southern tip of South Africa (McDowall, 1973a; Skelton, 1993; Waters and Cambray, 1997). It is the only African galaxiid, but it is a highly variable one that was described as four separate species (Waters and Cambray, 1997). It breeds in fresh water. Waters *et al.* (2000) considered the divergence of *G. zebratus* to be ancient, possibly dating back to the breakup of Gondwana. In their molecular phylogeny *G. zebratus* was sister to the nondiadromous *Brachygalaxias* + *Galaxiella*. No members of the family or subfamily occur in the Northern Hemisphere despite Day's description of *G. indicus*, which McDowall (1973b) regarded as a *nomen dubium*.

Galaxias neocaledonicus lives in the rivers at the southern end of the island of New Caledonia (Marquet and Mary, 1999). McDowall (1968) supported the use of *Nesogalaxias* for the genus of this species based on the absence of pleural ribs from vertebrae posterior to the pelvic girdle. Waters *et al.* (2000) presented mtDNA data that suggested *N. neocaldonicus* diverged from *Galaxias brevipinnis* 3–11 Mya and concluded that the New Caledonian species belongs in *Galaxias*. They agree with McDowall (1968) that marine dispersal is the best explanation for the presence of a galaxiid in New Caledonia. Most Pacific islands are of volcanic origin, but New Caledonia is of continental origin. It existed as an island since the Late Cretaceous when it separated from Australia (Watson and Pöllabauer, 1998).

Galaxias maculatus (Fig. 172) has one of the most widely disjunct distributions of any freshwater fish (Berra *et al.*, 1996), and this distribution pattern inspired one of the most interesting debates in biogeography. I present this controversy in detail. *Galaxias maculatus* occurs in eastern and western Australia, Tasmania, Lord Howe Island, New Zealand, Chatham Island, southern Chile, Argentina, and the Falkland Islands (McDowall, 1970). Two hypotheses have been advanced to explain this

FIGURE 172. *Galaxias maculatus* (Eigenmann, 1928, Plate X, Fig. 5).

disjunct distribution: dispersal (movement through the sea) and vicariance (continental drift).

McDowall (1970) attributed the wide geographic distribution to transoceanic dispersal of the marine whitebait stage. Juveniles have been taken at sea 700 km from the New Zealand mainland (McDowall *et al.*, 1975). The age at migration from the sea of juvenile *G. maculatus* is 100–200 days (McDowall *et al.*, 1994). McDowall (1970) suggested that *G. maculatus* originated in Australia and dispersed eastward past Tasmania to New Zealand and South America via the East Australian Current and the West Wind Drift. Because this dispersal would have occurred after the breakup of Gondwana, McDowall (1967, 1971b, 1978) hypothesized that significant phenotypic evolution should exist among present-day populations of *G. maculatus* if their distribution predates the breakup of Gondwana 65 Mya. Since the various populations are all quite similar phenotypically, they could not have been isolated for 65 million years.

Rosen (1974, 1978), on the other hand, presented a vicariant alternative to McDowall's dispersal hypothesis. He suggested that because galaxiids occur on all Gondwanan continents except India, their present distribution reflects an ancient Pangaean pattern followed by continental drift. Rosen (1974, 1978) and Croizat *et al.* (1974) considered galaxiid fishes to be part of a pan-austral Gondwanan biota that fragmented in the Mesozoic. McDowall (1969, 1978) argued that there was no evidence that *G. maculatus* was part of such a hypothetical biota, and that recent dispersal would easily explain both the widespread distribution of *G. maculatus* and how the speciation process among disjunct, diadromous populations might be retarded. McDowall (1978) also argued that generalized tracks, one component of a vicariant model, assume the existence of an ancestral biota that can be demonstrated only by the fossil record, and that the galaxiid fossil record dates to the early Miocene of New Zealand and is very limited (McDowall, 1976b; McDowall and Pole, 1977). Anderson (1998) assigned scaled fossils from the Late Cretaceous of South Africa to the Galaxiidae as defined by Johnson and Patterson (1996).

Ball (1975) and Craw (1979) entered the debate with comments on the nature of biogeographical hypotheses. In discussing distributional relationships between African and South American freshwater fish families, Lundberg (1993) wrote that "drift-related vicariance seemed unlikely" for the Galaxiidae. Likewise, Nelson (1994) thought it untenable that the distribution of *G. maculatus* dated to the breakup of Gondwana and believed that dispersal via ocean currents was a more parsimonious hypothesis.

Berra *et al.* (1996) used allozyme electrophoresis of muscle extracts of *G. maculatus* from eastern and western Australia, New Zealand, and Chile to test the hypothesis that populations from the western Pacific and the eastern Pacific do not differ genetically. They found only minor differentiation in allele frequency at some loci and no fixation of alternative alleles. The populations appeared to be part of the same gene pool indicating that gene flow via dispersal through the sea occurs today. It is unlikely that South American and Australasian populations would be conspecific if they have exchanged no migrants since the breakup of Gondwana at the end of the Mesozoic. Only a small amount of gene flow is necessary to prevent accumulation of genetic differences by random drift. The marine larval stage of *G. maculatus* has a 6-month period to traverse the distance between the southern continents (McDowall *et al.*, 1994) and could provide enough gene flow to deter fixation for alternative alleles. A study of mitochondrial DNA sequence divergence by Waters and Burridge (1999) supported the dispersal argument but reported greater population differentiation than detected by Berra *et al.* (1996) with allozymes.

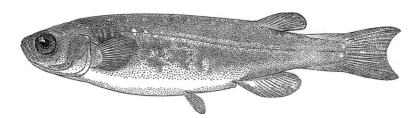

FIGURE 173. *Brachygalaxias bullocki* (Eigenmann, 1928, Plate X, Fig. 1).

Other South American galaxiids include two additional species of *Galaxias* and the small, colorful *Brachygalaxias bullocki* (McDowall, 1971b; Fig. 173). *Galaxias platei* occurs in Chile, Argentina, and the Falkland Islands, and the very rare, if not extinct, *G. globiceps* is known from only a few specimens taken near Puerto Montt, Chile (Berra and Ruiz, 1994; Berra and Barbour, 1998). *Brachygalaxias* occurs in central Chile from about Talca (35°S) to Chiloé (42°S) (Berra *et al.*, 1995). *Brachygalaxias bullocki* does not become much larger than 60 mm TL. It resembles *Galaxiella* of Australia, but whether this similarity is due to convergence or phylogeny is not clear.

Australia has about 20 species of galaxiids, with 13 species of *Galaxias*, 3 *Galaxiella*, and 4 *Paragalaxias* (McDowall and Frankenberg, 1981).

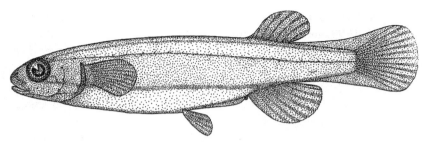

FIGURE 174. *Galaxiella pusilla* (reproduced with permission from McDowall, 1996, Fig. 10.15).

Galaxiella reach a maximum SL of about 45 mm. One species, *Galaxiella pusilla* (Fig. 174), occurs in southeastern Australia and Tasmania, and 2 species are sympatric in coastal drainages of southwest Western Australia (Berra and Allen, 1989b). Like *Brachygalaxias*, these small fishes have an orange band on their flank, black stripes, posterior dorsal fins, large eyes, and reduced supraneural bones. Waters *et al.* (2000), citing weak mtDNA support and strong morphological similarities, suggested that the three Australian species of *Galaxiella* be reassigned to *Brachygalaxias*, in which they were originally included. They consider *Galaxiella* and *Brachygalaxias* to represent an ancient Gondwanan radiation.

The four diminutive, stout-bodied species of *Paragalaxias* are confined to lakes at high elevation in the Central Plateau of Tasmania (McDowall, 1996, 1998a). Their large dorsal fin is forward, above the pelvic fin bases. *Paragalaxias julianus* (Fig. 175) is the largest species at 100 mm TL. MtDNA data suggest that *Paragalaxias* is closely related to the Tasmanina *G. parvus*–Victorian *G. olidus* clade (Waters *et al.*, 2000).

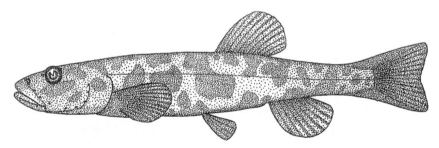

FIGURE 175. *Paragalaxias julianus* (reproduced with permission from McDowall, 1996, Fig. 10.20).

Galaxias olidus has a wide distribution in eastern Australia up to 1800 m in the Great Dividing Range. It occurs in coastal drainages as well as in the Murray–Darling river system from southern Queensland to South Australia. Like most Australian galaxiids, it is confined to fresh water. Its home range in a small mountain stream was estimated at 13–26 m of stream (Berra, 1973). The spotted galaxias, *G. truttaceus*, is a large, stout-bodied species that can reach more than 200 mm. Its brown to olive body is covered with round, ocellated spots, giving the fish a trout-like appearance. It occurs in coastal Victoria and Tasmania. *Galaxias occidentalis* is found only in coastal drainages of southwestern Western Australia.

New Zealand has about 19 galaxiids, 15 in *Galaxias* and 4 in *Neochanna*. Recent isozyme studies have shown that *Galaxias vulgaris* from South Island is actually a species complex (Allibone *et al.*, 1996; Waters and Wallis, 2001). Five new *Galaxias* have been described from that group (McDowall and Wallis, 1996; McDowall, 1997a; McDowall and Chadderton, 1999). Isozyme data indicated that populations of the *G. vulgaris* complex are highly structured in the southern part of South Island but relatively genetically homogeneous in the northern part. It is hypothesized that lack of diversity in the north reflects the recent formation of the Canterbury plain as a consequence of Pliocene and Pleistocene uplift and erosion, while the unstable braided river channels typical of the Canterbury plain have allowed continual genetic exchange among rivers. In the southern part of South Island the relatively stable geological history of this area allows for high diversity (Wallis *et al.*, 2001).

Galaxias argenteus is the largest galaxiid. This very stout fish, found throughout New Zealand, is reported to reach 580 mm and a weight of 2.7 kg (McDowall, 1990). *Galaxias gracilis*, known only from a few lakes in northern North Island, is a diminutive species reaching about 55 mm. *Galaxias brevipinnis* has a reputation as a climber. This species is able to negotiate damp, vertical rock faces above falls. Their downward pointing pectoral and pelvic fins form a gripping surface for moving up the damp rocks (McDowall, 1990). Five amphidromous species of New Zealand *Galaxias* contribute to the whitebait fisheries (McDowall, 1984b): *G. maculatus*, *G. brevipinnis*, *G. fasciatus*, *G. argenteus*, and *G. postvectis*. An amphidromous species is a diadromous fish whose migration from fresh water to the marine environment or vice versa is not for the purpose of spawning but occurs regularly as part of its life cycle (McDowall, 1988). Diadromy has been a major factor in the evolution of New Zealand's freshwater fish fauna (McDowall, 1998). Waters *et al.* (2000) concluded that species with a marine larval stage do not constitute a monophyletic group, and that migratory ability is distributed throughout the family. They further suggested that loss of this saltwater-tolerant, migratory phase could be a major cause of speciation within the family.

Mudfishes, *Neochanna apoda* (Fig. 176), *N. diversus*, and the newly described *N. heleios*, lack pelvic fins. *Neochanna burrowsius* has short pelvic fins of only four or five rays as opposed to the usual six or seven pelvic rays of *Galaxias*. *Neochanna* lives in swampy habitats that desiccate

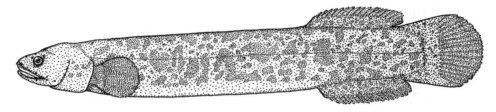

FIGURE 176. *Neochanna apoda* (reproduced with permission from McDowall, 1990, Fig. 7.57).

during the summer. These fish aestivate in the bottom mud of their habitat during the dry season. Morphological and molecular studies have placed the Tasmanian mudfish, *Galaxias cleaveri*, as a close relative of *Neochanna* (McDowall, 1997b; Waters and White, 1997). The Tasmanian mudfish is amphidromous, whereas the New Zealand mudfishes are freshwater species. The Tasmanian mudfish has small pelvic fins and small eyes. It is also capable of aestivation.

For life histories of the various galaxiid species see McDowall (1996, 1990) and Merrick and Schmida (1984). For descriptions of Australian galaxiid larvae see Gill and Neira (1994, 1998). For systematic information see McDowall and Frankenberg (1981), Waters (1996), and Waters *et al.* (2000).

Map references: Allen (1989),* Arratia (1981),* Berra *et al.* (1995, 1996),* McDowall (1971a,b), 1990, 1996),* McDowall and Frankenberg (1981),* Marquet and Mary (1999),* Skelton (1993),* Waters and Cambray (1997)*

Class Actinopterygii
Subclass Neopterygii

Order Salmoniformes
(Per) Family Salmonidae—**salmons, trouts** (sal-mon'-i-dē)

THE SALMONIDAE IS THE ONLY FAMILY INCLUDED IN THIS order according to Nelson (1994). Other arrangements include the suborders Salmonoidei and Osmeroidei together in the Salmoniformes (Johnson and Patterson, 1996). Fink and Weitzman (1982), Lauder and Liem (1983), Williams (1987), and Sanford (1990) each have differing interpretations of the taxonomic position of the Salmoniformes relative to the Ostariophysi, Osmeriformes, Esociformes, and the neoteleosts (Fig. 1).

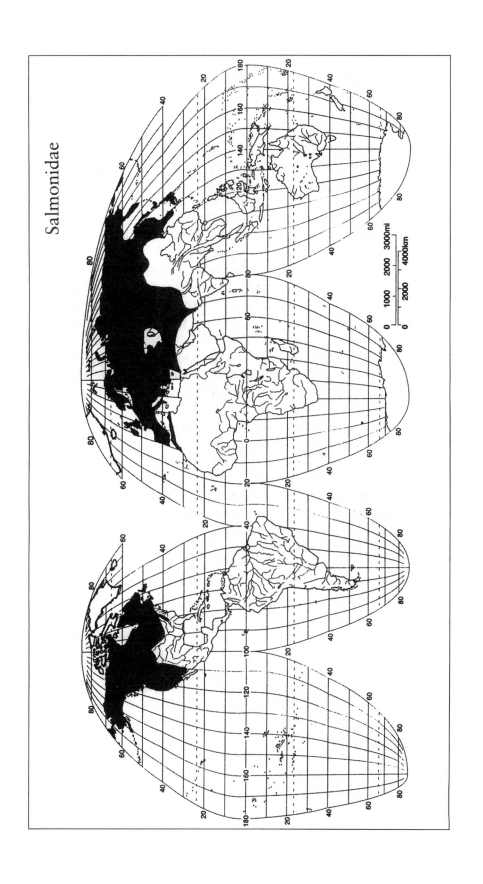

Salmonidae

Salmonids are distributed in a holarctic pattern in cool and cold waters of the north temperate zone, including northwest Mexico, Taiwan, and north Africa between 1000 and 2500 m in the Atlas mountains of Algeria and Morocco. The map shows only native distribution.

Salmonids are elongate, soft-rayed fishes with moderately to slightly indented caudal fins. An adipose fin is present, as is a pelvic axillary process. The swim bladder is connected to the gut (physostomous). Pyloric caeca may number up to about 210. The salmonid karyotype is tetraploid. The cranium contains a large proportion of cartilage, and the last three vertebrae turn upward toward the caudal fin. Most species have young with parr marks. The family includes both freshwater and anadromous species. Salmonids are greatly prized both for sport and for commercial fisheries. They have been introduced into cool waters of many parts of the world where they do not naturally occur, such as Australia, New Zealand, southern South America and Lake Titicaca near the equator, and southern Africa. Because of their great economic importance, they have been well studied and a great deal is known about their biology. Their taxonomy, however, remains controversial. There are 11 genera and about 66 species (Nelson, 1994). Three subfamilies are recognized: Coregoninae, Thymallinae, and Salmoninae. In the past these groups have been treated as separate, but related, families.

The Coregoninae include the whitefishes and the ciscoes. These fishes have relatively large scales, with fewer than 110 along the lateral line. The dorsal fin has less than 16 fin rays. No teeth are present on the maxilla in the adults, but maxillary teeth may be present in fry. A few species have teeth on the small-sized vomer (Behnke, 1972). There are three genera and about 32 species. *Prosopium* is the genus of the round whitefishes. There are 6 North American species, 1 of which, *P. cylindraceum*, the round whitefish, extends into northeastern Asia where it occurs from the Yenisei River to the Bering Strait and the Kamchatka peninsula (Scott and Crossman, 1973). Three species of *Prosopium* are only found in Bear Lake on the Utah–Idaho border (Rohde, 1980).

Coregonus is the genus of about 8 species of lake whitefishes and 17 ciscoes. Their distribution is circumpolar. The lake whitefish, *C. clupeaformis* (Fig. 177), ranges widely across northern North America from the Atlantic coast to Alaska. It has been an important commercial species, but yields have declined due to habitat destruction and overfishing (Scott and Crossman, 1973; Trautman, 1981). *Coregonus artedi*, the cisco or lake herring, is another species whose numbers have fluctuated tremendously from year to year. In the first two decades of the 1900s, 9–22 million kilograms were landed annually in Lake Erie alone (Scott and Crossman, 1973). By 1967, the catch had declined to a mere 2.3 million kilograms (Trautman, 1981). It is primarily a lake species and a plankton feeder.

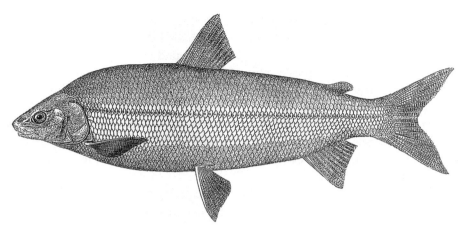

FIGURE 177. *Coregonus clupeaformis* as *C. clupeiformis* (Goode, 1884, Plate 196).

The inconnu, *Stenodus leucichthys*, represents the third genus within the subfamily Coregoninae. It occurs in northwestern North America west into the Arctic drainages of northern Asia. There is an isolated population in the Caspian Sea basin (Scott and Crossman, 1973).

The subfamily Thymallinae consists of *Thymallus* with about five species. These fishes are called graylings and are not related to the Southern Hemisphere graylings *Prototroctes*. Their generic name *Thymallus* refers to the fact that their flesh is said to smell like the herb thyme (Maitland, 1977). *Thymallus* occurs in northern North America and Eurasia. Graylings have a long, flowing dorsal fin with more than 17 rays. They also have teeth on the maxilla. The arctic grayling, *T. arcticus* (Fig. 178), is a strikingly beautiful fish with a purple dorsal surface, gray flanks

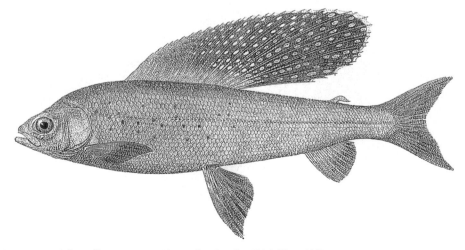

FIGURE 178. *Thymallus arcticus* as *T. signifer* (Goode, 1884, Plate 195).

with pink iridescence, and a white belly. Dark spots are scattered along the flanks anterior to the long, dark, speckled dorsal fin. The other fins are bronze in color. The arctic grayling is found in northwestern Canada from Hudson Bay to Alaska and across the Bering Sea into northern Eurasia. This species may reach 76 cm TL and 2.7 kg (IGFA, 1993), but specimens larger than 50 cm are very rare (R. Behnke, personal communication). Graylings live in clear, cold, large rivers and lakes. Adults feed on insects and other invertebrates. They leap when hooked and are sought after by fly fishers (Scott and Crossman, 1973). *Thymallus thymallus* is the European species.

Species in the subfamily Salmoninae have very small cycloid scales numbering more than 110 in the lateral line. The dorsal fin has fewer than 16 rays. This subfamily's distribution extends further north and south than the Coregoninae and Thymallinae. There are 7 genera and at least 30 species. The taxonomy of this group is complex and will provide work for ichthyologists for decades (Phillips and Pleyte, 1991; Stearley and Smith, 1993; Phillips and Oakley, 1997).

The genus *Salvelinus* includes the chars and some trouts, a group discussed in detail by Balon (1980). The arctic char, *S. alpinus*, has a circumpolar distribution and occurs further north than any North American freshwater fish (Scott and Crossman, 1973). The Great Lakes populations of the beautifully colored lake trout, *S. namaycush*, have been badly depleted, and in some cases exterminated, by the parasitic sea lamprey *Petromyzon marinus* (Scott and Crossman, 1973). The brook trout (*S. fontinalis*; Fig. 179), Dolly Varden (*S. malma*), and the bull trout (*S. confluentus*) are highly prized angling species.

FIGURE 179. *Salvelinus fontinalis* (Goode, 1884, Plate 192).

Salmo is the genus of Atlantic trouts. There are about five freshwater and anadromous species from northeastern North America and Europe.

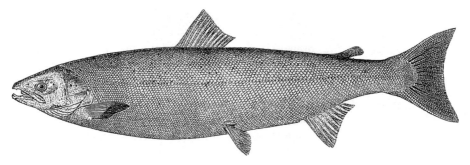

FIGURE 180. *Salmo salar* (Goode, 1884, Plate 186).

The Atlantic salmon, *Salmo salar* (Fig. 180), is an anadromous species that spawns in cold, rocky streams where the young remain for 2 or 3 years before they migrate to the sea. After 1 or more years at sea, the adults return to fresh water to spawn. Atlantic salmon do not usually die after spawning as do Pacific salmon. They can grow to 1.4 m. The life history and fisheries of the Atlantic salmon have been described by Netboy (1968, 1980). The brown trout, *Salmo trutta*, is native to Europe, north Africa, and Asia but has been introduced into North America and the Southern Hemisphere.

There are at least four species of Pacific trouts: Mexican golden trout (*Oncorhynchus chrysogaster*), cutthroat trout (*O. clarki*), Gila trout (*O. gilae*), and rainbow trout (*O. mykiss*) (Behnke, 1992). These trouts are found along the Pacific coast of North America south and inland to Arizona, New Mexico, and Mexico. There are freshwater as well as anadromous populations. The latter do not necessarily die after spawning. See Smith and Stearley (1989), Behnke (1992), and Stearley and Smith (1993) for a phylogenetic tree and discussions of the relationships within the genera of salmonid fishes.

The rainbow trout, *O. mykiss* (Fig. 181), is one of the most highly sought-after and one of the most widely cultured and introduced species in

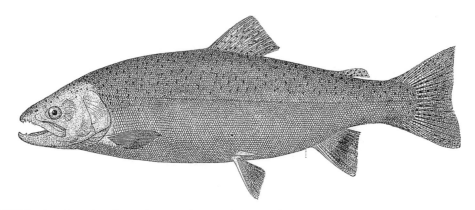

FIGURE 181. *Oncorhynchus mykiss* as *Salmo gairdneri* (Goode, 1884, Plate 187A).

the world. It was previously recognized as *Salmo gairdneri*. The scientific name change is a good example of the self-correcting nature of science. As we learn more, prevailing views change in the face of new evidence. This reevaluation and name change was explained by Smith and Stearley (1989). It was eventually realized that the rainbow trout was the same species as the Asian Kamchatka trout, *Salmo mykiss*. The specific epithet, *mykiss*, dated from 1792 and had priority over the 1836 usage of *gairdneri*. Osteological and molecular studies showed that the Pacific trouts, including the rainbow trout, were more closely related to *Oncorhynchus* than to *Salmo*. Therefore, *S. gairdneri* had to become *O. mykiss* according to the rules of zoological nomenclature.

Pacific salmon of the genus *Oncorhynchus* are composed of about six anadromous species that occur in North Pacific coastal areas from the Arctic Ocean to Japan and California. Most of these species make long spawning migrations from the sea into freshwater streams to lay their eggs and die. Their combined biomass can be so staggeringly large that the decomposing bodies provide an important source of nutrition for terrestrial predators such as bears, wolves, and eagles and fertilizer for riparian plant communities (Willson et al., 1998; Cederholm et al., 1999). After hatching, the young migrate to the sea, where they grow for 1–4 years before reappearing at their natal stream to complete the cycle. Experiments leading to an understanding of the homing of salmon were described by Hasler (1966). The males of some species develop an upturned lower jaw (kype) and a nuchal hump during the spawning migration (Fig. 182). These

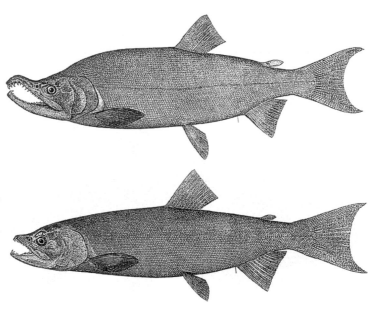

FIGURE 182. *Oncorhynchus nerka*, male (top) and female (bottom) (Goode, 1884, Plate 190).

structures are used in male-to-male combat. Some populations of normally anadromous species may be completely landlocked.

Some specialized terminology is applied to salmonid reproduction. Their gravel nest is termed a redd. An embryo with yolk sac attached is called an alevin. When the yolk sac is absorbed the small fish is known as a fry. A young fish with a series of bars on its sides is known as a parr, and the bars are called parr marks (Fig. 183). Eventually, parrs of anadromous species transform into smolts that are silvery. It is the smolts that migrate to the sea, where they grow into adults. Young, male adults that spawn at their small size are termed jacks. Spent fish are called kelts.

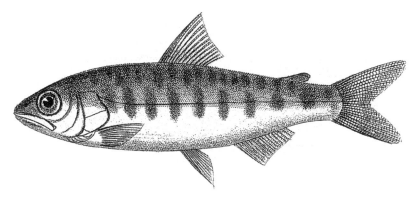

FIGURE 183. *Oncorhynchus mykiss*, parr, as *Fario aurora* (Girard, 1858, Plate LXVIII, Fig. 5).

The Pacific salmons are of great economic value and support both commercial and sport fisheries. The chinook salmon, *O. tshawytscha*, is one of the largest salmonids and may weigh 57 kg. It is now the most common species stocked in the Great Lakes. The coho salmon, *O. kisutch*, also has been introduced into the Great Lakes, where it feeds on the abundant alewife, *Alosa pseudoharengus*. Kokanee, a freshwater form of the sockeye salmon, *O. nerka*, has deep red flesh and feeds on crustaceans. It is one of the most expensive salmons to purchase fresh or canned. The pink salmon, *O. gorbuscha*, and the chum salmon, *O. keta*, are also important commercially and are raised in hatcheries.

The largest salmonid known is the Siberian *Hucho hucho taimen*. It reaches 70 kg and 2 m in length. A commercially caught specimen reportedly weighed 114 kg (Wheeler, 1975). This specimen is often referred to as *H. taimen*, but *H. hucho* has priority between these two closely related forms (R. Behnke, personal communication). Other Eurasian genera include *Brachymystax* (northern Asia to Korea), *Acantholingua* (former Yugoslavia), and *Salmothymus* (former Yugoslavia and Turkey).

The oldest salmonid is *Eosalmo driftwoodensis* known from Eocene freshwater sediments of British Columbia (Cavender, 1986; Wilson and

Williams, 1992; Wilson and Li, 1999). *Oncorhynchus* dates back at least 6 million years to the Miocene of Oregon and California (Nelson, 1994).

A great deal of life history information about various members of the Salmonidae can be found in Scott and Crossman (1973). Groot and Margolis (1991) profiled the lives of Pacific salmon. The phylogenetic implications of salmonid life histories are described by Hutchins and Morris (1985), and an unusual history of salmon in Australia can be found in Clements (1988). In fact, one of the classic studies in fisheries biology was done on trout in New Zealand (Allen, 1951). It is interesting to compare the elegant color drawings of some of these spectacularly beautiful fishes by fish artist Joseph Tomelleri (Behnke, 1992) with the watercolor paintings of Prosek (1996).

Map references: Behnke (1986, 1992*), Berg (1948/1949), Darlington (1957), Lee *et al.* (1980),* Page and Burr (1991),* Rosen (1974),* Scott and Crossman (1973)*

Class Actinopterygii
Subclass Neopterygii

Order Percopsiformes; Suborder Percopsoidei
(1st) Family Percopsidae—**trout–perches**   [per-kop'-si-dē]

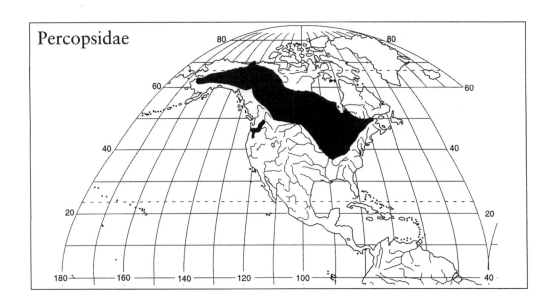

THE NEOTELEOSTEI (ALL FISHES CLASSIFIED ABOVE THE Salmoniformes) is composed of 7 superorders (Nelson, 1994). Most of these are deep-sea and pelagic marine fishes, but two of the superorders, Paracanthopterygii and Acanthopterygii, contain freshwater fish families (Fig. 1). Nelson does not give the Neoteleostei formal taxonomic rank, but its monophyly is reviewed by Fink and Weitzman (1982), Lauder and Liem (1983), Rosen (1985), Johnson (1992), Lecointre and Nelson (1996), and Johnson and Patterson (1996).

The superorder Paracanthopterygii was delimited by Patterson and Rosen (1989), and there is doubt that it is monophyletic. It is thought to be a side branch of teleost evolution, with the acanthopterygian either sharing a common ancestry or being derivatives of a related lineage (Nelson, 1994). Nelson's classification reflects the view that the Paracanthopterygii and the Acanthopterygii are sister groups. Parenti (1993) provided support for the idea that the Paracanthopterygii and Atherinomorpha are sister groups. Nelson included five orders within the Paracanthopterygii: Percopsiformes, Ophidiiformes, Gadiformes, Batrachoidiformes, and Lophiiformes. The latter group contains only marine fishes, most of which occur in deep water.

The order Percopsiformes contains three families of freshwater fishes endemic to North America. These families possess a mosaic of derived and ancestral characters. For example, weak spines in the dorsal and anal fins, cycloid and/or ctenoid scales, and an adipose fin may be present. The pelvic fins are close to but somewhat behind the pectoral fins (subthoracic), and the premaxilla is not protractile. See Gosline (1963) for a summary of the differences between percopsiform, amblyopsid, and gadiform fishes and Murray and Wilson (1999) for reasons to remove the amblyopsids from the Percopsiformes.

The Percopsidae consists of only two species. The trout–perch, *Percopsis omiscomaycus*, is widely distributed from the Yukon to the Atlantic coast and south in the Mississippi Valley to Missouri and Kentucky. The sand roller, *P. transmontana* (Fig. 184), has a much more restricted distribution in the Columbia River basin in western Idaho, southern Washington, and Oregon.

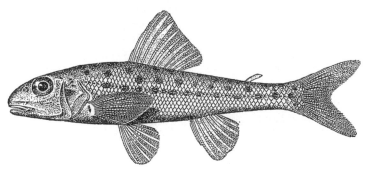

FIGURE 184. *Percopsis transmontana* as *Colombia transmontana* (Jordan and Evermann, 1905, p. 247).

The two species in this family have a large, naked head, subthoracic pelvic fins, cycloid and ctenoid scales, and an adipose fin. Weak spines are present in the dorsal, anal, and pelvic fins. The strange mixture of trout-like adipose fin with perch-like fin spines and ctenoid scales is responsible for the common name, trout–perch. Both species of *Percopsis* have chambers in the lower jaws and cheeks known as "pearl organs."

Percopsis omiscomaycus can reach 20 cm and inhabits lakes and deep pools of rivers usually over sand. It has a complete lateral line. The preopercle has few or no small spines. Trout–perch spawn in spring, and their diet consists of aquatic insect larvae. It is an important forage species for pike, trout, walleye, and perch. *Percopsis transmontana* prefers quiet backwaters of tributaries to the Columbia River with undercut banks, submerged tree roots, and debris. It often occurs in waters deeper than 15 m. The sand roller may occupy deeper waters during the day and shallow waters during the night (R. S. Wydoski, personal communication). It is smaller than the trout–perch and reaches only about 9.6 cm. Its back is humped, and the lateral line is incomplete. There are a few large spines on the preopercle. Its biology is poorly known. Fossil percopsids date from the late Paleocene of Alberta (Murray and Wilson, 1996), and two genera are known from the Eocene Green River Formation (Cavender, 1986; Grande, 1984).

See Magnuson and Smith (1963) and Scott and Crossman (1973) for life history details of P. omiscomaycus. Wydoski and Whitney (1979) discuss P. transmontana.

Map references: Lee *et al.* (1980)*, Miller (1958)*, Nelson (1976)*, Page and Burr (1991)*, Scott and Crossman (1973)*

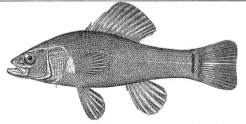

Class Actinopterygii

Subclass Neopterygii

FIGURE 185. *Aphredoderus sayanus* (Jordan and Evermann, 1900, Plate CXXII, Fig. 331).

Order Percopsiformes; Suborder Aphredoderoidei

(1st) Family Aphredoderidae—**pirate perch** [af'-rē-dō-der'-i-dē]

THIS MONOTYPIC FAMILY, COMPOSED OF *APHREDODERUS sayanus* (Fig. 185), is found in the Mississippi River valley and along the

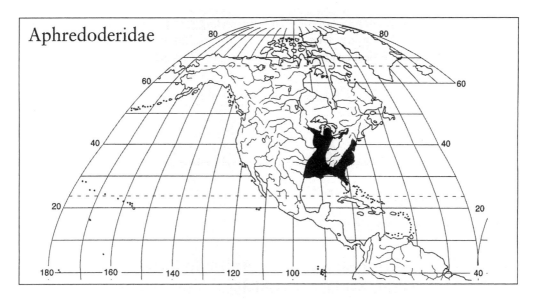

Aphredoderidae

Gulf and Atlantic coastal plains from eastern Texas to New York. There are disjunct populations in Lake Erie and Lake Ontario drainages. Boltz and Stauffer (1993) recognized two subspecies, an Atlantic slope form and a Gulf of Mexico–Mississippi Valley–Great Lakes form. Several extinct species date to the Oligocene or late Eocene of western North America (Cavender, 1986). The preferred habitat of *Aphredoderus* is sluggish water, ponds, oxbows, marshes, and other places that have bottoms of decomposing organic matter and are heavily vegetated.

Pirate perch have ctenoid scales, a large mouth, deep body anteriorly, three or four dorsal spines, two or three anal spines, and one spine in the subthoracic pelvic fin. During development the third soft anal ray becomes a hard spine. Thus, juveniles have two anal spines and adults have three. The most peculiar characteristic of pirate perch is the migration of the anal opening from the normal position (just anterior to the anal fin) in juveniles to the throat region between the gills (jugular) in adults. The name "Aphredoder-" means "excrement throat" (Jenkins and Burkhead, 1994). Mansueti (1963) studied the ontogeny of the young. Unlike the Percopsidae, they lack an adipose fin. The lateral line is incomplete or absent. There is a strong spine near the upper edge of the opercle. The swim bladder is not connected to the gut (physoclist). Maximum size is about 14 cm.

The reproductive biology of pirate perch is not well-known. Martin and Hubbs (1973) suggested that the eggs may be incubated in the gill cavity and noted that when eggs were artificially stripped from a female, they moved along a groove into the branchial chamber. The jugular location of the urogenital pore would facilitate this (Pflieger, 1975). In addition, some cavefish (Amblyopsidae) relatives of the pirate perch are buccal incubators. Bolts and Stauffer (1986) reported the presence of three eggs in the branchial chamber of a preserved female pirate perch. This remains a topic for further study. Diet consists of insects, small invertebrates, and small fishes. See Becker (1983) and Jenkins and Burkhead (1994) for life history details.

Map references: Boltz and Stauffer (1993)*, Lee *et al.* (1980)*, Miller (1958)*, Page and Burr (1991)*

Class Actinopterygii
Subclass Neopterygii

Order Percopsiformes; Suborder Aphredoderoidei
(1st) Family Amblyopsidae—**cavefishes**   (am'-blē-op'-si-dē)

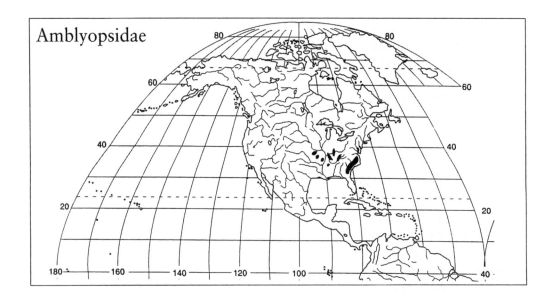

THE CAVE-, SPRING-, AND SWAMP-DWELLING FISHES IN THIS interesting family are endemic in fresh waters of southern and eastern United States in unglaciated, limestone regions. There are four or five genera and six species, and each has its own distribution disjunct from the others; however, *Amblyopsis spelaea* and *Typhlichthys subterraneus* occur syntopically (Page and Burr, 1991).

Amblyopsis rosae, the Ozark cavefish, is presently known from 21 caves distributed over seven counties in three states: northwestern Arkansas, northeastern Oklahoma, and the southern corner of Missouri. The verified historic range was slightly larger (Romero, 1998a; Cooper,

1980). *Amblyopsis spelaea,* the northern cavefish, is found in subterranean waters of thousands of caves from the Mammoth Cave area of Kentucky north into southern Indiana (Romero and Bennis, 1998). This is the largest species in the family at about 105 mm SL.

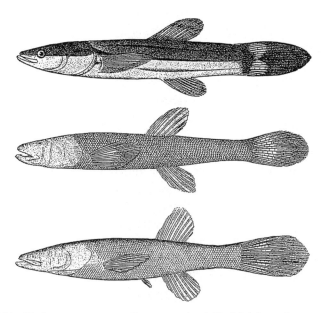

FIGURE 186. *Chologaster cornuta* as *C. cornutus* (top), *Typhlichthys subterraneus* (middle), and *Amblyopsis spelaeus* (bottom) (Jordan and Evermann, 1900, Plate CXV, Figs. 305–307.

Chologaster cornuta (Fig. 186), the swampfish, lives in swamps, ponds, ditches, and sluggish streams in the Atlantic coastal plain from southeast Virginia to east-central Georgia. *Chologaster agassizi,* the spring cavefish, lives in caves and springs from south-central Tennessee through south-central and western Kentucky to southern Illinois (Cooper, 1980). Page and Burr (1991), Mayden *et al.* (1992), and Etnier and Starnes (1993) referred to this species as *Forbesichthys agassizi.* These two surface-dwelling species are dark brown, whereas the other amblyopsids are pinkish due to the visibility of blood through the pigmentless skin.

Speoplatyrhinus poulsoni, the Alabama cavefish, is known only from subterranean waters in the north bank of the Tennessee River at the type locality west of Florence, Alabama, in Key Cave, Lauderdale County (Mettee et al., 1996). It was discovered in 1970 (Cooper and Kuehne, 1974). Only nine specimens are known, and it is protected as an endangered species since this is one of the rarest and most endangered species of fishes in the world (Romero, 1998c). *Typhlichthys subterraneus,* the southern cavefish, has a disjunct distribution both east and west of the Mississippi River in underground waters of the Ozark plateau of central and southeastern Missouri and northeastern Arkansas and in northwest

Alabama, northwest Georgia, central Tennessee, Kentucky, and southern Indiana (Cooper, 1980; Romero, 1998b).

Amblyopsids have small or rudimentary eyes. Four species are blind. Only the two species of *Chologaster* have functional eyes. Pelvic fins are absent except in *A. spelaea*, which has small, abdominal pelvics. Scales are cycloid and embedded. The depressed head is naked and has rows of sensory papilla that may extend onto the body and caudal fin. These tactile organs compensate for the lack of sight. The anus and urogenital openings are jugular as in the Aphredoderidae, and they may be buccal incubators (Poulson, 1963).

The species of amblyopsids represent an evolutionary sequence from surface dwelling to spring living and to cave living that shows a reduction in eyes, pigment, and pelvic fins. This seems to be correlated with the amount of time each species has been linked to cave dwelling (Poulson, 1963). Swofford *et al.* (1980) studied genetic differentiation in cavefishes, and Poulson (1986) speculated on evolutionary reductions by neutral mutations.

Jenkins and Burkhead (1994) suggested various hypothesis to explain the disjunct distribution of this family and gave life history information on *C. cornuta*. It feeds on midge larvae, amphipods, and cladocerans and lives only about 14 or 15 months. Pouslon and White (1969) reported that *C. cornuta* is a buccal incubator. Pflieger (1975) listed copepods as the principal food of *A. rosae* and reported that about 70 eggs are incubated in the gill chamber by the female. See references cited by Cooper (1980), Etnier and Starnes (1993), and Mettee *et al.* (1996) for further life history information.

Woods and Inger (1957), Rosen (1962), and Patterson and Rosen (1989) examined the systematics of this family. Amblyopsids have not been reported from the fossil record (Cavender, 1986; Wilson and Williams, 1992). Murray and Wilson (1999) recommended the removal of the Amblyopsidae from the Percopsiformes in order to make the latter monophyletic. They considered the Amblyopsidae to warrant recognition within its own order, Amblyopsiformes, within the Anacanthini.

Map references: Cooper (1980),* Etnier and Starnes (1993),* Jenkins and Burkhead (1994),* Mettee *et al.* (1996),* Page and Burr (1991),* Pflieger (1975),* Robinson and Buchanan (1989),* Woods and Inger (1957)*

Class Actinopterygii
Subclass Neopterygii

Order Ophidiiformes; Suborder Bythitoidei
(Per) Family Bythitidae—**viviparous brotulas**   [bī-thit'-i-dē]

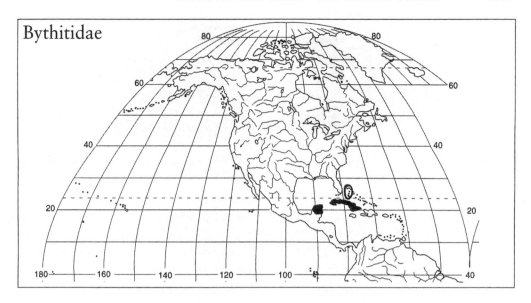

Bythitidae

THIS IS A MOSTLY MARINE FAMILY (ATLANTIC, INDIAN, AND Pacific Oceans) of 33 genera and about 100 species that has about 5 species in fresh or slightly brackish waters of caves, sinkholes, and blue holes in Cuba, the Bahamas, and the Yucatan peninsula of Mexico. It is this fresh-water distribution that is shown on the map.

This family gives birth to live young, which is unusual for a paracan-thopterygian group. The males have a clasper-like copulatory organ. The subfamily Brosmophycinae contains the few freshwater forms. Their caudal fin is separate from the long-based dorsal and anal fins, whereas in the other subfamily, Bythitinae, the fins are confluent.

Lucifuga is a genus of six or more species (Cohen and McCosker, 1998), four of which are blind or nearly blind species found in limestone caves and sinkholes in Cuba and the Bahamas (Burgess, 1983; Nelson, 1994). *Lucifuga subterranea*, *L. teresinarum*, *L. simile*, and *L. dentata* (Fig. 187) are Cuban and *L. spelaeotes* is from the Bahamas (Cohen and Robins, 1970; Cohen and McCosker, 1998). Eigenmann (1909b) reported

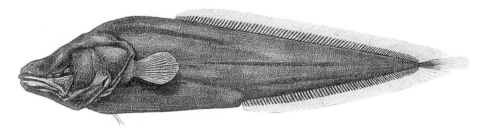

FIGURE 187. *Lucifuga dentata* as *Stygicola dentatus* (Eigenmann, 1909, Plate 13).

FIGURE 188. *Ogilbia pearsei* as *Typhlias pearsei* (reproduced with permission from Hubbs, 1938, Plate 3).

on the eyeless troglodytes from Cuba. *Ogilbia pearsei* (Fig. 188) occurs in freshwater caves of the Yucatan peninsula (Hubbs, 1938) and other species live in brackish water in the Galapagos Islands (Cohen and Nielsen, 1978). *Lucifuga* and *Ogilbia* are not known to occur together.

Map references: Burgess (1983b),* Cohen and McCosker (1998), Cohen and Nielsen (1978), Hubbs (1938)

Class Actinopterygii
Subclass Neopterygii

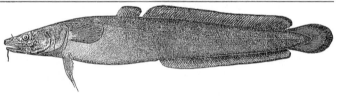

FIGURE 189. *Lota lota* as *L. maculosa* (Goode, 1884, Plate 61).

Order Gadiformes
(Per) Family Gadidae—cods   (gad'-i-dē) [gā'-di-dē]

THE COD FAMILY IS A MOSTLY BOTTOM-DWELLING, MARINE group from cold waters of the Arctic, Atlantic, and Pacific Oceans in both the Northern and Southern Hemispheres. There is one holarctic freshwater species, the burbot, *Lota lota* (Fig. 189). The burbot's distribution is shown on the map.

Many cods have an elongate body with a dorsal fin divided into two or three sections; the most posterior section is separate from the caudal fin. Others are not notably elongate and a few are deep bodied. There are no fin spines. A chin barbel is present in most species, and the swim bladder is physoclistous. The scales are cycloid, and the pelvic fins are jugular or thoracic. The classification of gadiform fishes is unsettled. See Cohen (1984), Markle (1989), Howes (1989, 1991b), and Cohen *et al.* (1990) for varying opinions.

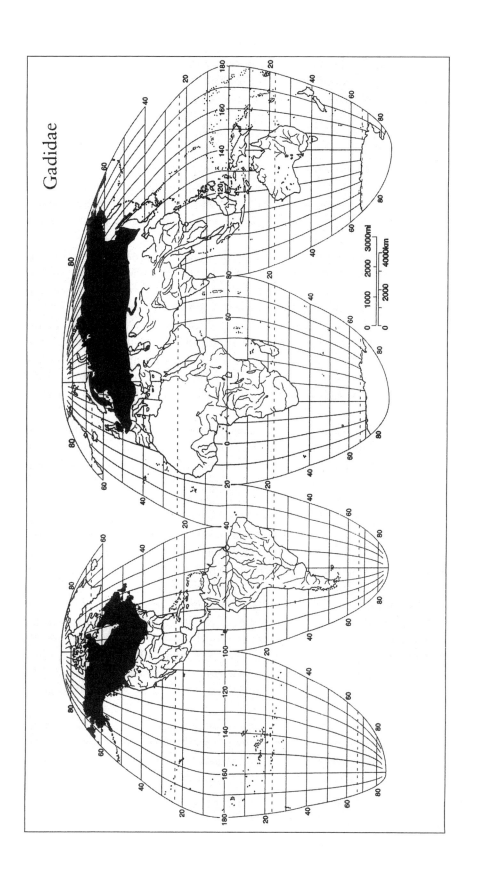

Gadidae

The cod family includes 15 genera with about 30 species (Nelson, 1994) and is, or was, of enormous economic importance. It is second only to the clupeids in the commercial value of its flesh and fish meal. *Gadus morhua*, the Atlantic cod, has been fished for more than 300 years by Europeans and North Americans. Although *G. morhua* can reach 1.8 m and 90 kg, fish more than 10 kg are rare today. Recent landing amounted to about 2 million metric tons each year. More than 6 million metric tons of Alaska pollock, *Theragra chalcogramma*, was landed annually. These fish produce vast numbers of eggs per female. For example, a single pollock female may spawn 15 million eggs, and a cod may produce up to 7 million eggs.

However, even this fantastic level of fecundity cannot support the intense overfishing that has occurred. Total catches and individual fish sizes are getting smaller, and the fisheries are collapsing (Hutchings and Myers, 1994). Cod on the northern Grand Banks off northeastern Newfoundland have declined below 1% of the levels observed in the late 1980s (Myers et al., 1997). The incredibly rich Grand Banks and Georges Bank fisheries of eastern North America are now essentially closed following their collapse (Safina, 1995). Safina (1997) sounded the alarm that the world's fisheries cannot be sustained at the current rate of exploitation.

The burbot, *L. lota*, is the only freshwater species of cod. Nelson (1994) considered it in the subfamily Lotinae, which Cohen (1984), Dunn (1989), and Markle (1989) treated as a family. Burbot live in cold, deep water in lakes and large rivers. They range throughout northern Eurasia south to about 45°N. In North America burbot occur across Alaska, Canada, and the northern United States dipping as far south as Kentucky in the Midwest (Page and Burr, 1991). Clay (1975) speculated that the burbot in the lower Ohio River system may be escapees from stocked lakes in Indiana and Ohio, strays from the Missouri River or upper Allegheny River system, or may represent a native population.

In lakes, burbot may be found as deep as 200 m. An unusual feature of burbot biology is that they spawn in midwinter, under ice, at night (Scott and Crossman, 1973). Maximum size is about 1.2 m and 34 kg. Young burbot feed on invertebrates and adults more than 50 cm feed voraciously on fishes. Its flesh is not as highly esteemed as other cods (Trautman, 1981). The oldest fossil *Lota* dates to the lower Pliocene of Austria (Cavender, 1986).

The Atlantic tomcod, *Microgadus tomcod*, enters rivers on the Atlantic coast of North America from Labrador to Virginia, where some populations have become landlocked in freshwater lakes. Like the burbot, tomcod spawn in midwinter under ice (Scott and Crossman, 1973).

Map references: Berg (1948/1949), Cohen *et al.* (1990),* Lee *et al.* (1980), Maitland (1977),* Nelson (1976),* Page and Burr (1991),* Scott and Crossman (1973),* Svetovidov (1962)*

Class Actinopterygii
Subclass Neopterygii

Order Batrachoidiformes
(Per) Family Batrachoididae—toadfishes   (bat'-ra-koi'-di-dē)

THE BATRACHOIDIDAE IS A COASTAL, BENTHIC MARINE FAMILY
that occurs in the Atlantic, Indian, and Pacific Oceans. There are 19 genera
and 69 species (Nelson, 1994). The greatest diversity is in tropical America.
It is one of the few tropical marine families that does not have its greatest
diversity in the Indo-Pacific region (Collette and Russo, 1981). There are 4
freshwater species in South America (Collette, 1995b), and a few other
species may enter river mouths (Greenfield, 1997).

The family is characterized by a large head with dorsally positioned
eyes and most genera have a scaleless body with multiple lateral lines, but
some genera, such as *Batrachoides*, are scaled. The soft dorsal and anal fins
are long and are preceded by a small spiny dorsal fin of two or three short,
sharp spines. Only three pairs of gills are present, whereas most other fishes
have four pairs. The pelvic fins are jugular. Fleshy flaps and barbels
surround the mouth of many species. Toadfishes are well camouflaged
against the bottom habitat. There are three subfamilies: Batrachoidinae,
Porichthyinae, and Thalassophryninae.

The Batrachoidinae is found off the coasts of the Americas, Africa,
Europe, southern Asia, and Australia. *Batrachoides goldmani* has a fresh-
water distribution in the same river systems in Mexico and Guatemala as
Hyporhampus mexicanus and *Strongylura hubbsi* (Belonidae) (Collette and
Russo, 1981; Collette, 1974a). *Batrachoides gilberti* frequently enters fresh
waters of Central America (Greenfield and Thomerson, 1997).
Potamobatrachus trispinosus occurs in the Rio Tocantins near Jatobal,
Pará, Brazil. It is the only genus of toadfishes with three subopercular spines
(Collette, 1995b). *Allenbatrachus grunniens* occurs in the Ganges River
(Greenfield, 1997) and *A. reticulatus* lives in fresh water near Singapore
(D. W. Greenfield, personal communication).

Porichthyinae lives in the eastern Pacific and western Atlantic. These
fishes are well-known for the noise emitted by males with their swim bladders during courtship. Fine (1997) reviewed sound production in *Opsanus
tau* and *Porichthys notatus*. The sustained low-frequency hum produced by
midshipmen (*P. notatus*) in San Francisco Bay was loud enough to disturb
houseboat dwellers (Ibara et al., 1983).

The Thalassophrynine from the eastern Pacific and western Atlantic is
distinguished from the other two subfamilies by the absence of a lateral line

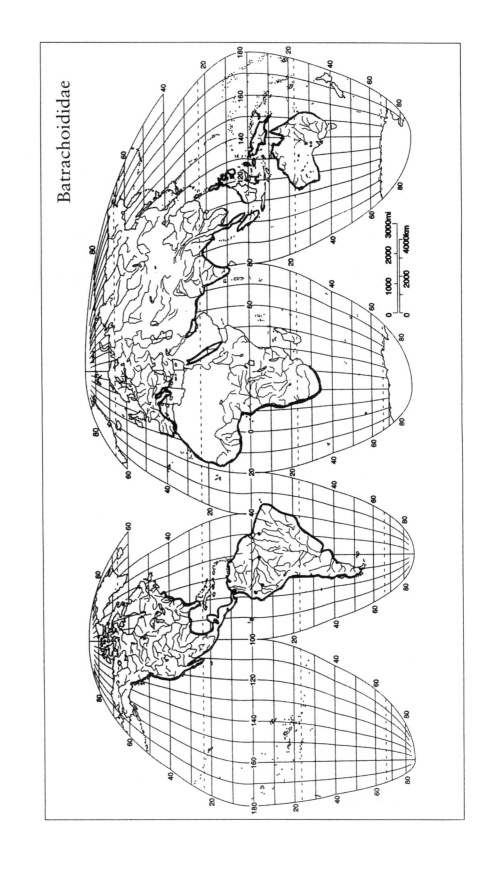

Batrachoididae

or by the presence of a single lateral line, two hollow dorsal spines, and a hollow opercular spine. These spines are connected to venom glands. This is the only group of venomous fishes to have hollow spines for the delivery of venom. Other groups have less advanced grooved spines. The Thalassophrynine is convergent with other venomous fishes such as the stonefishes (Scorpaeniformes) and weeverfishes (Perciformes) (Helfman et al., 1997). The Batrachoidinae and the Porichthyinae have three and two solid dorsal spines, respectively, and are nonvenomous.

There are two genera of venomous toadfishes: *Daector* (five species) and *Thalassophryne* (six species) (Collette, 1966a, 1973). *Daector* includes four marine species from the tropical eastern Pacific and *D. quadrizonatus* from freshwater tributaries of the Atrato River (Caribbean Sea drainage) in Colombia. *Thalassophryne* is found along the western Atlantic coast of Panama and South America except for *T. amazonica* (Fig. 190), which is known only from the Amazon River.

FIGURE 190. *Thalassophryne amazonica* (by Mildred H. Carrington from Collette, 1966; reproduced with permission).

Map references: Collette (1966a,* 1973, 1995b), Collette and Russo (1981),* Springer (1982)

Class Actinopterygii
Subclass Neopterygii

Order Mugiliformes
(Per) Family Mugilidae—**mullets** (mū-jil'-i-dē)

THE SUPERORDER ACANTHOPTERYGII, AS THE NAME DENOTES (acantho = spine, pterygion = fin), includes most of the spiny finned fishes.

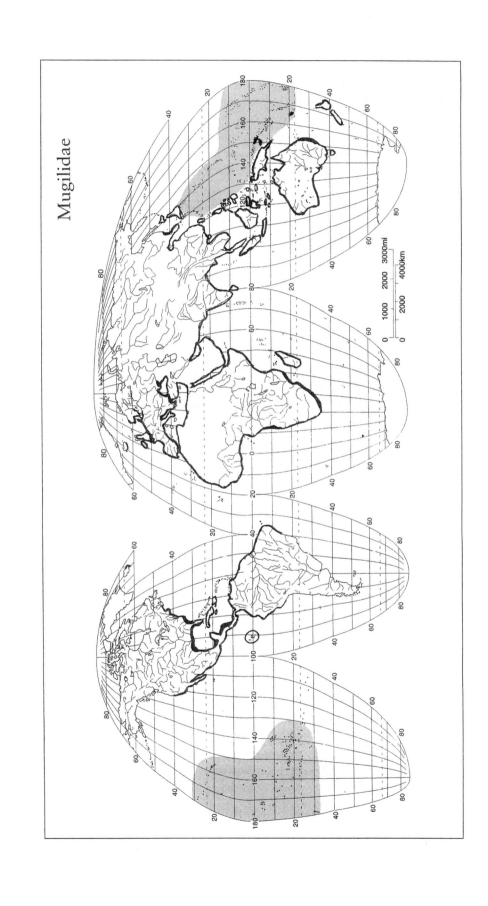

Mugilidae

Fishes in this superorder have a more mobil and/or protrusible upper jaw than fishes below this level of classification (Lauder and Liem, 1983). Thirteen orders and 251 families in three series—Mugilomorpha, Atherinomorpha, and Percomorpha—make up the Acanthopterygii (Nelson, 1994; Johnson and Patterson, 1993) (Fig. 1).

The order Mugiliformes includes only one family—Mugilidae, the mullets. Current thinking about mullet classification considers them related to the Atherinomorpha (Stiassny, 1990a, 1993; Johnson and Patterson, 1993). However, Parenti (1993) concluded that the atherinomorphs (without the mullets) are related to the Paracanthopterygii. Older classifications (Gosline, 1968, 1971) placed the mullets as the most primitive perciform family and in the same perciform suborder as the Polynemidae, Sphyraenidae (barracudas), Melanotaeniidae, and others.

Mullets are a catadromous, cosmopolitan, coastal marine family with several species that enter rivers throughout the world. There are about 17 genera and approximately 80 species (Harrison and Howes, 1991), but more than half of the species are assigned to either *Liza* or *Mugil*. Most of the other genera have one or two species each.

Mullets have two widely separated dorsal fins. The first dorsal consistently has four spines, one distinct and three very closely placed. The subabdominal pelvic fins have one spine, as does the anal fin. The typical mullet body is fusiform with a forked tail and large, weakly ctenoid scales. A few species have truncate or emarginate tails and some have cycloid scales. The lateral line is absent or very faint. Adipose eyelids may be present in some genera. Mullets usually travel in lively schools and feed on plankton and detritus which they extract and process from the bottom sediment via sieve-like gill rakers, a gizzard-like stomach, and very long intestinal tract. Adults range from 30 to 90 cm and may be of considerable commercial importance in some areas. In the Orient, they are raised in aquaculture.

The striped or gray mullet, *Mugil cephalus* (Fig. 191), has a worldwide circumtropical distribution in coastal waters. It often ascends rivers to the

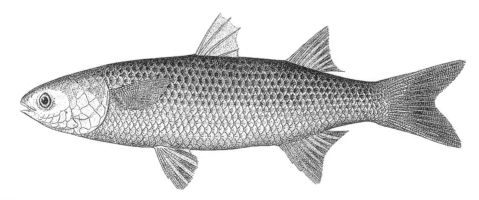

FIGURE 191. *Mugil cephalus* as *M. albula* (Goode, 1884, Plate 179).

fall line and is known to ascend the Mississippi River to Illinois (L. M. Page, personal communication). Most of the coastal distribution shown on the map is for this species. The mountain mullet, *Agonostomus monticola*, occurs along the American Atlantic coast from North Carolina to Venezuela. Adults and subadults often ascend coastal streams right to the headwaters. In Australia many species of mullet enter rivers. For example, the freshwater mullet, *Myxus petardi*, occurs in southeastern coastal drainages from the Burnett River of Queensland south to the Georges River south of Sydney (Allen, 1989; Thomson, 1996). In New Guinea, *Cestraeus goldiei* adults have been recorded far inland in fast-flowing streams at an altitude of 350 m and are presumed to breed there (Allen, 1991).

Liza abu is known from fresh waters and estuaries of Iraq, Pakistan, and the Indus River (Talwar and Jhingran, 1992). In addition to *Mugil cephalus*, *Myxus capensis* and *Liza macrolepis* enter east coastal rivers of southern Africa (Skelton, 1993). *Agonostomus telfairii* is only known from fresh waters of Madagascar, Mauritius, and Reunion Islands (Thomson, 1986). See Schultz (1946) for a taxonomic review and Thomson (1964) for a bibliography of systematic references.

Map references: Allen (1989,* 1991), Ben-Tuvia (1986b),* Berg (1948/1949), Bussing (1998)*, Lee *et al.* (1980),* Maitland (1977),* Skelton (1993),* Talwar and Jhingran (1992), Thomson (1986, 1996*)

Class Actinopterygii
Subclass Neopterygii

Order Atheriniformes; Suborder Bedotioidei
(2d) Family Bedotiidae—**bedotiids**
(be-dō-tī'-i-dē) [be dō-tē'-i-dē]

THE SERIES ATHERINOMORPHA INCLUDES THREE ORDERS: THE Atheriniformes, Beloniformes, and Cyprinodontiformes (Fig. 1). Nelson (1994) reviewed the history of this classification. The order Atheriniformes has not been demonstrated to be monophyletic (Rosen and Parenti, 1981; Parenti, 1984a, 1993; Lauder and Liem, 1983; Stiassny, 1990a). As currently structured, it includes eight families.

Atheriniform fishes are mostly freshwater, surface-dwelling, silvery fishes with two dorsal fins. The first dorsal has weak spines. Both cycloid and ctenoid, easily shed (deciduous) scales are present in this group. The lateral line may be entirely absent or weakly developed.

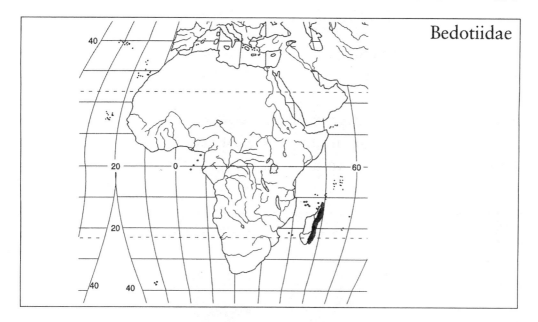

Bedotiidae

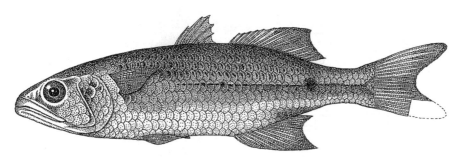

FIGURE 192. *Rheocles wrightae* (by Patricia Wynn from Stiassny, 1990; reproduced with permission).

The Bedotiidae is endemic to inland waters in the eastern forests of Madagascar (Stiassny and Raminosoa, 1994). Two genera, *Bedotia* (four species) and *Rheocles* (five species, represented by *R. wrightae*; Fig. 192), have been described (Stiassny, 1990a; Stiassny and Reinthal, 1992; Stiassny and Raminosoa, 1994). They occur from coastal lowlands up to an elevation of about 500 m (Allen, 1994). *Rheocles* has a restricted range limited to heavily forested streams of the central and eastern highlands (Stiassny, 1990a). Bedotids are small (about 80 mm), often colorful, and resemble melanotaeniids from Australia and New Guinea. This may be an example of convergent evolution between species of two areas (Madagascar and Australia) whose freshwater fish fauna is largely derived from marine fishes (Allen, 1994). A bedotiid characteristic is that the last six or seven vertebral centra and their neural and hemal spines are thickened (Stiassny, 1990a).

Map references: Stiassny (1990a),* Stiassny and Raminosoa (1994), Stiassny and Reinthal (1992)*

Class Actinopterygii

Subclass Neopterygii

Order Atheriniformes; Suborder Melanotaenioidei
(2d) Family Melanotaeniidae—**rainbowfishes**
(mel'-a-nō-tē-nī'-i-dē) [mel'-a-nō-tē'-nē-i-dē]

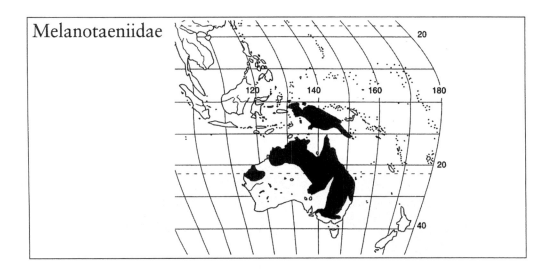

RAINBOWFISHES ARE SMALL (5–12 CM), COLORFUL FISHES OF the fresh waters of northern and eastern Australia and New Guinea. The Melanotaeniidae is the only freshwater fish family solely confined to this region. Rainbowfishes do not occur west of Weber's line, nor do they reach the Bismark Archipelago or the Solomon Islands to the east. Torres Strait, which separates Australia and New Guinea, is a recent feature. A land bridge occupied that area until about 7000 years ago when rising sea level created the water barrier (Allen and Cross, 1982). The Melanotaeniidae is the second largest atheriniform family with seven genera and 68 species, including 13 Australian species and 55 New Guinean species, of which 3 are shared by both regions (Allen, 1994, 1995; Allen and Renyaan, 1998).

Rainbowfishes have compressed bodies with a weakly developed or absent lateral line and large scales. The two dorsal fins are separated. There are three to seven spines in the first dorsal. The second or soft dorsal is long, with up to 22 rays. The anal fin is also long, with one spine and 10–30 rays. Melanotaeniids can be separated from atherinids by the presence of a membranous attachment along the length of the innermost pelvic fin ray

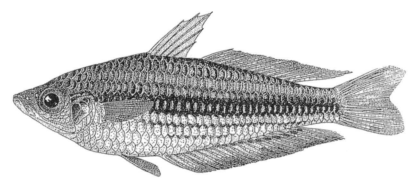

FIGURE 193. *Melanotaenia ogilbyi* (Weber, 1913, Fig. 28).

and the abdomen. Head and jaw shape and the fine conical teeth provide important diagnostic characteristics (Allen and Cross, 1982).

Melanotaenia, represented by *M. ogilbyi* (Fig. 193), is the largest genus with 46 species in Australia and New Guinea (Allen, 1996a, 1997; Allen and Renyaan, 1998). *Melanotaenia maccullochi* and *M. splendida* (with 5 subspecies) are shared by both regions, as is *Iriatherina werneri* (Allen, 1991; Allen and Renyaan, 1998). Males of the latter species possess strikingly elongated fin filaments. Crowley and Ivantsoff (1991) compared various populations of *Melanotaenia splendida* and found little genetic difference among them. *Cairnsichthys rhombosomoides* from northern Queensland and *Rhadinocentrus ornatus* from southern Queensland and northern New South Wales are endemic to eastern Australia (Allen, 1989). *Glossolepis* (7 species) and *Chilatherina* (10 species; Allen and Renyaan, 1996, 1998) are restricted to northern New Guinea except for *Chilatherina campsi*, which is found on both sides of the Central Dividing Range.

The recently described *Pelangia mbutaensis* is known only from the Mbuta basin near Etna Bay on the south coast of Irian Jaya. It is believed to be the sister group of *Glossolepis* from northern New Guinea and the ancestral population was probably isolated when the Vogelkop collided with the main part of New Guinea (Allen, 1998b).

Melanotaeniids are schooling fishes and eat a variety of foods, including algae, aquatic insects, and small crustaceans. They provide valuable mosquito control. In turn, they are eaten by waterfowl and larger fishes. Rainbowfish inhabit streams, lakes, and swamps, usually lower than 1500 m. Spawning is stimulated by the rainy season but occurs year-round. Eggs are laid in aquatic vegetation, and they hatch within 7–18 days. Most species attain sexual maturity within 1 year. Larval characteristics of two sympatric species have been studied by Aarn *et al.* (1997).

Rainbowfishes exhibit sexual dimorphism, with males having more elaborate fins and brighter colors. Because they are so beautiful and relatively easy to breed, melanotaeniids have become aquarium favorites. They are the only Australian native fishes to have an impact at the hobbyist level (Allen, 1989).

Allen (1980) reviewed the family and Allen and Cross (1982) described the biology of all species to that date. Allen (1995) discussed aquarium biology, and Allen (1996b) provided a list of melanotaeniids from Irian Jaya (formerly West Irian, the Indonesian-controlled half of the island of New Guinea). The 42 species of *Melanotaenia* recognized at the time of publication and their distribution are listed by Allen (1997).

There is a record of three *Melanotaenia* taken from a sandbar in the Missisippi River in Randolph County, Illinois, about 88 km upstream from St. Louis, in 1930 (Fuller et al., 1999). These fish probably escaped from a pet store or were released by a hobbyist.

Map references: Allen (1989, 1991, 1995,1996a, 1998b),* Allen and Cross (1982)*

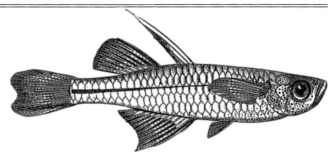

Class **Actinopterygii**
Subclass **Neopterygii**

FIGURE 194. *Pseudomugil novaeguineae* as *P. novae-guineae* (Weber, 1913, Fig. 27).

Order **Atheriniformes; Suborder Melanotaenioidei**
(2d) Family **Pseudomugilidae—blue eyes**   [sū'-dō-mū-jil'-i-dē]

PSEUDOMUGILIDS ARE TINY, DELICATE, COLORFUL FISHES THAT occur in fresh and brackish waters of southern New Guinea and eastern and northern Australia. Their body form is represented by *Pseudomugil novaeguineae* (Fig. 194). They are rarely found in marine waters (Allen, 1991). Blue eyes are so called because of the striking blue color of the iris. They are close relatives of the melanotaeniids and were previously classified with them but have been given family rank by Saeed *et al.* (1989). Dyer and Chernoff (1996) consider pseudomugilids as a subfamily of the Melanotaeniidae. Ivantsoff *et al.* (1996) distinguish the Telmatherinidae, Melanotaeniidae, and Pseudomugilidae. Most adults are less than 5 cm.

There are three genera with 16 species. *Pseudomugil* has 14 species. Eleven species are found in New Guinea and nearby islands (Allen, 1991;

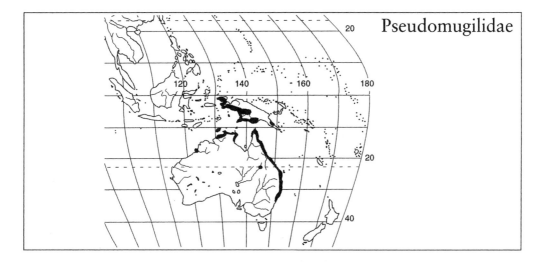

Allen *et al.*, 1998), and 6 species are known from Australia (Allen, 1989). Some species are shared by both New Guinea and Australia. The beautiful and widespread *Pseudomugil signifer* occurs in eastern coastal streams of Australia from Cape York peninsula to Narooma south of Sydney (W. Ivantsoff, personal communication). Aarn *et al.* (1997) reported on its larval development.

Kiunga ballochi is only known from rain forest streams in the upper Fly River of Papua New Guinea (Allen, 1991). It has a vestigial first dorsal fin with only two or three spines, and there is a keel between the pelvic and anal fins. *Scaturiginichthys varmeilipinnis* is known only from five spring-fed pools on a cattle station in arid Lake Eyre drainage, western Queensland (Ivantsoff et al., 1991). This tiny fish is only about 26 mm TL and has reduced or absent pelvic fins. Males have red-edged dorsal and anal fins. Spawning in captivity has been reported by Unmack and Brumley (1991).

In general, blue eyes resemble their melanotaeniid relatives, but blue eyes tend to have more slender bodies. Two dorsal fins are present, but the spines in the first dorsal and anal fin are not rigid. The terminal or oblique mouth is very small, and the pectoral fins are set high on the body. Cycloid scales are present, and a lateral line is absent. Males may be brightly colored, and their fins have elongate filaments that are utilized in elaborate courtship displays. Most pseudomugilids form large schools, and spawning occurs sporadically throughout the year in aquatic vegetation. Females deposit three to nine adhesive eggs that hatch in 10–21 days. Variation depends on temperature and which species is involved. Sexual maturity occurs in 1 year. Their habitat is similar to that of rainbow-fishes and includes drainage ditches, streams, lakes, and ponds. Blue eyes feed on microcrustaceans and insects (Allen, 1991). They are valued as aquarium fishes due to their beauty, small size, and peaceful disposition.

Allen and Cross (1982) and Allen (1995) discussed the biology of the various species.

Map references: Allen (1989, 1991)*, Allen and Cross (1982)*, Ivantsoff *et al.* (1991), Saeed *et al.* (1989)*

Class Actinopterygii
Subclass Neopterygii

Order Atheriniformes; Suborder Atherinoidei
(Per) Family Atherinidae—silversides
(ath'-er-in'-i-dē) [a-thur-rī'-ni-dē]

THE SILVERSIDES ARE A MARINE FAMILY WITH SOME SPECIES IN fresh water, especially in the Americas and Australia. There are about 25 genera and 160 species, with about 50 freshwater species (Nelson, 1994). White (1985) recognized six subfamilies, two of which are treated as families by Nelson (1994) and in this book. The other four subfamilies are the Atherinopsinae and Menidiinae (which make up the New World silversides) and the Atherioninae and Atherininae (Old World silversides). Saeed *et al.* (1994) separated the Old and New World atherinids, and Dyer (1998) treated the New World silversides (Menidiinae and Atherinopsinae) as a family—the Atherinopsidae.

Atherinids are compressed, elongate, silvery fishes that occur in large schools over shallow waters throughout the world. They have two separated dorsal fins. The first dorsal has flexible spines. There is one spine in the anal fin. Scales are usually cycloid, and the lateral line is absent. There is often a silvery lateral band. The mouth is terminal and almost always protrusible, and most species are planktivorous. The pelvic fins are abdominal, and the pectoral fins are high on the body. The eyes are large. Maximum size is about 60 cm but most are less than 15 cm. Silversides are egg layers. Males are generally smaller than females. Atherinids are important food, forage, and bait species in various parts of the world.

In North America, the brook silverside, *Labidesthes sicculus*, ranges from the upper Mississippi and Great Lakes basin to the Gulf coastal plain from Texas to Florida and up the Atlantic slope to South Carolina (Lee, 1980). It is the only true freshwater atherinopsid in North America. Grier *et al.* (1990) reported that brook silverside males have a genital palp that is used to transmit sperm to females, and that females were found carrying embryos in their ovarian cavities. This is the first report

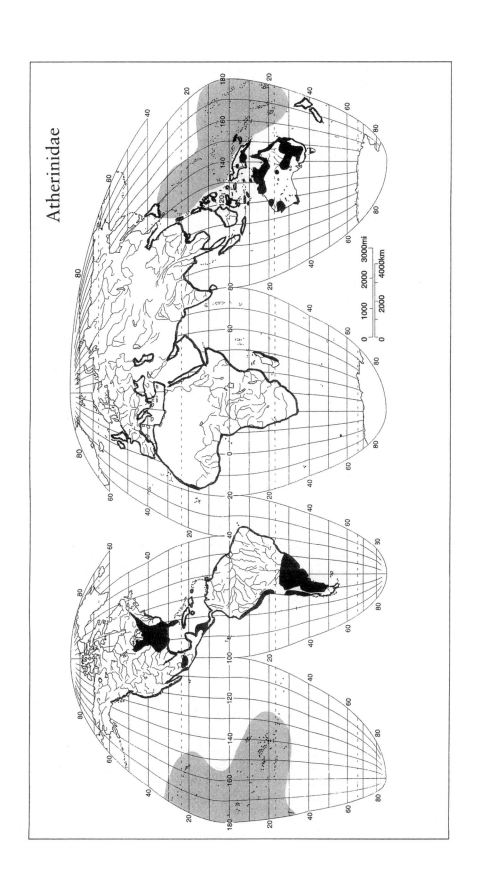

Atherinidae

of internal fertilization within the Atherinidae. The tidewater silverside, *Menidia beryllina*, occupies coastal habitats from Maine to Veracruz, Mexico, and ascends the lower Mississippi River to Missouri (Gilbert and Lee, 1980). *Menidia clarkhubbsi*, from coastal waters of southeastern United States, is a complex of unisexual "species" that arose by hybridization between *M. beryllina* and another *Menidia* species similar to *M. peninsulae*. *Menidia clarkhubbsi* reproduces by a process called gynogenesis. Sperm stimulates the egg to cleave but makes no genetic contribution to the offspring, which are all clones of the mother. See Echelle *et al.* (1989) for further details. *Alepidomus evermanni* is endemic to Cuban fresh waters.

The California grunion, *Leuresthes tenuis*, is a marine species remarkable for its nocturnal beach spawning during the peak of spring high tide from March to July. Two to 6 days after a new or full moon, hundreds of grunions wriggle up the sand beaches from Baja California to San Francisco. They run a gauntlet of people waiting to harvest them by hand. Females burrow into the sand as far up on the beach as high tide will take them and lay their eggs. Males curl around the partially burried females and fertilize the eggs. The adults then wriggle back into the sea. Two weeks later, the next high tide washes the eggs out of the sand and induces hatching within minutes. The fry are swept by the waves into the sea (Walker, 1959; Breder and Rosen, 1966). See Idyll (1969) for photographs of this striking event.

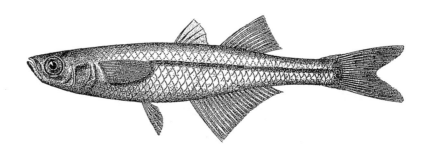

FIGURE 195. *Chirostoma mezquital* (Meek, 1904, Fig. 53).

In Mexico, 18 species of *Chirostoma* (represented by *C. mezquital*; Fig. 195) form a species flock in the Lerma–Santiago River basin and contiguous areas of the Mesa Central. These species have radiated into a variety of ecological niches including piscivores that resemble pikes and baracudas (Barbour, 1973; Echelle and Echelle, 1984).

In South America *Basilichthys* occurs in western Andean streams from northern Peru to Chiloé Island, Chile (Eigenmann, 1928; Dyer, 1997, 1998). Dyer listed 5 species of *Basilichthys*, 1 of which is undescribed. *Odontesthes* (about 20 species) is found in Atlantic coastal fresh waters from Patagonia to southeastern Brazil as well as in Chilean streams (Schultz, 1948; Arratia, 1981; López et al., 1987; Dyer, 1997, 1998). *Odontesthes* also has a coastal marine distribution in southern temperate South America. The latter two genera are sister groups according to Crabtree (1987) and Dyer (1997).

Australia and New Guinea have about 19 species of *Craterocephalus*, called hardyheads, in fresh water and another 5 species live in estuaries (Ivantsoff et al., 1987; Allen, 1991; Ivantsoff and Crowley, 1996). The fly-specked hardyhead, *C. stercusmuscarum*, is very widespread in eastern and northern Australia and southern New Guinea (Allen, 1989). *Craterocephalus eyresii* occurs in Lake Eyre drainages and *C. fluviatilis* occurs in the Murray–Darling system. These fishes are only about 6 cm SL.

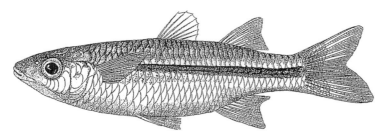

FIGURE 196. *Craterocephalus nouhuysi* as *Atherinichthys nouhuysi* (Weber, 1913, Fig. 26).

Craterocephalus nouhuysi (Fig. 196), from mountainous regions in the Lorentz and upper Fly–Strickland River systems of New Guinea, reaches 10 cm TL (Allen, 1991). The strawman, *Quirichthys stramineus*, is a beautiful, small atherinid from northern Australia that resembles a melanotaeniid in appearance and spawning behavior. The semitransparent males have a black anterior edge on the mast-like first dorsal fin, a black lateral stripe, and a yellow belly.

Schultz (1948) reviewed the family, and White (1985) examined the phylogeny and distribution of the New World silversides. Chernoff (1986a,b) reclassified some New World silversides, and White (1986) discussed the vicariance biogeography of the Atherinopsinae. Crabtree (1987) examined the phylogenetic relationships of the Atherinopsinae based on allozyme analysis, and Dyer (1997) provided a phylogenetic revision of the Atherinopsinae based on morphology and allozyme data. Ivantsoff and Crowley (1991) reviewed the Australian *Atherinomorus*.

Map references: Allen (1989, 1991),* Arratia (1981),* Arratia *et al.* (1983),* Barbour (1973),* Berg (1948/1949), Bussing (1998),* Dyer (1997, 1998),* Lee *et al.* (1980),* Ivantsoff and Crowley (1996),* Maitland (1977),* Page and Burr (1991),* Quignard and Pras (1986),* Scott and Crossman (1973),* Skelton (1993),* Stiassny and Reinthall (1992), White (1986)*

Class Actinopterygii
Subclass Neopterygii

Order Atheriniformes; Suborder Atherinoidei
(Per) Family **Telmatherinidae—sailfin silversides or Celebes rainbowfishes** (tel'-math-er-in'-i-dē) [tel'-ma-thur-rī'-ni-dē]

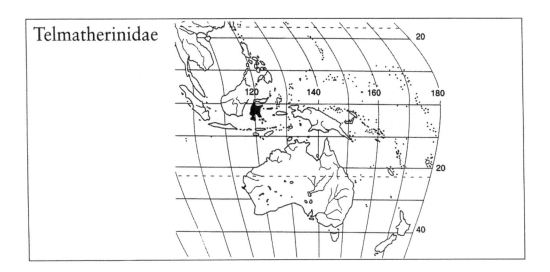

THE TELMATHERINIDAE IS FOUND IN FRESH AND BRACKISH waters of Sulawesi (= Celebes). One species, *Kalyptatherina helodes*, is known from brackish waters of Misool and Batanta islands off the extreme western end of New Guinea. This species was formerly assigned to the genus *Pseudomugil* (Ivantsoff and Allen, 1984), but Saeed *et al.* (1989) and Saeed and Ivantsoff (1991) created a new genus and placed it in the Telmatherinidae.

There are five genera and 17 species in the family, and most of these inhabit the Malili Lakes in the central area of eastern Sulawesi (Aarn et al., 1988; Kottelat, 1990b, 1991a; Allen, 1994). Sulawesi is considered a composite island both geologically and biologically (Parenti, 1991). It is thought that this part of Sulawesi was part of Australian Gondwana from about the mid-Miocene approximately 15 Mya (Powell et al., 1984), and this could have allowed fauna to disperse across Wallace's line. The distribution of the Telmatherinidae may provide confirmation of this former connection (Saeed and Ivantsoff, 1991; Allen, 1994).

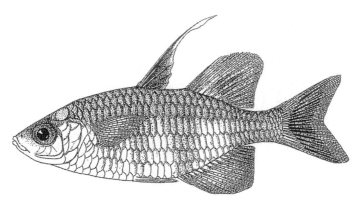

FIGURE 197. *Telmatherina celebensis* (Weber and DeBeaufort, 1922, Fig. 72).

Telmatherina is the largest genus with nine species, including *T. celebensis* (Fig. 197). *Paratherina* has four species, and *Tominanga* contains two species (Kottelat et al., 1993). Kottelat (1990b, 1991a) described about half of the species in the family. Aarn *et al.* (1988) described *Marosatherina* from Sulawesi for the species previously known as *Telmatherina ladigesi*. Sizes of telmatherinids range from about 3 to 20 cm. Telmatherinids resemble atherinids, pseudomugilids, and melanotaeniids. There are two dorsal fins, and the body is scaled. The midlateral band is poorly developed or absent. Males are more colorful than females, and males have a sail-like first dorsal fin.

Map references: Allen (1991),* Kottelat *et al.* (1993), Saeed and Ivantsoff (1991)

Class Actinopterygii
Subclass Neopterygii

Order Atheriniformes; Suborder Atherinoidei
(Per) Family Phallostethidae—**phallostethids or priapium fishes**
[fal-ō-steth'-i-dē]

THIS UNUSUAL FAMILY OCCURS THROUGHOUT COASTAL AND freshwater habitats in Thailand, the Malay Peninsula, northern Borneo, the Philippines, Sumatra, and Sulawesi (Parenti, 1989, 1996; Parenti and Louie, 1998; Kottelat et al., 1993). The Phallostethidae includes the fishes formerly placed in the Neostethidae. There are 21 species in four genera

Phallostethidae

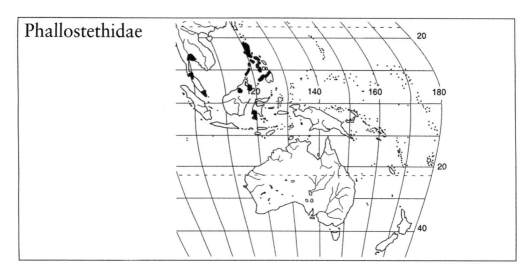

and two subfamilies (Parenti, 1989, 1996; Parenti and Louie, 1998). *Neostethus* (11 species), *Phallostethus* (2 species), and *Phenacostethus* (3 species) are assigned to the subfamily Phallostethinae found in Thailand, Malaysia, Indonesia, and the Philippines. Five species of *Gulaphallus* make up the subfamily Gulaphallinae in the Philippines. *Neostethus djajaorum*, from southwestern Sulawesi, is the only member of this family known from east of Wallace's line (Parenti and Louie, 1998).

A unique feature of this family is the presence of a complex, asymmetrical copulatory structure under the head of males. It is called the priapium and is formed from modifications of pelvic skeletal elements, pleural ribs, and the cleithra. Elongate, bony, curved projections called the toxactinium and the ctenactinium are used to clasp the female while the male impregnates her. Roberts (1971b) described the osteology of the priapium, and Parenti (1989) described this fleshy organ in detail and provided consistent terminology for this very intricate structure which contains ducts from the kidney, gonads, and intestine. *Neostethus bicornis* males have two long ctenactinia (Fig. 198), whereas other *Neostethus* males have a long and short one. Food and feeding adaptations of *N. bicornis* are reported by Mok and Munro (1991).

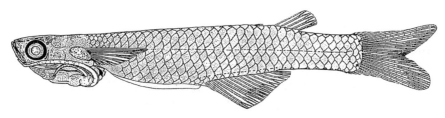

FIGURE 198. *Neostethus bicornis* as *Ceratostethus bicornis* (reproduced with permission from Roberts, 1971, Fig. 1).

Many fish with internal fertilization retain the fertilized eggs and give birth to live young. However, phallostethids lay fertilized eggs that have a filamentous process, presumably for attachment to vegetation. The anus and genital papilla of males open anteriorly and are on opposite sides of the body. Thus, males are either right- (dextral) or left- (sinistral) "handed" with respect to the genital opening (Parenti, 1986a). Both conditions may be found in males of the same species. Females have a vestigial or no pelvic fin except in *G. falcifer* (Parenti, 1986b, 1989).

Phallostethids have a fleshy keel on the belly that extends from the urogenital opening to the anal fin. The anal fin is long. The first dorsal fin is easily overlooked because it consists of only one or two small spines or it may be absent. Scales are cycloid.

Phallostethids are surface-feeding, schooling fishes like most atheriniforms and can be found in large numbers in certain areas (Roberts, 1971c). They are often transparent and the maximum size is about 37 mm SL (Parenti, 1996b). *Phenacostethus smithi*, at 15.6 mm SL, is one of the smallest atheriniform fishes known (Smith, 1945; Roberts, 1971c).

Map references: Parenti (1989, 1996),* Parenti and Louie (1998),* Roberts (1971c)*

Class Actinopterygii
Subclass Neopterygii

Order Beloniformes; Suborder Adrianichthyoidei
(2d) Family Adrianichthyidae—**ricefishes, medakas**
[ā-dri-an-ik'-thē-i-dē]

THE BELONIFORMES INCLUDE FIVE FAMILIES AND ABOUT 200 species, of which approximately 25% occur in fresh or slightly brackish water (Nelson, 1994). Members of this order have nonprotrusible upper jaws, and the lower caudal lobe has more principal fin rays than the upper lobe. Rosen and Parenti (1981) and Collette *et al.* (1984) provided taxonomic characters to support recognition of the ordinal status for this group.

The family Adrianichthyidae was previously linked with the killifishes (Cyprinodontiformes) but is now considered to be more closely related to the halfbeaks, flyingfishes, needlefishes, and sauries within the Beloniformes (Rosen and Parenti, 1981). The Adrianichthyidae includes fishes formerly placed in the Oryziidae and Horaichthyidae (Rosen and Parenti, 1981). The latter two groups are currently recognized as

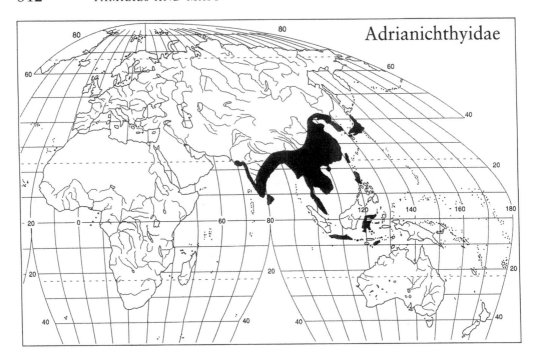

Adrianichthyidae

subfamilies in some classifications (Nelson, 1994). Adrianichthyids live in
fresh and brackish waters from India to Japan and south along the Indo-
Australian archipelago across Wallaces's line to Timor, Sulawesi (=
Celebes), and Luzon in the Philippines. They do not occur naturally on
Sumatra or Borneo according to Parenti (1991, 1994); however, Iwamatsu
et al. (1982) reported *Oryzias javanicus* from western Borneo. This may
represent an introduction (Roberts, 1989a). Adrianichthyids have a single
dorsal fin and no spines. The lateral line is absent on the body, and paired
nostrils are present. There are four genera and 18 species in three subfamilies (Nelson, 1994).

The subfamily Oryziinae includes the medakas (Japanese for killifish)
or ricefishes, so called because they occur in rice paddies as well as in
drainage ditches, streams, ponds, and lakes. The name of the subfamily's
single genus, *Oryzias*, is based on the generic name of the rice plant, *Oryza*.
There are about 13 species, and their distribution is similar to that of the
family range given previously. *Oryzias melastigma* is the species usually
listed for the Indian subcontinent (Talwar and Jhingran, 1992); however,
Roberts (1998) recognized additional species and demonstrated that
O. dancena is the correct name for the Indian ricefish. Ricefishes are small
(usually less than 3 cm SL), compressed fishes whose jaws are not greatly
enlarged. The dorsal fin is small, and the anal fin is elongated. Both are
posteriorly placed, and the fins of males are larger than those of females.
The females lay eggs among aquatic vegetation. Their mouth is terminal,
and they feed on mosquito larvae and other insects at the surface. Many of
the species are endemic to lakes on Sulawesi. See Uwa and Parenti (1988),

Uwa *et al.* (1988), Kottelat (1990c), and Kottelat *et al.* (1993) for a list of species.

These fishes, especially *O. latipes* (Fig. 199), widespread in eastern Asia, are widely used in genetic, embryological, and toxicological research because they are easy to rear and breed in captivity. The same features also

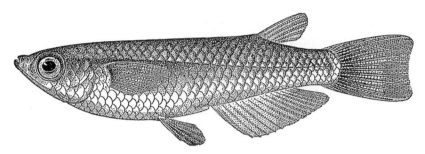

FIGURE 199. *Oryzias latipes* (Jordan and Metz, 1913, Fig. 21).

make them popular with aquarists. Parenti (1987) studied the teeth and jaw structure of *O. latipes*. *Oryzias javanicus* reaches Java and Lombok (Weber and De Beaufort, 1922; Kottelat et al., 1993).

The subfamily Adrianichthyinae is found only in Sulawesi, mainly in Lakes Poso and Lindu. Their mouth is shovel-like, and they range in size from 7 to 20 cm SL. Following the current classification, there are four species in two genera: *Adrianichthys kruyti* (Fig. 200) and three species of *Xenopoecilus* (Kottelat et al., 1993). Females of *X. poptae* carry clusters of

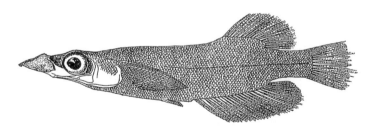

FIGURE 200. *Adrianichthys kruyti* (Weber and DeBeaufort, 1922, Fig. 101).

developing embryos against their belly, protected by the pelvic fin. The larvae hatch almost immediately upon egg deposition. Broken egg membranes reportedly cover extensive regions of the surface of Lake Poso. *Xenopoecilus poptae* was the object of a November to January hook-and-line fishery during their spawning season (Weber and De Beaufort, 1922), but it is now feared to be extinct (J. Albert, personal communication).

Horaichthys setnai is the only member of the subfamily Horaichthyinae. It is found in fresh and brackish waters of coastal western

India from near the Gulf of Kutch just south of the Tropic of Cancer to Trivandrum near the southern tip of India (Silas, 1959; Talwar and Jhingran, 1992). This very small species is elongate, compressed, and transparent. A small dorsal fin is inserted far back on the body, near the caudal fin. The anal fin is long and modified in both sexes. In males the first six rays form a gonopodium. In females the second to sixth fin rays are elongated. Pectoral fins are large and inserted high on the body. Both pelvic fins are present in the male, but in females the genital opening is deflected to the left, the right pelvic fin is absent, and the left pelvic fin is medial. *Horaichthys setnai* produces pointed, barbed spermatophores that are transmitted to the genital opening of the females via the gonopodium during courtship (Grier, 1984). Because the right pelvic fin of the female is missing, a larger surface area for attachment of the spermatophore is available (Parenti, 1994).

Horaichthys setnai is the smallest fish in India, with a maximum TL of about 2 cm (Talwar and Jhingran, 1992). It has a remarkable range of salt tolerance, related to monsoon flooding and summer evaporation. It feeds at the surface on copepods, diatoms, and insect larvae. Breeding occurs year-round but reaches a peak in July and August. The fertilized eggs are laid in weedy areas. See Kulkarni (1940) for a thorough study of morphology and life history and Hubbs (1941) for further comments.

Map references: Herre (1953), Jayaram (1981), Kottelat *et al.* (1993), Masuda *et al.* (1984), Parenti (1991), Roberts (1998),* Smith (1945), Una and Parenti (1988),* Weber and De Beaufort (1922)

Class Actinopterygii
Subclass Neopterygii

Order Beloniformes; Suborder Belonoidei
(Per) Family Belonidae—needlefishes
(be-lon'-i-dē) [be-lō'-ni-dē]

NEEDLEFISHES ARE A WORLDWIDE MARINE FAMILY OF 10 genera and 32 species, about one-third of which are confined to fresh waters in South America, the Indian subcontinent, Southeast Asia, and Australia and New Guinea (Nelson, 1994).

Belonids are thin, elongate fishes with teeth-studded jaws of equal length extended to form a beak. They resemble unarmored gars,

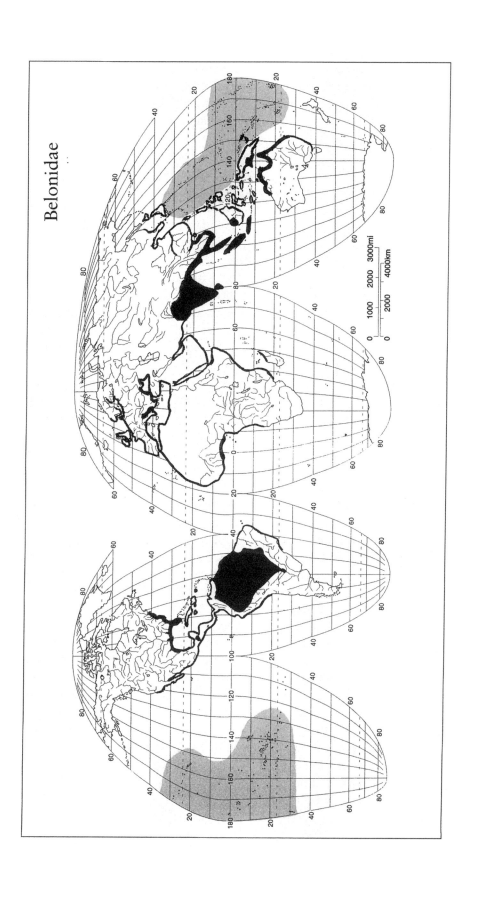

Belonidae

Lepisosteus. Their scales are small and cycloid. The dorsal and anal fins are long and posteriorly positioned opposite each other. The pelvic fins are abdominal, and the pectoral fins are short and located high on the body. None of the fins have spines. The lateral line is situated low on the body, and the lower lobe of the caudal fin may be elongated in some marine species. This latter feature is related to the leaping ability of this family, which is considerable. In fact, a fisherman attracting marine needlefish with lights was killed when a needlefish leaping at great speed pierced a vital organ or blood vessel (Munro, 1967). Needlefish are very fast and feed on other fishes and, in turn, are eaten by even faster tuna and swordfish. Their white flesh is consumed in many parts of their range. Oddly, they have green-colored bones (Collette and Parin, 1994).

Needlefishes have large, round eggs with attaching filaments (Collette et al., 1984). When the young hatch, their jaws are equal. As they grow the lower jaw elongates first, forming a "halfbeak" stage. Eventually, the upper jaw catches up, and the "needlenose" configuration results. This heterochrony of development is related to the change of diet from plankton-eating juveniles to fish-eating adults (Boughton et al., 1991).

In North America, the Atlantic needlefish, *Strongylura marina*, is the only belonid that commonly enters fresh waters. It travels upstream as far as the fall line. Along the coast, it ranges from Maine to Rio de Janeiro (Collette, 1974a; Burgess, 1980). *Strongylura hubbsi* occurs in fresh water in Guatemala and Mexico (Collette, 1974a). *Strongylura timucu* has been taken in fresh waters of Costa Rica, and it ranges along the coast from south Florida to Rio de Janeiro (Cressey and Collette, 1971).

In South American fresh waters, *Potamorrhaphis petersi* is found in the upper Orinoco, *P. guianensis* occurs throughout the Amazon and the Guianas, and *P. eigenmanni* is concentrated in the Paraquay–Paraná system (Collette, 1974b, 1982a). *Pseudotylosurus microps* occurs mainly in the Orinoco, the Guianas, and the lower Amazon, and *P. angusticeps* is found in the upper Amazon in Peru and Ecuador as well as in the Paraná and the Paraguay (Collette, 1974c).

Two species of diminutive *Belonion* also occur in South American fresh waters. *Belonion apodion*, which lacks the pelvic girdle and pelvic fins, and *B. dibranchodon* resemble a hemiramphid in that the upper jaw is substantially shorter than the lower jaw. Members of this genus are small, about 50 mm TL. The small size and unequal jaws are considered to be pedomorphic or neotenic conditions in which juvenile traits are retained into adulthood (Collette, 1966b). The largest member of this family is the worldwide marine species, *Tylosurus crocodilus*, known as the houndfish, which can reach 1.2 m SL (Cressey and Collette, 1971). There are newspaper reports that its leaping has injured people in Florida.

Xenentodon cancila (Fig. 201) is widely distributed in fresh waters of Pakistan, India, Bangladesh, Sri Lanka, Burma, Thailand, and the Malay Peninsula (Roberts, 1989a; Talwar and Jhingran, 1992). It ranges far inland

FIGURE 201. *Xenentodon cancila* as *Belone cancila* (Day, 1878b, Plate CXVIII, Fig. 5).

(Smith, 1945) and reaches about 40 cm TL. A second species, *X. canciloides*, occurs in Malaysia, Sumatra, and western Borneo (Roberts, 1989a).

In Australia and New Guinea, belonids are called longtoms. In central southern New Guinea, *Strongylura kreffti* has been taken as far as 830 km up the Fly River (Allen, 1991). In northern Australia, this species is found in rivers from the Fitzroy eastward to coastal streams of eastern Queensland as far south as the Dawson River (Allen, 1989). It appears to breed in fresh water.

Collette (1974a, 1982a) pointed out that there is an inverse relationship between the distribution of the Belonidae and the Hemiramphidae whereby the Belonidae predominates in American fresh waters and the Hemiramphidae is more numerous in the Indo-Australian region. Cressey and Collette (1971) provided a very interesting biogeographical study of needlefishes and copepods as hosts and parasites. Collette and Berry (1965) commented on three previous studies of needlefishes.

Map references: Allen (1989, 1991),* Burgess (1980),* Collette (1968, 1974a,b,c, 1982a),* Collette and Parin (1986a, 1994),* Cressey and Collette (1971),* Roberts (1989a), Talwar and Jhingran (1992), Weber and De Beaufort (1922)

Class Actinopterygii
Subclass Neopterygii

Order Beloniformes; Suborder Belonoidei
(Per) Family Hemiramphidae—**halfbeaks** (hem-i-ram'-fi-dē)

HEMIRAMPHIDS ARE A SURFACE-DWELLING MARINE FAMILY with a cosmopolitan distribution. There are 13 genera and more than 100 species, 25 of which occur in fresh waters, mostly in the Indo-Australian region (Nelson, 1994; Collette, 1995a; Meisner and Collette, 1999). Only 2 freshwater species occur in the Neotropical region (Collette, 1982a).

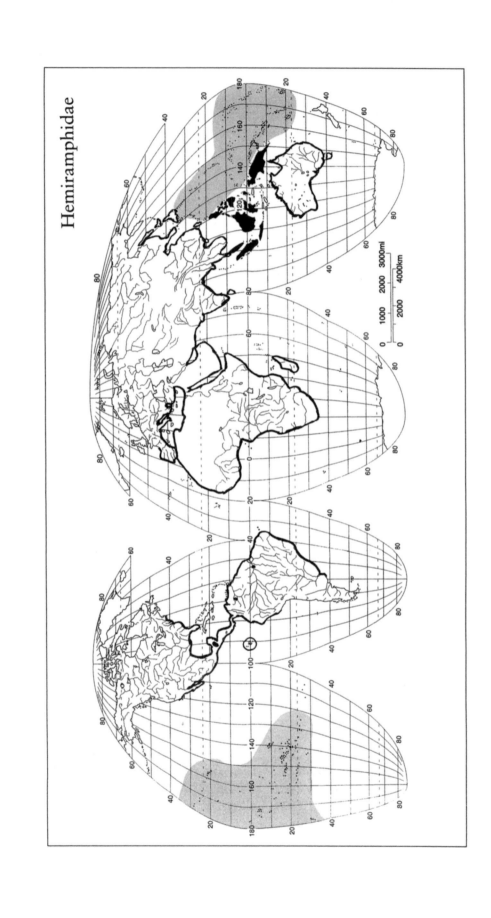

Hemiramphidae

Halfbeaks are slender, elongate fishes with the lower jaw extending far beyond the upper jaw in most species. The front margin of the upper jaw forms a prominent triangular projection. The dorsal, anal, and pelvic fins are set far back on the body. Large, cycloid scales are present, and the lateral line is low on the body. Fin spines are absent. The pectoral and pelvic fins are usually small. Maximum length is about 45 cm. Most halfbeaks are herbivores and feed on sea grass (*Zostera*), algae, and diatoms with some invertebrates ingested either deliberately or accidentally with the vegetation (Collette, 1974d). Freshwater species feed on insects (Anderson and Collette, 1991).

In some species the lower lobe of the caudal fin is enlarged, reflecting a relationship to the flying fishes (Exocoetidae) which use the expanded caudal fin for generation of sufficient propulsive force to leap over the water surface. A sculling motion of the tail allows halfbeaks to make short leaps, but most species do not have the expanded pectoral fins necessary for gliding like flying fishes. The term "flying halfbeak" is used for species with longer pectoral fins (Collette and Parin, 1994). The marine *Oxyporhamphus* has a more or less straight upper jaw margin that does not form a triangular projection. It also has long pectoral fins and shows characteristics intermediate between flyingfishes and halfbeaks.

Hyporhamphus mexicanus occurs in freshwater streams of Mexico and Guatemala that drain into Campeche Bay, Gulf of Mexico, and has a distribution similar to that of the needlefish *Strongylura hubbsi* (Collette, 1974a). *Hyporhamphus brederi* inhabits the Orinoco River between the mouths of the Meta and Apure Rivers of Venezuela and the Amazon River at Santarem, Estado Para, Brazil (Collette, 1982a). *Hyporhamphus xanthopterus* is known from lakes in Kerala, southwestern India (Collette, 1981), and *H. limbatus* (Fig. 202) is a common euryhaline species from coastal waters, estuaries, and fresh waters of Pakistan, India, and Burma to Thailand (Talwar and Jhingran, 1992).

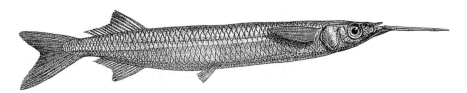

FIGURE 202. *Hyporhamphus limbatus* as *Hemiramphus limbatus* (Day, 1878b, Plate CXVIII, Fig. 3).

The other genera with freshwater and estuarine species include *Dermogenys*, *Hemirhamphodon*, *Nomorhampus*, *Tondanichthys*, and *Zenarchopterus*. Species within these genera have a modified anal fin called an andropodium. The five internally fertilized halfbeak genera listed previously represent a monophyletic group treated as a subfamily,

Zenarchopterinae, by Meisner and Collette (1999). Adult male *Tondanichthys* have not been collected, so the morphology of their anal fin is not known (Collette, 1995a). The freshwater halfbeaks are viviparous except for species of *Zenarchopterus* (and presumably *Tondanichthys*), which are oviparous (Collette et al., 1984; Grier and Collette, 1987; Anderson and Collette, 1991). Marine species are egg layers. Kottelat and Lim (1999) provided field observations of mating behavior of two species of *Zenarchopterus*. They hypothesized that in *Z. gilli* the highly modified dorsal and anal fins of males function as a clasping mechanism to hold the female in place during mating.

Dermogenys (about 21 species including undescribed forms) is the most widely distributed viviparous Southeast Asian halfbeak. It occurs from India, Burma, and Thailand, along the coast of the Malay Peninsula and Greater Sunda islands, to Sulawesi (Roberts, 1989a; Talwar and Jhingran, 1989; Meisner and Collette, 1998). Roberts (1989a) remarked that *Dermogenys* never occurs sympatrically with *Hemirhamphodon*, although their general distributions overlap. In Thailand, *Dermogenys pusillus* males are cultivated as a fighting fish. Its importance is second only to the Siamese fighting fish, *Betta*. Smith (1945) gave a vivid account of these fighting contests, in which the usual hold is an interlocking of jaws at their base with the long axis of the bodies at right angles.

Nomorhamphus is endemic to Sulawesi. Kottelat *et al.* (1993) listed eight nominal species but stated that the genus needs a rigorous revision. The lower jaw of this genus is longer than the upper jaw, but it does not form a slender beak like other halfbeaks. *Tondanichthys kottelati* is known only from Sulawesi, at the center of diversity of *Dermogenys* and *Nomorhampus* and just beyond the range of *Hemirhamphodon*. *Dermogenys* and *Nomorhampus* are considered sister groups and together they represent the sister group to *Hemirhamphodon* (Meisner and Collette, (1999).

Hemirhamphodon is an interesting genus of about six species that feeds predominantly on floating ants (Anderson and Collette, 1991). They have anteriorly directed conical teeth on the lower jaw beyond the length of the upper jaw. Most other hemiramphids have a toothless beak. In addition, the teeth on the upper jaw extend further posteriorly than in other halfbeaks. Anderson and Collette (1991) suggested that the unusual dentition of this group is utilized to entangle the legs of their prey which prevents the ants from escaping from the mouth of the halfbeak. These fish also have large sensory pores on the head that detect the activity of struggling ants in the water (Collette and Parin, 1994). *Hemirhamphodon* occurs on the Malay Peninsula, Sumatra, Bangka, Belitung, and Borneo (Roberts, 1989a; Kottelat et al., 1993). Many of the viviparous halfbeaks are small. The small size, combined with an elongate body, does not leave much room for developing embryos. This has resulted in superfetation taken to the extremes in *H. pogonognathus*. Females of this species mature at about

34 mm and produce small numbers of large young (7 or 8 mm) throughout the year. Stored sperm is used to fertilize a few eggs every couple of days. A pregnant female can have up to eight different broods in her body at once. With a gestation period of 6–8 weeks, the female gives birth at 2- to 8-day intervals (Roberts, 1989a). This makes room for more embryos (Collette and Parin, 1994).

Zenarchopterus, a genus of about 20 species, occurs throughout the Indo-Pacific region from East Africa to Samoa. Six species inhabit the fresh waters of New Guinea (Allen, 1991). See Collette (1982b, 1985) for descriptions of New Guinea species of freshwater *Zenarchopterus*. *Arrhamphus sclerolepis* inhabits estuaries in southern New Guinea and northern Australia, where it is called the snub-nosed garfish because of its very short lower jaw (Collette, 1974d). Landlocked populations are capable of breeding in reservoirs in eastern Queensland (Allen, 1989). *Zenarchopterus ectuntio* (Fig. 203) occurs in fresh water in India, Burma, Thailand, Malay Peninsula, Borneo, Hong Kong, and Taiwan and may grow to 18 cm TL (Talwar and Jhingran, 1992; Kottelat et al., 1993).

FIGURE 203. *Zenarchopterus ectuntio* as *Hemiramphus ectunctio* (Day, 1878b, Plate CXVIII, Fig. 5).

Map references: Allen (1989, 1991),* Anderson and Collette (1991),* Berg (1948/1949), Collette (1974d,* 1974e,* 1976,* 1986), Collette and Parin (1986b),* Collette and Su (1986),* Meisner and Collette (1998,* 1999), Roberts (1989a),* Talwar and Jhingran (1992)

Class Actinopterygii
Subclass Neopterygii

Order Cyprinodontiformes; Suborder Aplocheiloidei
(2d) Family Aplocheilidae—**Old World aplocheiloids** [ap'-lō-kī'-li-dē]

MEMBERS OF THE ORDER CYPRINODONTIFORMES ARE RELA-tively small, surface-dwelling species with some salt tolerance. They are

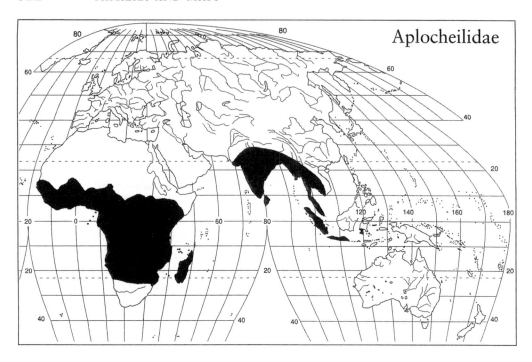

Aplocheilidae

found on all continents except Australia and Antarctica. Because of their surface orientation they are often referred to as topminnows. However, true minnows (Cyprinidae) lack jaw teeth, which cyprinodontiform fishes possess. Thus, they are also called tooth carps. The caudal fin of topminnows is often rounded but never forked. Their lateral line is well developed on the head but not the body. The dorsal surface of the head is flattened, and the mouth is terminal and often directed upward. The scales are usually cycloid, and the fins are soft-rayed. The dorsal fin originates more or less behind the middle of the body. The pelvic fins are abdominal and may be absent in some groups. Males are often brightly colored and may have elaborate fins. The group is important in the aquarium trade. Parenti (1981) provided a comprehensive phylogenetic analysis of the order. Costa (1998b) supported the monophyly of the families as proposed by Parenti (1981). Nelson (1994) recognized 8 families with about 800 species in 88 genera. Parker (1997) compared molecular and morphological phylogenies.

Parenti (1981) and Costa (1998b) treated the Old World and New World aplocheiloids in two related families, Aplocheilidae and Rivulidae. Robins *et al.* (1991), Nelson (1994), and Eschmeyer (1998) arranged them in two subfamilies, the Aplocheilinae (Old World rivulines) and the Rivulinae (New World rivulines). Some authors treat the rivulines as a subfamily of the Cyprinodontidae. I follow Parenti (1981) and Costa (1998b) and recognize both families. Aplocheiloids are the only cyprinodontiforms in which the pelvic fin bases are inserted close together.

The Aplocheilidae occurs in fresh water in Africa south of the Sahara Desert into South Africa, Madagascar, the Seychelles, the Indian subcontinent, Sri Lanka, and the Indo-Malaysian archipelago to Java (Parenti, 1981) and Sulawesi (Parenti and Louie, 1998). There are 8 genera and approximately 185 species (Nelson, 1994). The supracleithrum is fused to the posttemporal in the Aplocheilidae as opposed to the unfused condition in the Rivulidae, and the first postcleithrum is present. These fishes are egg layers.

The 35 species of *Nothobranchius* are called annual fishes because they complete their life cycle in 1 year. These fishes live in temporary pools or floodplains in tropical Africa. Lungfishes are often the only other fish found in these regions. *Nothobranchius* lays eggs in the bottom sediment that survive the desiccation of their habitat. The adults die when the pool dries out. Development is suspended until the rains come, and then the fertilized eggs hatch and another generation begins. Annual fishes and most rivulines feed on aquatic insects including mosquito larvae. Antimalarial and tsetse fly spraying programs are a threat to these fishes (Skelton, 1993). Many *Nothobranchius* species are popular aquarium animals (Jubb, 1981). *Nothobranchius guentheri* from eastern Africa achieves sexual maturity at 4 weeks of age and reproduces daily throughout its life. The rapid maturation, the fastest known in fishes, combined with the rampant mating promiscuity maximize reproductive success in an unpredictable environment (Haas, 1976a,b).

Aphyosemion is an African genus of about 60 species, many of which are small, such as *A. bivittatum* (Fig. 204), which reaches a TL of 45 mm. *Aphyosemion gardneri*, larger at 65 mm TL, is a beautifully colored species

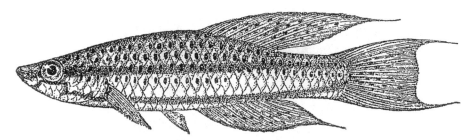

Fig. 204. *Amphyosemion bivittatum* as *Fundulus bivittatus* (Boulenger, 1915, Fig. 15).

from tropical west Africa that may survive after laying its drought-resistant eggs if water persists. For this reason it is considered a semiannual species. The males of some species of *Aphyosemion* have lyre-shaped tails and are valuable aquarium specimens. The eggs of *Aplocheilus annulatus*, also from west Africa, hatch within a few days of laying. This fish is not considered an annual species. Romand (1992) provided keys and color photographs of west African species, and Wildekamp *et al.* (1986) listed all known African species. Scheel (1990) summarized what is known of the biology of Old World aplocheiloids and presented many color photographs.

Aplocheilus panchax, one of about five species in the genus, is found from Pakistan to Sulawesi (Parenti and Louie, 1998). This colorful fish reaches a maximum size of about 55 mm and is an important agent of mosquito control (Smith, 1945; Talwar and Jhingran, 1992). It is also used as a fighting fish in Thailand. *Aplocheilus lineatus* (Fig. 205) from India is

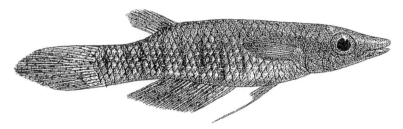

FIGURE 205. *Aplocheilus lineatus* as *Haplochilus lineatum* (Day, 1878b, Plate CXXI, Fig. 6).

the largest and most beautiful Asian member of the genus. It reaches 70 mm SL and is popular with tropical fish hobbyists.

Map references: Parenti (1981),* Romand (1992),* Skelton (1993)*

Class Actinopterygii
Subclass Neopterygii

Order Cyprinodontiformes; Suborder Aplocheiloidei
(2d) Family Rivulidae—New World aplocheiloids or rivulines (riv-ū'-li-dē)

THE NEW WORLD RIVULINES RANGE FROM SOUTHERN Florida, the Bahamas, Cuba, Hispaniola, and Trinidae to Middle America between central Mexico and Panama along Atlantic drainages and Costa Rica and Panama in Pacific drainages. In South America the family is widely distributed throughout most cis-Andean river basins from northern Venezuela to south of Buenos Aires Province, Argentina. The family is also present in trans-Andean drainages of Colombia and Venezuela and Pacific coastal basins of northwestern Colombia (Costa, 1998a; Parenti, 1981; Murphy *et al.* 1999). There are 27 genera and about 200 species (Costa, 1991, 1998a). In the Rivulidae, unlike its sister group the Aplocheilidae, the supracleithrum is not fused to the posttemporal, and the first postcleithrum

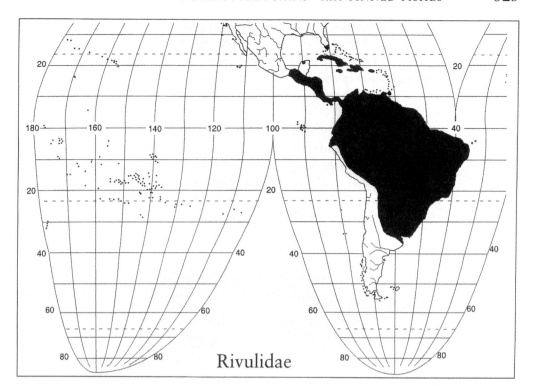

is absent. Costa (1998a) reported on the phylogeny and classification of the Rivulidae and listed the various genera and their relationships within the family including *Cynolebias*, *Rivulus*, *Pterolebias*, and *Austrofundulus*. Costa (1998b) presented phylogenetic hypotheses and classification of cyprinodontiform families.

Except for *Rivulus*, all rivulids are annual fishes (Costa, 1998a). *Cynolebias bellotii* (Fig. 206) is a typical representative well-known in the aquarium trade. Adults bury their fertilized eggs in the bottom mud subsequent to the dry season. The adults die, and only the eggs survive

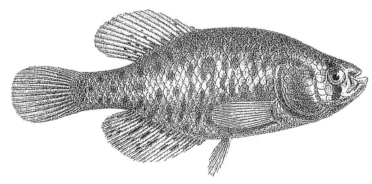

FIGURE 206. *Cynolebius bellottii* (Steindachner, 1882b, Plate V, Fig. 2).

and hatch months later at the beginning of the following wet season. This annual life cycle is similar to that of the Old World aplocheiloids, but it may have evolved independently in the two lineages. Costa considered annualism in rivulids to be a single evolutionary event (lost in *Rivulus*) and to have been achieved by successive colonization of peripheral aquatic habitats.

Costa (1998a) discussed miniaturization sensu Weitzman and Vari (1988) among rivulids. He reported 8 species that are less than 26 mm SL as adults. Since all the known miniatures occur in a single area of endemism, Costa suggested that something in this area favored miniaturization. He concluded that the degree of annual fish diversity in a temporary pool was the selecting force leading to miniaturization in this case.

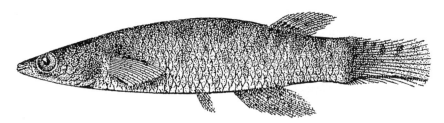

FIGURE 207. *Rivulus tenuis* as *Cynodonichthys tenuis* (Meek, 1904, Fig. 27).

Rivulus, represented by *R. tenuis* (Fig. 207), is the most widespread Neotropical genus, having a distribution coincident with the subfamily and including about 80 species (Costa and Brasil, 1991b; Huber, 1992; Costa, 1998a). It superficially resembles the African *Aphyosemion*. Huber suggested that this is due to convergence, and that the two genera had diverged before Africa and South America drifted apart. He stated that the characters shared by *Rivulus* and *Aphyosemion* are plesiomorphic (primitive), and that the distinctive characters are apomorphic (advanced). Murphy *et al.* (1999) considered *Rivulus* to be paraphyletic.

Rivulus marmoratus is the only member of the Rivulidae native to the United States. It is widely distributed in fresh and brackish water from southern Florida throughout the West Indies and from the Bahamas to the islands off Venezuela to southern Brazil (Gilbert and Burgess, 1980; Taylor et al., 1995). It is the most marine member of the otherwise freshwater genus (Davis et al., 1995), and it is tolerant of a wide range of salinities (0–68 ppt), temperatures (7–38°C), and high levels of hydrogen sulfide (Taylor et al., 1995).

Rivulus marmoratus has the distinction of being the only known fish species that is a synchronous, self-fertilizing hermaphrodite. It has both ovaries and testes and fertilizes its own eggs before laying, thus producing homozygous clones (Harrington, 1961; Harrington and Kallman, 1968).

Secondary males can occur by loss of ovarian tissue and the addition of male colors. Such males can supply cross-fertilization to the clonal population (Soto and Noakes, 1994). Males can be induced in the laboratory by incubation of eggs at low temperatures (Kallman and Harrington, 1964). Harrington (1971) reviewed the ecological influences such as day length and temperature and the genetic factors that interact to produce secondary males. In Belize populations, as many as 20% of the fish were males (Turner et al., 1992), but males are extremely rare in Florida populations (Davis et al., 1995). Kweon *et al.* (1998) showed that although *R. marmoratus* is internally self-fertilizing, its sperm is of a primitive form usually found in externally fertilizing fishes and quite different from sperm of internally non-self-fertilizing teleosts. They concluded that the mode of reproduction of this species may have recently shifted from external to internal fertilization.

Rivulus marmoratus reaches a maximum TL of 60 mm. It is maroon to brown dorsally and laterally with small black spots on the side. A whitish halo surrounds a large black spot on the upper half of the caudal base. This species inhabits drainage ditches, mangrove forests, salt marshes, and the burrows of land crabs, *Cardisoma guanhumi* (Taylor et al., 1995). It can survive the desiccation of its habitat by remaining in damp detritus, and it is capable of amphibious emersion from the water in pursuit of prey (Davis et al., 1995; Greenfield and Thomerson, 1997). *Rivulus marmoratus* feeds on a mixture of terrestrial and aquatic organisms, including insects, polychaetes, gastropods, and crustaceans (Taylor, 1992). Huber (1992) considered *R. marmoratus* to be a junior synomym of *R. ocellatus*.

Map references: Bussing (1998),* Costa (1998a), Gilbert and Burgess (1980),* Lee *et al.* (1983),* Parenti (1981)*

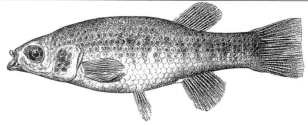

Class Actinopterygii

Subclass Neopterygii

FIGURE 208 *Profundulus oaxacae* as *Fundulus oaxacae* (Meek, 1904, Fig. 28).

Order Cyprinodontiformes; Suborder Cyprinodontoidei
(2d) Family Profundulidae—**Middle American killifishes**
[prō-fun-dū'-li-dē]

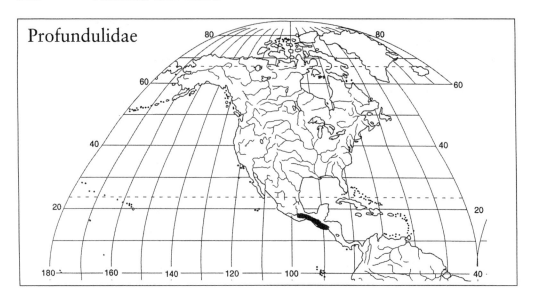

Profundulidae

THIS SMALL FAMILY CONSISTS OF *PROFUNDULUS* WITH FIVE species, including *P. oaxacae* (Fig. 208). Its distribution is centered on the Isthmus of Tehuantepec and it ranges from southern Mexico to Honduras, including both Atlantic and Pacific drainages (Parenti, 1981). These fishes typically inhabit mountain streams between 600 and 2440 m elevation (Miller, 1955a). *Profundulus* is considered to be the most primitive cyprinodontoid genus by Parenti. Costa (1998b) considered it to be closely related to the Goodeidae. It is characterized by a high number of gill rakers (14–23) on the anterior arm of the first gill arch, whereas *Fundulus* (Fundulidae) typically has less than 12 gill rakers (Miller, 1955a; Parenti, 1981). These fishes are nonannual egg layers with external fertilization. The anal fin of the male is not modified into a gonopodium. Miller (1955a) reviewed the group.

Map references: Costa (1998a),* Miller (1955a),* Parenti (1981)*

Class Actinopterygii
Subclass Neopterygii

Order Cyprinodontiformes; Suborder Cyprinodontoidei
(2d) Family Fundulidae—**topminnows and killifish**
[fun-dū'-li-dē]

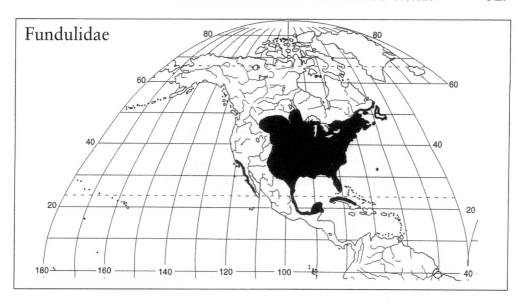

THE FUNDULIDAE IS DISTRIBUTED IN FRESH, BRACKISH, AND
coastal marine waters in the lowlands of North and Middle America
south to Yucatan and in Bermuda and Cuba. It is also represented on the
Pacific coast of California and Mexico. Parenti (1981) removed these
fishes from the Cyprinodontidae (Greenwood et al., 1966) and placed
them in a family of their own. There are four or five genera: *Adinia*,
Fundulus, *Leptolucania*, and *Lucania*. Parenti's (1981) action to resurrect
Plancterus for the plains killifish (*F. zebrinus*) has not been widely
adopted. Its major distinction is a highly convoluted intestine. Robins *et al.*
(1991), Page and Burr (1991), and Mayden *et al.* (1992) relegated it
to *Fundulus*. See Poss and Miller (1983) for a review of the taxonomic
status of *F. zebrinus* (Fig. 209). The oldest fossil fundulid dates to the early
middle Miocene of Montana and may belong to the genus *Plancterus*

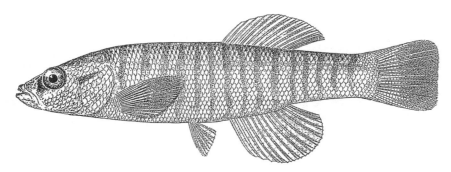

FIGURE 209. *Fundulus zebrinus* (Jordan and Evermann, 1900, Plate CIII, Fig. 276).

(Cavender, 1986). Mayden *et al.* (1992) listed 33 species in the family in continental North America; Parenti (1981) cited approximately 40 species. Most recent estimates recognize 5 genera and about 48 species (Smith-Vaniz et al., 1999). Bernardi (1997) provided a molecular phylogeny of the family.

The term "killifish" is derived from the Dutch word "killivisch." "Kills" are small waterways, and "visch" means fish (Rosen, 1973a). Killifish are exceptionally euryhaline. Some species of *Fundulus* from brackish habitats can tolerate salt concentrations higher than 70 ppt, which is twice that of sea water, whereas normally freshwater species can survive salinities up to 29 ppt (Griffith, 1974). The largest killifish is *F. grandissimus* from the Yucatan peninsula of Mexico that can reach about 180 mm SL, whereas the pygmy killifish, *Leptolucania ommata*, barely reaches 15 mm SL (Jenkins and Burkhead, 1994).

Topminnows (the other common name) feed at the surface with a protrusible, terminal mouth that includes jaws with teeth. Their head is flattened. They are carnivorous and consume many insects including mosquito larvae. Some species will also take small fishes. Topminnows prefer calm regions of fresh and brackish waters, including streams, marshes, swamps, ponds, and lakes. The single dorsal fin lacks spines and is inserted posteriorly. The pelvic fins are abdominal. The caudal fin is rounded or truncate. Males tend to have larger fins and more colorful pigmentation than females. Males develop breeding tubercles and do not have a gonopodium. Females are egg layers. The lateral line is absent from the body but well developed on the head. Scales are cycloid. There is no connection between the swim bladder and the gut (physoclist).

Fundulus is the largest genus with 30 or more species. Miller (1955b) provided an annotated list of species. Its evolutionary relationships have been studied by Wiley (1986), who concluded that *Fundulus* cannot be shown to be monophyletic. However, he did discuss four monophyletic subgenera. Wiley further opined that *Fundulus* + *Lucania* form the sister group to *Adinia* + *Leptolucania*. Many species of *Fundulus* are very colorful, and they are popular aquarium animals. The greatest area of diversity is eastern and central United States. *Fundulus olivaceus*, the black-spotted topminnow, is one of the more common and widespread fishes in the central and lower Mississippi River basin and Gulf coast drainages. *Fundulus notatus*, the black-stripe topminnow, is a very similar species with fewer and less discrete dorsolateral spots. *Fundulus notatus* (Fig. 210) occurs further north (into the lower Great Lakes area) than *F. olivaceus*, but they can be sympatric in the southern part of their ranges. *Fundulus diaphanus*, the banded killifish, occurs farther north than other freshwater killifish, ranging from South Carolina to the Maritime Provinces and Newfoundland. Its distribution extends west to Montana (Scott and Crossman, 1973).

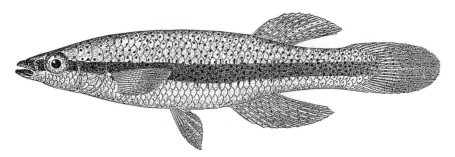

FIGURE 210. *Fundulus notatus* (Jordan and Evermann, 1900, Plate CVIII, Fig. 289).

The euryhaline mummichog, *F. heteroclitus*, occurs along the Atlantic coast from northeastern Florida to the Gulf of St. Lawrence region in Canada. This species is common in salt marshes, estuaries, and tidal areas and is a popular laboratory animal for physiological and embryological studies. More papers have been published on the biology of *F. heteroclitus* than any species of *Fundulus*, and this species ranks among the most popular fish species for scientific research. *Fundulus grandis* (Fig. 211), the Gulf

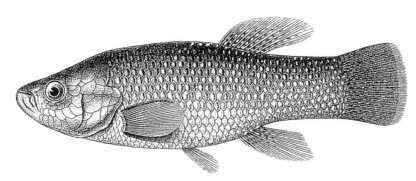

FIGURE 211. *Fundulus grandis* (Girard, 1859, Plate CXII, Fig. 5).

killifish, ranges from northeastern Florida to Cuba. *Fundulus bermudae* and *F. relictus* occur in inland habitats in Bermuda (Smith-Vaniz et al., 1999). How these little fishes colonized such small, isolated islands is not known, but Smith-Vaniz *et al.* (1999) speculate that waif dispersal via hurricanes or coastal flooding in association with large floating debris may have been possible. The Bermuda species are members of the *F. heteroclitus* species group, and several endemic species remain to be described from Bermuda (Smith-Vaniz et al., 1999). These new species will be recognized largely on the basis of mtDNA sequences (R. C. Cashner, personal communication). *Fundulus* does not occur in the Bahamas.

Fundulus bifax, the stippled studfish from the Tallapoosa River system of Alabama and Georgia, was recognized as unique almost solely on the basis of complete allelic differentiation at six loci from the southern studfish, *F. stellifer*. It was the first North American species to be described primarily on the basis of biochemical rather than morphological data (Cashner et al., 1998).

Fundulus parvipinnis, the California killifish, occurs in lagoons, bays, and estuaries from Morro Bay, California, to Magdalena Bay, Baja California. It can tolerate very high salinities. Moyle (1976) mentioned that populations also occur in streams of southern California. Distribution maps for many species of *Fundulus*, *Leptolucania*, and *Lucania* can be found in Lee *et al.* (1980) and Page and Burr (1991).

Fundulus lima, the Baja killifish, is known from pools in Baja California del Sur, near the tip of the peninsula in San Ignacio. Miller (1955b) regarded it as an odd freshwater endemic. It shares a disjunct Baja distribution with *F. parvipinnis*, which is primarily a marine species. *Fundulus lima* has a strikingly distinct breeding morphology and reproductive behavior that make it the most un-*Fundulus*-like member of the genus. Myers (1930) described the extraordinary development of the scale ctenni into elongate spines in breeding males. There may be as many as four spines at the posterior margin, but the middlemost spine is greatly elongated and may equal the length of the scale from which it extends. This condition gave rise to the specific epithet *lima*, which means "file." The coloration of breeding males is coal black (R. C. Cashner, personal communication). Brill (1982) reported that the eggs are laid on vegetation and splashed with water to keep them moist until hatching. Bernardi and Powers (1995) showed that *F. lima* and *F. parvipinnis* are close relatives.

The diamond killifish, *Adinia xenica*, is so named because of the profile of its body. This small (50 mm TL), compressed species occurs in brackish waters from Florida to Texas. Its life history has been studied by Hastings and Yerger (1971). It is the only member of its genus. *Leptolucania ommata*, from Georgia, central Florida, and Alabama, is also the only species in its genus. It has a large black spot surrounded by a ring of whitish-yellow on the caudal peduncle. *Lucania goodei* from fresh waters in peninsular Florida and *L. parva* from salt marshes of Cape Cod to Texas and from far up the Rio Grande and Pecos River in Texas and New Mexico are usually found in heavily vegetated areas.

Map references: Costa (1998b),* Lee *et al.* (1980),* Lagler *et al.* (1977),* Miller (1958, 1961),* Moyle (1976), Page and Burr (1991),* Parenti (1981),* Scott and Crossman (1973)*

Class Actinopterygii
Subclass Neopterygii

Order Cyprinodontiformes; Suborder Cyprinodontoidei
(2d) Family Valenciidae—**valenciids**
(va-len-shī'-i-dē) [va-len-shē'-i-dē]

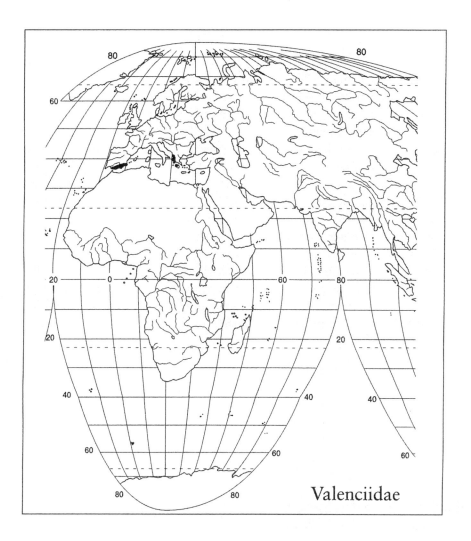

Valenciidae

THIS SMALL FAMILY OCCURS IN SOUTHEASTERN SPAIN, WEST-
ern Greece, and on the Greek Island of Corfu (Kerkira Island) and consists

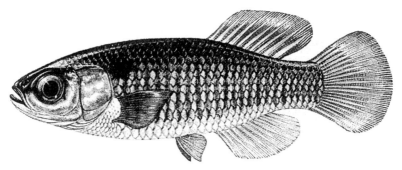

FIGURE 212. *Valencia hispanica* (reproduced with permission from Maitland, 1977, p. 197).

of two species, *Valencia hispanica* (Fig. 212) and *V. letourneuxi* (Bianco and Miller, 1989). Parenti (1981) examined an old specimen labeled "Italy" in the fish collection of the Academy of Natural Sciences of Philadelphia (L. Parenti, personal communication), but according to Bianco and Miller (1989) it does not occur there. Parenti (1981) placed this *Fundulus*-like genus in its own family. Its distinguishing feature is an elongate and attenuate dorsal process of the maxilla. These fish are egg layers and reach about 8 cm TL.

Valencia hispanica is found in coastal areas of southeast Spain in both fresh and brackish water (Maitland, 1977). Habitat destruction in the form of swamp drainage and coastal development and competition from the introduced *Gambusia affinis* pose threats to this species (Wheeler, 1975). *Valencia letourneuxi* is nearly extinct in its type locality on Corfu, but it has been found in several other localities in western Greece (Bianco and Miller, 1989). *Valencia letourneuxi* can tolerate salinities up to 46 ppm but usually lives in clean, standing fresh water among vegetation (Bianco and Miller, 1989). The two species are very similar.

Map references: Bianco and Miller (1989),* Maitland (1977)*

Class Actinopterygii
Subclass Neopterygii

Order Cyprinodontiformes; Suborder Cyprinodontoidei
(2d) Family Anablepidae—four-eyed fishes   (an'-a-blep'-i-dē)

THE ANABLEPIDAE HAS MEMBERS FROM SOUTHERN MEXICO to Nicaragua, along the northern coast of South America from Venezuela to Para, Brazil, and in southern Brazil, Argentina, and Uruguay (Parenti,

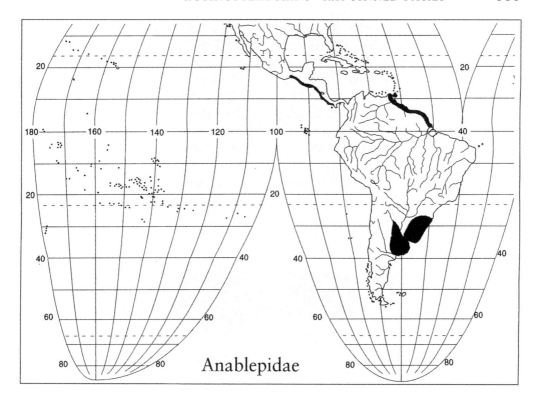

Anablepidae

1981). There are three genera: *Anableps* (three species), *Jenynsia* (nine species), and *Oxyzygonectes dowi*. Ghedotti (1998) discussed the relative merits of the specific epithet *dowi* vs *dovii* and recognized nine species of *Jenynsia*. Parenti united these three genera into one family. Previously, the Anablepidae included only *Anableps*. *Jenynsia* occupied its own family, Jenynsiidae, and *Oxyzygonectes* was included in the subfamily Fundulinae of the Cyprinodontidae.

Anableps and *Jenynsia* are placed in the subfamily Anablepinae. The males have thickened and elongated anal rays that are twisted around each other to form a fleshy, tubular gonopodium offset to the left or right. Fertilization is internal.

The three species of *Anableps* occur in fresh and coastal waters of the Pacific slope of Central America (*A. dowi*; Fig. 213) and northern South

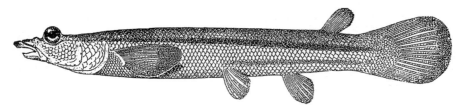

FIGURE 213. *Anableps dowi* as *A. dovii* (Jordan and Evermann, 1900, Plate CXIII, Fig. 300).

America (*A. anableps* and *A. microlepis*). Adult size is 15–30 cm. The body is elongate with a posteriorly positioned dorsal fin and a broad head. They are known as four-eyed fishes (cuatro ojos) because the eyes of these surface-dwelling animals are adapted for seeing in air and under water. The cornea is divided into an upper and lower half by a band of tissue. The iris is partially divided, resulting in two pupils, and a divided retina forms images of above-water and below-water objects simultaneously. Even the lens is differentially thickened to adjust for the different indices of refraction between air and water. Because of their visual acuity, they are very difficult to collect with a seine. They can see a collector approaching from 10 m or more, and they are excellent leapers (Miller, 1979).

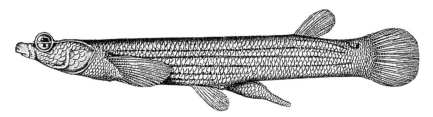

FIGURE 214. *Anableps anableps*, male, as *A. tetropthalmus* (Cunningham, 1912, Plate XXXI).

The male intromittent organ may be moved to the left or to the right (Fig. 214). The females are larger than the males and have a scale that occludes the genital opening on either the left or the right side. Consequently, it is thought that a "left-handed" male must copulate with a "right-handed" female and vice versa. Four-eyed fish are viviparous. See Herald (1962) for references to reproductive habits, Schwassmann and Kruger (1965) and Graham (1972) for information on the unique visual system, and Zahl *et al.* (1977), Zahl (1978), and Miller (1979) for natural history and illustrations.

Jenynsia includes about nine species with normal eyes from the lowlands of Brazil, Paraguay, Uruguay, and Argentina (Parenti, 1981; Ghedotti and Weitzman, 1996). The male's tubular gonopodium and the female's genital aperture are either dextral or sinistral as in *Anableps*. Fertilization is internal, and *Jenynsia* is viviparous. *Jenynsia lineata* females may reach 10 cm, but the males are only about 2.5 cm TL. See Breder and Rosen (1966) for references to reproductive structure and behavior.

The subfamily Oxyzygonectinae includes only one species—*Oxyzygonectes dowi* from the Pacific coast of Costa Rica. Males of this species do not have the anal fin modified as a gonopodium. Fertilization is external. The female is oviparous.

Map references: Ghedotti and Weitzman (1996),* Miller (1966, 1979), Parenti (1981),* Rosen (1973a)

Class Actinopterygii
Subclass Neopterygii

Order Cyprinodontiformes; Suborder Cyprinodontoidei
(2d) Family Poeciliidae—**poeciliids**
(pē'-si-lī'-i-dē) [pē'-si-lē'-i-dē)

THE POECILIIDS ARE A LARGE FAMILY OF ABOUT 30 GENERA
and 300 species found in North and Middle America, the Caribbean, South
America to southern Uruguay, the Congo basin and the African rift lakes,
Dar es Salaam, and Madagascar (Parenti, 1981). A considerable degree of
salt tolerance has allowed the spread of these secondary-division fishes to
some West Indian islands.

Poeciliids have pectoral fins set high on their sides, and the pelvic fins
are anteriorly positioned. Parenti (1981) recognized three subfamilies.
Previously, the family Poeciliidae included only the live-bearers, which now
compose the subfamily Poeciliinae. The other two subfamilies,
Fluviphylacinae and Aplocheilichthyinae, are egg layers and were formerly
placed in the Cyprinodontidae. See Rosen and Bailey (1963) for an earlier
view of the systematics and zoogeography of poeciliids.

The live-bearers or Poeciliinae occur in North America through
Central America, the Caribbean, and South America to Uruguay. There are
16 genera and about 200 species (Parenti, 1981; Rauchenberger, 1989a).
The males of these fishes have a gonopodium formed mainly from anal
rays 3–5. The pelvic fins of males also have curved rays which probably
function to hold the female during copulation. Fertilization is internal.
Females give birth to live young, except for 1 species from northeastern
South America, *Tomeurus gracilis*, that lays fertilized eggs. The
gonopodium of males of this odd species is located under the pectoral fins.
Meffe and Snelson (1989) provided details of the ecology and evolution of
poeciliids.

Live-bearers range in size from 1.5–18 cm and, in fact, the least
killifish, *Heterandria formosa* (Fig. 215), of the southeastern United States
may be the smallest live-bearing vertebrate in the world. The greatest
diversity of live-bearers is in Central and tropical America. *Poecilia*
(= *Limia*) makes up a significant portion of the freshwater fauna in the
Greater Antilles, probably because of its salt tolerance (Lee et al., 1983).
Rodriguez (1997) recognized *Limia* as a distinct genus from *Poecilia*
and considered *Xiphophorus* to be the sister taxon to the clade formed
by *Pamphorichthys*, *Poecilia*, and *Limia*. *Gambusia affinis* (Fig. 216)
and *G. holbrooki* have been introduced throughout the world for

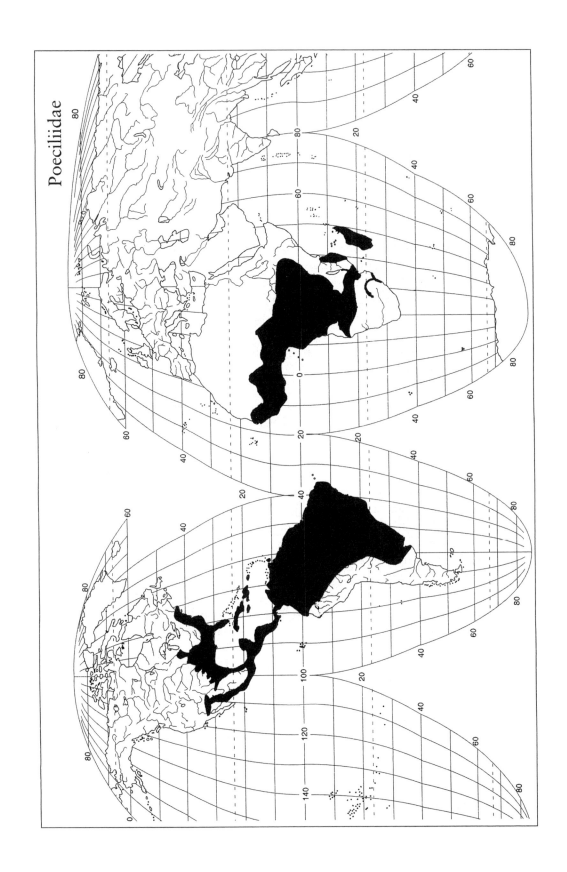

Poeciliidae

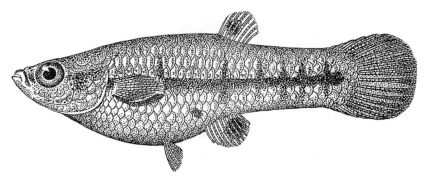

FIGURE 215. *Heterandria formosa*, female (Jordan and Evermann, 1900, Plate CXIV, Fig. 302).

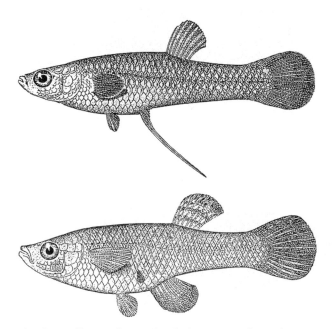

FIGURE 216. *Gambusia affinis* (male, top; female, bottom) (Jordan and Evermann, 1900, Plate CXIII, Figs. 299 and 299a).

mosquito control and are regarded as a pest that threatens native fish species, especially in Australia and New Guinea (Courtenay and Meffe, 1989; Arthington and Lloyd, 1989; Ivantsoff and Aarn, 1999). See Swanson *et al.* (1996) for information on *Gambusia* use in mosquito control and Rauchenberger (1989b) for a discussion of the systematics and biogeography of *Gambusia*.

Dawes (1991) reviewed the aquarium biology of many commonly kept live-bearers. Males are usually more colorful than females (e.g., guppies, *Poecilia reticulata*), have larger, more ornate fins (e.g., sailfin mollies, *P. latipinna*; Fig. 217), or have additional secondary sexual characteristics (e.g., swordtails, *Xiphophorus montezumae*; Fig. 218). These characteristics have attracted the attention of aquarists and have made this group one of the most important families in the aquarium industry. *Xiphophorus* has also been an important laboratory animal in genetic research. Marcus and McCune (1999) listed the 22 described species of *Xiphophorus* and discussed their ontogeny and phylogeny. They speculated on the selective forces that contributed to the origin of swords as well as to the loss of these structures in some species.

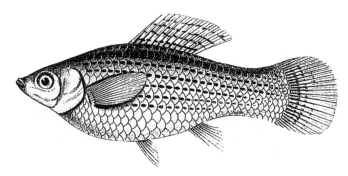

FIGURE 217. *Poecilia latipinna*, female, as *P. lineolata* (Girard, 1859, Plate CXI, Fig. 9).

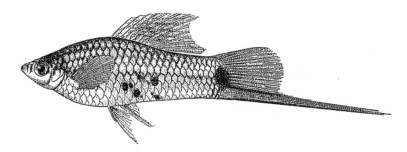

FIGURE 218. *Xiphophorus montezumae* (Meek, 1904, Fig. 50).

Internal fertilization and sperm storage are the rule in this subfamily. Males transmit spermatophores via a gonopodium. The females are able to store sperm for more than 10 months, and several successive broods may be fertilized from one mating. Most female poeciliids produce large eggs (2 mm) that require little supplementation by the female during development. The young are nourished by the yolk deposited before fertilization. This condition is termed ovoviviparity or lecithotrophy (Constantz, 1989). Species such as *Poecilia latipinna*, *P. formosa*, *P. reticulata*, *Xiphophorus helleri*, *Gambusia affinis*, and *Poeciliopsis monacha* are classified in this category. Other species, such as *Heterandria formosa* and several species of *Poeciliopsis*, produce tiny eggs (0.4–0.8 mm) with sparse yolk stores. Embryos of these species receive maternal nutrients during development. This condition is termed viviparity or matrotrophy (Constantz, 1989). See Breder and Rosen (1966) for reproductive details of various species.

The Amazon molly, *P. formosa*, is so called not because it lives in the Amazon River but because it is an all-female "species" reflecting the mythic race of female warriors. *Poecilia formosa* actually is found from the southern tip of Texas into Veracruz, Mexico. It mates with closely related males of other *Poecilia* species in the area. The sperm stimulates egg cleavage but does not fuse with the female's egg nucleus and therefore does not contribute to inheritance. This process is termed gynogenesis. All offspring are clones of the female and thereby genetically identical to their mother. It is thought that this peculiar reproductive pattern arose via hybridization between two *Poecilia* species, probably *P. mexicana* and *P. latipinna*. *Poecilia formosa* was the first discovered all-female fish. Unisexuality is now also recognized in another poeciliid, *Poeciliopsis*. Schultz (1989) discussed the origins and relationships of unisexual poeciliids. Six papers dealing with various aspects of the biology of unisexual *Poecilia* and *Poeciliopsis* are presented by Dawley and Bogart (1989).

Xenodexia ctenolepis from Guatemala is an unusual member of the Poeciliinae. It has ctenoid scales with very sharp ctenii (other poeciliids have cycloid scales), and males possess a very long and complex copulatory organ and clasper modified from the right pectoral fin (Hubbs, 1950).

The subfamily Fluviphylacinae consists of four species, the type of which is *Fluviphylax pygmaeus* (Costa, 1996). This minute (19 mm SL) species has very large eyes and a very small preorbital distance (Parenti, 1981). The males, which are larger than the females, lack a gonopodium, and the females are oviparous. It is found in the Amazon basin of Brazil. Roberts (1970b) discussed this species. Costa (1996) reviewed *Fluviphylax* and described three new species.

The third subfamily is the African Aplocheilichthyinae. It consists of about 100 species in seven genera. The males do not have a gonopodium, and fertilization is external. The females lay eggs. These fishes are found in savanna and forest lowland regions of west and central Africa south of the Sahara, the Rift lakes, portions of southern Africa, and Madagascar (Parenti, 1981; Romand, 1992; Skelton, 1993; Wildekamp et al., 1986). Genera include *Aplocheilichthys*, *Hypsopanchax*, *Lamprichthys*, *Pantanodon*, *Plataplochilus*, and *Procatopus*. *Pantanodon madagascariensis* occurs on Madagascar (Stiassny and Raminosoa, 1994). Wildekamp et al. (1986) listed the African species.

Map references: Bussing (1998),* Lee *et al.* (1980, 1983),* Page and Burr (1991),* Parenti (1981),* Romand (1992),* Rosen (1975),* Rosen and Bailey (1963),* Skelton (1993),* Wildekamp *et al.* (1986)

Class   Actinopterygii
Subclass   Neopterygii

Order   Cyprinodontiformes; Suborder Cyprinodontoidei
(2d) Family   Goodeidae—**goodeids**   [goo-dē'-i-dē]

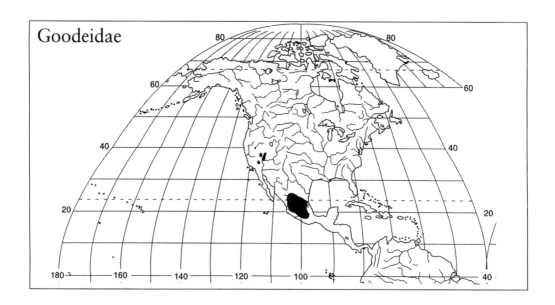

THE GOODEIDS HAVE A DISJUNCT DISTRIBUTION IN THE SOUTH-
ern Nevada Death Valley system and Mesa Central, Mexico. There are
about 19 genera and 40 species (Nelson, 1994). Parenti (1981) combined
the typical live-bearing goodeids from Mexico with 2 genera of egg layers
from Nevada that were formerly placed in the Fundulinae of the
Cyprinodontidae. She divided the newly rearranged Goodeidae into two
subfamilies, the Empetrichthyinae and the Goodeinae. Miller and Smith
(1986), however, assigned the 2 genera of the Empetrichthyinae to a separate family, Emptrichthyidae. Robins *et al.* (1991) retained the Mexican
forms within the Cyprinodontidae. I follow Nelson (1994), who utilized
Parenti's (1981) classification, as does Eschmeyer (1998). Costa (1998b)
considered the Goodeidae *sensu lato* to be the sister group of the
Profundulidae.

The Empetrichthyinae lacks pelvic fins and pelvic fin skeletons. Males
do not have an intromittent organ and fertilization is external. These 6-cm-
long fishes have large molariform pharyngeal teeth and fleshy bases of the
dorsal and anal fins (Parenti, 1981). There are two genera, *Empetrichthys*
and *Crenichthys*. *Empetrichthys merriami* from Ash Meadows, Nevada, is

reportedly extinct. The last specimen was seen in 1948. Competition and predation by introduced fishes have apparently exterminated all populations of the Ash Meadows killifish (Soltz and Naiman, 1978). *Empetrichthys latos*, known as the Pahrump killifish, is native to three springs in Pahrump Valley, Nye County, Nevada. However, removal of water for irrigation has eliminated *E. latos* from its native springs (Page and Burr, 1991). Of the three described subspecies (Miller, 1948), only *E. l. latos* survives. It now exists only in three refuge sites outside Pahrump Valley where it was transplanted to prevent its extinction (Soltz and Naiman, 1978). It is listed as an endangered species. Uyeno and Miller (1962) described a fossil species, *E. erdisi* from the Pleistocene of Nevada.

The genus *Crenichthys* is distinguished from *Empetrichthys* by the presence of bicuspid outer teeth, a scaled anal fin base, and 20 or more gill rakers on the first arch as opposed to 12 or 13 in *Empetrichthys*. There are two species. The White River springfish, *C. baileyi*, is an endangered species in warm springs of the White River system of Nevada. It is relatively common but threatened by human encroachment and introductions of nonnative fishes (Page and Burr, 1991). The Railroad Valley springfish, *C. nevadae*, is common in a very small area of warm (36–38°C) springs in Railroad Valley, Nye County, Nevada (Hubbs et al., 1974). The pool at the head of one of the springs measured only 4 × 6 m. It is listed as a threatened species. Reviews of topics related to desert fishes including hydrology, refuges, conservation, and life history can be found in Naiman and Soltz (1981) and Minckley and Deacon (1991).

The subfamily Goodeinae consists of about 17 genera and 36 species of live-bearers from the Mesa Central, Mexico, with most species concentrated in the Rio Lerma basin (Parenti, 1981). This adaptive radiation occurred in the cool, tropical highlands (915–2130 m) of western Mexico where there is a paucity of primary freshwater fishes (Miller and Fitzsimons, 1971). The anal fin of males is split and goodeids are sometimes called splitfins. The anterior rays of the anal fin are short and close to one another and somewhat separated by a notch from the rest of the fin as exhibited by *Skiffia lermae* (Fig. 219). Males possess an internal, muscular, gonopodium-like structure called a pseudophallus that connects the sperm ducts to the genital opening. It is thought that this structure transmits sperm to the female (Fitzsimons, 1972). The eggs are small and have little yolk. Goodeids give birth to live young (viviparity) that receive nutrition from the female through a structure analogous to the mammalian placenta. This structure is called a trophotaenia and consists of 2–12 ribbon-like extensions that hang from the anal region of developing embryos. The trophotaeniae presumably absorb nutrients from the ovarian fluid in which they are bathed. See Hubbs and Turner (1939) and Miller and Fitzsimons (1971) for descriptions of this strange arrangement. *Ataeniobius toweri* is the only live-bearing goodeid that does not have trophotaeniae. It is regarded as the

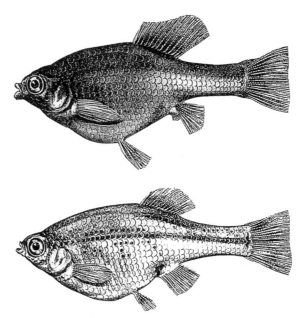

FIGURE 219. *Skiffia lermae* (male, top; female, bottom) (Meek, 1904, Plate VIII).

most primitive goodeid by Hubbs and Turner (1939). The caudal fin in embryos of *A. toweri* is much larger and more highly vascularized than that in other goodeids. These fishes also have relatively little yolk. It is assumed that the caudal fin is the site of nutrient absorption.

Body form ranges from streamlined to deep-bodied, and various species are carnivores, herbivores, or omnivores. Maximum size is about 20 cm TL, but most species are half that size (Fitzsimons, 1972). Females are larger than males. Goodeids are found in a variety of habitats, including warm springs, small and large lakes, swift-flowing streams, marshes, and ditches. Although primarily a highlands group, some species descend as low as 180 m (Miller and Fitzsimons, 1971).

Hubbs and Turner (1939) revised the family, and Turner (1946) added new information. Miller and Fitzsimons (1971) commented on the classification, and their paper is credited with introducing the goodeids to aquarists via *Ameca splendens* (Dawes, 1991). It is necessary to keep at least two fish together in the same tank because "one goodeid deserves another" (J. M. Fitzsimons, personal communication). Fitzsimons (1972, 1976) described courtship, karyotypes, and isolating mechanisms for several species. Uyeno *et al.* (1983) provided karyotypes for 35 species. Genera include *Allophorus, Allotoca, Characodon, Goodea* (represented by *G. atripinnis*; Fig. 220), *Skiffia, Xenoophorus,* and *Xenotoca* (Uyeno et al., 1983; Smith and Miller, 1987). The oldest known fossil goodeid, *Tapatia occidentalis*, was found in Miocene deposits in the state of Jalisco, Mexico

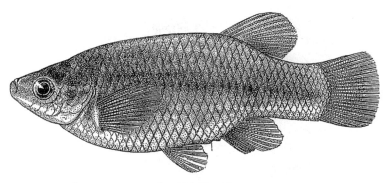

FIGURE 220. *Goodea atripinnis* (Meek, 1904, Fig. 43).

Map references: Costa (1998b),* Lee *et al.* (1980),* Page and Burr (1991),* Parenti (1981),* Uyeno and Miller (1962),* Uyeno *et al.* (1983)*

Class Actinopterygii
Subclass Neopterygii

Order Cyprinodontiformes; Suborder Cyprinodontoidei
(2d) Family Cyprinodontidae—**pupfishes, killifishes** (si-prin'-ō-don'-ti-dē)

THE CYPRINODONTIDAE LIVES IN FRESH, BRACKISH, AND coastal marine waters of North, Middle, and South America, the Caribbean, and the Mediterranean Anatolian regions including north Africa. Nine genera and about 100 species are recognized in the family as structured by Parenti (1981, 1984b), who removed *Profundulus, Fundulus, Valencia,* and other genera to their own families in her 1981 revision of the cyprinodontiform fishes. These small, egg-laying fishes are divided into two subfamilies, Cubanichthyinae and Cyprinodontinae.

The Cubanichthyinae, created by Parenti (1981), contains two species, *Cubanichthys cubensis* from Cuba and *C. pengelleyi* from Jamaica. An enlarged supraoccipital crest and several other osteological features separate this subfamily from other cyprinodontiforms. *Cubanichthys* had been considered an island form of *Lucania* (now Fundulidae) (Rosen, 1975). For field and aquarium observations on *C. pengelleyi,* consult Foster (1969).

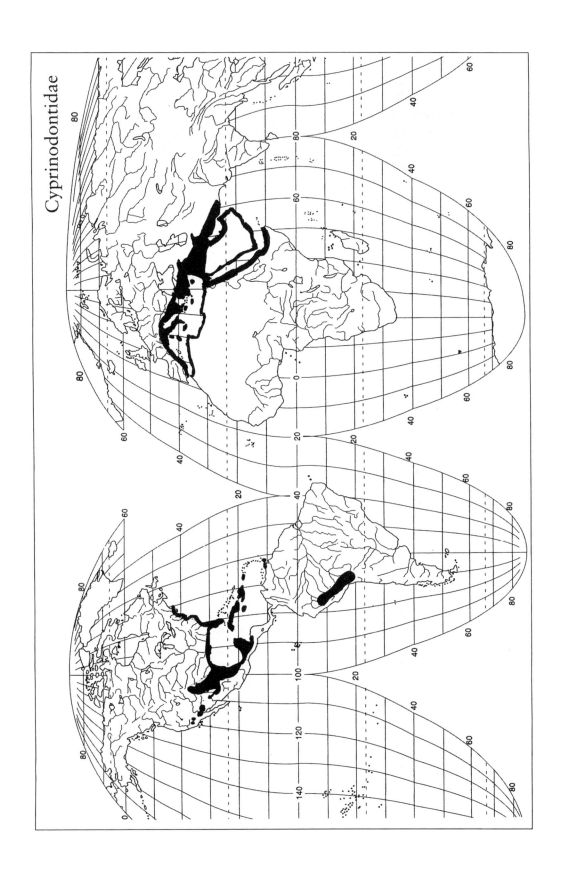

Cyprinodontidae

The subfamily Cyprinodontinae is divided into two tribes. The tribe Orestiini is composed of three genera and about 57 species (Parenti, 1981, 1984b). These fishes have an extremely robust lower jaw due to a medial extension of the dentary. The genus *Aphanius* consists of about 10 species that occur in fresh or brackish waters along the north coast of Africa, Spain, Italy, Greece, and Turkey and along the coast of the Arabian peninsula (Parenti, 1981; Krupp, 1983). Pelvic fins are absent in *A. apodus* but present in the other species. The genus is probably not monophyletic (Parenti, 1981). Females are larger than males, and the males often have enlarged dorsal and anal fins. *Aphanius dispar* (Fig. 221) reaches 70 mm TL and occurs in oasis pools of hypersaline to fresh water, landlock populations in Saudia Arabia and Iran, and coastal zones bordering northeast Africa and the Middle East (Wildekamp et al., 1986). The genus *Kosswigichthys* consists of 4 species from freshwater lakes of Turkey. Some species lack scales.

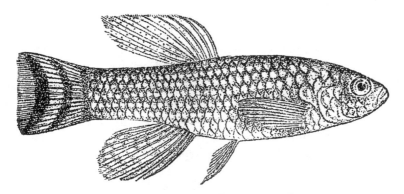

FIGURE 221. *Aphanius dispar*, male, as *Cyprinodon dispar* (Boulenger, 1915, Fig. 12).

The genus *Orestias* includes about 43 species of Andean killifish distributed throughout the central Andean highlands from Ancash Province of northern Peru to Antofagasta Province of northern Chile (Parenti, 1984b,c). More than half of these species are endemic to Lake Titicaca at 3810 m above sea level on the border between Peru and Bolivia. This is the highest freshwater lake inhabited by fishes. It is also incredibly deep at 281 m. This lake, the highest large navigable lake in the world, is 122 km long and 45 km wide, encompassing an area of 8284 km2. *Orestias* has radiated into a variety of forms and, in the absence of other fish families save one catfish (*Trichomycterus*), fills many ecological niches normally occupied by other groups. There are inshore forms with deep bodies, slender offshore forms, piscivores, planktivores, and miniature versions of normal killifish (Parenti, 1984b,c). The predacious *O. cuvieri* is the largest member of the genus at 22 cm SL, and members of the *O. gilsoni* complex are sexually mature at 30 mm SL. The *Orestias* in Lake Titicaca is considered an assemblage of several species flocks because *Orestias* is not monophyletic (Parenti, 1984c). There are several distinct groups of *Orestias* that have undergone adaptive radiation in Lake Titicacca and also Lake Poopó, Bolivia. Brown and rainbow trout and the silverside, *Basilichthys bonairensis*, have been introduced into Lake Titicacca and pose a threat to *Orestias* (Parenti, 1984b).

The remaining tribe, Cyprinodontini, consists of the pupfishes that inhabit North and Central America to Honduras and the West Indies in a southward arc to Venezuela (Parenti, 1981). There are 5 genera and about 42 species. Pupfishes tend to be deep bodied with a deep, strongly compressed caudal peduncle, unlike the slender topminnows and killifish (*Fundulus*). They have a depressed head and upturned mouth and no lateral line. Males may have an arched back. The name of the most specious genus (36 species), *Cyprinodon*, means toothed carp and refers to the jaw teeth that minnows lack. The common name pupfish reflects the reproductive behavior that was interpreted as "playing like puppies."

These fishes are noted for their remarkable abilities to tolerate a wide range of environmental conditions which are most extreme in the Death Valley system of southern California and western Nevada. The six subspecies of Amargosa pupfish, *Cyprinodon nevadensis*, are capable of spending various lengths of time in water up to 42°C as they cruise up and down their shallow desert streams and marshes (Soltz and Naiman, 1978). Temperatures may fluctuate as much as 20°C in one day. The Cotton Marsh pupfish, *C. salinus milleri*, has been reported living in water of 160 ppt, which is 4.5 times as salty as the sea (Soltz and Naiman, 1978). However, most pupfish prefer salinities between 22 and 45 ppt, which is still remarkable for a freshwater fish. The salinity of ocean water is typically 35 ppt. Simpson and Gunter (1956) reported sheepshead minnows, *C. variegatus* (Fig. 222), from waters of 142 ppt along the Texas coast during a drought. Desert and coastal *Cyprinodon*, faced with such high salinities, must

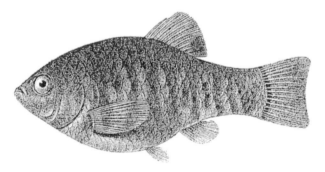

FIGURE 222. *Cyprinodon variegatus* as *C. gibboseus* (Girard, 1859, Plate CXIV, Fig. 5).

osmoregulate like saltwater fishes. That is, they ingest water and excrete monovalent ions via active transport of specialized chloride cells in the gills. Fertile hybrids of most *Cyprinodon* species have been obtained in laboratory experiments (Turner and Liu, 1977).

The story of the Devils Hole pupfish, *Cyprinodon diabolis*, is an interesting one and is documented by Soltz and Naiman (1978). This species is only found in a pool at Devils Hole, Ash Meadows, Nevada. This is the smallest range of any known vertebrate species. This pool is 15 m below the surface of the surrounding desert at an elevation of 730 m. It is the hydrostatic head of a large underground aquifer that has no surface outlet.

The pupfish rely on a shallow limestone shelf of about 20 m². Algae and small invertebrates live on the shelf and provide food for the pupfish. Spawning takes place over the shelf. In the early 1970s this shelf was becoming exposed due to the pumping of water from the underground aquifer for irrigation. This would have caused the extinction of the only population of this species. A legal battle led to the U.S. Supreme Court and a ruling was issued that pumping had to be curtailed to the extent that the shelf remained covered with water in order to ensure a stable population of pupfish. The *C. diabolis* population fluctuates around 200–400 individuals. An artificial refugium has been established below Hoover Dam as a precaution in case Devils Hole is disrupted.

Devils Hole has been isolated from surface waters for at least 10,000–20,000 years. Thus, *C. diabolis* has evolved in isolation for a long time, and this has resulted in substantial evolutionary changes from *C. nevadensis*, its closest relative. *Cyprinodon diabolis* is the smallest pupfish at about 25 mm SL. It lacks pelvic fins and has retained some juvenile characteristics such as large head and eye, long anal fin, and reduced pigmentation. These may be adaptations for survival in conditions of sparse nutrition.

For further information on the environments, physiological capabilities, life histories, and conservation of desert fishes see Miller (1948), Hubbs *et al.* (1974), Soltz and Naiman (1978), Naiman and Soltz (1981), Minckley and Deacon (1991), and Wright (1999).

The tribe Cyprinodontini includes four genera in addition to *Cyprinodon*. The flagfish, *Jordanella floridae* (Fig. 223), is found in vegetated ponds and slow-flowing steams and brackish waters in peninsular Florida north to St. Johns and Ochlockonee River drainages. It has a large black spot on the side of the body and an elongate dorsal fin of at least 15 rays. It reaches about 65 mm TL. *Jordanella pulchra* occurs in Yucatan peninsula south to Belize (Greenfield and Thomerson, 1997).

Cualac tessellatus is only known from San Luis Potosi, Mexico (Miller, 1956). *Floridichthys carpio* occurs along the Gulf coast from Key West to Cape San Blas, Gulf County, Florida (Stevenson, 1976). *Floridichthys polyommus* inhabits coastal waters in the Yucatan peninsula south to Honduras

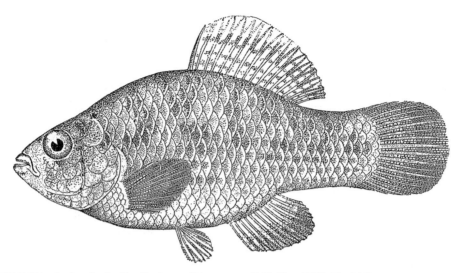

FIGURE 223. *Jordanella floridae* (Jordan and Evermann, 1900, Plate XCII, Fig. 298).

(Greenfield and Thomerson, 1997). Miller and Walters (1972) described *Megupsilon aporus* from Nuevo Leon, Mexico.

Map references: Krupp (1983),* Lee *et al.* (1980, 1983),* Maitland (1977),* Page and Burr (1991),* Parenti (1981, 1984b),* Tortonese (1986a),* Turner and Liu (1977)*

Class Actinopterygii
Subclass Neopterygii

Order   Gasterosteiformes; Suborder Gasterosteoidei
(Per) Family   Gasterosteidae—**sticklebacks**   (gas'-ter-ō-stē'-i-dē)

THE SERIES PERCOMORPHA INCLUDES THE MOST ADVANCED euteleostean fishes. Percomorphs have a connection between the pelvic girdle and the pectoral girdle, either directly or via a ligament. Rosen (1973b) recognized the Percomorpha and Johnson (1993) reviewed problems with the classification. I follow Nelson's (1994) treatment of this large and diverse group of 9 orders, 229 families, and more than 12,000 species. Most of these are marine fishes, but many major fresh-water fish lineages belong to the Percomorpha (Fig. 1). Johnson and Patterson (1993) presented an alternate arrangement of percomorph relationships.

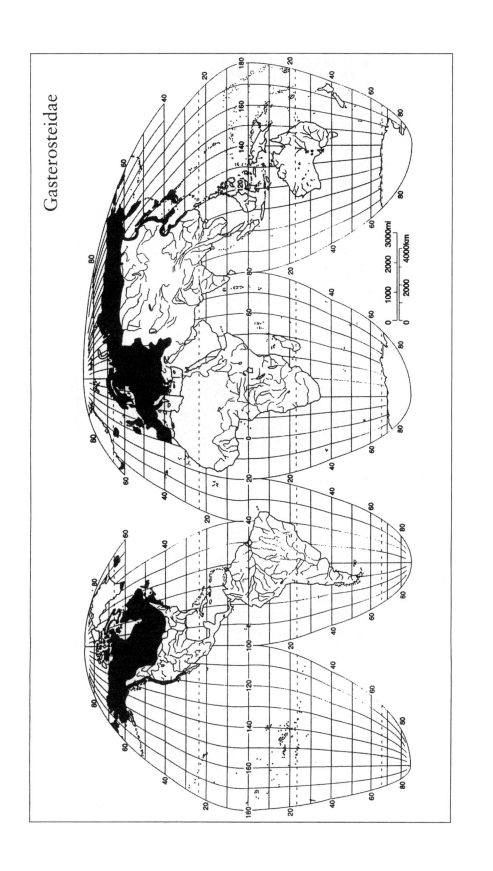

Gasterosteidae

The order Gasterosteiformes is composed of two suborders. The suborder Gasterosteoidei includes the stickleback family Gasterosteidae, and the suborder Syngnathoidei includes the pipefishes.

The Gasterosteidae is a family of small fishes found in marine, brackish, and fresh water of the Northern Hemisphere. The common name, stickleback, is derived from the 3–16 individual dorsal spines, each of which has its own membrane. These spines provide a defense mechanism against a wide variety of hungry predators such as *Esox*. The spines are followed by a normal dorsal fin. Pelvic fins also have a stout spine and up to three soft rays. A spine precedes the anal fin as well. The jaws are small and studded with teeth. The caudal peduncle is very narrow. There are five genera and only seven described species (Nelson, 1994). Several genera have species complexes of many different forms. There is a great deal of phenotypic plasticity within the freshwater species, and the stickleback group is very difficult taxonomically (Miller and Hubbs, 1969; Nelson, 1971; Blouw and Hagen, 1984).

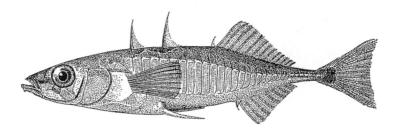

FIGURE 224. *Gasterosteus aculeatus* (Goode, 1884, Plate 181).

The *Gasterosteus aculeatus* (three-spine stickleback; Fig. 224) complex is found in fresh waters in a nearly circumpolar distribution. In some places it is anadromous. It also occurs in salt water along the Atlantic and Pacific coasts of both North America and coastal Eurasia (Scott and Crossman, 1973). It has been reported from north Africa near Algiers (Arnoult, 1986a). Marine forms may be heavily armored with bony plates, whereas freshwater forms are partially plated or lack plates and scales. The elaborate courtship behavior of sticklebacks has been well studied. In *G. aculeatus* the ventral surface of the male typically becomes bright red in the breeding season. The male constructs a tubular nest of plant material cemented together with a sticky kidney secretion. The female is lured to the nest and spawns. The male enters, fertilizes the eggs, and tends the eggs and young. He may entice several different females to spawn in the same nest. See Tinbergen (1952) for a classic study of stickleback instinctive behavior. Several books have been written about stickleback biology, including Wootton (1976, 1984) and Bell and Foster (1994). McLennan (1993) used behavioral characters to construct a phylogeny of the family. Coad (1981b) provided a bibliography of the sticklebacks.

A recent study of sympatric populations of G. *aculeatus* demonstrated that populations that evolved under different ecological conditions showed strong reproductive isolation from one another, whereas populations that evolved independently under similar ecological conditions lacked isolation (Rundle et al., 2000). For example, a large-bodied benthic form of three-spine stickleback that feeds on invertebrates in the littoral zone was reproductively isolated from a slender limnetic form that feeds on plankton in open waters of the same lake. Thus, the benthic and limnetic forms behave as biological species. However, the benthic form from a different lake did not develop reproductive isolating mechanisms from the benthic form from the original lake. The same applied to the limnetic forms from the two different lakes. This system provides a very informative window on the operation of natural selection and parallel speciation.

Gasterosteus wheatlandi, the black-spotted stickleback, is a marine species from the Atlantic coast from Newfoundland to New York (Scott and Crossman, 1973). There is also an undescribed marine species of *Gasterosteus*, known as the white stickleback, from the Atlantic coast of Canada (Blouw and Hagen, 1990; Jamieson et al., 1992). This form is very closely related to G. *aculeatus*. Haglund *et al.* (1990) found that sympatric populations of the white and red (G. *aculeatus*) sticklebacks were no more different with respect to allozymes than were allopatric populations of either form.

The nine-spine stickleback complex, *Pungitius pungitius*, including P. *occidentalis* in eastern North America, also has a circumpolar distribution in fresh and salt water (Haglund et al., 1992). Large bony lateral plates are absent from most populations except for a few populations on the Atlantic coast of North America. *Pungitius* spawns in fresh water (Scott and Crossman, 1973). The four-spine stickleback, *Apeltes quadracus*, occurs in coastal areas of eastern North America from the Gulf of St. Lawrence to Virginia. It has no bony plates on the sides, and it rarely enters fresh water. The brook stickleback, *Culaea inconstans*, is found in cool, clear, fresh waters throughout north-central North America. All of the sticklebacks mentioned previously are nest builders and are less than 10 cm TL. Life history information for the four previous genera can be found in Scott and Crossman (1973). Additional information on the brook stickleback is presented by Becker (1983).

Spinachia spinachia, the 15-spine stickleback, is an elongate marine species from the Atlantic of northern Europe. It is the largest stickleback at about 20 cm TL (Wooton, 1976). The oldest North American stickleback fossils are assigned to G. *aculeatus* from the late Miocene of California (Cavender, 1986).

Map references: Arnoult (1986a), Banister (1986),* Bell and Forster (1994),* Huglund *et al.* (1992),* Lee *et al.* (1980),* Maitland (1977),* Page and Burr (1991),* Morrow (1980),* Nelson (1976),* Scott and Crossman (1973),* Wootton (1976),* Ziuganov (1991)*

Class Actinopterygii
Subclass Neopterygii

Order Gasterosteiformes; Suborder Syngnathoidei
(Per) Family Syngnathidae—**pipefishes and seahorses**
(sing-nath'-i-dē) [sing-nā'-thi-dē]

THE SYNGNATHIDAE IS THE FAMILY OF PIPEFISHES AND seahorses which occur in shallow marine and brackish waters of the Atlantic, Indian, and Pacific Oceans. About 68% of the species are found in the Indo-Pacific region. A few species of pipefishes enter coastal rivers throughout the world. Family members have an elongate body encased in bony rings and a tubular snout with a small, toothless mouth. The gill openings are reduced to a pore on each side. Pelvic fins are absent. A primitive kidney without glomeruli is present on the right side only. The males incubate the eggs exposed or partly to completely concealed within a pouch under the trunk or tail (Dawson, 1982). There are two subfamilies, Syngnathinae and Hippocampinae.

The pipefishes are placed in the Syngnathinae. There are about 52 genera and 190 species (Dawson, 1982, 1985). About 17 species live in fresh water. Most of these are in the Indo-Pacific genus *Microphis* and 4 species are assigned to *Doryichthys*, also from the Indo-Pacific (Dawson, 1981). All 4 species of *Doryichthys* occur in western Broneo, and Roberts (1989a) mapped the distribution of this genus. Allen (1991) listed 3 species of *Hippichthys* and 8 species of *Microphis* from the fresh waters of New Guinea, including a pregnant male specimen of *M. spinachioides* taken 570 km upstream in a tributary of the Sepik River. An additional 35 pipefish species are euryhaline. For example, the Gulf pipefish, *Syngnathus scovelli*, can occur some distance upstream from Florida to Mexico. The 12-cm TL *Microphis brachyurus* (Fig. 225) inhabits freshwater streams, rivers, and estuaries in southern India (Talwar and Jhingran, 1992).

Pipefish species whose males have a pouch in which the female deposits eggs are considered more advanced than those in which the eggs develop and hatch while attached to a bare patch on the male's ventral surface. Embryos in the pouch receive nutrition from the male. The role reversal in

FIGURE 225. *Microphis brachyurus* as *Doryichthys bleekeri* (Day, 1878b, Plate CLXXIV, Fig. 3).

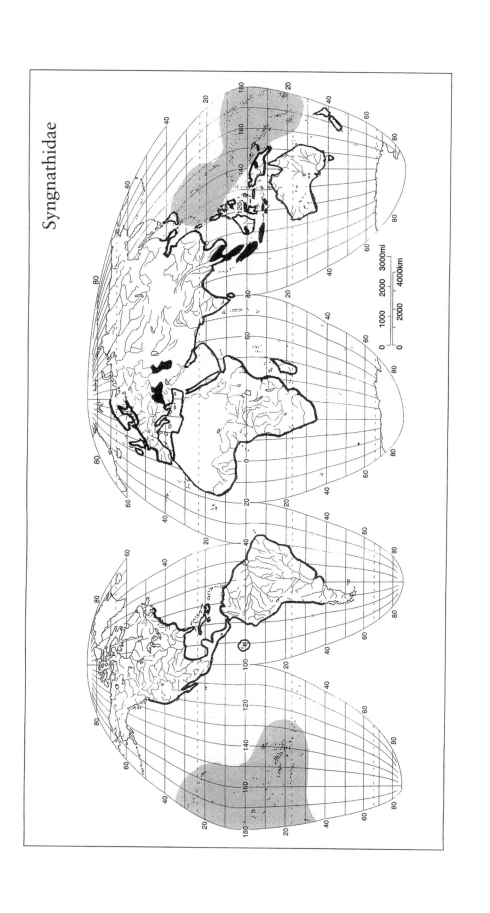

Syngnathidae

reproductive behavior is carried further by females that court and compete for males (Rosenqvist, 1990). Color photographs and descriptions of mating and pregnancy in the marine leafy sea dragon, *Phycodurus eques*, from Western Australia are provided by Groves (1998). Pipefish have a caudal fin and their tail is not prehensile. Their head is in line with the body like a normal fish.

The pipehorse, *Acentronura dendritica* (= *Amphelikturus dendriticus*), from the western Atlantic has a coiled, prehensile tail and a caudal fin, and its head is flexed at about 45° from the long axis of the body. It is a mixture of pipefish and seahorse characteristics; hence its common name.

The subfamily Hippocampinae includes about 25 species of seahorses in the genus *Hippocampus* (Vari, 1982b). Seahorses lack a caudal fin and their tail is prehensile. The long axis of the head is bent about 70–90°. They have a brood pouch under the tail that opens through a pore directly beneath the dorsal fin. Seahorses are marine and do not enter freshwater; however, Myers (1979) reported the possible presence of a seahorse from the Mekong River of Thailand. Consult Herald (1959) for a discussion of the relationship between seahorses and pipefishes. Herald's (1962) popular account of pipefishes and seahorses is still worth reading.

Map references: Allen (1991),* Berg (1948/1949), Dawson (1982, 1985, 1986),* Lee *et al.* (1980),* Maitland (1977),* Roberts (1989a)*

Class   Actinopterygii
Subclass   Neopterygii

Order   Gasterosteiformes; Suborder Syngnathoidei
(Per) Family   Indostomidae—**indostomids**   [in-dos-tō'-mi-dē]

THIS ODD FAMILY CONSISTS OF THREE SOUTHEAST ASIAN species. Until recently, the genus *Indostomus* was known only from a single widespread species, *I. paradoxus* (Fig. 226). This small (2.7-cm) pipefish-

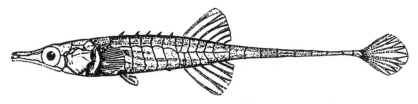

FIGURE 226. *Indostomus paradoxus* (reproduced with permission from Talwar and Jhingran, 1992, Fig. 243).

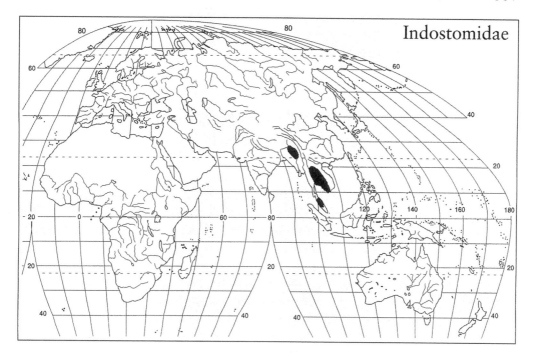

like species was first described from Lake Indawgyi in northern Myanmar (= Burma) about 100 km west of Myitkyina (Prashad and Mukerji, 1929). It also occurs in Kampuchea (= Cambodia) near Phnom Penh, western Malayasia in Trengganu province, and Thailand (Kottelat, 1982c, 1989). Britz and Kottelat (1999b) described two new species: *I. spinosus* from the Mekong basin in Thailand and Laos and *I. crocodilus* from the Malay Peninsula in Thailand.

The phylogenetic position of the Indostomidae is uncertain. Banister (1970) gave *Indostomus* ordinal status, but Pietsch (1978) argued to include the Indostomidae within the Gasterosteiformes, as did Johnson and Patterson (1993).

Indostomus has a thin, cylindrical body covered with bony scutes. The caudal peduncle is very long and slender, and there is a fan-like caudal fin. The small mouth is at the end of a tubular snout. The operculum is studded with six spines. Five isolated spines are anterior to the dorsal fin. The dorsal and anal fins are opposite each other and about the same size, with six rays each (Prashad and Mukerji, 1929; Talwar and Jhingran, 1992). *Indostomus* resembles a pipefish with the dorsal spines of a stickleback. Very little is known about the biology of these unusual fishes. They prefer swampy habitats with thick vegetation (Britz and Kottelat, 1999b).

Map references: Britz and Kottelat (1999b), Kottelat (1982b), Prashad and Mukerji (1929)

Class   Actinopterygii
Subclass   Neopterygii

Order   Synbranchiformes; Suborder Synbranchoidei
(Per) Family   Synbranchidae—swamp eels   [syn-bran'-ki-dē]

SWAMP EELS HAVE A REMARKABLY DISJUNCT DISTRIBUTION IN tropical and subtropical fresh waters of the New and Old Worlds. They occur in Mexico, Central and South America, Africa, Asia, and Australia (Rosen, 1975; Rosen and Greenwood, 1976). The largest number of species are in Asia and Australasia.

Synbranchids are not related to true eels (Anguilliformes), but they have an eel-shaped body. Fin reduction is the rule in this group. The dorsal and anal fins are reduced to rayless skin folds, and the caudal fin is reduced or absent except for a tiny fin in *Macrotrema*. Pelvic fins and girdles are absent, and pectoral fins are present only for about 2 weeks in larval synbranchids. During this time the fan-like larval pectoral fins direct a stream of oxygen-rich surface water across the thin, highly vascularized skin in a direction opposite blood flow (Liem, 1981). This countercurrent flow improves the oxygenation efficiency of the cutaneous blood vessels. As the young fish develop air-breathing capabilities, the pectoral fins are shed, usually in about 2 weeks (Liem, 1994). Scales are absent except in some species of *Monopterus*, and the eyes are small or vestigial. The gill openings unite to form a single pore or slit under the head or throat. This feature gave rise to the name Synbranchidae, which means "fused gills." The swim bladder is absent. Most species begin life as females and undergo a sex reversal up to 4 years later into males (Liem, 1968). Such a protogynous hermaphroditic condition is rare among freshwater fishes.

Swamp eels can live in water with very low oxygen due to the highly vascularized lining of the mouth and pharynx that allow air breathing. *Synbranchus* is a facultative air breather, whereas air breathing in *Monopterus* is obligatory (Graham, 1997). Synbranchids are also capable of extended amphibious excursions by snake-like undulations of their body. This is usually done during the wet season. Overland movements allows synbranchids to reach streams above waterfalls. *Monopterus* has been observed to feed while out of water (Liem, 1987). They hunt their prey by means of a well-developed sense of smell. During the dry season some synbranchids estivate in humid burrows that they dig by a cork-screwing motion of their body. Finlessness may be an adaptation for such movements (Liem, 1994). Laboratory experiments showed

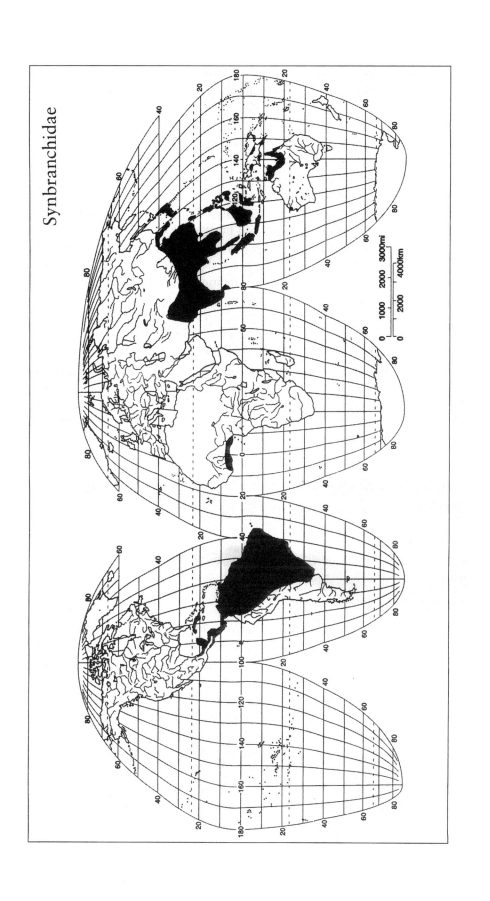

Synbranchidae

that *Synbranchus marmoratus* can survive up to 9 months in a drying burrow (Graham, 1997). Their enlarged urinary bladder may play a role in water retention during estivation. The Synbranchidae is among the most specialized assemblage of fishes, and Lauder and Liem (1983) consider the snakeheads, *Channa* (Channidae), to be closely related to the synbranchids.

Rosen and Greenwood (1976) revised this family of four genera and 15 species (Nelson, 1994). They recognized two subfamilies: Macrotreminae and Synbranchinae. The Macrotreminae consists of a single species, *Macrotrema caligans*, from fresh and brackish waters of Thailand and the Malay Peninsula. This is a small species reaching only about 20 cm TL. The eyes are reduced and sunken below the skin. The gill openings are large. This is thought to be the primitive condition. It is not an air breather and does not undergo sex reversal.

The Synbranchinae is composed of three genera. *Ophisternon* includes six species, two of which are from the New World and four from the Old World. This is a very unusual distribution pattern and may reflect continental drift (Rosen 1975). *Ophisternon infernale* is a blind form known from caves of Yucatan, Mexico (Hubbs, 1938). *Ophisternon aenigmaticum* occurs in northern South America, Guatemala, Mexico, and Cuba (Rosen and Greenwood, 1976). *Ophisternon bengalense* (Fig. 227) inhabits the

FIGURE 227. *Ophisternon bengalense* as *Symbranchus bengalensis* (Day, 1878b, Plate CLXVII, Fig. 2).

Indo-Malaysian region, the Philippines, and New Guinea. *Ophisternon gutturale* is from northern Australia and southern New Guinea, and *O. candidum* is only known from four specimens from a well in the Northwest Cape region of Western Australia (Allen, 1989). It is a white, depigmented form with no eyes. *Ophisternon afrum* occurs along the west African coast from Guinea Bissau to the Niger delta (Daget, 1992b).

There are two species of *Synbranchus* from the Neotropical region. *Synbranchus marmoratus* is the largest species in the family at 1.5 m TL. It is sympatric with *O. aenigmaticum* over much of its range from Mexico and Guatemala, both slopes of Central America, and Atlantic drainages of South America to Argentina (Rosen and Greenwood, 1976). *Synbranchus madeirae* is known from the Río Madeira in the Bolivian Amazon.

Monopterus contains six widely distributed, highly predacious species, including three species that were formerly included in *Amphipnous* that some authorities placed in a separate family, Amphipnoidae. Rosen and Greenwood (1976) synonymized *Amphipnous* with *Monopterus*. *Monopterus albus* occurs from India through Southeast Asia north to Japan. Inger and Kong (1962) reported it from north Borneo. *Monopterus albus* reaches the northeast coast of Australia. Lake (1971) thought it was introduced to Australia, but Rosen and Greenwood suggested that it may be native. There are only two reports of its occurrence in Australia (Merrick and Schmida, 1984). This species, known as the rice eel, reaches 90 cm TL and is utilized for food. It can survive the seasonal draining of rice paddies by burrowing deep into the mud. Its gills are vestigial. *Monopterus boueti* is known from Sierra Leone, Liberia, and Ivory Coast (Rosen and Greenwood, 1976; Daget, 1992b). *Monopterus cuchia* inhabits fresh and brackish waters of Pakistan, northern and northeastern India, Nepal, and Burma (Talwar and Jhingran, 1992). This species and M. *fossorius* and M. *indicus*, both from western India, possess pharyngeal pouches for air breathing. These species were formerly in the Amphipnoidae. *Monopterus eapeni* is a cavernicolous species that inhabits subterranean waters and wells in the southwest Indian state of Kerala (Talwar and Jhingran, 1992).

Monopterus albus is established in Florida and Georgia probably due to an aquarium release or a fish farm escape (Fuller et al., 1999). It was brought to Hawaii as a food fish before 1900.

Map references: Allen (1989, 1991),* Bussing (1998),* Daget (1992b),* Kottelat *et al.* (1993), Lee *et al.* (1983),* Masuda *et al.* (1984), Roberts (1989a), Rosen (1975),* Talwar and Jhingran (1992)

Class Actinopterygii
Subclass Neopterygii

Order Synbranchiformes; Suborder Mastacembeloidei
(1st) Family Chaudhuriidae—**earthworm eels**
(chowd-hū-rī'-i-dē) [chowd-hū-rē'-i-dē]

THE CHAUDHURIIDAE AND THE MASTACEMBELIDAE WERE formerly thought to be derived from a perciform ancestor (Greenwood

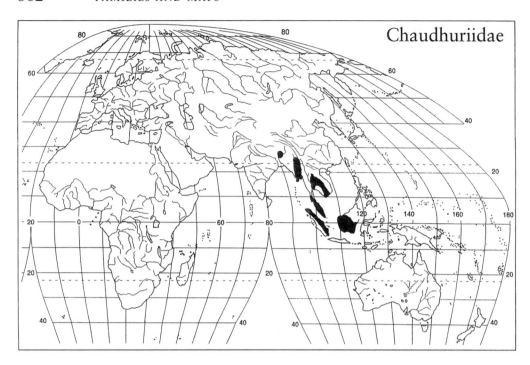

Chaudhuriidae

et al., 1966), but recent studies place them in the suborder Mastacembeloidei of the Synbranchiformes (Gosline, 1983; Travers, 1984a,b; Johnson and Patterson, 1993).

Members of the Chaudhuriidae have a brown, worm-like body. The long dorsal and anal fins are continuous to or confluent with the caudal fin. There are no dorsal and anal fin spines, and pelvic fins are absent. *Rhynchobdella sinensis* has dorsal and anal spines and is placed is this family by Travers (1984b); however, Kottelat and Lim (1994) consider it a mastacembelid. The pectoral fins are small in some species. The snout is blunt without a rostral appendage as is present in the Mastacembelidae. The body is naked or covered with minute scales embedded in the skin (Annandale, 1918; Kottelat and Lim, 1994). They are small fishes reaching a maximum TL of about 75 mm SL and are found in dense vegetation in still or slowly moving waters. They are extremely slippery and difficult to handle, even when preserved (Kottelat and Lim, 1994).

The family distribution ranges from northeastern India through Thailand, Malaysia, and to Borneo. There are five genera with seven species (Travers, 1984b; Kottelat, 1989, 1991b; Kottelat and Lim, 1994). *Chaudhuria caudata* (Fig. 228) is known from Inle Lake in Burma and from near Bangkok and a Mekong tributary in Thailand (Annandale, 1918; Roberts, 1971c, 1980). Its caudal fin is distinct from the dorsal and anal fins. The Indian *Chaudhuria khajuriai* (Fig. 229) has confluent caudal and anal fins. *Nagaichthys filipes* occurs in western Borneo and western

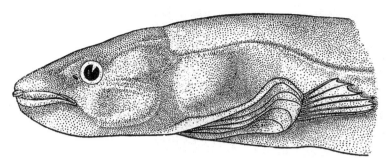

FIGURE 228. *Chaudhuria cauda* (Annandale, 1918, Plate IV, Fig. 1).

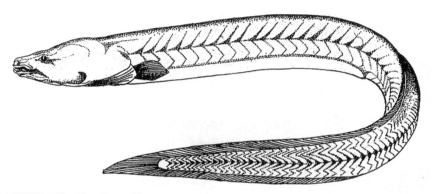

FIGURE 229. *Chaudhuria khajuriai* (reproduced with permission from Talwar and Jhingran, 1992, Fig. 296).

Malaysia (Kottelat, 1991b). Its dorsal, anal, and caudal fins are joined, and the rudimentary pectoral fin has a single filamentous ray.

Pillaia indica and *P. khajuriai* are found in Assam in northeast India. Talwar and Jhingran (1992) list these species as *Chaudhuria*. The caudal fin of *Pillaia* is confluent with the dorsal and anal fins, and this genus possesses a fleshy rostrum (Yazdani, 1972). Yazdani (1976) assigned *Pillaia* to its own family, the Pillaiidae. These fishes are found in heavily vegetated, slow-moving streams and spend most of their time buried in the bottom mud (Yazdani, 1978).

Kottelat and Lim (1994) described *Chendol*, a new genus characterized by fused dorsal, caudal, and anal fins. *Chendol keelini*, from the Malay Peninsula, is the only chaudhuriid with scales, lateral line, and a cephalic line system. *Chendol lubricus* occurs in east Borneo and is naked and has very small eyes. *Bihunichthys monopteroides* from the Malay Peninsula is minute, reaching only 36 mm SL. Its caudal fin, consisting of a single ray, is not fused with the dorsal and anal fin, and the pectoral fin is composed of only two rays (Kottelat and Lim, 1994).

Map references: Kottelat (1989, 1991b), Kottelat and Lim (1994), Nichols (1943), Roberts (1971c, 1980), Travers (1984b)*

Class Actinopterygii
Subclass Neopterygii

Order Synbranchiformes; Suborder Mastacembeloidei
(1st) Family Mastacembelidae—**spiny eels**
(mas'-ta-sem-bel'-i-dē)

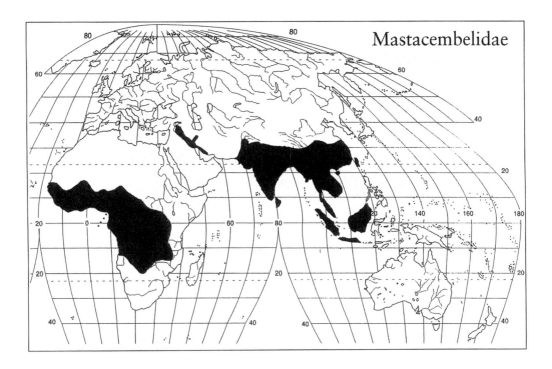

SPINY EELS ARE FOUND IN FRESH WATERS OF TROPICAL AFRICA and in Asia from the Euphrates area of Syria through the Indian region, the Malay Archipelago, and China. Some Asian, but not African, species are found in brackish waters (Seegers, 1996). Mastacembelids are called spiny eels because they have an eel-like body with 9–42 isolated, depressible dorsal spines anterior to the long dorsal fin that has 52–131 soft rays. The anal fin is also long with 2 or 3 spines and 30–130 rays. Talwar and Jhingran (1992) reported that some species will wiggle backward when handled, thus spiking a fisher's hand with the suddenly erect dorsal spines. Smith (1945) stated that the stout spines can inflict painful wounds.

Mastacembelids have a fleshy rostral appendage of varying length, at the end of which are a pair of tubular, anterior nares. The posterior nares are near the small eyes (Gosline, 1983). This specialized rostrum presumably functions in the location of aquatic insects and worms (Roberts, 1980). Pelvic fins are absent, and the body of most species is covered with small, cycloid scales. The caudal fin is reduced. Some mastacembelids are brightly colored and kept as aquarium animals. All are utilized for food. Maximum size is about 90 cm TL for the Asian *Mastacembelus erythrotaenia* (Rainboth, 1996), whereas some African and Asian species reach only about 14 cm (Roberts, 1980). Some species burrow into the bottom mud during the dry season, but they apparently lack an accessory air-breathing organ (Graham, 1997). They do have thin, vascularized skin that may be suitable for aerial gas exchange.

There are 5 genera and about 70 species in two subfamilies: Mastacembelinae and Afromastacembelinae (Travers, 1984a,b, 1992b; Roberts, 1980, 1986b, 1989a; Kottelat, 1989; Kottelat et al., 1993; Kottelat and Lim, 1994; Vreven and Teugels, 1996, 1997). The Mastacembelinae is a south Asian subfamily that ranges from Iran to China and Borneo and contains 2 genera, *Macrognathus* (approximately 12 species) and *Mastacembelus* (approximately 13 species). Members of this subfamily have a distinct caudal fin, not confluent with the dorsal and anal fins. The rim of the anterior nostril of *Macrognathus* has six finger-like fimbriae (except in 1 species), whereas the nostril rim of *Mastacembelus* has two fimbriae and two broad flaps (Roberts, 1986b). Roberts (1980) speculated that the rostrum of *Macrognathus* functions in food gathering in addition to its sensory function. *Macrognathus zebrinus* (Fig. 230) from Burma displays the typical spiny eel shape and flexible rostrum. *Mastacembelus*

FIGURE 230. *Macrognathus zebrinus* as *Mastacembelus zebrinus* (Day, 1878b, Plate LXXII, Fig. 3).

mastacembelus is the Middle Eastern species. Kottelat and Lim (1994) reviewed the nomenclature of *Rhynchobdella sinensis*, which they placed in the Mastacembellidae, whereas Travers (1984b) considered it in the Chaudhuriidae. Kottelat and Lim (1994) proposed that *Sinobdella* replace *Rhynchobdella*.

The African species of mastacembelids are assigned to the subfamily Afromastacembelinae. In these fishes, the dorsal and anal fins are confluent with the caudal fin. Before Travers' (1984b) revision all African species except *Caecomastacembelus brichardi* were described and placed in *Mastacembelus*. Vreven and Teugels (1996) and Seegers (1996) discussed the confusion in the diagnoses of the African genera of spiny eels. A generic revision is needed. Currently, two African genera are recognized: *Caecomastacembelus* and *Aethiomastacembelus* (Travers, 1988). The latter genus contains most of the species Travers (1984b) assigned to *Afromastacembelus*. *Caecomastacembelus* includes 24 species that usually have more than 100 anal rays, whereas *Aethiomastacembelus* has about 20 species and less than 95 anal rays (Travers, 1992b; Vreven and Teugels, 1996, 1997). *Caecomastacembelus ansorgii* (Fig. 231) from Angola is a typical subfamily representative. It reaches 445 mm TL.

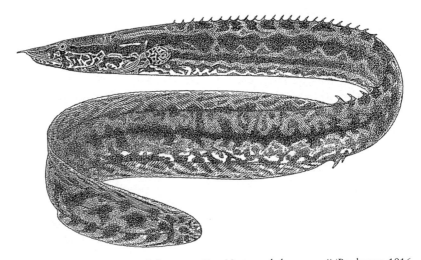

FIGURE 231. *Caecomastacembelus ansorgii* as *Mastacembelus ansorgii* (Boulenger, 1916, Fig. 75).

Afromastacembelids are found in tropical Africa, but not in northern or southern Africa (Travers, 1984a,b, 1988, 1992a,b; Travers et al., 1986). The Congo basin including Lake Tanganyika is a center of diversity for this group in Africa (Roberts, 1980; Travers et al., 1986). Copley (1941) reported spiny eels from the Athi River basin in Kenya.

Map references: Copley (1941), Roberts (1980,* 1989a), Skelton (1993),* Travers (1984b, 1992b),* Vreven and Teugels (1996, 1997)*

Class Actinopterygii
Subclass Neopterygii

Order Scorpaeniformes; Suborder Cottoidei
(per) Family Cottidae—**sculpins** (kot'-i-dē)

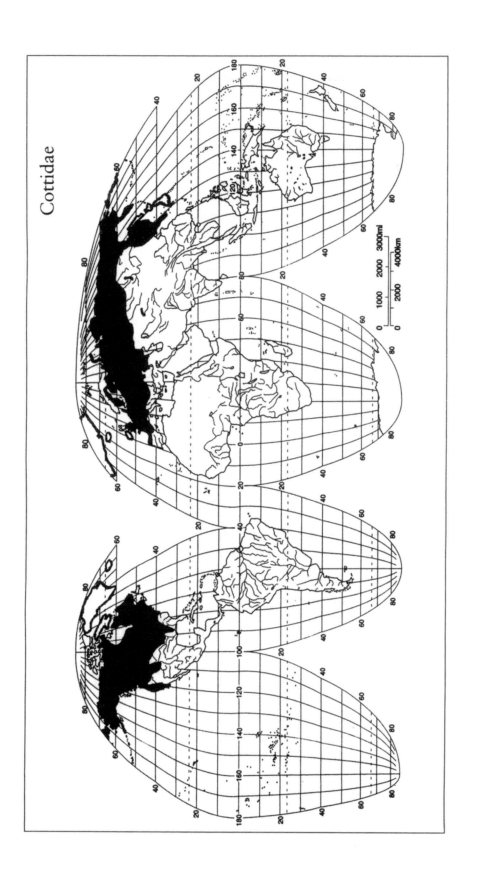

Cottidae

FISHES OF THE ORDER SCORPAENIFORMES ARE CALLED "MAIL-cheeked" fishes because they have a bony strut that runs under the eye, across the cheek to the preoperculum. Many species have head spines attached to this suborbital stay that protect the eye. Scorpaeniform fishes tend to be benthic with rounded pectoral and caudal fins. There are 25 families, 266 genera, and about 1271 species (Nelson, 1994). This order is overwhelmingly marine. The only freshwater scorpaeniforms are about 52 species in the suborder Cottoidei (Yabe, 1985). Washington *et al.* (1984) discussed scorpaeniform relationships.

The taxonomic position of the Scorpaeniformes is controversial. The suborbital stay is the one unifying characteristic, and it is not certain that the group is monophyletic. Nelson (1994) treated the Scorpaeniformes as a preperciform sister group, as did Eschmeyer (1998). Lauder and Liem (1983) and Roberts (1993) considered this group to be a perciform derivative and listed it after the perciforms.

The Cottidae is a marine and freshwater family of about 70 genera and 300 species that has a circumpolar distribution in the Northern Hemisphere (Nelson, 1994) and 4 deep-water species of the genus *Antipodocottus* from the eastern Australia–New Guinea–New Zealand region (DeWitt, 1969; Nelson, 1990). These Southern Hemisphere marine cottids are not represented on the map. The greatest diversity is in North Pacific coastal areas.

Cottids are big-mouthed, bottom-dwelling fishes with large heads and fan-like pectoral fins. Their bodies are often covered with tubercles or prickles. A few scales may be present. The first dorsal fin is spiny, but the anal fin lacks spines. The pelvic fins have one spine. The gill covers are often studded with strong spines. A swim bladder is absent. Cottids are egg layers. Marine species can reach a larger size—78 cm TL in *Scorpaenichthys marmoratus*—whereas freshwater species tend to be smaller and rarely exceed 17 cm TL.

The genus *Cottus* has a circumpolar distribution and includes about 42 species (counting several undescribed forms) in fresh water, including *C. kessleri* from Lake Baikal in Siberia. The slimy sculpin, *C. cognatus*, occurs in both North America and northeast Asia (Scott and Crossman, 1973). Thirteen species are Eurasian, and 28 are North American (Jenkins and Burkhead, 1994). The Klamath Lake sculpin, *C. princeps* (Fig. 232), live along the rocky and sandy shores of Upper Klamath and Agency Lakes in Oregon. It has been extripated from Lost River, Oregon (Page and Burr, 1990). *Cottus bairdi*, the mottled sculpin, is a widespread, common inhabitant of streams with a disjunct distribution in eastern and western North America. *Cottus pygmaeus*, the pygmy sculpin, is known only from Coldwater Spring within the Coosa River system near Anniston, Alabama. It only reaches 38 mm SL as an adult, and it is a threatened species (Mettee et al., 1996). *Cottus gobio*, known as the bullhead, is widespread in European streams. The largest *Cottus* is the prickly sculpin, *C. asper*, of the

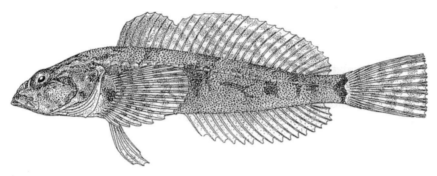

FIGURE 232. *Cottus princeps* as *C. evermanni* (Jordan and Evermann, 1900, Plate CCXCII, Fig. 708).

Pacific slope that reaches 192 mm TL, although there is a dubious record of 305 mm (Scott and Crossman, 1973).

Most *Cottus* are sedentary and well camouflaged against the bottom with a blotched or mottled color pattern. They are ambush predators that lie and wait for passing prey, such as small crustaceans, insects, and fishes, that they inhale into their gaping mouth. Sculpins are often accused without much justification of preying on salmonid eggs and fry, but Moyle (1977a,b) pointed out their beneficial effects such as serving as a food source for salmonids.

Previously, some sculpins from Lake Baikal were classified in the family Cottocomephoridae (Sideleva, 1980). These genera—*Cottocomephorus, Paracottus*, and *Batrachocottus*—are now placed in the subfamily Cottocomephorinae of the Cottidae (Nelson, 1994). Lake Baikal is the deepest lake in the world at 1741 m (Kozhov, 1963). It is also one of the most ancient. It forms a giant crescent 636 km long and up to 80 km wide at an elevation of 456 m amid the mountains of central Asia. It covers an area of 34,173 km². It is famous for the high degree of endemism of its biota. The cottocomephorids, Comephoridae, and Abyssocottidae

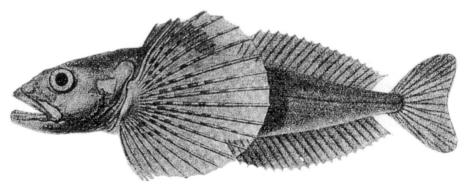

FIGURE 233. *Cottomephorus grewingki*, male (Berg, 1948/1949, Fig. 933).

represent an excellent example of an endemic freshwater fauna derived from marine ancestors via adaptive radiation (Smith and Todd, 1984). Some species are typical sculpin-like bottom feeders, whereas others feed in open-water habitats. *Cottocomephorus grewingki* (Fig. 233) is a pelagic species found along the entire coastline of Lake Bikal from shore to a depth of 300 m. Although it only reaches 19 cm, it was so plentiful that it supported a commercial fisheries (Berg, 1948/1949).

The oldest cottids are from late Miocene deposits of Lake Idaho in southeast Oregon (Cavender, 1986). Consult Jenkins and Burkhead (1994), Moyle (2000), Scott and Crossman (1973), and Berg (1948/1949) for life history information.

Map references: Berg (1948/1949), Lee *et al.* (1980),* Fedorov (1986),* Maitland (1977),* Page and Burr (1991),* Scott and Crossman (1973)*

Class Actinopterygii
Subclass Neopterygii

Order Scorpaeniformes; Suborder Cottoidei
(per) Family Comephoridae—**baikal oilfishes** [cō-mē-fōr'-i-dē]

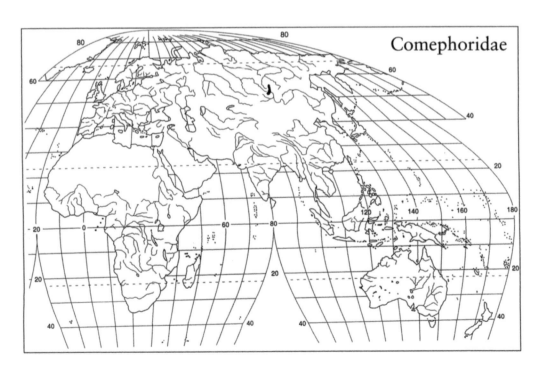

Comephoridae

THIS SMALL FAMILY CONSISTS OF TWO SPECIES: *COMEPHORUS baicalensis* (Fig. 234) and *C. dybowskii*. They are found only in Lake Baikal over deep water (1000 m) (Berg, 1948/1949; Sideleva, 1980). They are called oilfishes because of their very high fat content.

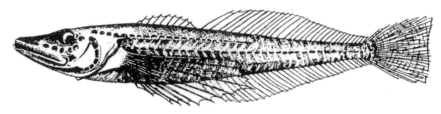

FIGURE 234. *Comephorus baicalensis*, female (Berg, 1948/1949, Fig. 936).

The body is naked and covered with thin skin, and living fish are translucent. The pectoral fins are very long, and pelvic fins are absent but pelvic bones are present. The head is not armed but has well-developed lateral line pores. The caudal fin is truncate, an unusual characteristic for a member of the Cottoidei. The preopercular and dentary bones are cavernous. This skeletal weight reduction and the high oil content are probably related to the pelagic habits of a fish derived from benthic cottid ancestors. Comephorids are ovoviviparous. They feed on the endemic amphipods of Lake Baikal (Kozhov, 1963). Maximum size is about 20 cm TL.

The fat content of *C. baicalensis* is as much as 25% of its body weight (Berg, 1948/1949). Dead fish are so buoyant that they freeze into the ice during winter. When the ice melts in spring, their bodies are washed up on shore by the waves. Spawning occurs in the summer months. The larvae spend the day at great depth but ascend to about 10 m during the night (Berg, 1948/1949). The females may die after giving birth to live young. Wheeler (1974) wrote that males comprise only 3% of the adult population. See Kozhov (1963) for information about Lake Baikal as a habitat.

Map references: Berg (1948/1949), Kozhov (1963)

Class **Actinopterygii**
Subclass **Neopterygii**

Order **Scorpaeniformes; Suborder Cottoidei**
(per) Family **Abyssocottidae—abyssocottids** [a-bis'-ō-kot'-i-dē]

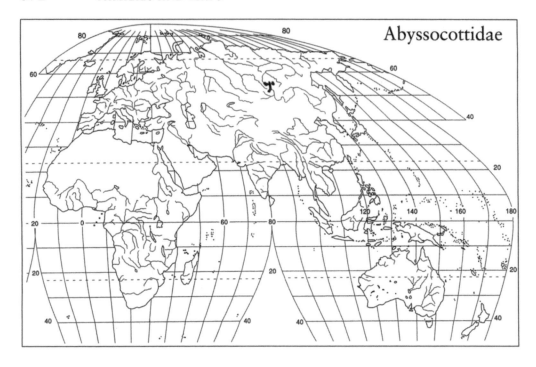

Abyssocottidae

THE ABYSSOCOTTIDS ARE FOUND PRIMARILY IN LAKE BAIKAL, but some species also occur in rivers near the Lake Baikal drainage system in Siberia such as the Vitim and Lena. There are 6 genera with 20 species (Nelson, 1994). These were formerly classified in the Cottocomephoridae and the Cottidae (Berg, 1948/1949).

Unlike the Comephoridae, these fishes have pelvic fins, each with one spine. Otherwise, abyssocottids resemble typical cottids. They are egg layers. The genera are *Asprocottus* and *Limnocottus*, with six species each, and *Abyssocottus*, *Cottinella*, *Neocottus*, and *Procottus* have one to three species each. *Neocottus werestschagini* (Fig. 235) lives on the silted sandy bottom of Lake Baikal at a depth of 877 m (Berg, 1948/1949). Nelson (1994) cited Russian language references for species descriptions. Many of these species were discussed by Berg in English.

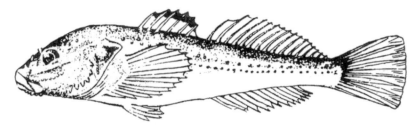

FIGURE 235. *Neocottus werestschagini* as *Abyssocottus werestschagini* (Berg, 1948/1949, Fig. 916).

Fishes of the Cottidae, Comephoridae, and Abyssocottidae comprise more than half of the 50 species of fishes found in Lake Baikal (Nelson, 1994). See Kozhav (1963) and Smith and Todd (1984) for a discussion of the species flock of cottoids found in Lake Baikal.

Map references: Berg (1948/1949), Kozhav (1963)

Class   Actinopterygii
Subclass   Neopterygii

Order   Perciformes; Suborder Percoidei
(per) Family   Centropomidae—**snooks**   (cen'-trō-pom'-i-dē)

THE PERCIFORMES IS THE LARGEST ORDER OF VERTEBRATES with about 9300 species (Nelson, 1994). There is enormous diversity of body form within this order, ranging from a tiny goby, *Trimmatom namus*, at 1 cm TL to the black marlin, *Makaira indica*, that can reach 4.5 m TL and weigh 900 kg (Johnson and Gill, 1994). To accommodate such diversity in one order, taxonomists have created 18 suborders, 148 families, and about 1500 genera of perciform fishes (Nelson, 1994). Needless to say, there is controversy in perciform classification. As presented here and by Nelson (1994), the order is not monoplyletic. However, Johnson and Patterson (1993) considered the Perciformes to be monophyletic if the Scorpaeniformes, Pleuronectiformes, and Tetraodontiformes are included. Johnson (1993) provided an overview of perciform classification.

The suborder Percoidei is the largest of the 18 suborders of Perciformes. It includes the most generalized perciformes in 71 families and about 2900 species. Only seven families contain about half of the species. One of these speciose families is a freshwater family (Percidae), and the rest are primarily marine. Only about 12% of percoids are found in fresh water, and about half of these are in the family Percidae (Nelson, 1994). Percoids are probably ancestral to the remaining Perciformes and perhaps even to the Pleuronectiformes and Tetraodontiformes.

Percoids are distinguished from typical lower teleosts such as the Ostariophysi (minnows, characins, and catfishes) and the Protacanthopterygii (pikes, smelts, and trouts) by the following characters: (i) Spines are present in the dorsal, anal, and pelvic fins; (ii) there are two dorsal fins; (iii) there is no adipose fin; (iv) scales are usually ctenoid (a few percoid species have cycloid scales); (v) pelvic fins are thoracic instead of abdominal; (vi) there is one spine and five soft rays in the pelvic fin; (vii) the

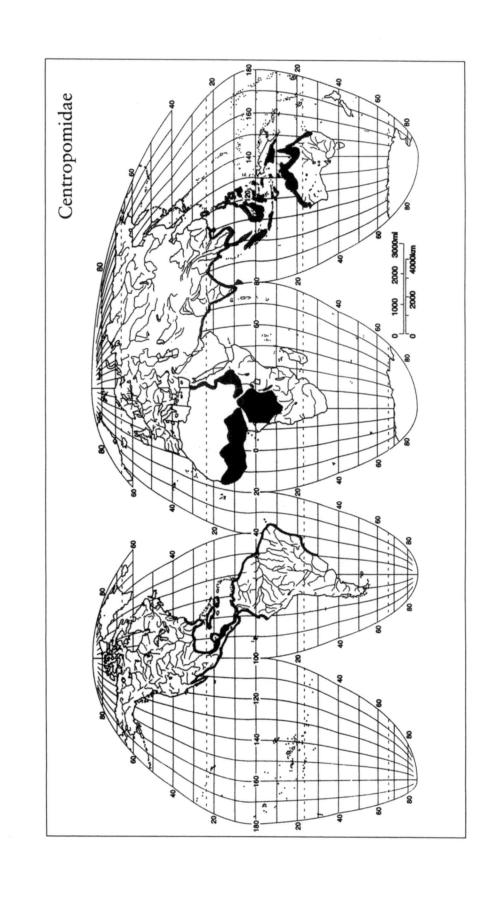

Centropomidae

upper jaw is bordered by the premaxilla, not the maxilla; (viii) the swim bladder is physoclistic and not connected to the gut by a duct as in physostomes; and (ix) there are 17 or fewer principal caudal fin rays (Nelson, 1994).

The Centropomidae is a widespread marine family that has some freshwater species, especially in Africa. Members of this family have a lateral line that reaches the posterior margin of the caudal fin, except in 1 species. Some species have three rows of lateral line canals on the tail. There are 22 species in three genera that are divided into two subfamilies (Nelson, 1994). Fossils of this family date to the lower Eocene of northern Italy. Greenwood (1976) revised the family.

The subfamily Centropomidae is composed of the genus *Centropomus* with 12 species. The dorsal fins of *Centropomus* are separated by a small gap, and their caudal fin is forked. These fishes occur in tropical and subtropical waters of the New World, with 6 species from the Atlantic and 6 from the Pacific. None of the species occurs in both oceans, but transisthmian pairs have been reported (Rivas, 1986). Results of a recent allozyme analysis (Tringali et al., 1999) are in good agreement with the taxonomic revision based on morphology (Rivas, 1986). The genus occurs along the Atlantic coast from Pamlico Sound, North Carolina, to Porto Alegre, Brazil, and along the Pacific coast from Baja California to Paita, Peru, and the Galapagos Islands. Some individuals may enter coastal fresh waters seasonally. Four species frequent coastal rivers of Florida including *C. undecimalis* (Fig. 236), which can reach 1.4 m TL and 24.3 kg (Burgess, 1980d; IGFA, 1993). The life history of *C. undecimalis* was studied by Gilmore *et al.* (1983).

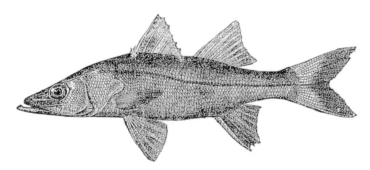

FIGURE 236. *Centropomus undecimalis* (Jordan and Everman, 1905, p. 369).

The subfamily Latinae consists of 9 species of *Lates* (7 of which are confined to African fresh waters) and *Psammoperca waigiensis* from the Indo-West Pacific. In this subfamily the dorsal fins are not completely separated or have one or two isolated spines between them. The caudal fin is usually rounded.

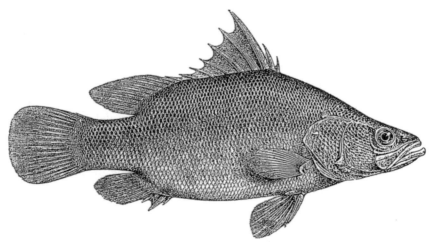

FIGURE 237. *Lates niloticus* (Boulenger, 1915, Plate 106, Fig. 82).

The Nile perch, *L. niloticus* (Fig. 237), can exceed 2 m TL and 200 kg and is a voracious predator (Kaufman, 1992). It occurs commonly in the major African river basins, including the Nile, Chad, Senegal, Volta, and Congo (Daget, 1986). *Lates niloticus* is an important food species in Africa and has been introduced into Lake Victoria, where it has done extensive damage to the endemic species flocks of haplochromine cichlids (Pitcher and Hart, 1995). About 200 species (65%) of the endemic cichlids of Lake Victoria have been exterminated (Goldschmidt et al., 1993). The destruction of native cichlid species by Nile perch has occurred in smaller African lakes as well (Ogutu-Ohwayo, 1993). *Lates niloticus* was proposed for introduction into Australia (Midgley, 1968), but fortunately the plan was recognized as much too dangerous (Barlow and Lisle, 1987). The other African species of *Lates* (= *Luciolates*) occur in Lakes Rudolf, Albert, and Tanganyika (Daget, 1986). Nile perch were intentionally stocked in some Texas reservoirs by Texas Parks and Wildlife, but fortunately the fish died due to cold temperatures (Fuller et al., 1999).

Australia has its own *Lates* in the form of the barramundi, *L. calcarifer*. This important game and food species is widely distributed in coastal, estuarine, and fresh waters from the Persian Gulf to China and the Indo-Australian archipelago. It is found in southern New Guinea and northern Australia, and it can reach 1.8 m and 60 kg (Lake, 1978). In New Guinea, barramundi have been taken more than 800 km up the Fly River (Allen, 1991). It is the most important commercial fish in Australian fresh waters. It is a protandrous hermaphrodite. Males transform into females at a TL between 500 and 1000 mm. This species is enormously fecund. A 1.2-m female can produce up to 40 million eggs. Barramundi undertake a catadromous migration and spawn in brackish water from August to February.

(African *Lates* spawn in fresh water.) Life history and aquaculture of barramundi were reviewed by Merrick and Schmida (1984) and Copeland and Grey (1987).

Map references: Allen (1989, 1991),* Bussing (1998),* Daget (1986), Greenwood (1976),* Grey (1987),* Paugy (1992),* Rivas (1986)*

Class Actinopterygii
Subclass Neopterygii

Order Perciformes; Suborder Percoidei
(per) Family Ambassidae—**Asiatic glassfishes** (am-bas'-i-dē)

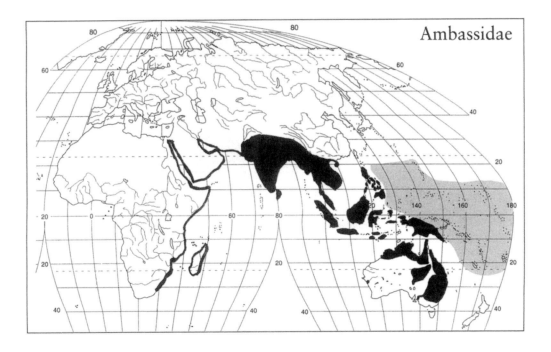

THE AMBASSIDAE (ALLEN AND BURGESS, 1990; ESCHMEYER, 1998), also referred to as Chandidae (Nelson, 1994), occurs from southern Africa through India to New Guinea and Australia. They were once classified with the centropomids, but Greenwood (1976) removed them. There has been much confusion regarding the use of *Chanda* versus *Ambassis*, and this is detailed by Smith (1945). *Ambassis* is the largest genus in the family and includes about 24 species (Allen and Burgess, 1990).

The Asiatic glassfishes have greater diversity in fresh water than in the sea. This is unusual for a percoid family. There are eight genera in the family, seven of which are found in fresh water. Likewise, of the 41 species, 21 are confined to fresh water, mostly in India, Southeast Asia, and the Indo-Australian archipelago (Roberts, 1989a; Allen and Burgess, 1990). Many of the species are transparent, with the vertebral column and swim bladder visible, hence the common name glassfishes. As such, they make interesting aquarium animals, especially *Chanda ranga*, a small schooling species from rivers of India and Burma. Most glassfishes are small, under 10 cm, but *Parambassis gulliveri* (Fig. 238) of northern Australia and southern New Guinea can reach 30 cm TL (Lake, 1978). Glassfishes are generalized perch-like fish with a deep notch between the two dorsal fins in most species and a forked tail. There are usually seven or eight dorsal and three anal spines.

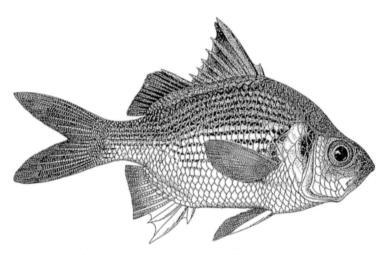

FIGURE 238. *Parambassis gulliveri* as *Ambassis gigas* (Weber, 1913, Fig. 31).

Fraser-Brunner's (1955) systematic review of the family was brought up to date by Roberts (1989a) and Allen and Burgess (1990). Maugé (1986a) listed 3 species of *Ambassis* that range from the east coast of Africa (Natal northward) to the Philippines and 2 other species that are only found in rivers of the east coast of Madagascar and South Africa. Talwar and Jhingran (1992) cited 16 species of *Ambassis*, *Chanda*, and *Parambassis* from rivers and estuaries of the Indian region. They make the point that these fishes are helpful in the biological control of guinea worms and malaria. In this area, glassfishes are eaten locally and provide a cheap source of protein. They are also utilized as bait and fertilizer.

Roberts (1989a) described a scale-eating glassfish, *Paradoxodacna piratica*, from Western Borneo. It feeds predominantly on scales of cyprinids by means of a shortened lower jaw and modified teeth. Roberts considered *Chanda* to be monotypic, containing only *C. nama*, a highly specialized scale eater. He placed other *Chanda* species in the next avaialble genus, *Parambassis*. One of these, *P. apogonoides*, is an oral brooder. Males of this species have been found with two size classes totaling 81 young in the orobranchial cavity, most likely representing two broods. *Gymnochanda filamentosa* is a scaleless glassperch from the Malay Peninsula and western Borneo (Roberts, 1989a). Sixteen species in four genera are listed by Kottelat *et al.* (1993) for Indonesia.

In Australia and New Guinea, a total of 22 species in 4 genera are recorded (Allen and Burgess, 1990). Fifteen of these are in the genus *Ambassis*. *Ambassis agassizii* extends throughout the Murray–Darling River system of eastern Australia. *Denariusa bandata* is shared by northern Australia and south-central New Guinea (Allen, 1989), as are several species of *Ambassis*. The genus *Tetracentrum* (3 species) from New Guinea has a continuous spiny and soft dorsal fin, and *T. apogonoides* has four instead of the usual three anal fin spines.

Little is known of the biology of these fishes, but several species have been spawned in captivity. They are thought to feed at night on microcrustaceans and aquatic insects. Most are egg scatters, and they may congregate in large schools (Allen, 1991).

Map references: Allen (1989, 1991),* Allen and Burgess (1990),* Kottelat (1993), Maugé (1986), Roberts (1989a), Skelton (1993),* Smith (1945), Talawar and Jhingran (1992)

Class Actinopterygii
Subclass Neopterygii

Order Perciformes; Suborder Percoidei
(per) Family Moronidae—**temperate basses**   (mō-ron'-i-dē)

THE TEMPERATE BASSES MAKE UP A SMALL FAMILY OF TWO genera and six species with a disjunct distribution in North American Gulf coast and Atlantic drainages and in Europe and northern Africa. Gosline (1966) separated the freshwater percichthyids from the marine Serranidae, and Johnson (1984) removed this group from the Percichthyidae. This history reflects the taxonomic difficulty of dealing with generalized percoids.

Moronids have 8–10 stout spines in the first dorsal and 1 spine in the soft dorsal fin. There are 3 spines in the anal fin. The operculum is armed with 2 spines, and the preopercle is saw-toothed. The lateral line nearly

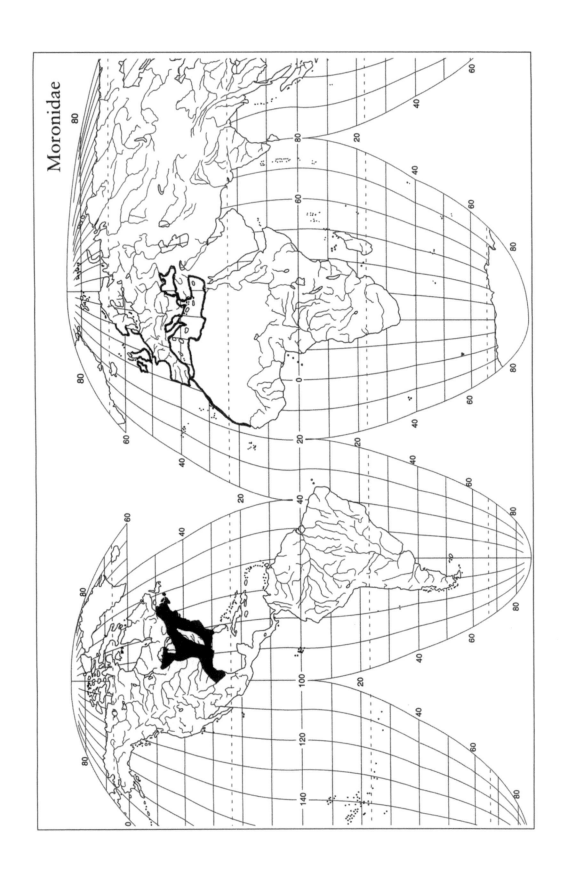

Moronidae

extends to the posterior edge of the caudal fin. There are also supplementary lateral lines on the caudal fin dorsal and ventral to the main canal. Temperate basses are medium to large fishes with a relatively elongate and deep body covered with ctenoid scales. Moronids superficially resemble the basses of the Centrarchidae, but centrarchids are nest builders and moronids scatter eggs more or less at random and do not brood the eggs or young. These fishes are silvery and three of the six species are striped, which is why moronids are sometimes called striped basses. This family is important to anglers and commercial fishers.

The genus *Morone* (= *Roccus*) is from eastern North America and includes four species. *Morone saxatilis*, the striped bass, is native to the Atlantic and Gulf coasts from the St. Lawrence River to eastern Texas, but it has been extensively introduced elsewhere including the Pacific slope of North America and Eurasia (Jenkins and Burkhead, 1994). It is the largest member of the genus. This species may reach nearly 2 m TL and 51 kg and is widely sought by anglers. It is an anadromous species that may ascend far upstream to spawn in spring in large aggregations near the surface. Males mature as early as age 2, whereas females of some populations first spawn at age 4 or 5. Females may release eggs numbered in the millions. After hatching, young fish typically remain in the rivers for 2 or more years before moving to the sea, where they move along the coast feeding on fishes. In some years, many juveniles may leave the river as young-of-the-year (J. Waldman, personal communication). Mature striped bass apparently return home to their natal stream, and at least 11 stocks or populations have been identified within their native range. Reservoir fisheries have also been successfully established. Longevity can reach into the 20s. *Morone chrysops* (Fig. 239), the white bass, is a strictly freshwater species of streams, rivers, ponds, and lakes from the

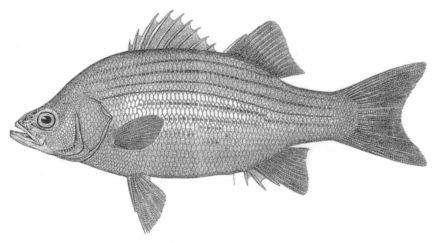

FIGURE 239. *Morone chrysops* as *Roccus chrysops* (Goode, 1884, Plate 171).

Great Lakes–St. Lawrence and Mississippi basins and the western Gulf coast. It is closely related to *M. saxatilis*, but only reaches about one-fourth of its size at 45 cm TL. Spawning occurs at the heads of lakes and reservoirs and in rivers.

The other North American species pair of moronids consists of the euryhaline white perch, *M. americana*, and the stenohaline yellow bass, *M. mississippiiensis*. *Morone americana* is the only American family member without stripes. It occurs in estuaries from Nova Scotia to South Carolina. It has been established in Lake Erie since 1953, but it is not clear if this is the result of natural dispersal or introduction (Trautman, 1981). *Morone mississippiensis* is a completely freshwater species from Lake Michigan and the Mississippi basin south to the Gulf. Life history details of all species of *Morone* have been reviewed in great detail by Becker (1983) and Jenkins and Burkhead (1994). Waldman (1986) reviewed the systematics of *Morone* and suggested that the two Asian species of *Lateolabrax* may be related to the Moronidae. He assigned the European–North African *Dicentrarchus* to the Moronidae.

Dicentrarchus labrax is a sporting and commercial species that occurs in estuaries of the Atlantic from Norway to Senegal and in the Mediterranean and Black Sea (Daget and Smith, 1986). It occasionally enters rivers and may grow to 1 m TL and 12 kg (Maitland, 1977). The biology of this species was discussed by Kennedy and Fitzmaurice (1972). Also see the synoptic book on sea bass by Pickett and Pawson (1994). *Dicentrarchus punctatus* (Fig. 240) is also a coastal species that ranges from

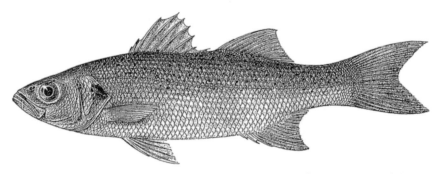

FIGURE 240. *Dicentrarchus punctatus* as *Morone punctata* (Boulenger, 1915, Fig. 81).

Spain south to Senegal and the southwestern Mediterranean. It reaches 70 cm TL (Bauchot, 1992). European ichthyologists tend to favor the two-genera classification, whereas Americans often unite all moronids in *Morone* (J. Waldman, personal communication).

Map references: Bauchot (1992), Daget and Smith (1986), Lee *et al.* (1980),* Maitland (1977),* Page and Burr (1991),* Scott and Crossman (1973),* Tortonese (1986b)*

Class Actinopterygii
Subclass Neopterygii

Order   Perciformes; Suborder Percoidei
(per) Family   Percichthyidae—**temperate perches**   [pers-ik-thē'-i-dē]

THIS FRESHWATER BASAL PERCOID GROUP WAS SEPARATED from the marine Serranidae by Gosline (1966). Johnson (1984) removed the Moronidae from the percichthyids and added several Australian genera that were formerly placed in other families. As currently constituted, it is not clear that the percichthyids are monophyletic. These freshwater fishes occur in southern South America (Argentina and Chile) and Australia. There are 9 genera and about 20 species.

· Percichthyids have a double dorsal fin that may or may not have a deep notch between the spiny and soft dorsal. There are three anal spines. The posterior field of the ctenoid scales is covered with simple needle-like ctenii. Some genera have secondarily cycloid scales. Fishes in this assemblage can range in size from a few centimeters to 1.8 m TL.

The South American percichthyids include two genera: *Percichthys* (two extant species and three fossil species) and *Percilia* (two extant species) (Arratia, 1982). These genera are restricted to the south of Chile and Argentina, but in the Upper Tertiary a fossil species, *Santosius antiquus*, occurred in Brazil (Arratia, 1982). *Percichthys trucha* is the largest of the South American percichthyids, and it is one of the few native species sought after by anglers in Chile and Argentina (introduced trout attract most anglers). It reaches 35 cm TL and feeds primarily on chironomids (Ruiz and Berra, 1994). *Percichthys melanops*, found only in Chile from about Valparaiso to Concepcion, reaches 20 cm TL.

Percilia gillissi and *Percilia irwini* (Fig. 241) are only found in Chile. *Percilia gillissi* is more widespread, occurring from Valparaiso to Porto

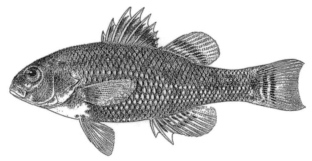

FIGURE 241. *Percilia irwini* (Eigenmann, 1928, Plate XVI, Fig. 1).

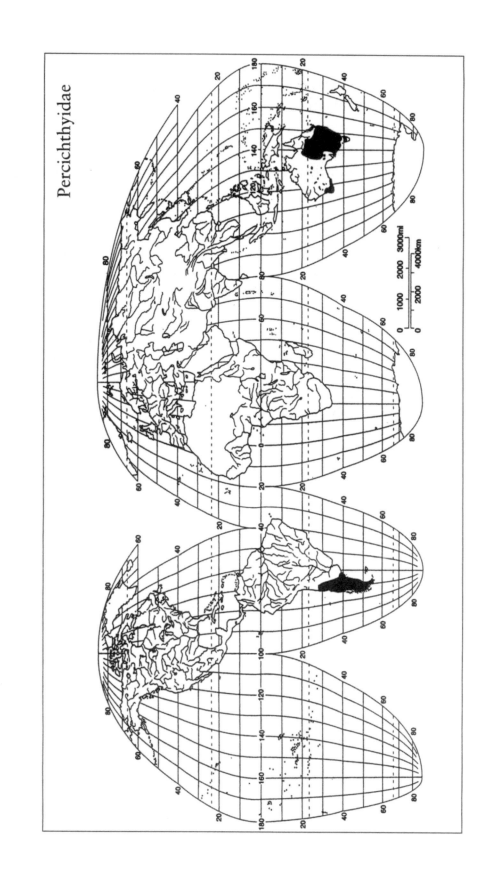

Percichthyidae

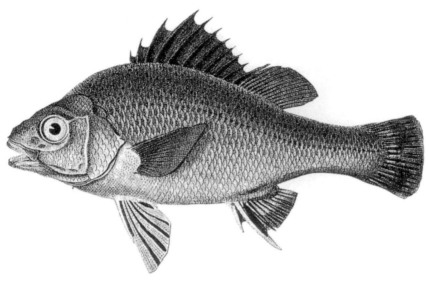

FIGURE 242. *Macquaria australasica* (Waite, 1923, p. 117).

Montt, whereas *P. irwini* is found near Concepcion (Arratia, 1981). These small fishes (maximum TL 90 mm) resemble miniature *Pericichthys* and are reminiscent of North American darters (*Etheostoma*) because of their coloration, appearance, and habits (Eigenmann, 1921).

There are seven percichthyid genera in Australia. *Macquaria* is the most generalized genus. MacDonald (1978) synonymized *Percalates* and *Plectroplites* within *Macquaria*. The Macquarie perch, *M. australasica* (Fig. 242), was once found throughout the cooler upper reaches of the Murray–Darling River system. Today its distribution is fragmented, and only small, discrete populations remain (Ingram et al., 1990). Populations also occur in several river systems in the southeastern coastal drainage (Cadwallader, 1981). Its conservation status is vulnerable (Jackson, 1998). Habitat degradation, particularly siltation, makes existence difficult for this species. *Macquaria australasica* can reach 46 cm and 3.5 kg (Harris and Rowland, 1996). It is an excellent sporting fish, but populations are now protected in most places. Artificial breeding attempts have not been very successful. Insects form a substantial part of its diet.

The golden perch, *Macquaria* (= *Plectroplites*) *ambigua* (Fig. 243), is a moderately large percichthyid that has been recorded up to 76 cm and 23 kg, but anglers are happy to catch 5-kg fish. It is also the object of a small commercial drum and gill net fishery that is due to be terminated in 2001 (S. Rowland, personal communication). It prefers warm, sluggish waters. The dorsal profile of golden perch is strongly convex. Its bronze coloration accounts for the common name. It is found throughout the Murray–Darling River system and some internal drainages such as Lake

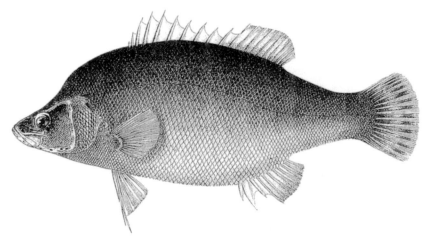

FIGURE 243. *Macquaria ambigua* as *Ctenolates ambiguus* (McCoy, 1884, Plate 84).

Eyre (Harris and Rowland, 1996). Electrophoretic evidence suggests that there are many genetically distinct populations, and that the Lake Eyre population should be considered a distinct species (Musyl and Keenan, 1992). However, like many other Australian native fishes, golden perch is disappearing from many areas, and its numbers are declining due to dams and altered stream flow and temperatures. Adults are migratory and move great distances upstream during spring and summer. Females are very fecund and spawn semibuoyant eggs during spring and summer floods. The pelagic eggs develop in the floodplain. Golden perch are propagated artificially and young are released into impoundments and farm ponds.

The Australian bass and the estuary perch, *Macquaria* (= *Percalates*) *novemaculeata* and *M. colonorum*, respectively, inhabit coastal rivers and estuaries in southeastern Australia. *Macquaria novemaculeata* is catadromous and moves far upstream but migrates downstream to estuaries to breed. It reaches 60 cm and 3.8 kg and is the object of a substantial recreational fishery. *Macquaria colonorum* is larger at 75 cm TL and 10 kg, but fish more than 3 kg are rare. It spawns in seawater at estuary mouths in winter.

The largest percichthyids belong to the genus *Maccullochella* and are called cods by Australians. The largest Murray cod, *M. peelii* (Fig. 244), ever recorded was 113.5 kg; however, most specimens taken today are approximately 10 kg, but anglers occasionally catch 20- to 40-kg fish (Harris and Rowland, 1996; Rowland, 1998a). The species can reach 1.8 m TL. Murray cod have a greenish background color with a mottled pattern, equal jaws or lower jaw protruding, and a concave head slope (Berra and Weatherley, 1972). This species was originally abundant throughout the Murray–Darling River system in eastern Australia, but now its abundance and range are seriously reduced because of overfishing by

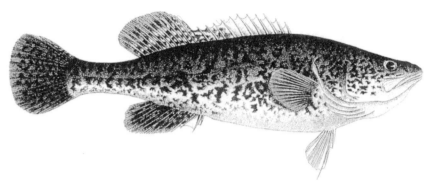

FIGURE 244. *Maccullochella peelii* as *Oligorus macquariensis* (McCoy, 1884, Plate 86).

commercial fisheries between the late 1800s and the 1930s and by dams that have reduced the frequency, magnitude, and length of flooding (Rowland, 1989; Ingram et al., 1990; Harris and Rowland, 1996). Murray cod generally prefer deep pools with cover such as fallen trees, stumps, and boulders. Females produce about 7000–90,000 eggs that are spawned onto a hard surface such as the inside of a hollow log (Rowland, 1998b). The male guards and fans the eggs during incubation. They have been bred in captivity, and young are released into farm ponds and rivers within their native range. As the top predator of Australia's inland waters, Murray cod consume fish, crayfish, mollusks, aquatic birds, turtles, frogs, and any terrestrial animals unlucky enough to fall into a Murray cod pool. This species is still highly valued by anglers and was an important part of the life of inland Aborigines (Rowland, 1989).

A smaller subspecies of cod, known as the Mary River cod, *M. peelii mariensis*, is endemic to the Mary River system in southeastern Queensland (Rowland, 1993). Its conservation status is endangered (Jackson, 1998) because it is now restricted to only a few small tributaries (Harris and Rowland, 1996). The eastern freshwater cod, *M. ikei*, is also endangered. It only occurs in the Clarence and Richmond Rivers in northeastern New South Wales (Rowland, 1993, 1996).

Although different in appearance from the Murray cod, the trout cod, *M. macquariensis*, was confused with *M. peelii* for many years (Berra and Weatherley, 1972). It was thought to be a juvenile Murray cod, or the two forms were said to be different sexes of the same species. Compared to the Murray cod, trout cod have a gray background color with speckled pattern, overhanging upper jaw, and a straight head slope. It is a smaller species, reaching only about 85 cm and 16 kg. It was once widespread throughout the southern tributaries of the Murray–Darling system where it was sympatric with Murray cod (Berra, 1974; Harris and Rowland, 1996). Today only two self-sustaining populations are known. One is in the Murray River between Yarrawonga and Tocumwal, New South Wales, and the other is in Seven Creeks, a tributary of the Goulburn River near Euroa, Victoria. This population was introduced into Seven Creeks many years ago. Another translocated trout cod population in Lake Sambell near Beechworth, Victoria, was exterminated in a fish kill in the late 1970s. I expressed fears of such a disaster (Berra, 1974) and wrote, "If some natural or man-made disaster were to ruin Lake Sambell and Seven Creeks, the danger of extinction would be very great." Trout cod are now considered endangered (Jackson, 1998; Ingram and Douglas, 1995), and a recovery plan has been developed (Douglas et al., 1994). I urged artificial breeding and stocking, and fortunately this has occurred. More than 200,000 juvenile trout cod have been released throughout their historic range since 1986 (Ingram and Rimmer, 1992; Harris and Rowland, 1996). It remains to be determined if these stockings result in established populations and spawning. Douglas *et al.* (1995) reported a natural hybrid between Murray cod and trout cod, but they showed that wild populations are genetically distinct and there is no introgression between the species.

In the past, *Gadopsis*, called blackfish in Australia, has been placed in its own family, Gadopsidae, and even in its own order, Gadopsiformes. At various times this enigmatic fish has been allied with blennioid or ophidioid fishes. Johnson (1984) treated it as a percoid in the Percichthyidae. He maintained that it shares some features with *Maccullochella*, and that the closest genus to *Gadopsis* is the monotypic *Bostockia*. Ovenden *et al.* (1988) reported on its evolutionary relationships.

Gadopsis is an elongate fish with a long dorsal fin and jugular pelvic fins that are reduced to one branched ray each and positioned anterior to the pectoral fins. It has tiny, secondarily cycloid scales and a rounded tail. There are two species. *Gadopsis marmoratus* (Fig. 245) has 6–13 stout dorsal spines and 22–31 dorsal rays. It occurs throughout Victoria and into eastern New South Wales. There are small populations in South Australia and near the New South Wales–Queensland border. Like all Australian percichthyids, its numbers are declining, probably due to siltation which smothers the eggs (Jackson et al., 1996). Its maximum size is up to 60 cm and 5.5 kg, but most specimens taken today are much smaller (45 cm). *Gadopsis marmoratus* is an opportunistic carnivore and feeds on many of the same prey items as introduced brown trout, such as aquatic insects and

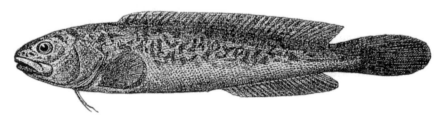

FIGURE 245. *Gadopsis marmoratus* (Waite, 1923, p. 161).

crustaceans. Blackfish tend to be benthic and nocturnal feeders, whereas the trout are midwater and diurnal feeders. The two-spined blackfish, *G. bispinosus*, from the higher altitude streams of northeastern Victoria at 200–700 m, has only 1–3 slender dorsal spines and 35–38 dorsal rays (Jackson et al., 1996). It prefers woody debris habitat in cool streams, but it is often sympatric with *G. marmoratus* in lower altitude regions. Sanger (1984, 1990) described and studied the biology of this species. For a description of the larvae of *Gadopsis*, *Maccullochella*, and *Macquaria* see Neira *et al.* (1998).

The nightfish, *Bostockia porosa*, is a small species (maximum SL about 13.5 cm) from coastal streams and lakes in southwestern Australia. This species has a large mouth, secondarily cycloid scales, and a series of large pores around the eye and on the snout and head. As its common name denotes, *Bostockia* is nocturnal. It feeds on crustaceans, insects, and small fishes (Allen, 1989).

There are six species of small pygmy perches in southwestern and southeastern Australia classified in three derived genera: *Edelia*, *Nannatherina*, and *Nannoperca*. These fishes were previously classified in the Kuhliidae, but Johnson (1984) placed them in the Percichthyidae. They are all less than 10 cm TL and have a deeply notched dorsal fin. The lateral line has two parts and is often incomplete or absent. Males are colorful and are kept as aquarium animals. *Edelia obscura* occurs in coastal drainages of Victoria, and *E. vittata* is from southwestern Australia. Kuiter *et al.* (1996) consided *Edelia* to be a synonym of *Nannoperca*. *Nannatherina balstoni* has a maximum size of 6 cm SL and inhabits coastal streams, ponds, and swamps in southwestern Australia. *Nannoperca australis* (Fig. 246) occurs in the Murray–Murrumbidgee River system in southeastern Australia. The other two species, *N. oxleyana* (classified as endangered) and *N. variegata* (vulnerable), have very restricted distributions around the Queensland–New South Wales border and the South

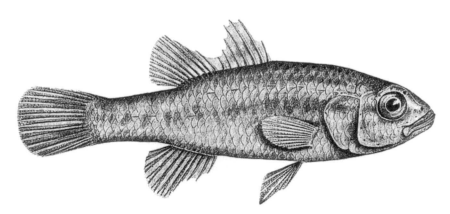

FIGURE 246. *Nannoperca australis* (Waite, 1923, p. 117).

Australia–Victoria border, respectively (Allen, 1989; Kuiter et al., 1996; Jackson, 1998).

Map references: Allen (1989),* Arratia (1982),* Arratia *et al.* (1985),* Berra and Weatherley (1972),* McDowall (1996)*

Class Actinopterygii
Subclass Neopterygii

Order Perciformes; Suborder Percoidei
(1st) Family Centrarchidae—**sunfishes**   (sen-trar'-ki-dē)

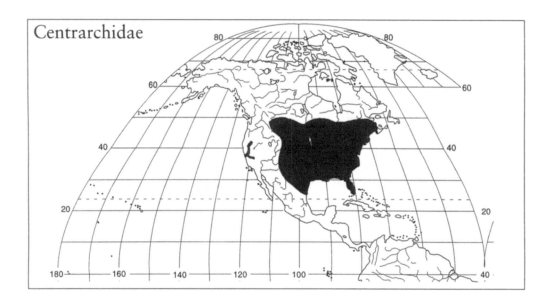

THE CENTRARCHIDAE IS THE SECOND LARGEST FRESHWATER fish family endemic only to North America. (The Percidae, with all its darters, is also found in Eruasia.) Only the catfishes, Ictaluridae, have more species. Centrarchids are native east of the Rocky Mountains from southern Canada to northern Mexico. Only one species, *Archoplites interruptus*, the Sacramento perch, occurs west of the Rocky Mountains in the Sacramento and San Joaquin River systems of California. The range of fossil centrarchids was more extensive, including Alaska and southern Mexico (Miller, 1961; Cavender, 1986; Smith et al., 1975). The earliest centrarchids date to the Eocene of northwestern Montana. There are 8 genera and 30 species (Mayden et al., 1992). This does not include the pygmy sunfishes, *Elassoma*, excluded from the Centrarchidae by Johnson (1984). Centrarchids have been extensively introduced throughout North America because of their importance as sport fishes (Fuller et al., 1999). They have even been introduced into Europe, Africa, South America, Japan, and elsewhere (Lever, 1966).

Centrarchids have two basic body styles: elongate (basses, *Micropterus*) and deep-bodied (sunfishes, *Lepomis* and others). The spiny and soft dorsal fins are usually broadly joined into one. There are three to eight anal spines and ctenoid scales. A lateral line is present. The pelvic fins are thoracic, and the pectoral fins are high on the body. The males of all centrarchids, including the primitive *Archoplites*, build circular nests by moving gravel with a fan-like motion of their caudal fin (Moyle, 2000). Males guard the eggs and young. The banded sunfishes, *Enneacanthus*, are the smallest at about 8 cm TL, whereas the largemouth bass, *Micropterus salmoides*, can reach more than 90 cm and 10 kg.

Branson and Moore (1962) presented a phylogeny of sunfishes based on a marvelously detailed study of the acousticolateralis system. Avise *et al.* (1977) obtained similar results from a study of allozymes. *Archoplites* seems to be the most basal genus followed by *Acantharcus*. *Pomoxis* and *Centrarchus* are sister groups, as are *Lepomis* and *Micropterus*. Another view, based on supraneural and predorsal bones presented by Mabee (1988), considered *Micropterus* to be the primitive sister group to the other living centrarchids. Wainwright and Lauder (1992) discussed sunfish phylogeny in relation to feeding biology.

The Sacramento perch, *Archoplites interruptus* (Fig. 247), is the one centrarchid to survive extinction west of the Rocky Mountains while other centrarchids died out, probably in the Miocene (Miller, 1958, 1961).

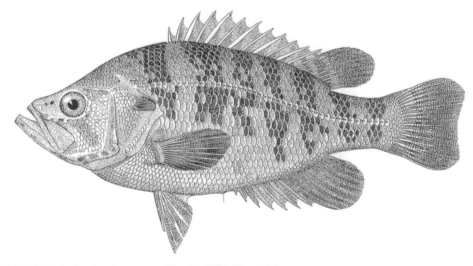

FIGURE 247. *Archoplites interruptus* (Goode, 1884, Plate 151).

Because of its long isolation and lack of competition from other centrarchids (until recent introductions), it has retained some primitive features and may be viewed as an example of the ancestral sunfish stock (Moyle, 1976). It may be related to *Centrarchus* and *Pomoxis* (Moyle, 2000). Its social and reproductive behaviors seem to be less complex than those of other sunfishes. Older literature reported that *Archoplites* is the only sunfish that does not build a nest or have an elaborate courtship ritual (Moyle, 1976), but recent observations indicate that males create a shallow depression for spawning by "digging" with their tail fin (Moyle, 2000). Sacramento perch have more dorsal spines (12 or 13) than any other centrarchid and 6 or 7 anal spines. *Archoplites* is a fairly deep-bodied fish with a large, oblique mouth. It is a piscivorous, ambush predator. It was once widely distributed in the Sacramento and San Joaquin Rivers, the Pajaro and Salinas Rivers, and Clear Lake. This species was common enough to support a fishery at the turn of the century. Today, it exists mostly in farm ponds and reservoirs. It has been nearly extirpated from its sluggish, native streams by competition from 11 species of more advanced, introduced sunfishes combined with the deleterious effects of habitat alteration (Moyle, 1976). However, Sacramento perch can live at high temperatures, salinities, and alkalinities where other fishes cannot. This may be its salvation as a species, and *Archoplites* has been introduced into alkaline lakes in California, Nevada, and Utah (Moyle, 1980). Maximum size is 61 cm TL, but this record is from the nineteenth century. Large fish today are half that size. See Moyle (1976, 2000) for a summary of life history information.

Acantharchus pomotis (Fig. 248), the mud sunfish, is widely distributed along the Atlantic coastal plain from southern New York to the St. Johns

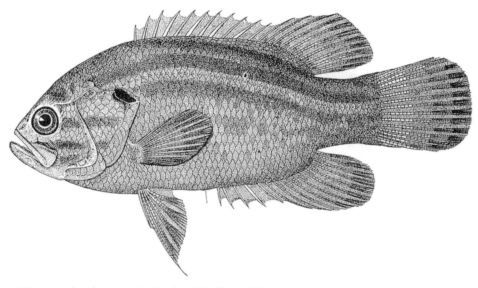

Fig. 248. *Acantharchus pomotis* (Goode, 1884, Plate 150).

River, Florida. It also occurs along the extreme eastern Gulf coastal plain of north Florida and southern Georgia, but it is uncommon there (Cashner et al., 1989). This species is the only centrarchid with exclusively cycloid scales. (Many centrachid species have cycloid scales on the cheek.) It has a chunky, compressed body with a rounded caudal fin and usually five (four to six) anal spines. Maximum size is about 21 cm TL (Jenkins and Burkhead, 1993). This secretive, nocturnal, mud-loving sunfish inhabits sluggish lowland streams, ponds, and swamps. Amphipods and decapods are the most frequent prey items found in the gut of *Acantharachus*, and this species seems to spawn earlier (December–May) than other centrarchids (Pardue, 1993).

FIGURE 249. *Pomoxis nigromaculatus* as *Pomoxys sparoides* (Goode, 1884, Plate 159).

There are two species of crappies, *Pomoxis*. The black crappie, *P. nigromaculatus* (Fig. 249), is distinguished from the white crappie, *P. annularis*, by the presence of seven or eight dorsal spines as opposed to the six dorsal spines of white crappies. Both have a long predorsal region and six or seven anal spines. Both species are so widely introduced that their true native range in the eastern United States is difficult to determine. Lee *et al.* (1980) provided interesting maps of former and present distribution for each species. Both species reach about 50 cm TL and are highly prized by anglers. White crappies appear to be more tolerant of turbid waters than black crappies. The common name "crappie" is often pronounced "croppie" and appears to be a corruption of a French Canadian word "crapet," but the etymology is obscure (Jenkins and Burkhead, 1993). The generic name *Pomoxis* means "sharp opercle," but the opercle is not particularly sharp in crappies. Both species spawn in March–July and are very fecund, producing eggs numbering in the hundreds of thousands. Jenkins and Burkhead summarized life history information. See Carlander (1997) for summaries of biological parameters such as age and growth and Williamson *et al.* (1993) for culture information on centrarchids.

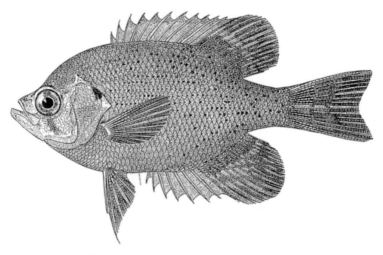

FIGURE 250. *Centrarchus macropterus* (Goode, 1884, Plate 158).

The flier, *Centrarchus macropterus* (Fig. 250), has a deep, extremely compressed body and large dorsal and anal fins. There are 11–13 dorsal fin spines and 7 or 8 anal fin spines. There is a large, black teardrop under the eye. In small specimens an eye spot or ocellus consisting of a black spot surrounded by a red halo is present in the soft dorsal. The body is dotted with dark spots arranged in interrupted lines along the sides of the fish. The mouth is small. Insects and microcrustaceans form most of the diet. The maximum reported size is 36 cm TL (Jenkins and Burkhead, 1993), but most adult fliers are about 11–17 cm. Spawning occurs in March–May. Fliers inhabit slow-moving, clear, heavily vegetated streams, swamps, ponds, and lakes along the coastal lowlands from Maryland to Florida, along the Gulf slope, and up the central Mississippi Valley. Anglers consider fliers excellent light fly rod quarry, but their small size limits interest.

There are four species of rock basses, *Ambloplites*, in eastern North America. These fishes usually have six (five to seven) anal spines and large red eyes. They are sometimes called redeyes. The young are compressed, but as they grow the body becomes more robust. *Ambloplites rupestris* (Fig. 251), the rock bass, is the most widely distributed and best known species. It is native to the Mississippi and Great Lakes–St. Lawrence region

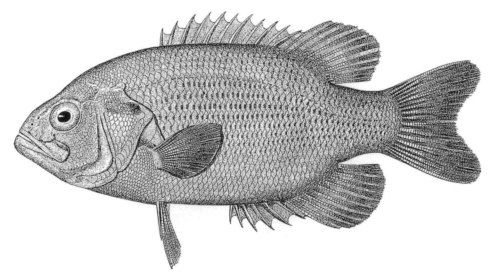

FIGURE 251. *Ambloplites rupestris* (Goode, 1884, Plate 149).

from southern Canada to Missouri. Like other centrarchids, it has been widely introduced outside of its native range. It has a large mouth and is a predator upon many different food items. Crayfish are a favorite food of adults. Maximum size is approximately 37 cm TL and 1.6 kg, but fish larger than 25 cm are uncommon (Jenkins and Burkhead, 1993). Rock bass prefer clear, moderate-gradient streams and are associated with cover such as undercut banks and tree stumps.

The Roanoke bass, *A. cavifrons*, has a limited distribution in the Chowan and Roanoke drainages of Virginia and the Tar and Neuse drainages of North Carolina (Cashner and Jenkins, 1982; Jenkins and Cashner, 1983). This species is similar to *A. rupestris* but grows to a larger average size (Jenkins and Burkhead, 1993). *Ambloplites constellatus*, the Ozark bass, has a limited distribution in the White River of Missouri and Arkansas (Cashner and Suttkus, 1977). *Ambloplites ariommus*, the shadow bass, has a disjunct distribution in Gulf slope drainages and in Arkansas and southeastern Missouri (Cashner, 1980).

There are three species of banded sunfish, *Enneacanthus*. They are small, compressed species whose banded color pattern provides disruptive camouflage in vegetated habitats. They can tolerate relatively acidic habitats (pH 4–6). The maximum size in this genus is about 9.5 cm TL. They usually have nine (7–11) dorsal fin spines. In fact, the name *Enneacanthus* means "nine-spined." Three anal fin spines unite this genus with *Lepomis* and *Micropterus*. *Enneacanthus* occurs along the Atlantic and Gulf slope from New Hampshire to Mississippi. These pretty fishes make fine aquarium animals, and the black-banded sunfish, *E. chaetodon* (Fig. 252), is even known to aquarists in Europe and Asia. The other species are the blue-spotted sunfish, *E. gloriosus*, and the banded

FIGURE 252. *Enneacanthus chaetodon* as *Mesagonistius chaetodon* (Goode, 1884, Plate 161).

sunfish, *E. obesus*. Jenkins and Burkhead (1993) summarized life history information.

Lepomis is the largest centrarchid genus with 12 species. Most of these are deep-bodied and compressed, but some are more elongate than others. All have three anal spines and a slightly forked caudal fin. This genus is distributed throughout eastern North America and has been widely introduced west of the Rocky Mountains. Adult male *Lepomis* are some of the most beautiful fishes in North America. Color is important in courtship and species recognition in this genus. In disturbed areas such as turbid, polluted, or overpopulated waters hybridization readily occurs. Juvenile sunfish are often difficult to identify. Hybridization and the difficulty of identifying small specimens are two reasons why 15 generic and 75 specific synonyms have appeared in the literature for this group (Jenkins and Burkhead, 1993). Bailey (1938) did much to clarify the nomenclatural mess, but some problems remain.

Even today, the warmouth, herein referred to as *Lepomis gulosus* (Fig. 253), is placed by some researchers in the genus *Chaenobryttus*. The nomenclatorial history of this species is interesting and instructive for ichthyology students. This dark brown, elongate, thick-bodied sunfish with a huge mouth was first recognized from Florida by William Bartram in 1791 in his *Travels* (Berra, 1997b). He named it *Cyprinus coronarius*. However, because he was not consistently binomial in his nomenclature, his names were rejected by the International Commission on Zoological Nomenclature. The earliest accepted name is *Pomotis gulosus* assigned to a specimen from Lake Pontchartrain, Louisiana, by Cuvier and Valenciennes in 1829. *Pomotis* was assigned to *Chaenobryttus* so the warmouth became generally

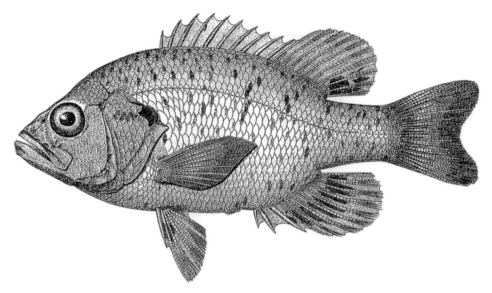

FIGURE 253. *Lepomis gulosus* as *Chaenobryttus gulosus* (Goode, 1884, Plate 152).

accepted as *Chaenobryttus gulosus* in the 1960 edition of the American Fisheries Society's checklist (Bailey et al., 1960). In the 1970 checklist, Bailey *et al.* (1970) considered *Chaenobryttus* to be a subgenus of *Lepomis*. They reasoned that the differences between *L. gulosus* and *L. cyanellus* were no greater than those between *L. cyanellus* and other *Lepomis*. Birdsong and Yerger (1967) provided evidence of a natural hybrid population of warmouth and *L. macrochirus* which supported the inclusion of warmouth in *Lepomis*. Smith and Lundberg (1972) reported a unique bony feature for warmouth and a fossil centrarchid that they say justified the retention of *Chaenobryttus*. Wainwright and Lauder (1992) recognized *Chaenobryttus* as distinct from *Lepomis* based on a study of feeding biology. Mayden *et al.* (1992) retained *Chaenobryttus* in order for *Lepomis* to be considered monophyletic. I would not be surprised if *Chaenobryttus* is used in the next edition of the Common Names List. Molecular studies have not clarified the situation (Avise and Smith, 1974, 1977). Warmouth prefer pools in low-gradient streams, swamps, ponds, and lakes. Fish and crayfish are the main prey of adults. Larimore (1957) provided a detailed life history of this species.

The green sunfish, *L. cyanellus*, is elongate with a large mouth and thick body. Males are beautiful and often have bright yellow or orange edges on the dorsal, anal, and caudal fins and a dark spot at the rear of the soft dorsal and anal fins. Their pectoral fins are short and rounded and the gill rakers are long and slender. The opercular flap is short. Green sunfish can tolerate a low pH and are often stocked in areas subject to acid mine drainage. They are pollution tolerant (Reash and Berra, 1987, 1989).

The longear sunfish, *L. megalotis*, is one of the most beautiful fishes in North America. Breeding males have a deep orange belly, deep blue wavy lines on the cheek and opercle, and a white edge on the long, black opercular flap. This species can reach about 24 cm TL. Its diet consists of invertebrates, mostly insects. It can repopulate decimated stream segments within 1 year of depopulation (Berra and Gunning, 1970), and it occupies a limited home range of 20–60 m of stream during most of the year in Louisiana (Berra and Gunnning, 1972). The dollar sunfish, *L. marginatus*, has similar colors to the longear, but its maximum size is about half that of the longear.

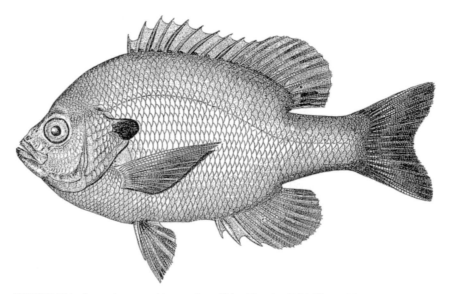

FIGURE 254. *Lepomis macrochirus* as *L. pallidus* (Goode, 1884, Plate 155).

The bluegill, *L. macrochirus* (Fig. 254), is probably the best known *Lepomis*. It has been introduced all over North America. It is the largest *Lepomis*, reaching about 40 cm and 2.15 kg (IGFA, 1993). It is deep-bodied and very compressed, with a small mouth and long, pointed pectoral fins. There is usually a dark spot at the posterior end of the soft dorsal fin. It feeds on insects with the help of very long gill rakers. It hybridizes with most other species of *Lepomis*. The pumpkinseed, *L. gibbosus*, also reaches about 40 cm TL and is widely introduced as a pan fish throughout North America. It lives in cooler waters than any other sunfish (O'Hara, 1968). The smallest member of the genus is *L. symmetricus*, the bantam sunfish, which reaches only about 9 cm. It is the only sunfish with an incomplete lateral line. The red-spotted sunfish, *L. miniatus*, was recently separated from the spotted sunfish, *L. punctatus* (Warren, 1992). *Lepomis auritus*, the red-breast sunfish, has an extremely long and narrow ear flap, and *L. microlophus*, the red-ear sunfish, has a red spot on the opercular flap. The latter species is specialized for feeding on snails by virtue of its large molariform pharyngeal teeth and is widely known as "shellcracker."

Most *Lepomis* have been well studied because they are very important as game species. Life history information and related references as well as color photographs can be found in Jenkins and Burkhead (1993) and Etnier and Starnes (1993).

The basses, sometimes called black basses, *Micropterus*, are the most elongate centrarchids. Their caudal fin is more deeply forked than in *Lepomis*, and the dorsal and anal fins of basses are smaller than those of sunfishes. The generic name *Micropterus* means "small fin" and is based on the fact that the original description of the genus was from a single specimen with a deformed dorsal fin (Jordan and Evermanm, 1896). Basses have a large mouth and a greater preference for current than do sunfishes. Like *Lepomis*, basses have three anal fin spines. There are seven described species (Page and Burr, 1991; Williams and Burgess, 1999).

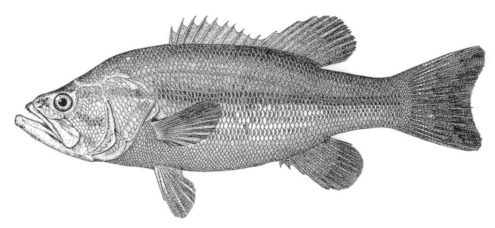

FIGURE 255. *Micropterus salmoides* (Goode, 1884, Plate 147).

The largemouth bass, *Micropterus salmoides* (Fig. 255), is known to almost all anglers in the United States. It is one of the most important game species in most states, and it has been stocked in South America, Africa, Europe, and Russia. The largest fish caught by an angler weighed 10.09 kg (IGFA, 1993) and was 82 cm TL (Becker, 1983). It was taken in a lake in Georgia. Adults prey on fishes, and bluegill sunfish are usually stocked in lakes as food for largemouth bass. *Micropterus salmoides* is found in warm water in rivers, streams, lakes, and ponds.

The smallmouth bass, *M. dolomieu*, has a more northerly distribution in cooler waters of rivers, streams, and lakes with rocky or gravelly bottoms. The angling record is 5.41 kg. Crayfish and fish are important items of diet. Other basses include the spotted bass, *M. punctulatus*, of the Mississippi basin and Gulf slope; the Suwannee bass, *M. notius*, from north Florida and Georgia; the Guadalupe bass, *M. treculi*, from central Texas; and the red-eye bass, *M. coosae*, from above the fall line in the Savannah, Chattahoochee, and Mobil Bay basins (Page and Burr, 1991). The most recent bass to be described is the shoal bass, *M. cataractae*, from the Apalachicola, Chattahoochee, and Flint River drainages of Alabama, Florida, and Georgia (Williams and Burgess, 1999). Previously, this bass was aligned with *M. coosae*, but it is most similar to *M. punctulatus*.

The reproductive behavior of sunfishes and basses has been well studied and is similar for most species. Males excavate a shallow, circular depression in the gravel or sand by fanning with the caudal fin from a nearly vertical position. In crowded ponds, these nests are often close together in shallow water. They can be easily observed by someone walking along the shore. Spawning occurs in midspring to early summer and involves a ritual courtship. A male and a female may circle over the nest with their heads and ventral surfaces touching, and eventually the female swoops down to the nest on her side and releases the eggs. The male follows and fertilizes the eggs. Several females may spawn in the same nest. Where colonial species nest in dense groups, some hybridization may take place when a female deposits her eggs in the "wrong" male's nest. The male guards the eggs. In addition to the male that builds and defends the nest, smaller males, called "sneakers," may dart in and add sperm to the mix, and intermediate-sized males in female colors ("female mimics" or "satellites") may gain admission to a nest by posing as a receptive female (Gross, 1984; Dominey, 1988).

Centrarchids are infamous for their proclivity to hybridize. Intergeneric hybrids are possible, and within *Lepomis* almost any combination is likely. Hubbs (1955) reviewed hybridization in fishes. *Lepomis* hybrids are predominately males and are stocked in ponds with bass to prevent the inevitable overpopulation of fecund *Lepomis* (Bennett, 1971). The hybrids are frequently fertile (Williamson et al., 1993). Carlander (1977) provided a wealth of fish biology information on centrarchids.

Map references: Lagler *et al.* (1977),* Lee *et al.* (1980),* Miller (1958, 1961),* Page and Burr (1991),* Rostlund (1952),* Scott and Crossman (1973)*

Class Actinopterygii
Subclass Neopterygii

Order Perciformes; Suborder Percoidei
(1st) Family Percidae—**perches, darters**   (pur'-si-dē)

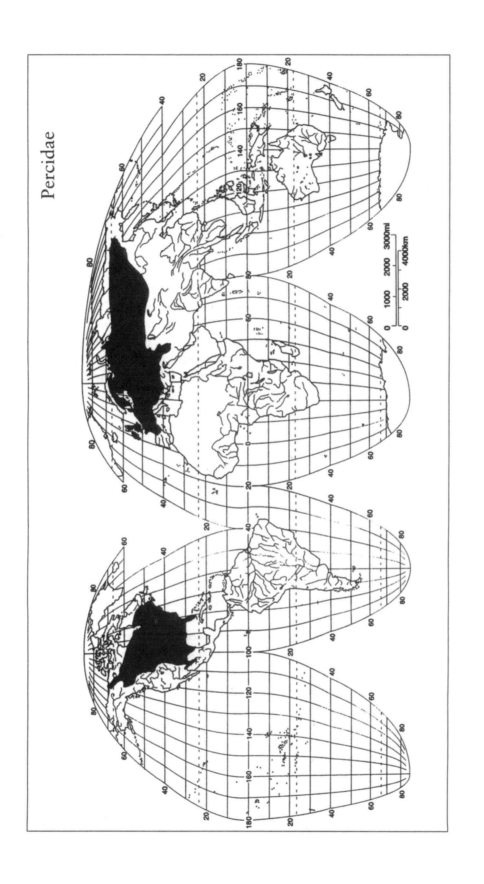

Percidae

THIS LARGE FAMILY OF ABOUT 195 SPECIES AND 10 GENERA HAS a holarctic distribution from eastern North America through Europe and northern Asia. Percids are absent from eastern Asia and western North America. The Percidae is the second largest freshwater fish family in North America (after the Cyprinidae) (Wood and Mayden, 1997), and its greatest diversity is in eastern North America. Almost all of the 180 North American percid species are darters (Jenkins et al., 1994; Wood and Mayden, 1997; Song et al., 1998.). The southeastern United States is the center of this diversity. Tennessee alone has at least 97 species of darters (Etnier and Starnes, 1993). Only 1 species of darter lives west of the Rocky Mountains in Mexico. Eurasia has 14 percid species and 4 endemic genera.

Percids are unusual among percoid families in having fewer than three anal fin spines. Most percids have two anal spines, and some have one. The two dorsal fins are separate or narrowly fused in all genera. The pelvic fins are thoracic with one spine and five rays. Pectoral fins are set high on the sides. Scales are ctenoid, and the body is elongate. The opercle has a single posterior spine. The lateral line is usually complete. Most darters are sexually dimorphic, and the males can be extremely colorful. Darters are small, usually less than 10 cm TL. The fountain darter, *Etheostoma fonticola*, is the smallest at 4 cm TL. Some of the piscivorous percids can reach large size such as the walleye, *Stizostedion vitreum*, at 90 cm and 11.3 kg. The Eurasian pikeperch, *S. lucioperca*, can reach 1.3 m TL and 15 kg (Jenkins et al., 1994).

Two subfamilies are recognized by Collette (1963) and Collette and Banarescu (1977): Percinae and Luciopercinae. Their arrangement is followed by Nelson (1994) and will be presented here. Wiley (1992) and Coburn and Gaglione (1992) presented alternative interpretations of subfamily relationships. A recent DNA sequence study divided the Percidae into three monophyletic subfamilies: Etheostomatinae (*Ammocrypta*, *Crystallaria*, *Etheostoma*, and *Percina*), Percinae (*Perca* and *Gymnocephalus*), and Luciopercinae (*Stizostedion*, *Zingel*, and *Romanichthys*) (Song et al., 1998).

The Percinae (*sensu* Nelson, 1994) possess well-developed anal spines, an enlarged anterior interhaemal bone, and a lateral line that ends before reaching the caudal fin. This subfamily includes about 167 species in seven genera. Most of these species are darters. There are two tribes within the Percinae: Percini (perches) and Etheostomatini (darters).

Members of the tribe Percini are spiny with a strongly serrate preopercle and a well-developed swim bladder. The genus *Perca* has a circumpolar distribution with yellow perch, *P. flavescens* (Fig. 256), in northern North America, *P. fluviatilis* in Eurasia, and *P. schrenki* in the Ballkhash and Alakul lakes region of Asia. *Perca flavescens* is an important commercial and sporting fish in the Great Lakes region of North America, and it has been widely introduced in other parts of the United States. *Perca fluviatilis*

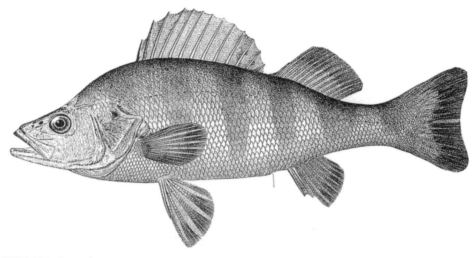

FIGURE 256. *Perca flavescens* as *P. americana* (Goode, 1884, Plate 168).

was the first nonnative fish species introduced into Australian fresh waters, where it is known as redfin. It has become a pest in stunted, dense populations and it competes with native species (McKay, 1984; Lever, 1996). *Perca fluviatilis* has also been introduced into South Africa and New Zealand. The biology of *Perca* and related species was summarized by Craig (1987).

There are four species of *Gymnocephalus* in Europe and western Asia. The ruffe, *G. cernuus*, which is widespread in Europe, has become established in Lakes Superior, Michigan, and Huron of North America, apparently as an escapee from the freshwater balast of oceangoing freighters. It was first detected in 1986 and has spread rapidly. Yellow perch and ruffe eat different food items with only slight overlap, so the effect of ruffe on perch is not expected to be great. *Percarina demidoffi* inhabits the northern Black Sea region and the Sea of Azov (Maitland, 1977).

The tribe Etheostomatini is characterized by smooth or only partly serrate preopercle margins and a reduced or absent swim bladder. These are small fishes, usually less than 80 mm TL. There are approximately 170 species of darters and three or four genera depending on the classification scheme one adopts. *Percina* has about 40 species, *Ammocrypta* includes 6 species, *Crystallaria* is monotypic, and *Etheostoma* accounts for about 123 species. Simons (1992) included *Ammocrypta* within *Etheostoma* and resurrected *Crystallaria* as distinct from *Ammocrypta*. Darter relationships have also been discussed by Page (1981), Bailey and Etnier (1988), Wood (1996), Wood and Mayden (1997), and Song *et al.* (1998).

These small fishes are sexually dimorphic, and the males, during spring and summer, are some of the most beautifully colored fishes in the world.

Their small size, brilliant colors, and darting motions make them the hummingbirds of the fish world. Kuehne and Barbour (1983) and Page (1983) included color photographs of most species. Etnier and Starnes (1993) and Jenkins *et al.* (1994) are rich sources of biological information and color photographs of these interesting fishes.

Darters generally prefer a benthic existence in flowing, shallow waters. The reduction or loss of the swim bladder as well as an emarginate tail and large pectoral fins facilitate life in this habitat. Darters feed on aquatic insects.

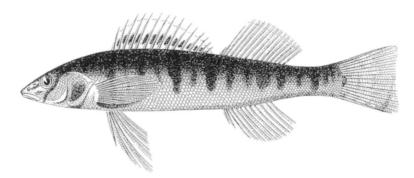

FIGURE 257. *Percina caprodes* as *P. carbonaria* (Girard, 1859, Plate LXXXIV).

Percina is thought by some ichthyologists to contain the most primitive darters. They have a small swim bladder and spend more time swimming in the water column than other darters. Nest building and parental care of eggs are not practiced by *Percina*. They bury their eggs in the gravel substrate of streams. Some *Percina* attain a larger size than other darters. The logperch, *P. caprodes* (Fig. 257), can reach 18 cm TL. It has the largest range of any darter—from Hudson Bay to southern Texas (Kuehne and Barbour, 1983). Most *Percina* have a row of dark blotches along the sides. Most are not as colorful as members of the genus *Etheostoma*. *Percina* has a row of enlarged scales along the midline of the breast and belly. These scales are absent in *Etheostoma*. Etnier and Starnes (1993) listed nine subgenera of *Percina* based on Page (1974): *Alvordius*, *Cottogaster*, *Ericosoma*, *Hadropterus*, *Hypohomus*, *Imostoma*, *Odontopholis*, *Percina*, and *Swainia*.

The crystal darter, *Crystallaria asprella*, from the Ohio, Missouri, and Mississippi River system and from several Gulf coast drainages is very slender and occurs over sand or gravel in rapidly flowing regions of small to medium-sized rivers. It can reach a relatively large size at 16 cm TL (Etnier and Starnes, 1993). There is a great deal of genetic divergence among populations of *C. asprella* (Wood and Raley, 2000).

The six species of *Ammocrypta* are called sand darters because they tend to burrow into sandy bottoms of rivers. They have a long, slender

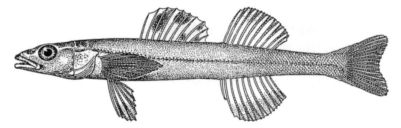

FIGURE 258. *Ammocrypta beani* (Jordan and Evermann, 1900, Plate CLXXII, Fig. 455).

body and are more or less translucent with a yellowish cast. Their eyes are located dorsally rather than laterally, which is an adaptation for their burrowing lifestyle. They often lie on the bottom with only their eyes exposed. *Ammocrypta beani* (Fig. 258) shows the typical morphology of the genus. Wiley and Hagen (1997) placed the sand darters in *Etheostoma* and discussed their mitochondrial DNA sequence variation.

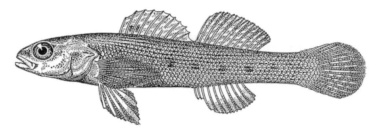

FIGURE 259. *Etheostoma virgatum* (Jordan and Evermann, 1900, Plate CLXXV, Fig. 465).

Etheostoma is the largest genus of North American freshwater fishes with about 123 species. The striped darter, *E. virgatum* (Fig. 259), models the usual body plan for this genus. New species continue to be described either as new discoveries or as taxonomic revisions. Species of *Etheostoma* have adapted to a variety of habitats including swamps and lake shores as well as the more typical riffle habitat of streams. Eggs are deposited on specific sites such as the surface of rocks or plant leaves. Some species bury their eggs in sandy bottoms of streams. Some species practice parental care of eggs (Page, 1985).

Etheostoma exile, the Iowa darter, occurs further north (northern Alberta) and west (western Alberta, Montana, and Colorado) than any other darter. *Etheostoma pottsi*, the Mexican darter, is the only darter to occur in a Pacific drainage (Rio Mesquital), and its distribution is further south than that of any other darter (Page, 1983; Smith et al., 1984). Other species of darters from the Chihuahuan desert of Mexico include *E. grahami*, *E. australe*, *E. lugoi*, and *E. segrex* (Norris and Minckley, 1997). Some darters have a very small range, such as the endangered *E. nuchale*, the watercress darter, which is found in only a few springs of the Black

Warrior River system near Bessemer, Alabama (Kuehne and Barbour, 1983; Mettee, 1996). Others, such as *E. nigrum*, the Johnny darter, have a very large range from Hudson Bay drainages to the Mobile basin. *Etheostoma nigrum* is thought to be more tolerant of a variety of conditions than most darters (Trautman, 1981), and it does not require the swiftly moving waters typical of darter habitats. It can occur in lakes.

Page (1983) provided an analysis of darter distribution by drainage systems and selected rivers. The greatest concentration of darter species is in central and eastern Tennessee and northern Alabama. A second center of diversity is in the Ouachita Mountains in western Arkansas and eastern Oklahoma. These areas lie between the maximum extension of Pleistocene glaciation and the coastal plain (Page, 1983). Such regions have been relatively undisturbed for a long time and are of great topographic diversity. Time and ecological opportunities have conspired to allow the evolution of impressive species diversity.

For keys to species, life history information, and color plates refer to Kuehne and Barbour (1983), Page (1983), Robison and Buchanan (1989), Etnier and Starnes (1993), Jenkins *et al.* (1994), and Mette *et al.* (1996).

Etheostoma taxonomy is complex, many species are similar, and the total number of species is large. Eighteen subgenera have been recognized, which helps to reduce the confusion to a manageable level. The following subgenera have been summarized by Etnier and Starnes (1993) from Bailey *et al.* (1954), Bailey and Gosline (1955), Collette and Knapp (1976), Page (1981), and Bailey and Etnier (1988): *Etheostoma, Ulocentra, Litocara, Doration, Boleosoma, Vaillantia, Psychromaster, Allohistium, Nothonotus, Oligocephalus, Ozarka, Fuscatelum, Catonatus, Hololepis, Microperca, Ioa, Belophlox,* and *Villora*. The relationships among these subgenera are not known with certainty, although there are some hypotheses such as presented by Wood and Mayden (1997), and even the species composition of some of the subgenera is not clear. Etnier and Starnes (1993) provided a list of Tennessee species currently assigned to the 18 subgenera. Porterfield *et al.* (1999) discussed phylogenetic relationships among the fan-tail darters of the subgenus *Catonotus*.

The subfamily Luciopercinae is characterized by weak anal spines, no enlargement of the anterior interhemal bone, and a lateral line that extends onto the tail. There may be supplementary canals on the caudal fin. As in the Percinae, parallel evolution has produced large predaceous species (tribe Luciopercini) and small bottom dwellers that lack a swim bladder (tribe Romanichthyini).

There are five species of piscivorous pikeperches, *Stizostedion*. Three species occur in Eurasia and two species are North American. These fishes have an elongate body, prominent canine teeth, and serrated preopercular bones. The walleye, *S. vitreum* (Fig. 260), is one of the most important sport fishes in North American fresh waters. Its firm, white, flaky flesh is

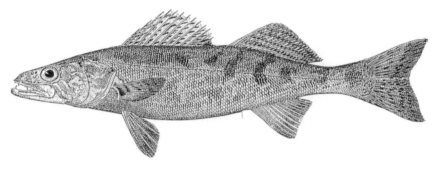

FIGURE 260. *Stizostedion vitreum* (Goode, 1884, Plate 169).

highly esteemed for its mild flavor. Walleye occupy a wide variety of cool-water habitats in rivers and lakes from the Mackenzie River of northern Canada and the southern half of Hudson Bay basin south into Alabama and Arkansas. It has been widely introduced outside of its native range. Tourist officials around Lake Erie refer to their lake as the "walleye capital of the world." Walleye are the object of important ice-fishing activity in the Great Lakes region during winter. Walleye reach about 90 cm and 11.3 kg (IGFA, 1993). Specimens larger than 4 kg are always females (Hackney and Holbrook, 1978). The name "walleye" refers to the large, glassy eye due to the oversized cornea of this nocturnal hunter. The specific epithet "vitreum" refers to the same feature. This effect is enhanced by the presence of a reflective tapetum lucidum behind the retina. Because of its financial importance as a premier sport fish, a great deal is known about the biology of walleye. For life history information see Scott and Crossman (1973), Becker (1983), Etnier and Starnes (1993), and Jenkins *et al.* (1994). Other sources of biological information include Collette *et al.* (1977), Hackney and Holbrook (1978), Ney (1978), and Colby *et al.* (1979).

A color variety of the walleye, known as the blue pike, was described as a separate subspecies, *S. v. glaucum*. This color morph was endemic to Lakes Erie and Ontario and the lower Niagara River, but it is now considered extinct (Trautman, 1981). However, blue-colored fish have been taken recently in small lakes in Canada. DNA testing should determine if blue pike are still extant and if they are a distinct species from the walleye. Faber and Stepien (1997) utilized mitochondrial DNA control region sequences to compared populations, species, and genera of percids including *Stizostedion*.

The walleye differs from its slightly smaller close relative, the sauger (*S. canadense*), by the presence of a white-tipped lower caudal lobe and a black blotch on the last two membranes of the dorsal fin. The sauger's caudal lobe is not white, and the black blotch is absent from the last dorsal fin membranes; however, its spiny dorsal fin membranes have many black markings. Sauger and walleye ranges overlap extensively, but sauger tend to be less northern and more southern in distribution. The native range of sauger is west of the Appalachian Mountains, whereas walleye occur on both sides of the Appalachians (Barila, 1980). Sauger can tolerate more turbidity than walleye. Hybrids between sauger and walleye have been produced. They are called saugeyes. They look like saugers but have longer gill rakers and lack the dorsal fin spots.

Stizostedion lucioperca, the pikeperch of Europe, occurs in slow-flowing rivers and lakes from The Netherlands to the Caspian Sea (Maitland, 1977). It too is valued as a sport fish. This species is the largest member of the Percidae. *Stizostedion volgensis* inhabits the Volga and Danube Rivers, and *S. marina* is found in brackish waters of the Black and Caspian Seas and the lower reaches of rivers. Kottelat (1997) referred the European *Stizostedion* to the genus *Sander*.

The darter-like fishes of the Luciopercinae include three species of *Zingel* (= *Aspro*) and *Romanichthys valsanicola* from swift-flowing water in the Danube basin of Romania. The two dorsal fins of *Zingel* are widely separated. *Zingel asper* is found in the River Rhône and its tributaries, and *Z. streber* and *Z. zingel* occur in the Danube (Maitland, 1977). These fishes lack a swim bladder, as do the North American darters, but the Eurasians are much larger. *Zingel zingel* may reach 48 cm TL.

The fossil record of the Percidae is unclear and has many gaps. *Perca*, *Percina*, and *Etheostoma* date to the Pleistocene in North America (Cavender, 1986), but their Eurasian origins must predate that by many millions of years (Collette and Banarescu, 1977; Patterson, 1981).

Map references: Banarescu (1990),* Berg (1948/1949), Collette and Banarescu (1977),* Darlington (1957),* Kuehne and Barbour (1983),* Lee *et al.* (1980),* Maitland (1977),* Norman and Greenwood (1975),* Norris and Minckley (1997),* Page (1983),* Page and Burr (1991),* Scott and Crossman (1973),* Smith *et al.* (1984)*

Class Actinopterygii
Subclass Neopterygii

Order Perciformes; Suborder Percoidei
(per) Family Apogonidae—**cardinalfishes, mouth-almighty**
[ap'-ō-gon'-i-dē]

THE APOGONIDAE IS A WORLDWIDE TROPICAL AND SUBTROPical family of small, reef-dwelling, colorful fishes. The group is overwhelmingly marine and most speciose in the Indo-Pacific, but 1 genus lives in fresh waters of New Guinea and northern Australia. The

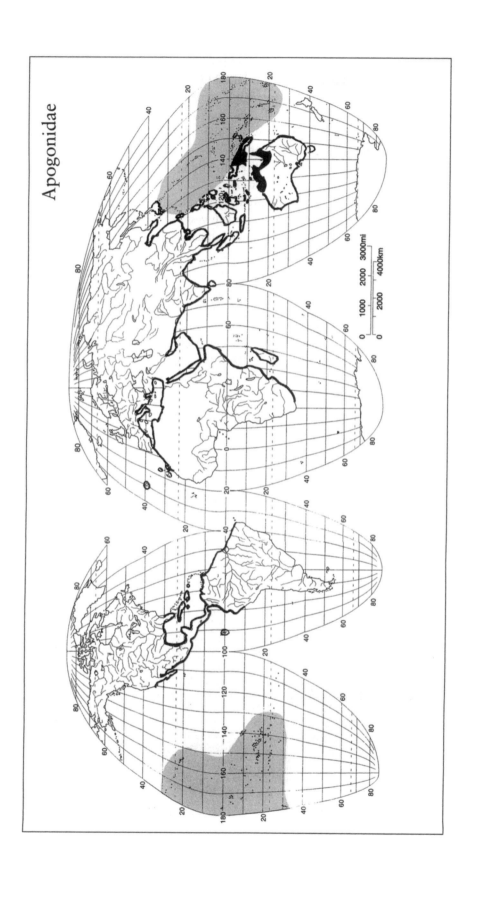

Apogonidae

distribution in the United States and elsewhere is coastal, not inland. There are 23 genera of apogonids and about 284 species, of which 89% occur in the Indo-west and central Pacific region (G. R. Allen, personal communication). About 180 species are in the cosmopolitan genus *Apogon*. The genus *Glossamia* is confined to fresh water and contains seven species, six of which are found in New Guinea and the seventh is shared by New Guinea and northern Australia (Allen, 1991). In addition, two species of *Apogon* also enter fresh water in New Guinea.

Apogonids have two separated dorsal fins except for the recently described *Paxton concilians* from northwestern Australia, which has a continuous dorsal fin (Baldwin and Johnson, 1999). The spiny dorsal of apogonids has six to eight spines, and the soft dorsal has one spine. Two anal spines are present. The caudal peduncle is relatively long. Ctenoid scales are the rule, but some species have cycloid scales. Many cardinalfish species are red, nocturnal hunters with large eyes and thereby resemble squirrelfishes (Holocentridae). The common name "mouth-almighty" is a reflection of their huge gape. Apogonids prey on small fishes, crustaceans, and, in fresh water, insect larvae. The family is a mouth-brooding group. Several hundred relatively large (4-mm) eggs are expelled by the female and scooped up into the male's mouth. The male broods the egg mass for several days or weeks until the well-developed young hatch. The egg-carrying parent does not feed during the period of incubation. In some species it is thought that the female incubates the eggs. Most cardinalfishes are less than about 12 cm TL, but some species grow to 20 cm TL.

Glossamia trifasciata (Fig. 261) reaches 10 cm SL and inhabits the Fly–Strickland and Lorentz Rivers of New Guinea (Allen, 1991). *Glossamia aprion* occurs among vegetation in streams and rivers of southern New

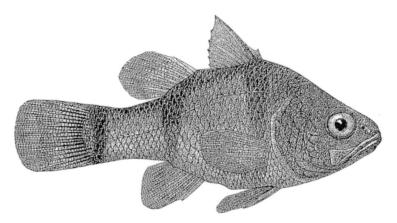

FIGURE 261. *Glossamia trifasciata* as *Apogon trifasciatus* (Weber, 1913, Fig. 32).

Guinea and northern Australia. It makes an interesting aquarium species; however, it is highly voracious and smaller tankmates will disappear from the aquarium. Merrick and Schmida (1984) included a color photograph of a male with a mouth full of eggs.

Fraser (1972) studied the comparative osteology of cardinalfishes and Fraser and Lachner (1985) revised two Indo-Pacific subgenera. Fraser and Struhsaker (1991) provided a key to the 19 genera of the subfamily Apogoninae. Roberts (1978) and Allen (1991) gave keys to the New Guinean freshwater species.

Map references: Allen (1989, 1991),* Allen and Robertson (1994), Fraser and Lachner (1985),* Lythgoe and Lythgoe (1992), Robins and Ray (1986), Tortonese (1986c)*

Class Actinopterygii
Subclass Neopterygii

Order Perciformes; Suborder Percoidei a̅ı
(per) Family Coiidae—**tigerperches** (kō-ī'-i-dē) [kō-ē'-i-dē]

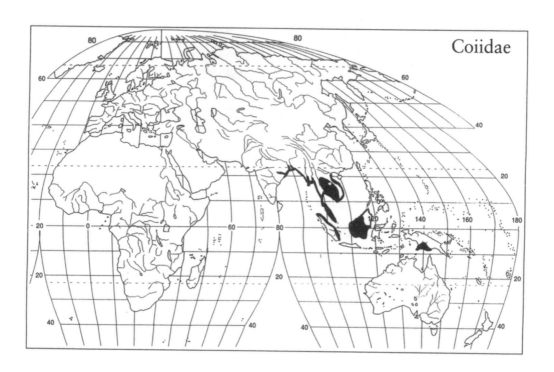

THE NOMENCLATURAL HISTORY OF THIS FAMILY HAS BEEN turbulent. Vari (1978) excluded the Southeast Asian *Datnioides* from the Terapontidae. Roberts (1989) placed the genus in its own family, Datnioididae. Nelson (1994) used the family name Lobotidae and included the widespread marine fish, *Lobotes surinamensis*, and *Datnioides* from fresh and brackish waters. Roberts and Kottelat (1994) reviewed the taxonomic history and decided that *Coius* was a senior synonym for *Datnioides*. They recognized the family name *Coiidae* for five species of *Coius* (formerly *Datnioides*). The relationships of this small perciform family are not clearly understood (Johnson, 1993). Lobotidae (the tripletail family) is retained for *L. surinamensis* (Schmid and Randall, 1997).

Coius is a deep-bodied, compressed, generalized percoid genus with a vertically barred color pattern. Of the three stout anal spines, the second is the largest. The edge of the opercle has fine teeth, and the lateral line scales number more than 50 (Kottelat et al., 1993). They are moderately large predatory fishes found throughout much of tropical Asia and New Guinea (Roberts and Kottelat, 1994).

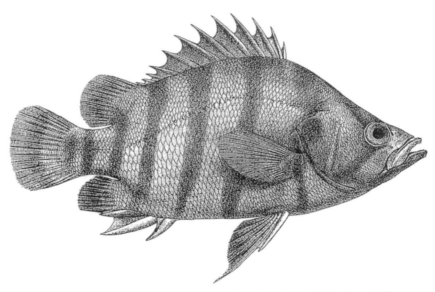

FIGURE 262. *Coius quadrifasciatus* as *Dantnioides polota* (Day, 1878b, Plate XXIV, Fig. 6).

Coius quadrifasciatus (Fig. 262) has the most extensive range. It occurs in the lower reaches of large rivers and brackish waters from India to Indonesia (Roberts and Kottelat, 1994). It has up to seven dark vertical bars on the body, not four as the specific epithet implies. In larger specimens the bars tend to coalesce. In New Guinea *C. campbelli* inhabits the Gulf of Papua drainages in both brackish and fresh waters (Roberts and Kottelat, 1994; Allen, 1991). It has been taken as far as 900 km upstream in the Fly River (Roberts, 1978). These two euryhaline species have relatively large scales and are probably phyletically related (Roberts and Kottelat, 1994).

A second group of *Coius* is restricted to fresh waters and has small scales. *Coius microlepis* is known from the Kapuas basin of western Borneo and the Musi basin in Sumatra (Roberts and Kottelat, 1994). This colorful species was heavily exploited for the aquarium trade and is now rare. It can reach 490 mm SL. Kottelat (1998) described C. *pulcher* from the Chao Phraya and Mekong basins based on material previously identified as C. *microlepis*. The two species cannot be distinguished by their external morphology but are separated by their color patterns. The other freshwater species is C. *undecimradiatus* from the middle and lower Mekong basin in Thailand, Laos, Cambodia, and Vietnam (Roberts and Kottelat, 1994).

Map references: Allen (1991), Kottelat (1998), Roberts (1989), Roberts and Kottelat (1994), Smith (1945), Talwar and Jhingram (1992)

Class **Actinopterygii**
Subclass **Neopterygii**

Order **Perciformes; Suborder Percoidei**
(per) Family **Polynemidae—threadfins**
(pol'-i-nem'-i-dē) [po-lē-nē'-mi-dē]

THE POLYNEMIDS ARE A CIRCUMTROPICAL MARINE FAMILY that occur in warm, shallow continental waters, with their greatest diversity in the Indo-Australian region. Some species enter coastal rivers throughout the world. Previously, this family was associated with the mullets and barracudas (Nelson, 1984). Recent studies suggested that the threadfins are closer to some Percoidei such as the drums (Feltes, 1986; Johnson, 1993).

The divided pectoral fin of these fishes provides them with their namesake, threadfins. The upper part of the pectoral fin has rays attached by membranes as in a normal fin. The lower portion of the pectoral fin has 3–16 long, unattached rays. The fish can fan out these filamentous feelers and receive sensory information about the substrate in turbid, coastal waters and rivers. Threadfins usually occur over sand and mud flats. Adipose eyelids are present, as in mullets, which is probably an additional adaptation to the murky habitat. The spiny and soft dorsal fins are widely

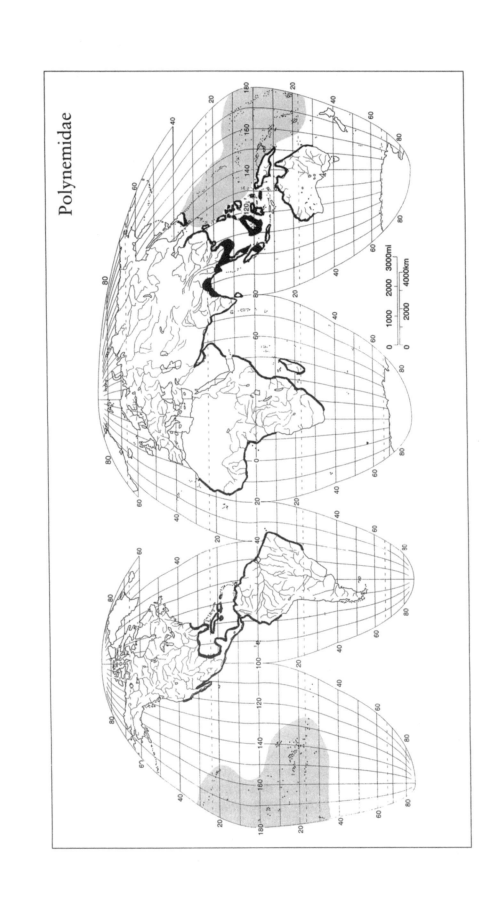

Polynemidae

separated. The pelvic fins are subabdominal and have one spine and five rays. The caudal fin is deeply forked. The snout overhangs a subterminal mouth. They feed on arthropods and fishes.

Feltes (1986, 1991, 1993) recognized seven genera and 33 species: *Eleutheronema* (2 species, northern Indian Ocean through Malay archipelago and China Sea), *Filimanus* (6 species, Pakistan to Solomon Islands, represented by *F. heptadactyla*; Fig. 263), *Galeoides decadactylus* (west coast of Africa), *Parapolynemus verekeri* (southern New Guinea and northern Australia), *Pentanemus quinquarius* (west coast of Africa; Fig. 264), *Polydactylus* (17 extant species, circumtropical, represented by *Polydactylus sextarius* and 1 extinct species, *P. fossilis*; Fig. 265), and

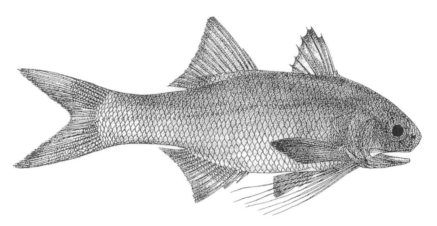

FIGURE 263. *Filimanus heptadactyla* as *Polynemus heptadactylus* (Day, 1878b, Plate XLII, Fig. 5).

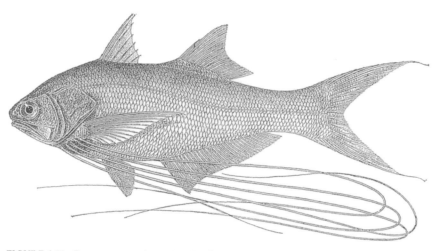

FIGURE 264. *Pentanemus quinquarius* (Boulenger, 1916, Fig. 61).

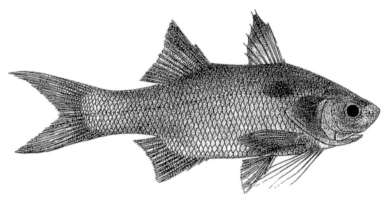

FIGURE 265. *Polydactylus sextarius* as *Polynemus sextarius* (Day, 1878b, Plate XLII, Fig. 6).

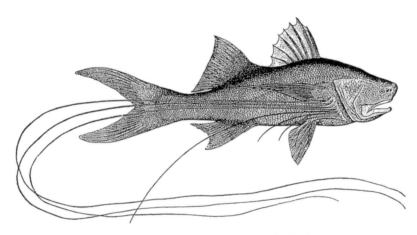

FIGURE 266. *Polynemus hornadayi* (Hornaday, 1885, pp. 386–387).

Polynemus (5 species, eastern India to Borneo, represented by *Polynemus hornadayi*; Fig. 266). *Polydactylus nigripinnis* from southern New Guinea and northern Australia deserves a closer look and may represent an undescribed genus (Feltes, 1986).

Polydactylus has the most extensive distribution. It is circumtropical and extends into temperate regions. *Polydactylus quadrifilis* makes spawning runs far up the rivers of west Africa. *Polydactylus macrochir* frequents rivers and estuaries of southern New Guinea and northern Australia, and *Polydactylus macrophthalmus* has only been taken in rivers of Borneo and Sumatra where *Polynemus multifilis* also occurs (Roberts, 1989a). A few species of the genus *Polynemus* enter coastal rivers, especially in Borneo (Roberts, 1989a). *Eleutheronema terradactylum* may be taken in lower stretches of rivers with tidal influence.

Threadfins are important commercially in the tropics and are taken in net fisheries, especially off the coasts of Nigeria, India, and Indonesia (Feltes, 1991). Most species are 20–60 cm TL; however, *Eleutheronema tetradactylum* of India may reach 2 m TL.

Map references: Daget (1992c), Daget and Njock (1986), Feltes (1986, 1991, 1993),* Hureau (1986),* Roberts (1989a), Talwar and Jhingran (1992)

Class Actinopterygii
Subclass Neopterygii

Order Perciformes; Suborder Percoidei
(per) Family Sciaenidae—**drums, croakers**
(sī-en'-i-dē) [sī-ē'-ni-dē]

SCIAENIDS ARE MOSTLY MARINE FISHES FOUND ON THE CONTI-
nental shelf of tropical and temperate regions throughout the world. There are about 70 genera and 270 species (Nelson, 1994). One species is confined to fresh waters in North and Central America and several species live in South American fresh waters. They are also represented in fresh waters in New Guinea. Sciaenids tend to be absent from the clear water and reefs around islands in the mid-Indian and Pacific Oceans (Nelson, 1994). More than half of the species occur in the New World rather than in the highly diverse Indo-Pacific (Springer, 1982). This unusual pattern is similar to that of the Batrachoididae. Trewavas (1977) reviewed the Indo-West Pacific species, and Sasaki (1989) revised the family, which he considered to have arisen in the New World and dispersed to the Old World across the Atlantic Ocean.

Drums have only two anal spines, and their lateral line extends onto the caudal fin as it does in the Centropomidae. The caudal fin may be emarginate to pointed. The two dorsal fins are separated by a deep notch, and the soft dorsal fin is long. The pelvic fins are thoracic, and the scales are ctenoid. Many species are deep-bodied with an arched back. One or more barbels may be present under the chin. Enlarged lateral line canals are present on the head. The gas bladder is oval or carrot shaped, often with many appendages of taxonomic significance (Chao, 1986a). The common names (drum or croaker) are derived from the fish's ability to make resonating sounds with the gas bladder and its muscles. These sounds may be used in courtship and navigation in murky waters.

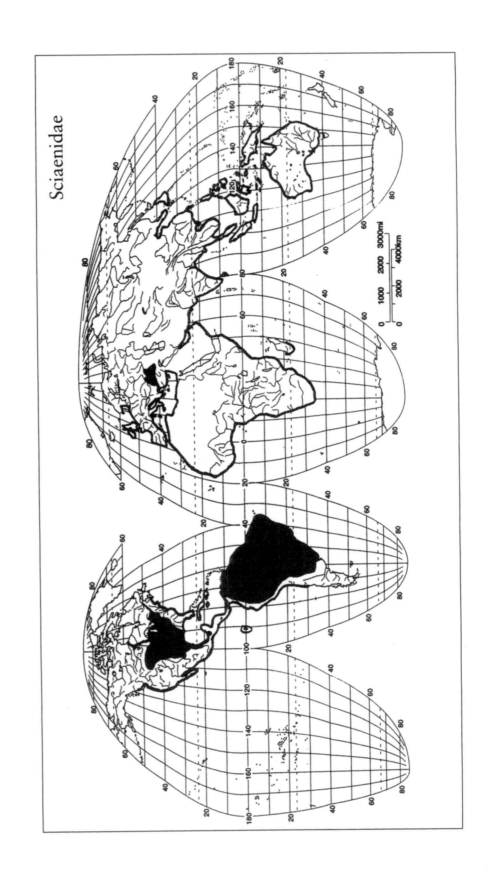

Sciaenidae

The resulting underwater noise was a source of confusion to navy sonar operators until its nature was understood. The sagitta (one of the otoliths) is exceptionally large in this family. This is not uncommon in fishes that produce sound. Some sciaenids also have an enlarged lapillus (another otolith). Schwarzhans (1993) reviewed the morphology of recent and fossil sciaenid otoliths. Otolith size and shape convey taxonomic information that is not always in agreement with the data derived from swim bladders (Chao, 1986a). Many sciaenid species are important food and sport fishes.

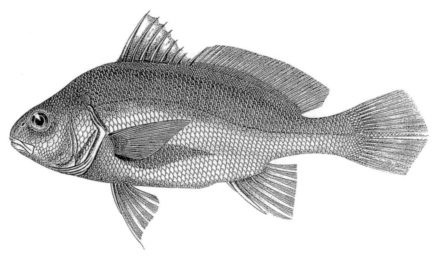

FIGURE 267. *Aplodinotus grunniens* as *Amblodon grunniens* (Girard, 1858, Plate XXIII, Fig. 1).

Some marine species enter estuaries and about 28 species are confined to fresh waters in the Atlantic drainages of the New World (Nelson, 1994). The freshwater drum, *Aplodinotus grunniens* (Fig. 267), occupies the largest latitudinal range of any North American freshwater fish from Hudson Bay to Guatemala (Fremling, 1980). This species is a bottom feeder in rivers and lakes and consumes fishes, crustaceans, and mollusks which it crushes with molar-like pharyngeal teeth. The crushed shells are excreted with the feces. It is one of the few North American freshwater fish with planktonic eggs and fry. (The Hiodontidae and some *Hybognathus* (Cyprinidae) also have planktonic eggs.) Planktonic eggs are common among marine fishes. The life history of *Aplodinotus*, mostly from Lake Erie populations, was reviewed by Scott and Crossman (1973). It can reach 89 cm TL and 18.2 kg, although most specimens seen by anglers are less than 2 kg. The freshwater drum is edible, but its coarse flesh is not highly esteemed. Beachcombers along Lake Erie often find the large otoliths of freshwater drums which tourists call "lucky stones."

These ivory-like bones can be 20–30 mm or larger in their longest dimension, and they are sometimes made into earrings or pendants. A laboratory manual of freshwater drum anatomy was produced by Fremling (1978).

Three genera are restricted to Atlantic freshwater drainages of South America: *Pachyurus* (8 species), *Pachypops* (4 species), and *Plagioscion* (9 species) (Sasaki, 1989). Chao (1986a) estimated that five to seven genera with 25–30 species live in Neotropical fresh waters.

Two species of *Nibea* are known from as far up the Fly River of Papua New Guinea as 515 km (Roberts, 1978; Allen, 1991).

Seatrouts, *Cynoscion*, are fast-swimming piscivores that are popular with surf and pier anglers. They are important sport and food fishes throughout their range. The red drum, *Sciaenops ocellatus*, from the Atlantic and Gulf coast of North America, is popular in restaurants as "blackened redfish."

The fossil record of the Sciaenidae dates back to at least the Eocene of North America. A Late Cretaceous fossil from Wyoming was assigned to the Sciaenidae, but Cavender (1986) considered that specimen to be closer to the Cichlidae.

Map references: Chao (1986b),* Freming (1978),* Lee *et al.* (1980),* Page and Burr (1991),* Scott and Crossman (1973),* Springer (1982)*

Class   Actinopterygii
Subclass   Neopterygii

Order   Perciformes; Suborder Percoidei
(per) Family   Monodactylidae—**moonfishes**
(mon'-ō-dak'-til'-i-dē)

THE MONODACTYLIDAE IS A SMALL MARINE FAMILY WHOSE members can ascend rivers and live in fresh water. The family ranges from the west coast of Africa through the Indo-Pacific.

Monodactylids have a very deep, strongly compressed body with small ctenoid scales. The dorsal and anal fins are undivided and long and accentuate the disk-like body. The base of the dorsal and anal fins is scaled. There are three anal spines. Pelvic fins are well developed in juveniles, but they become reduced to spines in adults. The mouth is small (Talwar and Jhingran, 1992).

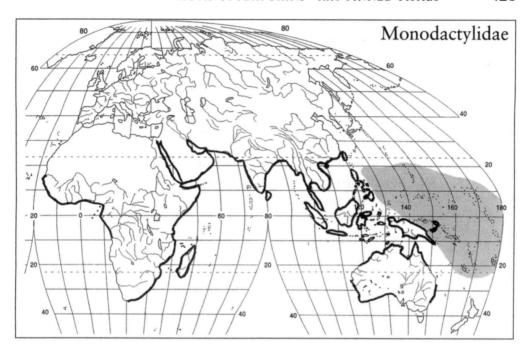

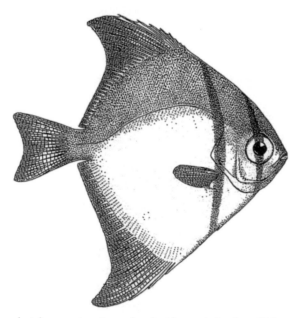

FIGURE 268. *Monodactylus argenteus* (reproduced with permission from Weber and DeBeaufort, 1936, Fig. 54).

There are four species in the genus *Monodactylus*. *Monodactylus argenteus* (Fig. 268), the silver moonfish, inhabits mangrove estuaries and river mouths, and it often moves into freshwater streams in eastern

Africa, Indonesia, New Guinea, and northern Australia (Allen, 1991; Kottelat et al., 1993; Skelton, 1993). It is a schooling fish, and small schools of juveniles make a graceful freshwater aquarium display. The young are mostly black posteriorly with two dark vertical bars through the eyes and operculum. The adults are more silvery. This species can reach 25 cm TL.

Monodactylus falciformis also ranges from the east coast of Africa through the Indo-Pacific. It can live in fresh water (Desoutter, 1986a). *Monodactylus sebae* is found along the west coast of Africa from Senegal to Congo Rivers in lagoons and estuaries. It has an extraordinarily high body. The distance from the tip of the dorsal fin to the tip of the anal fin is nearly twice the standard length of the fish. It rarely enters fresh water. Pethiyagoda (1991a) described *M. kottelati* from coastal Sri Lanka.

The genus *Schuettea*, with two Australian marine species (east and west coast), is sometimes placed in this family. These fishes have normally developed pelvic fins and cycloid scales. Tominaga (1968) placed them in a family of their own, but Eschmeyer (1998) retained *Schuettea* in the Monodactylidae.

Map references: Allen and Swainston (1988), Desoutter (1986a), Gommon *et al.* (1994), Kuiter (1993), Pethiyagoda (1991a),* Skelton (1993),* Sterba (1966)*

Class Actinopterygii
Subclass Neopterygii

Order Perciformes; Suborder Percoidei
(per) Family Toxotidae—**archerfishes**   (toks-ot'-i-dē)

ARCHERFISHES OCCUR IN COASTAL MANGROVE SWAMPS, brackish estuaries, and freshwater streams from India to Melanesia including New Guinea, the Solomon Islands, and Vanuatu (Allen, 1978, 1991). There are six species in a single genus, *Toxotes*, distributed as follows: *T. blythi*, Burma; *T. chatareus*, widespread from India to Australia/New Guinea (Fig. 269); *T. jaculatrix*, widespread from India to Vanuatu; *T. lorentzi*, northern Australia and New Guinea; *T. microlepis*, Thailand, Sumatra, and Borneo; and *T. oligolepis*, eastern Indonesia, New Guinea, and northern Australia (Allen, 1978).

Archerfishes have a strongly compressed, rhomboidal body and a straight slope from snout to dorsal fin. There are usually four or five

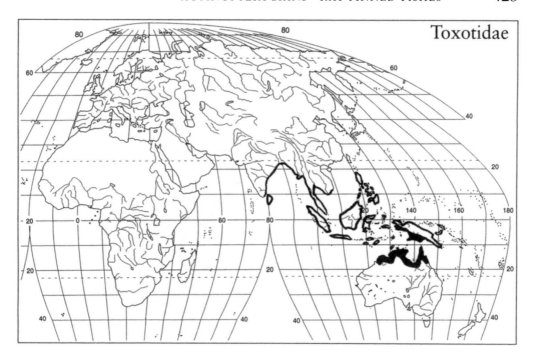

FIGURE 269. *Toxotes chatareus* (Day, 1878b, Plate XXIX, Fig. 6).

dorsal spines, and the spiny and soft dorsal fin are undivided and posteriorly set. The undivided anal fin has three sharp, stout spines and is positioned under the dorsal fin. The soft dorsal is smaller than the soft anal fin. The eye is large. The mouth is large and slanted upward, with the lower jaw protruding. The body usually has a pattern of black spots or bars.

These fishes are known for their remarkable ability to "shoot down" insects from vegetation overhanging the water surface. This ability was first explained by Smith (1936). *Toxotes jaculatrix* and, presumably, the other species have a palatine groove that, together with the tongue, forms a tube through which water drops are forcefully ejected by stong compression of the operculum. *Toxotes* can accurately hit an insect 1.6 m above the water's surface. When prey is knocked into the water, it is quickly devoured. Herald (1962) included action photographs, and Lake (1971) reported that his cigarette was extinguished from 90 cm by an aquarium-based *Toxotes*. Archerfish are even able to compensate for light refraction at the water's surface and for the curved trajectory of the expelled water "bullets" (Dill, 1977). See Smith (1936, 1945) and Allen (1973) for more absorbing details of this unique mode of existence. Archerfish also feed on crustaceans, aquatic insect larvae, flower buds, and other food in addition to what they can shoot into the water (Allen, 1978, 1991).

Toxotes chatareus and *T. jaculatrix* are the most widespread species. *Toxotes jaculatrix* usually occurs in brackish mangrove estuaries. It is the least likely archerfish to be found in fresh waters. The other 5 species are frequently found in fresh water and probably breed there (Allen, 1978). *Toxotes chatareus* is recorded from 800 km up the Fly River of Papua New Guinea and grows to about 40 cm TL, twice the size of any other archerfish (Roberts, 1978). Allen (1978) reviewed the family and provided keys and photographs. Fossils date to the lower Tertiary of central Sumatra.

Toxotes is an exciting aquarium animal. I have seen clever displays that feature a bull's eye target with a cricket in the center that the archerfish knocks into its tank with a jet of water droplets.

Map references: Allen (1978, 1989, 1991)*

Class Actinopterygii
Subclass Neopterygii

Order Perciformes; Suborder Percoidei
(1st) Family Nandidae—**Asian leaffishes**   (nan'-di-dē)

THE NANDIDAE AS RECOGNIZED BY NELSON (1994) CONSISTED of three groups that will be treated as distinct families here: Nandidae

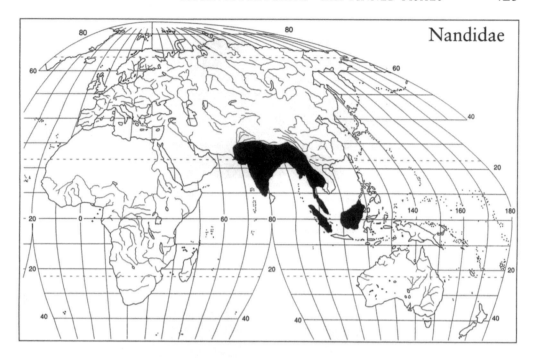

Nandidae

(*Nandus* + *Badis*), Polycentridae, and Pristolepididae. Barlow *et al.* (1968) placed *Badis* in its own family, Badidae. Gosline (1971) accepted three families (Badidae, Nandidae, and Pristolepidae) and listed them at the beginning of his Percoidei which was preceded by the Anabantoidei. Liem (1963, 1970) gave a detailed account of the osteology and functional anatomy of the Nandinae and, in the latter paper, explained why he considered that nandids and anabantoids are not closely related as suggested by Gosline (1971). Britz (1997) reviewed the tortured taxonomic history of these fishes, discussed egg surface morphology, and commented on phylogeny and biogeography. He concluded that the Nandidae and Badidae are not closely related. The monophyly of the Nandidae and its phylogenetic inter- and intrarelationships are still problematic.

The Nandidae as recognized here is a small family of two genera and three species. *Nandus nandus* (Fig. 270) occurs in India, Burma, and Thailand, and *N. nebulosus* is found in Thailand, the Malay Peninsula, Sumatra, and Borneo (Roberts, 1989a). Nandids typically have an enormous, widely protrusible mouth. Their compressed body form and camouflage coloration resemble a leaf. They ambush their unsuspecting prey in the vegetation of slow-moving streams. The dorsal fin of *Nandus* is long, with 12–14 spines and 11–13 soft rays. Three anal spines are present. The caudal fin and posterior part of the dorsal and anal fins can be so transparent as

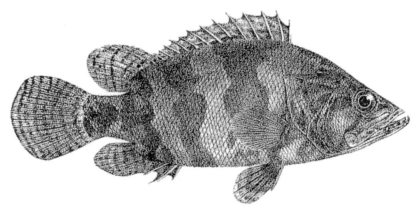

FIGURE 270. *Nandus nandus* as *N. marmoratus* (Day, 1878b, Plate XXXII, Fig. 1).

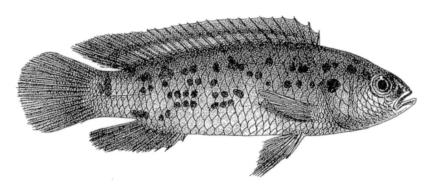

FIGURE 271. *Badis badis* as *B. buchanani* (Day, 1878b, Plate XXXI, Fig. 6).

to be hardly visible in living fishes (Talwar and Jhingran, 1992). The lateral line is incomplete. *Nandus nandus* reaches 20 cm TL and is utilized as a food fish. The two species of *Nandus* comprise the subfamily Nandinae.

The subfamily Badinae includes only one species, *Badis badis* (Fig. 271), from ponds, ditches, and rivers of Pakistan, India, Nepal, Bangladesh, and Burma (Talwar and Jhingran, 1992). This species has a small mouth, large eye, and three anal spines. It is the most beautifully pigmented of the nandids, and it can change colors and patterns rapidly. Barlow *et al.* (1968) described its behavior. It is an ambush predator. *Badis* reaches 8 cm TL and is often kept by tropical fish hobbyists. Males are territorial and aggressive.

Map references: Darlington (1957), Roberts (1989a), Talwar and Jhingran (1992)

Class Actinopterygii
Subclass Neopterygii

Order   Perciformes; Suborder Percoidei
(1st) Family   Polycentridae—**African and South American leaffishes**   (pol'-i-sen'tri-dē)

TWO GENERA FROM WEST AFRICA AND TWO GENERA FROM northeast South America comprise this small family. Like the Nandidae, polycentrids have very large, highly protrusible mouths and are voracious predators. They have compressed bodies and mottled coloration which give them a leaf-like appearance. Their coloration is variable depending on their surroundings and degree of excitement. Britz (1997) provided synapomorphies based on breeding behavior, egg structure, and larval morphology for Polycentridae, distinct from Nandidae and Pristolepidae.

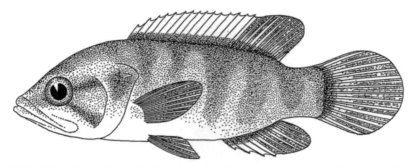

FIGURE 272. *Afronandus sheljuzhkoi* (reproduced with permission from Lévêque *et al.*, 1990, Fig. 48.2).

Afronandus sheljuzhkoi (Fig. 272) lives in small, forest rivers in southwest Ivory Coast (Thys van den Audenaerde and Breine, 1986). It is elongate, reaches 48 mm TL, and has four anal spines and no lateral line (Teugels, 1992c). *Polycentropsis abbreviata* (Fig. 273) occurs in rain forest areas of Nigeria, Cameroon, and Gaboon (Thys van den Audenaerde and Breine, 1986). It is deep-bodied, reaches 80 mm TL, and has 9–12 anal spines and an incomplete lateral line (Teugels, 1992c). Its coloration is a blend of greens and browns which serve to conceal it among vegetation. It builds a bubble nest beneath floating leaves at the water surface. The male tends the nest and herds the hatchlings into a depression in the stream bottom (Sterba, 1966).

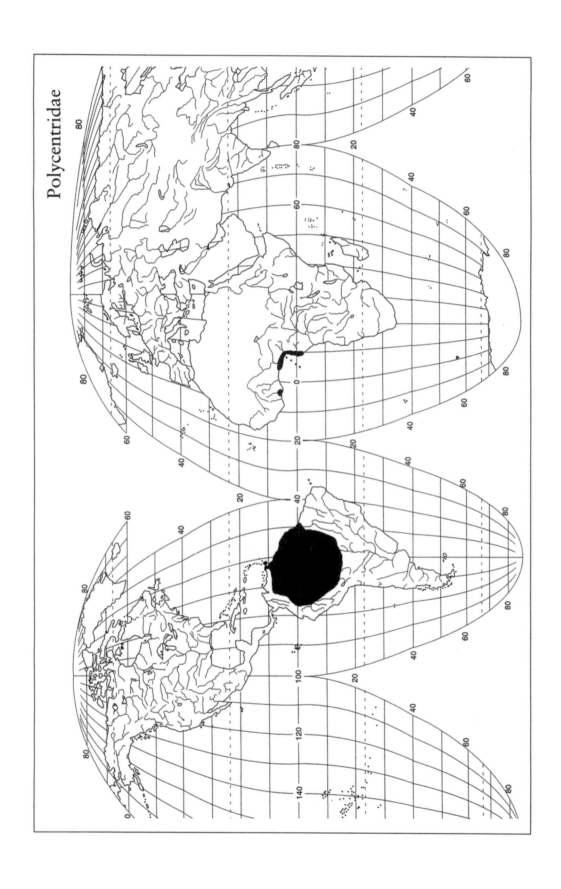

Polycentridae

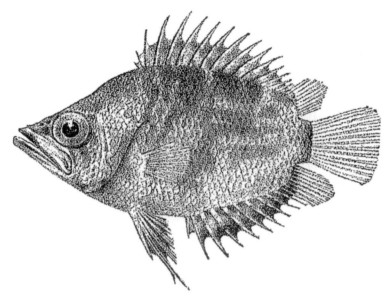

FIGURE 273. *Polycentropsis abbreviata* (Boulenger, 1915, Fig. 79).

In South America *Monocirrhus polyacanthus* occurs in Guiana and the Amazon lowlands into tropical lowland Peru, up the Rio Negro, and into the upper Orinoco drainage (Darlington, 1957; S. Kullander, personal communication). It reaches 10 cm TL and is the most leaf-like of the leaffishes. It has a deep body, pointed head, and huge mouth, with which it can "inhale" prey fishes two-thirds to three-fourths its own size. Some individuals possess a prominent chin flap that resembles a "stem" and heightens the leaf mimicry. This species advances toward potential prey with its head at a downward angle without any apparent fin movement. The stealthy approach is followed by a lightning-quick strike at the prey. The dorsal fin is very long with 16–17 spines, and the spiny and soft portions are joined. The anal fin has 12 or 13 spines and 11–14 soft rays. The caudal fin is rounded and the lateral line is incomplete. Because of their interesting appearance and behavior these fish are popular aquarium animals, but they require a large supply of live food.

Polycentrus punctatus (= *schomburgki*) is a lowland species that ranges from the Amapá swamps between French Guiana and Brazil to Trinidad and the lower Orinoco River (S. Kullander, personal communication). It has 16–18 dorsal fin spines and 8 or 9 rays. The anal fin has 13 spines and 6 to eight rays. It can reach about 10 cm TL in the wild, but captive specimens remain smaller. It too is a leaf mimic.

Map references: Darlington (1957), Sterba (1966), Teugels (1992c),* Thys van den Audenaerde and Breine (1986)

Class Actinopterygii
Subclass Neopterygii

Order Perciformes; Suborder Percoidei
(1st) Family Pristolepididae—**pristolepidids**
(pris'-tō-lep-id'-i-dē) [pris'-tō-lē-pid'-i-dē]

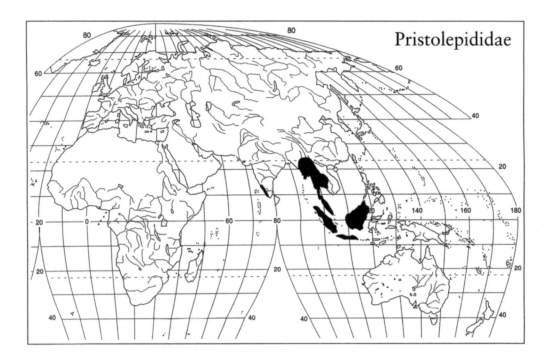

LIEM (1970) CONSIDERED THAT THIS GROUP WAS NOT ESPE-
cially closely related to the nandids. Roberts (1989a) wrote that *Pristolepis*
is more generalized than the nandids and is probably not closely related to
them. Roberts speculated that it may be related to the anabantoids or to the
"protoanabantoid stock." Barlow *et al.* (1968) wrote that *Pristolepis*
should not be included in the same family as *Badis* for which he created a
separate family, the Badidae, which he considered a protoanabantoid.
Pristolepis was placed in the Nandidae by Greenwood *et al.* (1966) and
considered a subfamily of the Nandidae by Nelson (1994).

The family Pristolepididae consists of three species of *Pristolepis* which
range from peninsular India to Borneo. These deep-bodied fishes have a
relatively short-jawed, small mouth that is only slightly protrusible and
three anal spines. The dorsal fin has 12–16 spines and 14–16 soft rays. The

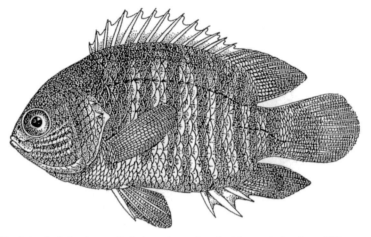

FIGURE 274. *Pristolepis fasciata* as *P. fasiatus* (reproduced with permission from Weber and DeBeaufort, 1936, Fig. 95).

lateral line is interrupted at the 20th scale (Kottelat et al., 1993), and pored scales of the lower lateral line extend onto the caudal fin (Roberts, 1989a). *Pristolepis fasciata* (Fig. 274) occurs in Burma, Thailand, and the Malay Peninsula (Roberts, 1989a; Kottelat et al., 1993). Other species include *P. marginata* from southwestern India (Talwar and Jhingran, 1992) and *P. grootii* from the Malay Peninsula, Java, Sumatra, and Borneo (Roberts, 1989a; Kottelat et al., 1993).

Map references: Darlington (1957), Kottelat *et al.* (1993), Roberts (1989a), Talwar and Jhingran (1992)

Class Actinopterygii
Subclass Neopterygii

Order Perciformes; Suborder Percoidei
(per) Family Terapontidae—**grunters** [ter-a-pon'-ti-dē]

GRUNTERS OCCUR IN COASTAL, BRACKISH, AND FRESH WATER of the Indo-West Pacific from the east coast of Africa through India, Southeast Asia to Japan, New Guinea, and Australia to Fiji and Samoa (Vari, 1978). Two species have invaded the eastern Mediterranean Sea via the Suez Canal from the Red Sea (Ben-Tuvia, 1986a). Their distribution is not shown on the map. There are 16 genera and about 45 species (Nelson,

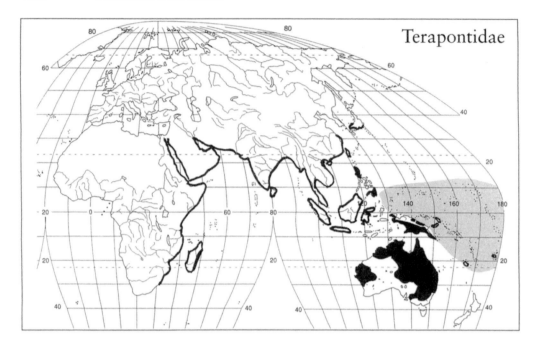

1994). Terapontids are unusual for a marine family in that most of their members are found in fresh water. Six genera and 33 species are restricted to the fresh waters of New Guinea and Australia (Allen, 1991). There are also freshwater species in the Philippines, Sulawesi (Vari, 1978), and Madagascar (Vari, 1992c).

These moderately compressed, oblong, perch-like fishes are of small to moderate size (65–800 mm SL) and have two spines on the operculum. The dorsal fin is fairly deeply notched, and a scaled sheath is present at the base of the spinous dorsal and anal fin. The spinous dorsal fin is depressible into the groove formed by the basal scaly sheath. Ctenoid scales are present. The lateral line extends onto the caudal fin. There are three anal spines. Some species emit a grunting noise produced by extrinsic swim bladder muscles which vibrate the resonant, transversely divided swim bladder. These muscles arise from the rear of the skull or posttemporal bone or both and attach to the anterodorsal surface of the anterior chamber of the swim bladder (Vari, 1978). No other family has this arrangement combined with a transversely divided swim bladder.

The family name is often written as "Theraponidae." Vari (1978), in his comprehensive review of the family, explained that although *Terapon* is an incorrect transliteration it is the original spelling and must apply to both generic and familiar names. Springer (1982) used Terapontidae, as did Vari (1992c) and Eschmeyer (1998). Nelson (1994) suggested that this spelling is correct.

In the absence of most primary-division freshwater fishes in Australia, terapontids have radiated into a diverse assemblage that constitutes an important group in that continent's fresh waters. A fossil teraponid from the Oligocene of Queensland indicates that this family has been present in Australia for at least 30 million years (Allen, 1989). Allen listed 23 species from Australian fresh waters and estuaries. *Bidyanus bidyanus*, called silver perch in Australia, is a widespread, important sporting fish in the Murray–Darling system of eastern Australia. It is also an excellent species for pond culture, with the potential to form a large aquaculture industry (Rowland et al., 1995). It commonly reaches 30–40 cm TL and 0.5–1.5 kg, but it has been known to reach 8.0 kg (Lake, 1978). A female may produce 300,000–500,000 pelagic eggs (Merrick and Schmida, 1984; Merrick, 1996) after a lengthy upstream migration beyond the peak of spring and summer flooding (Reynolds, 1983). This species, like the percichthyid, *Macquaria ambigua*, is one of the few freshwater fishes to produce planktonic eggs. The eggs hatch in about 30 hr, and the larvae begin feeding in about 6 days as they drift downstream. Despite its tremendous fecundity, the silver perch has declined dramatically, and it is classified as vulnerable (Jackson, 1998).

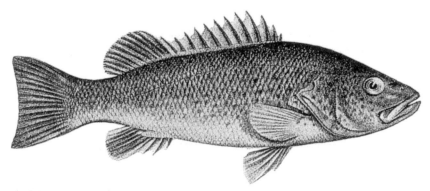

FIGURE 275. *Leiopotherapon unicolor* as *Therapon unicolor* (Waite, 1923, p. 119).

Leiopotherapon unicolor (Fig. 275), the spangled perch, may be the most widespread freshwater fish in Australia. Its range includes all the warmer fresh waters across the northern two-thirds of Australia. It is also known from the inland drainages of Lake Eyre and Bulloo-Bancannia (Midgley et al., 1991). It is remarkably hardy and can tolerate a wide range of temperatures and salinities (Merrick and Schmida, 1984). It may even estivate in leaf litter or the bottom mud when its habitat desiccates. It reappears soon after rain rehydrates its pools (Merrick, 1996). Spawning and temperature tolerance have been studied by Llewellyn (1973) and Beumer (1979a,b). See Merrick and Schmida (1984) and Allen (1989) for descriptions of other Australian grunters.

The southern half of New Guinea shares six genera with Australia (*Amniataba, Hephaestus, Mesopristes, Pingalla, Terpon,* and *Variichthys*)

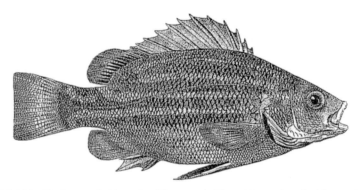

FIGURE 276. *Hephaestus habbemai* as *Therapon habbemai* (reproduced with permission from Weber and DeBeaufort, 1931, Fig. 28).

(Allen, 1991, 1993). *Hephaestus habbemai* (Fig. 276) is common in the upper montane regions of the Fly–Strickland and Lorentz systems of New Guinea 970 km from the sea (Allen, 1991). There are 16 species from New Guinea. The New Guinean terapontids were reviewed by Mees and Kailola (1977).

Lagusia micracanthus is endemic to the Lagusi River on the southern peninsula of Sulawesi (Vari, 1978). *Mesopristes elongatus* occurs in rivers on the east coast of Madagascar (Vari, 1992c). Other members of this genus occur on the islands of Indonesia, the Philippines, and Fiji.

Map references: Allen (1989, 1991),* Merrick (1996),* Springer (1982),* Vari (1978,* 1992c)

Class   Actinopterygii
Subclass   Neopterygii

Order   Perciformes; Suborder Percoidei
(per) Family   Kuhliidae—flagtails   (kōō-lī'-i-dē) [kōō-lē'-i-dē]

THIS MARINE FAMILY OF ONE GENUS, *KUHLIA*, AND ABOUT eight species ranges from Africa to the tropical eastern Pacific with at least two species occurring in fresh water (Nelson, 1994).

Flagtails are so called because the tail is usually conspicuously barred. Their elongate, compressed body form is similar to that of the centrarchid basses. The dorsal fin is deeply notched with 10 spines and 9–12 rays. The anal fin has 3 spines. Both dorsal and anal fins have a well-developed scaly sheath. The opercle has 2 flattened spines. The body is covered with ctenoid scales. Maximum size is about 50 cm TL.

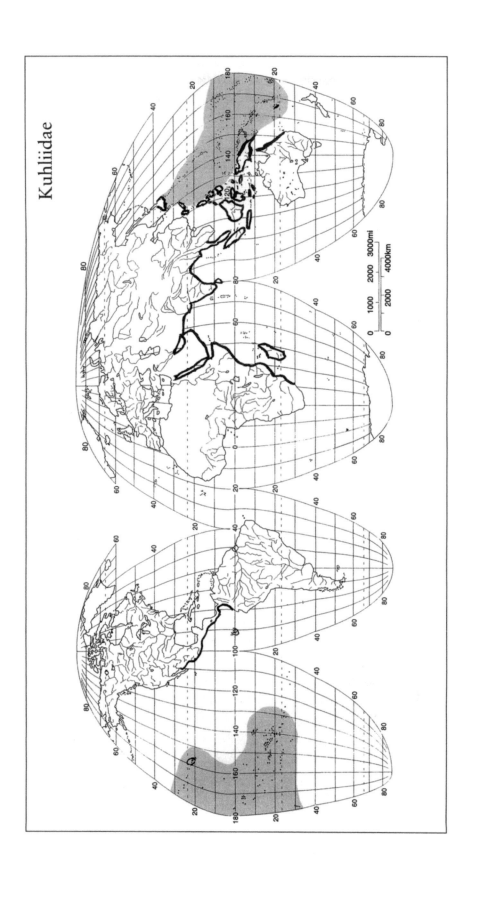

Kuhliidae

Formerly, six species of pygmy perches (*Edelia*, *Nanatherina*, and *Nannoperca*) were classified in the Kuhlidae, but current thinking places them in the Percichthyidae (Johnson, 1984). *Parakuhlia* mentioned by Allen (1991) from the Atlantic is a member of the family Haemulidae (Eschmeyer, 1998).

Marine kuhlids occur on coral reefs and in estuaries of the tropical Indo-Pacific extending to the tropical eastern Pacific. *Kuhlia taeniura* is the most widespread species and the only one to reach the Americas. Its range extends from the east coast of Africa to the Gulf of California south to Colombia and the Galapagos Islands (Thomson et al., 1979). Freshwater kuhlids such as *K. marginata* (Fig. 277) occur in coastal streams from New Guinea to Japan and the Society Islands and can penetrate far upstream (Berra et al., 1975).

The jungle perch, *K. rupestris* (Fig. 278), is widely distributed in the tropical Indo-Pacific from East Africa to Melanesia and eastern Australia.

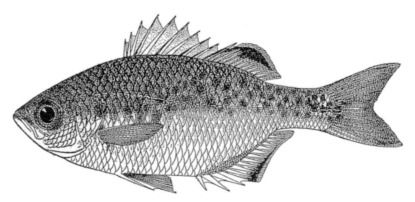

FIGURE 277. *Kuhlia marginata* (reproduced with permission from Weber and DeBeaufort, 1929, Fig. 72).

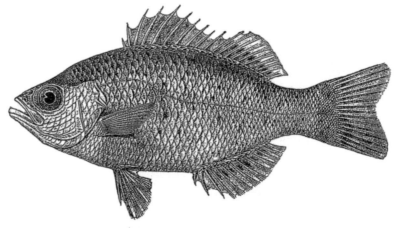

FIGURE. 278. *Kulia rupestris* (Boulenger, 1915, Fig. 76).

It commonly occurs in fresh water on islands such as Madagascar, Mauritius, Seychelles, India, Indonesia, Philippines, Palau, Caroline Islands, Solomon Islands, and New Caledonia (Allen, 1989; Marquet and Mary, 1999). It inhabits fast-moving, clear, rain forest streams. In Australia it is restricted to coastal drainages of eastern Queensland, where it is a popular angling species. It can reach 3 kg and feeds on fishes, crustaceans, insects, and figs (Merrick and Schmida, 1984).

Map references: Allen (1989, 1991),* Marquet and Mary (1999),* Masuda *et al.* (1988), Maugé (1986), Thomson *et al.* (1979), Weber and DeBeaufort (1929)

Class Actinopterygii
Subclass Neopterygii

Order Perciformes; Suborder Elassomatoidei
(1st) Family Elassomatidae—**pygmy sunfishes**
[ē-las'-sō-mat'-i-dē]

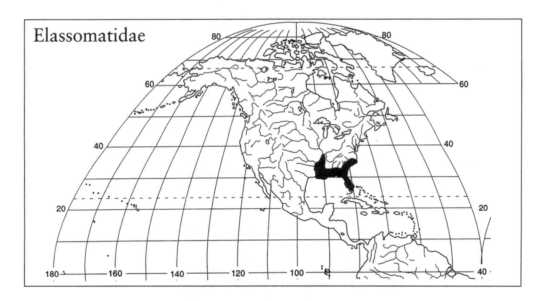

THE PYGMY SUNFISHES HAVE LONG BEEN CONSIDERED A subfamily of the Centrarchidae (Greenwood et al., 1966; Nelson, 1984). Evidence began accumulating in the 1960s that these small fishes were not just diminutive or neotenic sunfishes. Branson and Moore (1962) provided morphological data and Avise and Smith (1977) presented biochemical evidence that pygmy sunfishes were not very closely related to centrarchids. Johnson (1984, 1993) and Johnson and Patterson (1993) considered them to be more closely allied to atherinomorphs and their relatives than to perciform fishes and excluded the Elassomatidae from the Perciformes. Nelson (1994) retained them in the Perciformes but classified them as the only family in their own suborder Elassomatoidei. Johnson and Springer (1997) suggested that they may be most closely related to gasterosteiforms. A recent mitochondrial DNA study (Jones and Quattro, 1999) suggested a sister-group relationship between *Elassoma* and the Centrarchidae plus the Moronidae. These data did not support a sister-group relationship between *Elassoma* and atherinomorphs or cichlids. Grier and Parenti (1999) reported that *Elassoma* is unique among teleosts in having spermatogonia restricted to the distal ends of testicular lobules rather than being distributed along the lengths of lobules. Based on testicular morphology, Grier and Parenti asserted that *Elassoma* is not an atherinomorph.

There are six described species in a single genus, *Elassoma*, endemic to the southeastern United States. Two of these species are common, and four are relatively rare. These are small fishes with a maximum TL of 45 mm, but most specimens are smaller than 35 mm. The dorsal fin is long with three to five spines and 8–13 rays. The anal fin has three spines. The body is elongate and covered with cycloid scales. There is no lateral line. The caudal fin is rounded. The eye is large and the lower jaw protrudes beyond the upper jaw. The mouth is upturned. The males may be beautifully colored and various species of *Elassoma* are kept as aquarium animals by hobbyists throughout the world.

Elassoma is usually associated with spring and swamp habitats. Two species have wide-ranging distributions. The other four species have very restricted ranges occupying portions of one or two river systems (Mayden, 1993). *Elassoma zonatum*, the banded pygmy sunfish, is found throughout the coastal plain from eastern Texas to North Carolina and north along the Mississippi embayment to southern Illinois. Its distribution is basically that of the genus. *Elassoma evergladei* (Fig. 279), the Everglades pygmy sunfish,

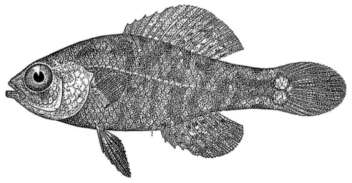

FIGURE 279. *Elassoma evergladei* (Jordan and Evermann, 1900, Fig. 414).

occurs on the Atlantic and Gulf coastal plain from North Carolina to Mobile Bay and south to the northern edge of the Everglades.

Elassoma okefenokee occurs in swamps in southern Georgia and northern Florida (Böhlke and Rohde, 1980). *Elassoma boehlkei* inhabits heavily vegetated creeks and ditches in the Waccamaw and Santee River drainages of North and South Carolina. *Elassoma okatie* is endemic to the southern triangle of South Carolina in the Lower Edisto, New, and Savannah River drainages (Rohde and Arndt, 1987). *Elassoma alabamae* has been recorded from three springs in the Tennessee River drainage of northern Alabama. A native population survives in the Beaverdam Creek system in Limestone County, Alabama (Mayden, 1993; Mettee et al., 1996). Some specimens from Beaverdam Creek have been successfully transplanted to the Pryor Spring system as a precaution (Mettee et al.,1996). This species is listed as endangered by the U.S. Fish and Wildlife Service.

Spawning behavior, embryology, and development of *Elassoma* have been studied by Mettee (1974). Walsh and Burr (1984) reported on the life history of *E. zonatum*. Female *Elassoma* deposit eggs in aquatic vegetation rather than in gravel nests as centrarchids do. Males guard the eggs for up to 48 hr. Small crustaceans, mollusks, and aquatic insects make up their diet (Walsh and Burr, 1984). Pygmy sunfish live for only 1 or 2 years.

Map references: Böhlke and Rohde (1980),* Mayden (1993),* Page and Burr (1991),* Rohde and Arndt (1987)*

Class Actinopterygii
Subclass Neopterygii

Order Perciformes; Suborder Labroidei
(2nd) Family Cichlidae—**cichlids** (sik'-li-dē)

THE CICHLIDS ARE FRESHWATER MEMBERS OF AN OTHERWISE marine suborder, the Labroidei. Recognition of the Labroidei is primarily based on characteristics of the pharyngeal jaw apparatus (Kauffman and Liem, 1982; Liem, 1986; Stiassny and Jensen, 1987). The position of the cichlids within the suborder is far from certain. The closest relatives of the cichlids (the remaining members of the suborder Labroidei) are the marine surfperches (Embiotocidae), damselfishes (Pomacentridae), wrasses (Labridae), and parrotfishes (Scaridae).

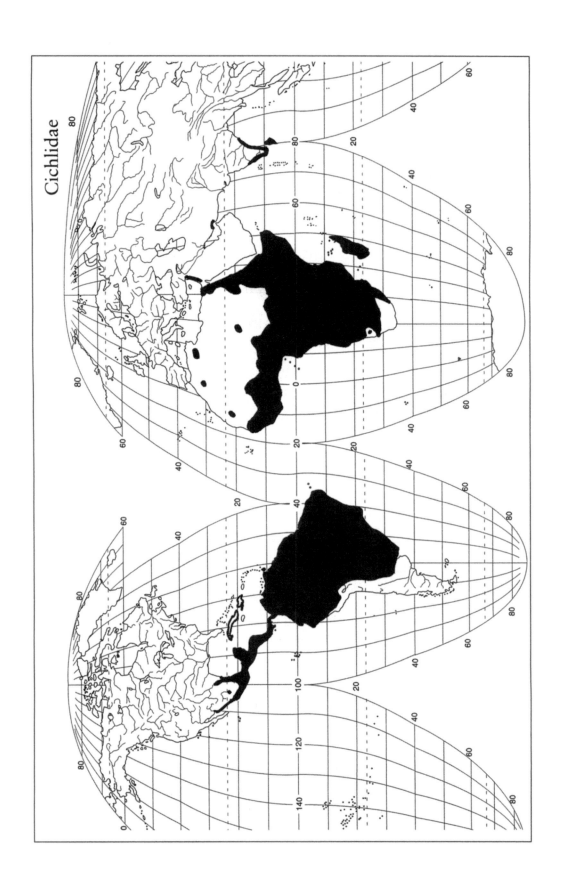

Cichlidae

The distribution of cichlids includes tropical and subtropical lowland regions of Mexico, Central America, West Indies (Cuba and Hispaniola), South America, Africa, Madagascar, the Middle East (Iran, Syria, Israel, and Palestine), coastal India, and Sri Lanka (Greenwood, 1994b). In North America one species reaches southern Texas. A few cichlids even occur in isolated Saharan oases (Roberts, 1975; Lévêque, 1990c).

The exact number of genera and species is not precisely known because there are many scientists currently working on various revisions, and they do not necessarily agree on generic limits. Whatever the total, many species (and genera) are undescribed. Nelson (1994) estimated that there were 105 genera and 1300 species but cautioned that this approximation may be too conservative. Greenwood and Stiassny (1998) suggested about 200 genera and 2000 species. The Cichlidae is probably the third largest family of bony fishes after Cyprinidae and Gobiidae. Cichlid jaws, behavior, family life, and threats to their continued existence are reviewed by Barlow (2000).

Daget *et al.* (1991) recorded 143 genera and 870 species from Africa. About 150 of these species are from rivers (Greenwood and Stiassny, 1998), and the remainder are from African lakes. South America, which lacks large lakes, has an estimated 350 cichlids, mostly in rivers according to Greenwood and Stiassny. Kullander (1998) reported that South American cichlids comprise about 50 genera and 450 species, and there are an additional 100 species in Middle America (S. Kullander, personal communication). The South American *Crenicichla* is the largest genus with more than 100 species. *Haplochromis* was considered the largest genus in older classifications with 206 species, but their taxonomy and phylogeny are problematic (Van Oijen et al., 1991; Greenwood, 1981). None of the genera occur on more than one continent (Greenwood, 1994b).

Cichlids come in a dazzling array of body shapes and colors, which makes them one of the most important groups of aquarium animals. They superficially resemble centrarchids in aspects of morphology, behavior, and ecology, but cichlids can easily be distinguished in having only one nostril on each side instead of the usual two, and their lateral line is interrupted (i.e., it consists of an anterior, longer, upper section and a posterior, shorter, lower section of pored scales). Some cichlids have a compressed, disc-shaped body, such as the South American discus fishes of the genus *Symphysodon*, but other fast-moving fish eaters, such as the pike cichlids of the genus *Crenicichla*, are elongate and streamlined. The freshwater angelfishes from South America, *Pterophyllum*, have exceptionally high dorsal and anal fins. Most cichlids have 3 anal fin spines, some have 4–9 anal spines, and the Asian *Etroplus* (represented by *E. maculatus*; Fig. 280) has 12–15 anal spines. Dorsal fin spines generally number 7–25. Most cichlids are relatively small species, but *Boulengerochromis microlepis* of Lake Tanganyika can reach 80–90 cm TL (Fryer and Iles, 1972; Greenwood and Stiassny, 1998) and the Neotropical

FIGURE 280. *Etroplus maculatus* (Day, 1878b, Plate LXXXIX, Fig. 4).

Cichla temensis apparently reaches about 1 m TL (S. Kullander, personal communication).

Cichlids are well-known among evolutionary biologists for their extraordinary ability to form species flocks in the African Great Lakes (Brooks, 1950; Fryer and Iles, 1972; Greenwood, 1974a, 1981; Meyer et al., 1990). Daget *et al.* (1991) provided a cichlid bibliography of 3000 references. A special section of the journal *Conservation Biology* [Vol. 7(3), 1993] included 10 papers on the cichlids of the Great Lakes of Africa. Other papers that discuss cichlid species flocks are included in Echelle and Kornfield (1984). Resolving relationships among members of rapidly evolving species flocks is difficult, even for molecular techniques (Kornfield and Parker, 1997).

A species flock can be defined as a monophyletic group of distinct, ecologically diverse species that have evolved in an isolated macrohabitat. Greenwood (1984c) discussed various definitions of species flocks. Lake Victoria (250,000–750,000 years old) in East Africa has at least 200 and perhaps more than 400 cichlid species, Lake Tanganyika (9–12 million years old) has about 170–200 cichlids, and Lake Malawi (4 million years old) has more than 450 species (Greenwood, 1994b; Stiassny and Meyer, 1999). Stauffer *et al.* (1997) estimate that there are more than 1500 cichlid species in Lake Malawi. Up to 99% of these fishes are endemic to their particular lake and are found nowhere else. Eccles and Trewavas (1989) and Konings (1990) reviewed the cichlids of Lake Malawi. Poll (1986) and Brichard (1989) discussed Lake Tanganyika cichlids. Teugels and Thys Van Den Audenaerde (1992) reviewed 41 species of west African cichlids in 14 genera. Meyer (1993) demonstrated via mitochondrial DNA analysis that the cichlids of Lake Victoria are genetically more similar to one another than they are to morphologically similar cichlids from Lakes Tanganyika and Malawi. Kocher *et al.* (1993) reported that the similar morphologies of cichlids in Lakes Tanganyika and Malawi are due to convergence.

Sturmbauer *et al.* (1997) studied phylogeographic patterns of cichlids from rocky habitats in Lake Tanganyika.

Liem (1973) credited the pharyngeal jaw apparatus and the great diversity of their highly modified teeth with allowing cichlids to specialize in feeding on a single type of food, thus partitioning food resources and facilitating explosive speciation in lakes. Jaw teeth are specialized for food capture, whereas the pharyngeal teeth process what is swallowed. Every conceivable food type from bacteria to algae, plankton to plants, snails to scales, and eggs to fish is eaten by some species of cichlid. One species, *Haplochromis* (= *Dimidiochromis*) *compressiceps* of Lake Malawi, has been suggested to specializes in biting the eyes out of other fishes (Fryer and Iles, 1972), but this has yet to be documented and eyes have not been found in its stomach (M. L. J. Stiassny and J. R. Stauffer, Jr., personal communication).

Hori (1993) reported that the scale-eating *Perissodus microlepis* of Lake Tanganyika comes in two forms: one with head and jaws curved to the right and another with head and jaws curved to the left. Right-headed fish scrape scales from the left side of their prey, and left-headed fish feed on the right side. Over an 11-year period the ratio between right- and left-"handed" scale eaters oscillated around 1:1. Cross-breeding experiments demonstrated that right- or left-"handedness" is a simple Mendelian trait with right dominant to left. *Perissodus microlepis* approaches its food source from behind and grabs several scales from the side of the prey. When one of the two phenotypes (e.g., right-handed individuals) is more abundant in the population, prey fishes will tend to be more alert against attacks to their left side. This results in left-handed individuals gaining greater hunting success and therefore greater fitness because the victims are not guarding their right flank as carefully. In other words, the rarer phenotype enjoys more success than the common phenotype. Thus, this frequency-dependent selection allows the polymorphism to reach an evolutionary stable state that oscillates around unity (Hori, 1993). Surely this is one of the ultimate feeding niche specializations.

As a general rule, the species of the "tilapiine" lineage, represented by *Oreochromis niloticus* (Fig. 281), are herbivores or detritivores, whereas those of the "haplochromine" lineage, such as *Astatotilapia burtoni*, are predators. Of course, there are exceptions, such as the sand-dwelling haplochromine planktivores reported by Stauffer *et al.* (1993). Even more astounding than the sheer number of cichlid species is the apparent rapidity of their evolution from an ancestral species. Seismic studies and core samples indicate that Lake Victoria dried up completely during the late Pleistocene approximately 12,000–15,000 years ago (Johnson et al., 1996). This implies that the rate of speciation of cichlids in Lake Victoria was extremely rapid—in fact, the fastest ever reported for vertebrates. Repeated isolation due to fluctuation of lake levels has been suggested as

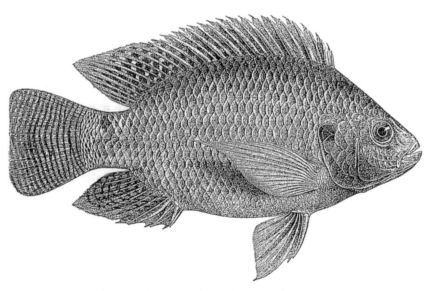

FIGURE 281. *Oreochromis niloticus* as *Tilapia nilotica* (Boulenger, 1915, Fig. 106).

an important speciation mechanism in lake cichlids (Stiassny and Meyer, 1999).

Many of Lake Victoria's cichlids have not been described and, unfortunately, never will be because they are being forced into extinction. This tragic waste of evolutionary resources began with overfishing of cichlids in the lake. In the mid-1950s Nile perch, *Lates niloticus*, a voracious predator, from Lake Albert (= Lake Uganda) was stocked into Lake Victoria as a food source for local people (Goldschmidt, 1996). Since that time, this species has eaten its way through most of the cichlid populations, and much of the species flock is in grave danger of extinction. This loss of so many endemic species qualifies as the largest mass extinction in historical times. Today, many zoos and aquariums are propagating Lake Victoria cichlids in an attempt to keep desperately endangered species extant, even if only in captivity. There is a voluminous literature on this tragedy. Kaufman (1992) provided an overview of catastrophic changes in Lake Victoria. Goldschmidt *et al.* (1993) and Ogutu-Ohwayo (1993) described the effects of the introduced Nile perch, *L. niloticus*, in Lake Victoria and Lake Nabugabo, respectively. Pitcher and Hart (1995) edited an important book on the impact of species changes in African lakes, and Goldschmidt (1996) wrote an absorbing, first-person narrative of the drama unfolding in Lake Victoria. Since many of the lake cichlids were algae eaters, their elimination by Nile perch, combined with an influx of nutrients as a result of deforestation, resulted in algal blooms which eventually die and decay and cause an oxygen drain on the lake water. This further degrades the habitat. The large Nile perch cannot be dried like small cichlids. *Lates* must be preserved by smoking. This has resulted in the increased felling of trees to supply firewood, and the resulting deforestation has led to erosion that further pollutes the lake and increases turbidity (Stiassny and Meyer, 1999). This dramatically demonstrates one of the principles of ecology—that everything is connected to everything else. Stiassny (1996) discussed the loss of freshwater biodiversity in Africa.

Cichlid reproduction is complex, with an elaborate courtship display and intense parental care (Keenleyside, 1991; Clutton-Brock, 1991; Stiassny and Gerstner, 1992; Barlow, 2000). Most of the American and Asian cichlids lay eggs on the substratum and both males and females tend the eggs and young. They are usually monogamous. This is considered the phylogenetically primitive condition. The more advanced African cichlids, particularly the lake species, are mouthbrooders and are highly sexually dimorphic with very colorful males and dull females. The importance of male coloration in cichlid courtship has been highlighted by recent events in Lake Victoria. Human activities around the lake have increased turbidity, which makes mate choice on the basis of coloration difficult. Drab coloration and low diversity are now found in regions of the lake that have become turbid due to eutrophication (Seehausen et al., 1997).

In nearly all mouthbrooders the eggs are carried by the female. In African pseudocrenilabrines during courtship the female nudges the male's vent, and he releases sperm, which fertilizes the eggs in her mouth. This system is aided by the presence of dummy egg spots (as shown on *Astatotilapia burtoni*; Fig. 282) on the male's anal fin which likely stimulate the female to nudge the male in an attempt to retrieve the "eggs." The fertilized eggs and, later, the newly hatched young are carried in the female's mouth. The mouthbrooding cichlids are the only species with egg spots, are

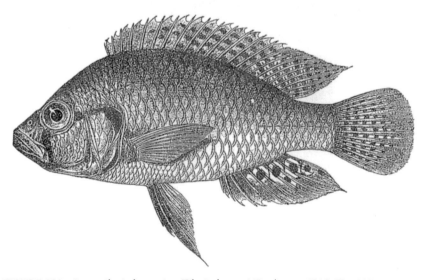

FIGURE 282. *Astatotilapia burtoni* as *Tilapia burtoni* (Boulenger, 1915, Fig. 140).

usually polygamous, and are mostly African. Some species of South American mouthbrooders are behaviorally intermediate and lay eggs on the substrate and brood only the young orally. The substrate spawning *Symphysodon discus* secrete mucus which the young pick from the sides of the adults. Wimberger *et al.* (1998) combined the study of parental care, mating systems, and biogeography with mitochondrial phylogenetics in the South American genus *Gymnogeophagus*. Stiassny and Mezey (1993) provided a review of cichlid egg structure.

Cichlid taxonomy, as one might expect, is very difficult and confusing. Important studies include Poll (1986), Greenwood (1987b), Stiassny (1987, 1990b, 1991, 1992), Casciotta and Arratia (1993a,b), Sültmann and Wayer (1997), Kullander (1998), and Farias *et al.* (1999). The 13 Madagascan species, in genera such as *Paratilapia*, *Paretroplus*, and *Ptychochromis*, are thought to form the most primitive sister groups to the rest of the family (Stiassny, 1991; Stiassny and Raminosa, 1994). *Heterochromis* (represented by *H. multidens*; Fig. 283) may be the sister group to the American and African species, and the Neotropical species probably form a monophyletic group (Stiassny, 1991; Casciotta and Arratia, 1993a). American and African cichlids, minus *Heterochromis*, seem to form a monophyletic group (Stiassny, 1992).

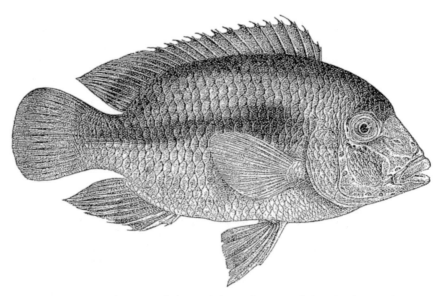

FIGURE 283. *Heterochromis multidens* as *Pelmatochromis multidens* (Boulenger, 1915, Fig. 277).

Kullander (1998) proposed a new phylogeny and classification of cichlids with eight subfamilies: Etroplinae, Pseudocrenilabrinae, Retroculinae, Cichlinae, Heterochromidinae, Astronotinae, Geophaginae, and Cichlasomatinae. According to this scheme, Old World cichlids from

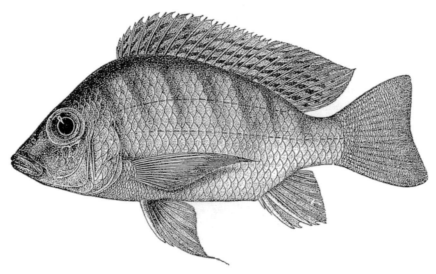

FIGURE 284. *Tylochromis lateralis* as *Pelmatochromis lateralis* (Boulenger, 1915, Fig. 260).

India (*Etroplus*) and Madagascar (*Ptychochromis*) are sister groups and, grouped as the Etroplinae, form the sister group of an African assemblage including tilapiines, haplochromines, hemichromines, and *Tylochromis* (represented by *T. lateralis*; Fig. 284) as the Pseudocrenilabrinae. *Retroculus* is the sister group of all remaining Neotropical cichlids, followed by *Cichla* and *Crenicichla*, which are sister groups within the Cichlinae. The African genus, *Heterochromis*, is the sister group of the remaining Neotropical cichlids, of which the Astronotinae (*Astronotus* + *Chaetobranchus*) forms the basal lineage, followed by the two major polygeneric subfamilies Geophaginae and Cichlasomatinae. Kullander (1998) maintained that the split between the Old and New World cichlids is well supported, even with *Heterochromis* among the Neotropical taxa. Farias *et al.* (1998) reported on a molecular phylogeny of Neotropical cichlids and organized them in five major lineage: heroines, cichlasomines, crenicichlines, geophagines, and chaetobranchines. Farias *et al.* (1999) put *Heterochromis* back among the Africans but confirmed *Retroculus* as a basal South American taxon. Thompson (1979) described the cytotaxonomy of 41 species of Neotropical cichlids. The earliest fossils date from the Eocene of the Neotropics, whereas east African fossils were deposited in the late Oligocene (Greenwood, 1994b).

Cichlids have been introduced to many countries where they are not native (Lever, 1996). This is especially true for various species of tilapiine cichlids that are utilized extensively in aquaculture, such as *Oreochromis mossambicus*, *O. niloticus*, and *O. aureus*. Lever (1996) provided maps showing the principal introductions outside Africa.

As secondary-division fishes, many cichlid species exhibit a fair degree of salt tolerance. Various species of tilapiine cichlids (*Oreochromis*,

Sarotherodon, and *Tilapia*) can move up and down the coastline between river systems. *Oreochromis mossambicus* has even become established in brackish and marine waters of islands in the South Pacific and Micronesia (Nelson and Eldredge, 1991). Fuller *et al.* (1999) documented many intentional and unintentional introductions of cichlids into the United States. The confusing nomenclature of *Oreochromis*, *Sarotherodon*, and *Tilapia* was reviewed by Trewavas (1973, 1981), Ivoylov (1981), and Stiassny (1991).

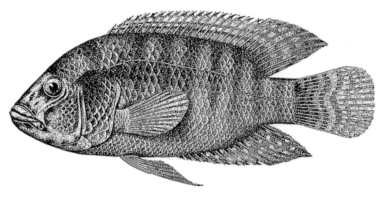

FIGURE 285. *Cichlasoma octofasciatum* as *C. hedricki* (Meek, 1904, Fig. 66).

The only member of this family native to United States waters is the Rio Grande cichlid, "*Cichlasoma*" *cyanoguttatum*, from the lower reaches of the Rio Grande system in Texas. It has been introduced to other localities. Because of Kullander's (1983) restriction of the genus *Cichlasoma* and the resulting confusion caused by not assigning generic names to the "orphaned" Middle American species, it has been suggested that Mexican and Central American representatives of the former broadly defined *Cichlasoma*, such as *C. octofasciatum* (Fig. 285), receive the next available valid generic name, *Herichthys* (Burgess and Walls, 1993). This move would convert the Rio Grande cichlid to *Herichthys cyanoguttatus*. The taxonomy and nomenclature of the Central American cichlasomine cichlids are currently probematical and undergoing revision. Kullander and Hartel (1997) and Kullander (1996) provided justification for use of particular generic names for Central American ex-*Cichlasoma* and diagnosed *Herichthys* as a northern Mexican endemic group. See Martin and Bermingham (1998) for molecular confirmation of these arrangements.

At least 44 cichlid species have been introduced in the United States (Fuller et al., 1999; Robins et al., 1991). Some of these are escaped aquarium fishes, and others are deliberate introductions for vegetation control, recreational fisheries, or aquaculture. Florida, because of its warm climate and abundant water, has the most exotic cichlid species. A few of the exotic cichlids in the southern United States are the oscar (*Astronotus ocellatus*),

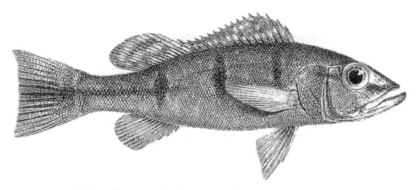

FIGURE 286. *Cichla ocellaris* (Steindachner, 1883, Plate I, Fig. 2).

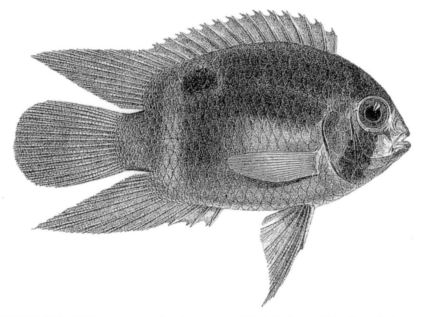

FIGURE 287. *Cleithracara maronii* as *Acara maronii* (Steindachner, 1882a, Plate II, Fig. 4).

peacock cichlid (*Cichla ocellaris*; Fig. 286), and Jack Dempsey (*Cichlasoma octofasciatum*).

Some Central and South American genera include *Acaronia*, *Aequidens*, *Apistogramma*, *Astronotus*, *Cichla*, *Cichlasoma*, *Cleithracara* (represented by *C. maronii*; Fig. 287), *Crenicichla*, *Geophagus*, *Gymnogeophagus*, *Heros*, *Pterophyllum*, *Retroculus*, and *Symphysodon*. Some African genera include *Haplochromis*, *Hemichromis*, *Heterochromis*, *Julidochromis*, *Lamprologus*, *Oreochromis*, *Pelmatochromis*, *Pelvicachromis*, *Pseudocrenilabrus*, *Pseudotropheus*, *Sarotherodon*, *Tilapia*, and *Tylochromis*.

In addition to the basic continental African species, Old World cichlids can be found in other areas. *Sarotherodon* extends from Africa to Syria.

Tristramella is endemic to Israel in the Sea of Galilee and waters around Damascus (Trewavas, 1942; Coad, 1982). *Iranocichla hormuzensis* is endemic to southern Iran in waters draining into the Persian Gulf at the Straits of Hormuz (Coad, 1982). This apparently represents a relict of a once wider distribution either across peninsular Arabia or the Tigris–Euphrates basin. *Tristramella* appears to be the closest relative of *Iranocichla* (Coad, 1982). Three species of *Etroplus* inhabit fresh and brackish waters of peninsular India and Sri Lanka (Talwar and Jhingran, 1992; Pethiyagoda, 1991b). Cichlids are absent from the fauna of Pakistan and northern India (Coad, 1982).

Map refrences: Bussing (1998),* Casciotta and Arratia (1993),* Coad (1982),* Darlington (1957), Eigenmann (1909a),* Lagler *et al.* (1977),* Lévêque (1990c),* Miller (1996),* Norman and Greenwood (1975),* Roberts (1975), Rosen (1975),* Skelton (1993),* Trewavas (1942)

Class Actinopterygii
Subclass Neopterygii

Order Perciformes; Suborder Labroidei
(per) Family Embiotocidae—**surfperches** (em'-bi-ō-tos'-i-dē)

THE EMBIOTOCIDAE ARE VIVIPAROUS FISHES THAT LIVE IN shallow waters along the northern Pacific coasts of North America and Asia. There are 13 genera and 24 species (Nelson, 1994). Twenty species are restricted to the western coast of North America and 1 of these, the tule perch, *Hysterocarpus traski* (Fig. 288), is confined to fresh water in

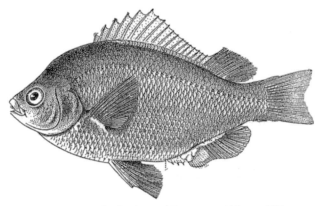

FIGURE 288. *Hysterocarpus traski* (Jordan and Evermann, 1905, p. 470).

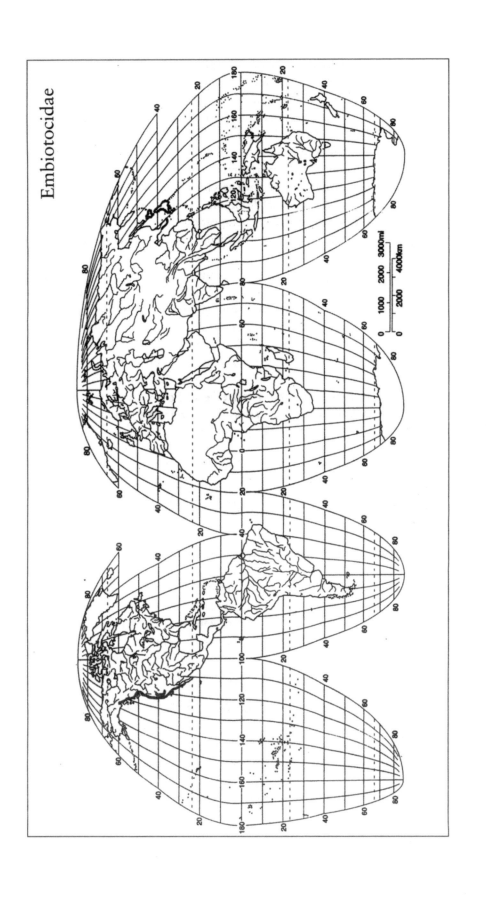

Embiotocidae

California. Three species occur off Japan and the southern coast of Korea (Masuda et al., 1984). The family is absent from the Aleutian Islands between North America and Asia.

Surfperches are deep-bodied and have a scaled ridge that runs along the base of the continuous dorsal fin. There are three anal spines, and the caudal fin is forked. They have cycloid scales and small mouths suitable for plucking small invertebrates from the substrate (DeMartini, 1969). Feeding behavior of surfperches was discussed by Hixon (1980), Schmitt and Cover (1982), and Laur and Ebling (1983). A functional and evolutionary perspective of their single lower pharyngeal jaw apparatus is given by Liem (1986). Maximum TL is about 45 cm.

Surfperches are unusual in that they give birth to young (viviparity) rather than lay eggs like most marine fishes. The anal fin of males is elongated and modified to facilitate sperm transmission. Embryonic development takes place in the female's enlarged ovaries, and the young absorb nourishment from the ovarian fluid in which they are immersed. The embryo has greatly enlarged and highly vascularized dorsal, anal, and pelvic fins that lie close to the highly folded, richly vascularized ovarian wall. This arrangement allows oxygen and nutrients from the maternal blood to reach the developing embryo. At birth the young are 3–5 cm. Some species become sexually mature in a few weeks after birth. Well-developed testes are reported in some newborn males. See Breder and Rosen (1966) for discussion of surfperch reproductive biology. Species living in unpredictable environments usually give birth to more young than those in a stable environment (Baltz, 1984).

The tule perch, *Hysterocarpus traski*, is native to large, low-elevation streams of the Sacramento–San Joaquin River system, Clear Lake, Coyote Creek, and the Russian, Napa, Pajaro, and Salinas Rivers in northern California (Moyle, 1976). Unfortunately, it is now apparently extinct in the Pajaro and Salinas Rivers and rare in the San Joaquin River, but it is abundant elsewhere. The tule perch prefers emergent aquatic vegetation and overhanging banks. It rarely enters brackish waters. The shiner perch, *Cymatogaster aggregata*, is tolerant of low salinities. During winter, it is found in the Pacific Ocean, often at depths of 70 m (Wydoski and Whitney, 1979). During summer it moves into estuaries and coastal rivers. It can be found in coastal streams from Port Wrangel, Alaska, to Baja California (Moyle, 1976).

There is a Pliocene fossil surfperch from California (Cavender, 1986). Tarp (1952) revised the family.

Map references: Eschmeyer and Herald (1983), Lee *et al.* (1980),* Masuda *et al.* (1984), Morrow (1980),* Moyle (1976), Page and Burr (1991)*

Class   Actinopterygii
Subclass   Neopterygii

Order   Perciformes; Suborder Notothenioidei
(per) Family   Bovichthyidae—**bovichthyids**   [bō-vik-thē'-i-dē]

BOVICHTHYIDS BELONG TO THE PERCIFORM SUBORDER Notothenioidei, a group of five marine families whose species occur in coastal regions of the Antarctic. Some of these fishes actually live in waters in which the temperature decreases below freezing (–1.9 C). Notothenioid fishes in this subzero water have glycoproteins in their blood that act as antifreeze and lower the freezing temperature of body fluids (Scott et al., 1986). Oxygen is abundant in very cold water and some of these species lack red blood cells and hemoglobin. Cold-water species tend to have aglomerular kidneys. A swim bladder is absent and most notothenioid fishes are benthic, but some achieve neutral buoyancy via reduction of skeletal bone and an abundance of triglycerides (Eastman and DeVries, 1986; Eastman, 1991). Gon and Heemstra (1990) provided keys to Antarctic fishes, and two books included an extensive summary of Antarctic fish biology and review of notothenoid interrelationships (Eastman, 1993; Miller, 1993).

The Bovichthyidae is a small marine family of 3 genera and about 11 species from southern South America, southern Australia, and New Zealand (Eastman, 1991, 1993). There is 1 freshwater species in southeastern Australia. This family is atypical for a notothenioid group in having a largely non-Antarctic distribution. It is also an unusual notothenioid family in that it has only one lateral line, teeth present on the palatine and vomer, and glomerular kidneys. Most notothenioid fishes have two or three lateral lines and no teeth on the palatine and vomer. Bovichthyids are the most primitive of the notothenioids (Eastman, 1991). The body is elongate, and a small spiny dorsal fin and a long soft dorsal fin are present. Only one nostril is present on each side of the head. Gon and Heemstra (1990) reviewed the nomenclature of this family, which led them to advocate the spelling Bovichtidae instead of Bovichthyidae as used by Nelson (1994).

The three genera are *Bovichtus*, *Cottoperca*, and *Pseudaphritis*. There are nine species of *Bovichthus* from around the Southern Hemisphere, including isolated islands such as Juan Fernandez and Tristan da Cunha.

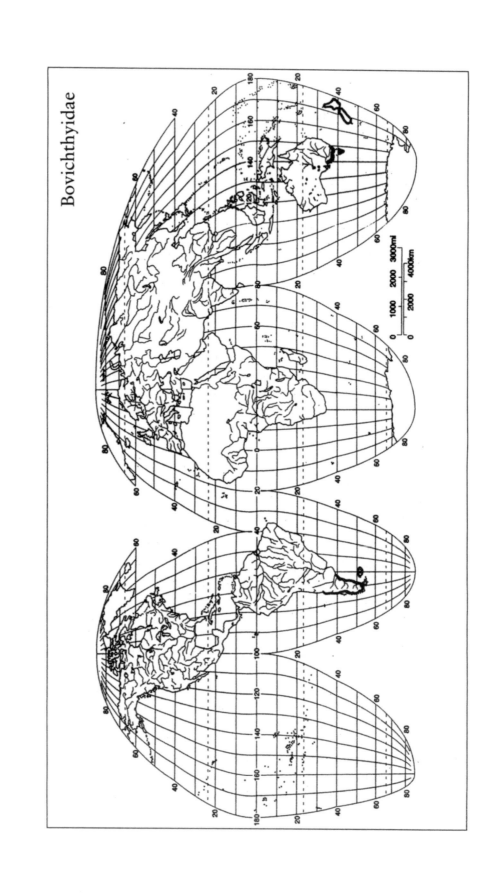

Bovichthyidae

FIGURE 289. *Pseudaphritis urvillii* (Waite, 1923, p. 164).

Cottoperca gobio occurs in deep waters around the tip of South America, including the Falkland Islands (Eastman, 1993).

Pseudaphritis urvillii (Fig. 289) is a catadromous species that occurs in coastal waters, estuaries, and rivers south of Bega, New South Wales, through Victoria, Tasmania, and South Australia. It can reach 34 cm TL, but most specimens are in the 10- to 15-cm range (Andrews, 1996). It is called congolli in New South Wales and tupong in Victoria. *Pseudaphritis urvillii* has a depressed head and pointy snout. This species ranges far upstream in southeastern Australia. It has been recorded 120 km inland. Life history details are not clear, but of 619 specimens I collected in the Tambo River of Victoria, all were female (Berra, 1982). Apparently, females migrate from the upper reaches of coastal rivers downstream to estuaries. Males are probably at the river mouths or in the estuaries where spawning takes place in autumn and winter. Scott *et al.* (1974) remarked that specimens could be transferred from salt water directly to fresh water without noticeable ill effects. *Pseudophritis urvillii* is an opportunistic carnivore. Diet includes benthic insects, snails, amphipods, shrimp, and fishes (Hortle and White, 1980). Eastman and DeVries (1997) studied the morphology of the digestive system of *Pseudaphritis*. Small specimens make interesting aquarium animals.

Eastman and Clarke (1998) stated that the Bovichthyidae is para- phyletic and that *Pseudaphritis* is more closely related to notothenioids than are *Cottoperca* and *Bovichthus*. This hypothesis is based on nucleotide sequence data gathered by Lecointre *et al.* (1997). Following Balushkin (1992), Eastman and Clarke (1998) adopted the monotypic family Pseudaphritidae for *P. urvillii*. They compared lacustrine species flocks of east African cichlids and Lake Baikal cottoids with the radiation of Antarctic fishes.

Map references: Allen (1989),* Andrews (1996),* Eastman (1993)

Class Actinopterygii
Subclass Neopterygii

Order Perciformes; Suborder Trachinoidei
(per) Family Cheimarrhichthyidae—torrentfish
[kī'-mar-rik-thē'-i-dē]

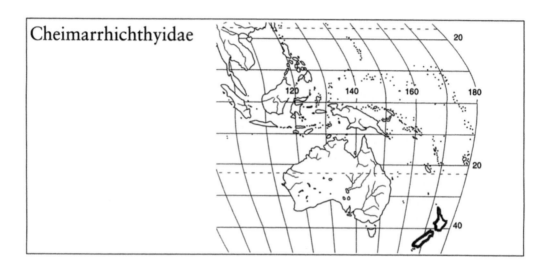

Cheimarrhichthyidae

THE PERCIFORM SUBORDER TRACHINOIDEI AS CONSTITUTED
in Nelson (1994) contains 13 families of mostly tropical marine fishes.
Pietsch (1989) and Pietsch and Zabctian (1990) discussed the family relationships within this suborder. Johnson (1993) considered the suborder to
be paraphyletic. Of the core trachinoids, the only family to occur in fresh
water, Cheimarrhichthyidae, is considered to be the most primitive (Pietsch
and Zabetian, 1990).

The Cheimarrhichthyidae is a monotypic family consisting of
Cheimarrichthys fosteri (Fig. 290), which is found in fast-flowing rivers
throughout coastal New Zealand (McDowall, 1990, 2000). It is called the
torrentfish because it lives in tumbling white waters usually in large rivers
with gravel and boulders and a broad bed. Torrentfish occupy the spaces
between boulders. Such rivers are very unstable, and their beds shift during
floods. Much of this habitat is difficult to reach, so torrentfish are not easily
observed and relatively little is known about them. Torrentfish may reach
an elevation of 700m and penetrate 300 km inland from the coast
(McDowall, 2000)

The torrentfish has a heavy body and broad head that is flattened on
the ventral surface. Three or four short, isolated dorsal spines precede the

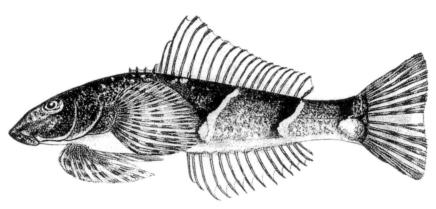

FIGURE 290. *Cheimarrichthys fosteri* (McCulloch and Phillipps, 1928, Plate IV, Fig. 2).

long, low soft dorsal fin. The anal fin usually has one spine and 15 soft rays. The pelvic fins are under the head, anterior to the broad pectoral fins. The caudal fin is slightly forked. The mouth is small and nonprotractile, and the snout overhangs the lower jaw. A lateral line is present with about 50 scales along its length. Maximum TL is about 160 mm, but most specimens are approximately 100–125 mm. It most likely spawns in the spring and has a marine larval stage, but the actual site of spawning is unknown. Juveniles enter fresh water in spring and spend the rest of their lives there. They feed on aquatic insects (McDowall, 1990, 2000).

Torrentfish are well adapted for their torrential habitat. The depressed head and its flattened ventral surface combined with the broad pectoral and pelvic fins are hydrodynamically attuned to the swift-flowing currents. The subterminal mouth is very effective for grazing invertebrates from rock surfaces.

McDowall (1973c) studied the osteology of *Cheimarrichthys* and concluded that it is closely related to *Parapercis* in the trachinoid family Pinguipedidae. However, Pietsch (1989) and Pietsch and Zabatian (1990) considered Cheimarrhichthyidae as an ancestral sister group to other trachinoids.

Map reference: McDowall (1990, 2000)*

Class Actinopterygii
Subclass Neopterygii

Order Perciformes; Suborder Gobioidei
(per) Family Rhyacichthyidae—**loach gobies** [rī'-a-sik-thē'-i-dē]

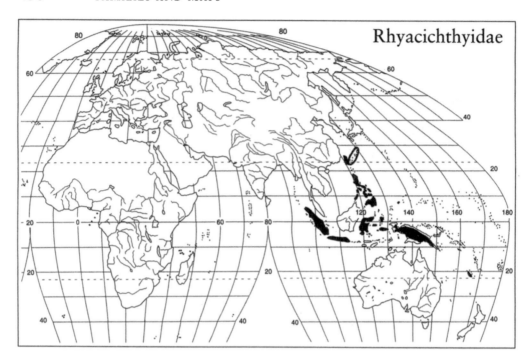

Rhyacichthyidae

THE SUBORDER GOBIOIDEI IS A VERY LARGE MARINE GROUP with about 268 genera and more than 2100 species. About 10% of these fishes occur in fresh waters. Nelson (1994) recognized eight families, including the six families indicated by Hoese (1984). Other arrangements consider only two families, Rhyacichthyidae and Gobiidae (divided into seven subfamilies) (Miller, 1973). Gobioid fishes are usually small, most lack a lateral line on the body, most lack a swim bladder, and many have the pelvic fins fused into a cup-shaped disk. The following are recent important taxonomic and phylogenetic studies: Winterbottom and Emery (1986), Harrison (1989), Birdsong *et al.* (1988), Winterbottom (1993a), Hoese and Gill (1993), and Pezold (1993).

The Rhyacichthyidae consists of two genera and three species found in freshwater streams of Sumatra, Java, Bali, Sulawesi, Moluccas, New Guinea, Palau, the Philippines, Taiwan, Solomon Islands, Ryukyu Islands of Japan, and New Caledonia (Watanabe, 1972; Koumans, 1953; Miller, 1973; Masuda et al., 1984; Dingerkus and Seret, 1992; Kottelat et al., 1993; Watson and Pöllabauer, 1998). *Rhyacichthys aspro* (Fig. 291) is widespread throughout the range listed previously except in New Caledonia. *Rhyacichthys guilberti* is only known from northeastern New Caledonia (Dingerkus and Seret, 1992). The recently described *Protogobius attiti* is known only from southeastern New Caledonia and is provisionally assigned to the Rhyacichthyidae because it has a complete lateral line (Watson and Pöllabauer, 1998).

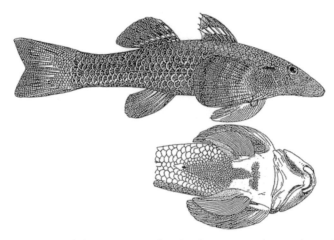

FIGURE 291. *Rhyacichthys aspro* (reproduced with permission from Weber and DeBeaufort, 1953, Fig. 95).

Loach gobies possess a complete lateral line on the head and body, indicating that the Rhyacichthyidae is probably the most primitive gobioid family as hypothesized by Hoese and Gill (1993). In *Rhyacichthys* the caudal fin is forked instead of rounded—an unusual characteristic for adult gobioid fish but common in larval and postlarval fry. The two dorsal fins are widely separated. The first dorsal has seven weak spines, and the second has one spine and eight or nine rays. The anal fin has one weak spine. The ventral surface of the head and belly is flattened and, when combined with widely separated pelvic fins and the very broad and obliquely oriented pectoral fins, forms an efficient hydrodynamic profile that holds the animal firmly in place in swift currents. This arrangement is an adaptation to life in torrential mountain streams which allows the fish to cling to rocks. The body shape superficially resembles the Balitoridae (loaches) or the Cheimarrhichthyidae, which also live in torrential mountain streams. This is an excellent example of convergent evolution. Diatoms, grazed from rocks with the small subterminal fleshy-lipped mouth, compose a large part of the diet of loach gobies. Miller (1973) discussed the osteology and adaptive features of *Rhyacichthys*. Maximum size is about 32 cm TL. Further descriptive information can be found in Herre (1927), Koumans (1953), and Miller (1973).

The other genus in the family, *Protogobius*, is less specialized and more "goby-like" in appearance (Watson and Pöllabauer, 1998). It is known only from a few specimens from swift, clear streams, living over gravel substrate. Osteological and mtDNA studies confirm it belongs to the Rhyacichthyidae. Unlike *Rhyacichthys*, *Protogobius* is a carnivore that feeds almost exclusively on small, transparent shrimp that are common in its environment (R. E. Watson, personal communication).

Map references: Allen (1991), Dingerkus and Seret (1992), Herre (1927, 1953), Kottelat *et al.* (1993), Koumans (1953), Masuda *et al.* (1984), Miller (1973), Wantanabe (1972)

Class Actinopterygii
Subclass Neopterygii

Order Perciformes; Suborder Gobioidei
(per) Family Odontobutidae—**odontobutids** [ō-don'-tō-bū'-ti-dē]

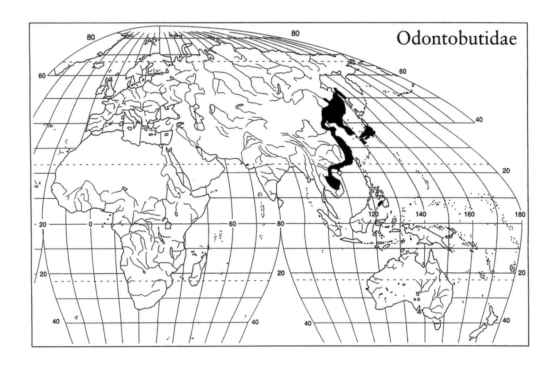

HOESE AND GILL (1993) ERECTED THE FAMILY Odontobutidae for three genera that were formerly contained within the Eleotridae. Odontobutids are found in fresh waters of the northwest Pacific, including China, Korea, Japan, Russia, northern Vietnam, northeastern Thailand, and Laos (Iwata et al., 1985; Hoese and Gill, 1993; Sakai et al., 1996; Vidthayanon, 1995; Kottelat, 1998). Two specimens have been reported from the Philippines at Zamboanga (Herre, 1927) but have not yet been confirmed. Three plesiomorphic characters are shared with the Rhyacichthyidae and synapomorphies involving the pectoral girdle, dorsal fin, and scale morphology suggest monophyly of a group consisting of all gobioids except rhyacichthyids and odontobutids (Hoese and Gill, 1993). Although Hoese and Gill utilized a characteristic of the caudal skeleton—specifically the procurrent cartilages—as an important feature in establishing the Odontobutidae, it

FIGURE 292. *Odontobutis obscura* (reproduced with permission from Weber and DeBeaufort, 1953, Fig. 75).

occurs only in *Micropercops*. The morphology of the procurrent cartilages in *Odontobutis* and *Percottus* is unremarkable (R. E. Watson, personal communication).

There are about eight species in the genera *Micropercops, Odontobutis*, and *Perccottus* (Hoese and Gill, 1993; Saki et al., 1996; Vidthayanon, 1995). The body form is that of an eleotrid. The lateral line is absent, and the pelvic fins are separate and do not form a sucking disk. The caudal fin is rounded. Nichols (1943) listed *Micropercops cinctus, M. swinhonis*, and *M. dabryi* from eastern China. Masuda *et al.* (1984) recorded *Odontobutis obscura* (Fig. 292) from central and southern Japan. *Odontobutis obscura*, with several subspecies, occurs widely throughout the range of the family (Iwata et al., 1985). Sakai *et al.* recorded *O. interrupta* and *O. platycephala* from Korea. *Odontobutis aurarmus* was recently described from northeastern Thailand (Vidthayanon, 1995) and *O. aspro* from Laos (Kottelat, 1998). Berg (1948/1949) included *Perccottus glehni* from the Amur basin, China, and northeast Korea.

Map references: Berg (1948/1949), Herre (1927), Hoese and Gill (1993), Iwata *et al.* (1985),* Kottelat (1998), Sakai *et al.* (1996),* Vidthayanon (1995)*

Class Actinopterygii
Subclass Neopterygii

Order Perciformes; Suborder Gobioidei
(per) Family Eleotridae—**sleepers, gudgeons**   [ē-lē-ō'-tri-dē]

ELEOTRIDS ARE FOUND IN TROPICAL, SUBTROPICAL, AND occasionally temperate, marine, brackish, and fresh waters throughout the world. They form an important component of the freshwater fauna of Australia, New Guinea, New Zealand, Hawaii, and islands of the

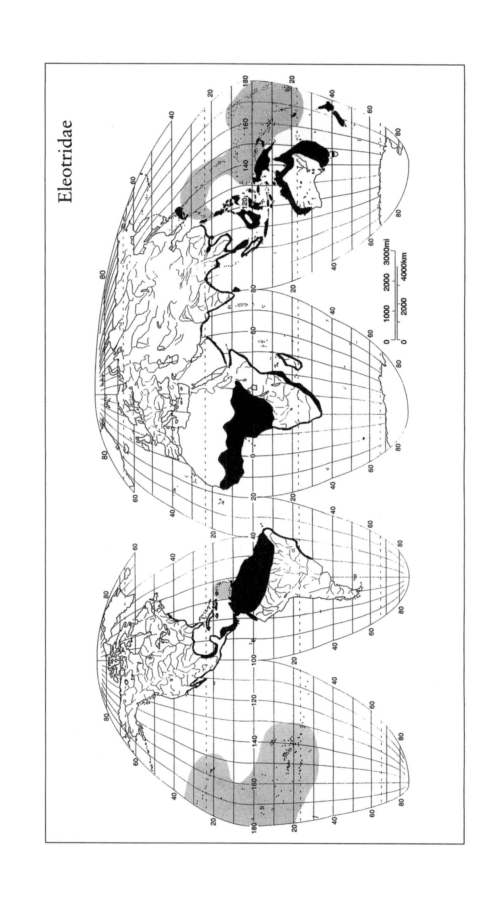

Eleotridae

Indo-Pacific. There are approximately 35 genera and 150 species of sleepers (Nelson. 1994). More of these species occur in fresh water than in brackish or marine waters.

The pelvic fins of eleotrids are separate and do not form a sucking disc as do the fins of most gobies. The mouth of sleepers is usually upturned or terminal but never inferior (Allen, 1991). The body may be covered with cycloid or ctenoid scales. There are no lateral line canals along the body.

The two eleotrid subfamilies recognized by Nelson (1994) (Butinae and Eleotrinae) are treated as subfamilies within the Gobiidae by Hoese and Gill (1993). The Butinae is a tropical group of 13 genera found in estuaries and freshwater streams of the Indo-Pacific and west Africa. The Eleotrinae is cosmopolitan in fresh water and mangrove habitats. It contains 22 genera.

Many sleepers are colorful and make excellent aquarium animals. They are easy to breed in captivity. Eleotrids attach their eggs to the substrate or vegetation. It is thought that the young of most freshwater eleotrids are swept downstream into brackish waters, and that the juveniles eventually migrate back into fresh waters to complete their life cycle. This suggests a marine origin for the family. Many eleotrids are benthic, but a large number have well-developed swim bladders. All eleotrids are carnivorous and feed primarily on insects, crustaceans, and small fishes. The common name "sleeper" reflects the fact that these fish spend a great deal of time resting on the bottom or positioned parallel to a stick or other object waiting for a potential meal. However, they can move very quickly and stop suddenly when chasing prey.

The fat sleeper, *Dormitator maculatus*, occurs in streams along the Atlantic and Gulf coasts of North America from as far north as the Hudson River of New York through the West Indes to Brazil (Lindquist, 1980a). This species approaches 60 cm TL and is used as a food fish. The largest sleeper, and probably the largest gobioid species, is *Oxyeleotris marmorata* at 66 cm TL (Roberts, 1989a). This species is from fresh water in Thailand, the Malay Peninsula, Sumatra, and Borneo and is considered a delicacy. Most eleotrids are much smaller at about 30–100 mm TL.

Géry (1969) wrote that eleotrids enter, more or less occasionally or for spawning, into South American rivers. Tropical America has four genera restricted to fresh waters: *Microphilypnus* (Amazon), *Leptophilypnus* (Atlantic and Pacific slopes of Central America), *Hemieleotris* (Pacific slope from Costa Rica to Colombia), and *Gobiomorus* (rivers of the Antilles and Central and South America) (Banarescu, 1990).

Allen (1991) recorded 41 species of gudgeons (as eleotrids are called in Australia and New Guinea) from the fresh waters of New Guinea. *Morgunda* is a prominent part of the freshwater fish fauna of New Guinea and Australia, with 16 species recorded from New Guinea and at least 6 species from Australia. Half of the New Guinean species are from the Lake Kutubu region in the Southern Highlands of Papua New Guinea.

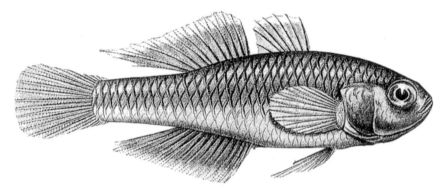

FIGURE 293. *Hypseleotris galii* as *Carassiops galii* (Waite, 1904a, Plate XXXIV, Fig. 2).

Oxyeleotris fimbriata is widely distributed throughout New Guinea on both sides of the central divide (Allen, 1991).

In Australia, *Hypseleotris* is an important component of the freshwater fish fauna, with about 9 of the 25 eleotrid species (Allen, 1989). *Hypseleotris galii* (Fig. 293) occurs in coastal streams from Fraser Island, Queensland, to Eden in southern New South Wales (Larson and Hoese, 1996). The species is very small, but males (55 mm SL) are larger than females (40 mm) and develop a hump on the head behind the eyes during breeding season. Normally gray or bronze, breeding males may become black with orange fins. *Hypseleotris klunzingeri* is widespread in streams and ponds of the Murray–Darling system of eastern Australia as well as in coastal drainages from the Fitzroy River in central Queensland to the Hunter River of New South Wales (Larson and Hoese, 1996). It often forms dense schools in midwater. It prefers vegetated habitats. Maximum TL is only about 45 mm. *Hypseleotris compressa* from coastal streams around the northern two-thirds of Australia and south-central New Guinea is a beautiful, little, sexually dimorphic fish. The male's dorsal and anal fins become intensely blue, black, and orange during the breeding season, and the ventral surface also develops a bright orange coloration. They readily breed in captivity, and the males guard the nest. The female may deposit up to 3000 eggs (Merrick and Schmida, 1984). Maximum TL is 100 mm, but most specimens are only about half this size.

Philypnodon grandiceps (Fig. 294) occurs throughout the Murray–Darling River system and coastal stream of southeastern Australia, including the north coast of Tasmania. It can reach 115 mm TL. *Mogurnda adspera* lives in Pacific coast drainages from Cape York peninsula southward to the Clarence River of northern New South Wales. *Mogurnda mogurnda* occurs in freshwater streams of northern Australia from the Kimberley region to Cape York peninsula and in central southern New Guinea. *Mogurnda larapintae* is found in the Finke River system of central Australia. Other desert *Mogurnda* remain to be described (G. Allen, personal communication). *Milyeringa veritas* is an eyeless, white or pinkish

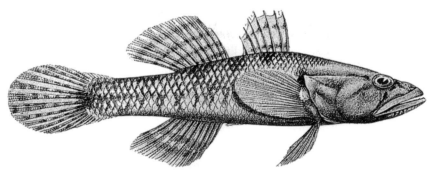

FIGURE 294. *Philypnodon grandiceps* (Waite, 1904a, Plate XXXVI, Fig. 2).

gudgeon from wells and sink holes of the North West Cape region of Western Australia. Two species of *Kimberleyeleotris* are known only from the Drysdale and Mitchelle Rivers of Western Australia. Other genera found in Australia include *Butis* and *Gobiomorphus*.

Eleotrids are called bullies in New Zealand, and they compose a significant part of a depauperate freshwater fish fauna. There are six species of *Gobiomorphus* in New Zealand's rivers, and all occur on both North and South Island (McDowall, 1990). *Gobiomorphus basalis* is only known from one population in the alpine reaches of the Awatere River system on South Island, but it is relatively widespread throughout the North Island. *Gobiomorphus huttoni* (Fig. 295) is the only eleotrid to occur on Stewart Island at the southern tip of South Island. This is the most southerly occurrence of any eleotrid. This colorful species is common throughout New Zealand. *Gobiomorphus cotidianus* is the most widespread bully in New Zealand. *Gobiomorphus gobioides* is the largest New Zealand bully at 22 cm TL.

In Hawaii, the endemic *Eleotris sandwicensis* is one of the few prominent stream fishes (Gosline and Brock, 1965). Kottelat *et al.* (1993) list 10

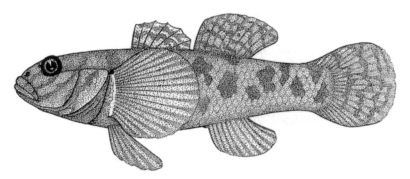

FIGURE 295. *Gobiomorphus huttoni*, female (reproduced with permission from McDowall, 1990, Fig. 18.1A).

genera and 19 species from the islands of Indonesia including Sulawesi. Herre (1927) presented an account of 18 genera and 32 species of eleotrids from the Philippines. Talwar and Jhingran (1992) reported *Eleotris, Butis, Ophiocara, Ophieleotris,* and *Odonteleotris* from fresh waters in the Indian region. *Ophiocara porocephala* (Fig. 296) reaches 3 cm TL and enters fresh waters throughout the Indo-Pacific region (Talwar and Jhingran, 1992).

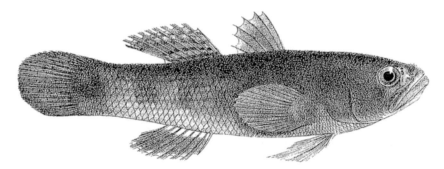

FIGURE 296. *Ophiocara porocephala* as *Eleotris porocephalus* (Day, 1878b, Plate LXVII, Fig. 1).

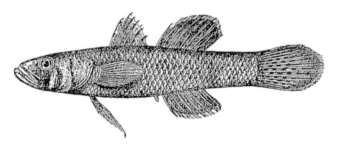

FIGURE 297. *Kribia kribensis* as *Eleotris kribensis* (Boulenger, 1916, Fig. 9).

There are 11 genera and 25 species of Eleotridae recorded from Africa (Maugé, 1986c). *Kribia* is an African genus restricted to fresh water. *Kribia kribensis* (Fig. 297) is found from Guinea to the Congo, *K. leonensis* is restricted to Sierra Leone, *K. nana* occurs in Lake Chad and the Congo River system, and *K. uelensis* is endemic to the Uelle River basin of central Africa (Harrison and Miller, 1992a; Maugé, 1986c). These species are less than 50 mm TL. *Eleotris, Dormitator,* and *Hypseleotris* also occur in African fresh waters. *Ratsirakia* and two species of the subterranean *Typhleotris* are restricted to fresh waters of Madagascar (Stiassny and Raminosoa, 1994).

Map references: Allen (1989, 1991),* Bartholomew (1911),* Berg (1948/1949), Bussing (1998),* Eschmeyer and Herald (1983), Harrison and Miller (1992a),* Lindquist (1980),* McDowall (1990),* Masuda *et al.* (1984), Maugé (1986c), Pethiyagoda (1991),* Skelton (1993),* Talwar and Jhingran (1992)

Class Actinopterygii
Subclass Neopterygii

Order Perciformes; Suborder Gobioidei
(per) Family Gobiidae—**gobies** (gō-bī'-i-dē) [gō-bē'-i-dē]

THE GOBIIDAE, AS CONSTITUTED ACCORDING TO NELSON (1994), is the largest marine fish family with about 212 genera and approximately 1875 species. If one were to include the eleotrids, as recommended by Hoese and Gill (1993), the number of species would exceed 2000 and the Gobiidae would rival the Cyprinidae as the largest fish family (or vertebrate family) in the world. The Cichlidae would not be far behind. The cyprinids are fairly well-known, but there are many more marine and estuarine gobies yet to be described. The Gobiidae is the marine family with the largest number of recently described new species (Berra, 1997a; Berra and Berra, 1977).

Gobies are distributed throughout the world in marine, brackish, and fresh water, primarily in the tropics and subtropics and especially in the Indo-West Pacific, but marine species are known from the subarctic of Norway and Iceland and subarctic streams of southern Siberia. Gobiids, like eleotrids, are common in freshwater streams of islands. Gobies usually have united pelvic fins that form a sucking disc, and all freshwater species have united pelvic fins. Some reef gobies may have separate pelvics. However, the degree of separation of gobiid pelvic fins is highly variable, and it is not always easy to distinguish an eleotrid from a gobiid. There are two dorsal fins. They lack a lateral line on the body, and scales may be cycloid, ctenoid, or absent. Most gobies are small under 10 cm, and their small size makes identification difficult. The smallest known vertebrate is a marine goby, *Trimmatom nanus*, from the Chagos archipelago in the Indian Ocean. This minute, scaleless fish is sexually mature at about 8–10 mm SL (Winterbottom and Emery, 1981; Kottelat and Vidthayanon, 1993), but an even smaller genus is being described from southern Japan (R. E. Watson, personal communication). In the Philippines, two freshwater species, *Pandaka pygmaea* and *Mistichthys luzonensis*, are nearly as tiny at 10 or 11 mm. Some species, such as *Gobioides broussenetii* from the Caribbean, may reach 50 cm TL.

With such a large number of species comes a great diversity of lifestyles, especially in marine gobies. Some gobies (*Gobiosoma*) function as cleaner fishes that remove external parasites from coral reef fishes (Limbaugh, 1961; Losey, 1987). Other gobies (*Crytocentrus steinitzni*) have a symbiotic relationship with shrimp and share a burrow with them (Preston, 1978;

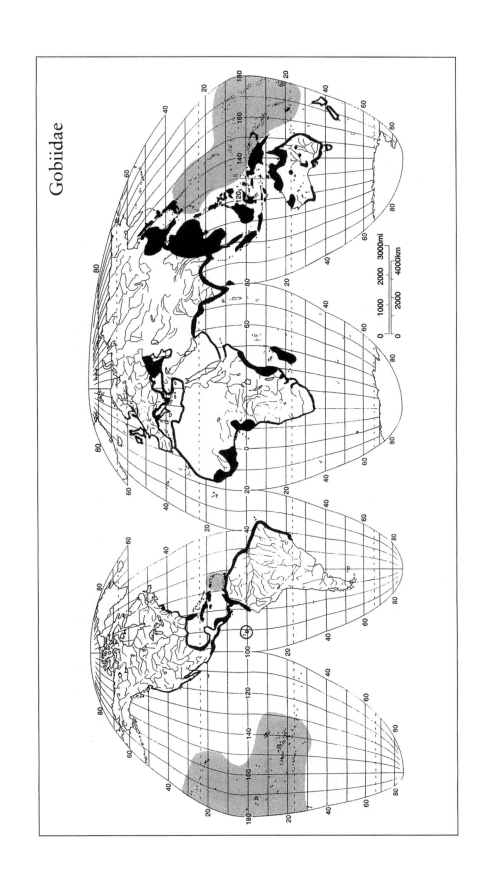

Gobiidae

Karplus, 1979). Other gobies live in contact with sponges, corals, and sea urchins.

Mudskippers (*Boleophthalmus*, *Periophthalmus*, *Periophthalmodon*, and *Scartelaos*) are amphibious, euryhaline gobies that leave the water and dart about on mudflats in search of food such as insects and crustaceans. Some respiration takes place cutaneously and via water retained in the branchial chambers. Their eyes are on moveable short stalks and are adapted for seeing in air as well as in water. They can even climb the prop roots of mangrove trees. See Polunin (1972) and Nursall (1981) for photographs and descriptions of their remarkable behavior. Murdy (1989) provided a taxonomic revision of mudskippers and their relatives (Oxudercinae). Consult Graham (1997) for an extensive review of mudskipper adaptations for aerial respiration. Horn *et al.* (1999) reviewed the biology of intertidal fishes.

Hoese (1984) subdivided the Gobiidae into four subfamilies, and this arrangement was followed by Birdsong *et al.* (1988). Pezold (1993) added a fifth subfamily. These are Oxudercinae, Amblyopinae, Sicydiinae, Gobionellinae, and Gobiinae. The Sicydiinae comprise freshwater gobies (larvae are euryhaline), and the Gobionellinae includes mostly estuarine gobies as well as many species inhabiting fresh waters. The total number of goby species in fresh water is not known, but it is approximately 200 (Hoese, 1994). Of these, about 100 species are sicydiine gobies that make up a group of tropical and subtropical stream fishes that exhibit a high degree of island endemism, especially in the Pacific (Parenti and Maciolek, 1993, 1996). These amphidromous gobies spend most of their life in fresh water, where they breed. The larvae are swept downstream to the sea, where they grow before ascending streams. Most Pacific islands with sufficient elevation to maintain perennial stream systems are populated by sicydiine gobies, as are high-gradient, coastal streams.

Currently, the six genera of the Sicydiinae are *Cotylopus*, *Lentipes*, *Sicydium*, *Sicyopterus*, *Sicyopus*, and *Stiphodon* (Watson, 1995; Parenti and Maciolek, 1996). New genera from Vanuatu and west Africa remain to be described (R. E. Watson, personal communication). *Cotylopus* is only known from mountainous streams of Réunion and is retained as a synonym of *Sicyopterus* in most accounts (Watson, 1995). *Sicydium* occurs in insular and coastal streams of the tropical eastern Pacific and the Atlantic, including the Caribbean Sea. The other genera occur in the central Pacific. According to Parenti and Maciolek (1996), *Sicyopus* is the most plesiomorphic genus, and *Sicyopterus* is one of the most derived sicydiines. Both genera also inhabit the Indian Ocean. The polarization of genera as stated by Parenti and Maciolek (1993) is challenged by Watson based on teeth characteristics of *Sicyopus* (Watson, 1999a, personal communication). Parenti and Thomas (1998), using pharyngeal jaw morphology, explored hypotheses of relationships of sicydiine gobies. *Stiphodon* from various islands has been reviewed by Watson (1996a, 1998, 1999b) and Watson *et al.* (1998).

Rhinogobius is considered the most speciose genus of freshwater gobies in western Pacific drainages. It contains about 40 species with a wide distribution from the Amur basin in Russia to Thailand and Japan, Taiwan, and the Philippines (Chen et al., 1998).

Allen (1991) reported that there are more than 300 species of gobies from New Guinea and that about 50 species are found in fresh water. Common species-rich genera include *Awaous, Glossogobius, Stenogobius, Sicyopterus, Sicyopus,* and *Stiphodon.* There are about 400 goby species in Australian waters, but only about 6 tropical species are restricted to fresh waters (Larson and Hoese, 1996b). *Glossogobius giurus* (Fig. 298) has been successful at penetrating fresh waters along the northern coast of Australia (Allen, 1989). *Redigobius macrostoma* is found in coastal rivers from the Queensland border south to the South Australian border (Larson and Hoese, 1996b). The desert gobies, *Chlamydogobius,* with 5 freshwater species, have a remarkable distribution for fishes from a marine family. They live in the middle of the Australian continent in pools associated with artesian springs and wells (Larson, 1995). One species, *C. eremius* (Fig. 299), occurs throughout the Lake Eyre system (Scott et al., 1974; Allen, 1989) and can withstand wide and rapid fluctuations of salinity

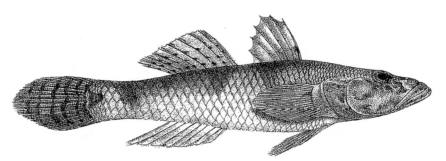

FIGURE 298. *Glossogobius giuris* as *Gobius giurus* (Day, 1878b, Plate LXVI, Fig. 1).

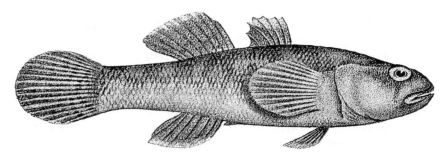

FIGURE 299. *Chlamydogobius eremius* as *Gobius eremius* (Waite, 1923, p. 171).

(Merrick and Schmida, 1984). Its adaptive features are discussed by Miller (1987). *Chlamydogobius japalpa* is restricted to the Finke River in central Australia. *Chlamydogobius micropterus* and *C. squamigenus* are considered endangered (Larson, 1995).

Freshwater gobies are probably most diverse in the Southeast Asian region, with some euryhaline species entering fresh water to well upstream. For example, Kottelat (1989) listed 76 species of gobies from Indo-Chinese inland waters, and Kottelat *et al.* (1993) listed 119 gobies from Indonesian fresh and estuarine waters. Many of these are endemic freshwater species. The island of Sulawesi has small species flocks of endemic *Glossogobius* and *Mugilogobius* (Larson and Kottelat, 1992; Kottelat et al., 1993). *Weberogobius amadi* (Fig. 300), from Lake Poso, Sulawesi, is possibly extinct (Kottelat et al., 1993).

Gobies appear to be poorly represented in the fresh waters of the Indian region, but this may be due to lack of collecting. Talwar and Jhingran (1992) recorded the genera *Awaous, Brachygobius, Glossogobius, Stigmatogobius,* and *Apocryptodon* from Indian fresh waters. *Stigmatogobius sadamundio* (Fig. 301) reaches 85 mm TL in fresh and

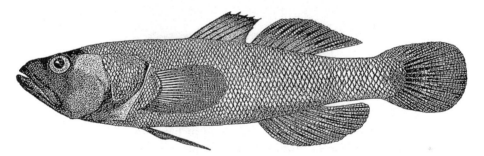

FIGURE 300. *Weberogobius amadi* (reproduced with permission from Weber and DeBeaufort, 1953, Fig. 42).

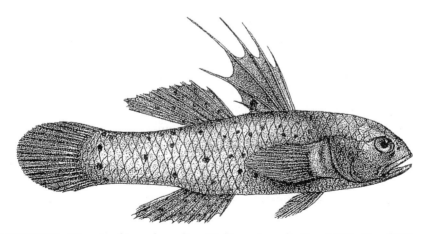

FIGURE 301. *Stigmatogobius sadanundio* as *Gobius sadanundio* (Day, 1878b, Plate LXIII, Fig. 10).

brackish waters of the Indian region and Thailand. It is one of the few gobies popular as aquarium animals (Kottelat et al., 1993). The freshwater gobies of Sri Lanka seem to have more in common with those of Indonesia than with those of India (Watson, 1998). Maugé (1986d) listed 41 genera and 90 African species, some of which occur in rivers. Harrison and Miller (1992b) reviewed the west African gobies. Two gobies are endemic to fresh waters of Madagascar (Stiassny and Raminosoa, 1994). *Sicyopterus lagocephalus* (Fig. 302) grows to 90 mm TL on Reunion, Mauritius, Comores, and Madagascar. Hatching and larval stages are found at sea, whereas postlarvae and adults occur in fresh water (Maugé, 1986d).

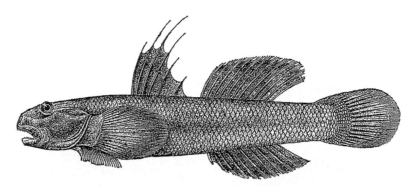

FIGURE 302. *Sicyopterus lagocephalus* as *Sicydium laticeps* (Boulenger, 1916, Fig. 26).

In North America, several gobies such as *Awaous tajasica*, *Evorthodus lyricus*, *Gobioides broussoneti*, *Gobionellus shufeldti*, *Gobiosoma bosci*, and *Microgobius gulosus* enter fresh water along the Atlantic and Gulf coasts (Lindguist, 1980b). The estuarine species *Acanthogobius flavimanus*, a native of Japan, is established in the Sacramento–San Joaquin River system and elsewhere in California as well as in southeastern Australia. *Neogobius melanostomus*, a native of the Black and Caspian Seas, was introduced into the Great Lakes region of North America via ballast waters of transoceanic cargo ships around 1990 (Jude et al., 1992). Today, there are populations in all five Great Lakes, with the heaviest concentration in the central basin of Lake Erie (Charlebois et al., 1997).

Gobies inhabit rivers in tropical South America (Géry, 1969). *Evorthodus* is a largely freshwater genus from Central America and the Guianas. *Euctenogobius* (a monotypic subgenus of *Awaous*), represented by A. *flavus*, is found in fresh and brackish waters from Colombia (near the Panama border) southeastward to Belém, Brazil, near the mouth of the Amazon River. Watson and Horsthemke (1995) considered A. *flavus* to be the most advanced form of *Awaous* and suggested that it is derived from Pacific stocks. They discussed aspects of its natural history. Watson (1996b) reviewed other South American *Awaous*. There are many endemic gobies in the Black Sea and Caspian region of Eurasia (Berg, 1948/1949; Darlington, 1957; Maitland, 1977).

Map references: Allen (1989, 1991),* Bartholomew (1911),* Berg (1948/1949), Bussing (1998),* Eschmeyer and Herald (1983), Harrison and Miller (1992b),* Koumans (1953), Larson (1995),* Larson and Hoese (1996b),* Lindquist (1980),* Maitland (1977),* Masuda *et al.* (1984), Maugé (1986d), Miller (1986, 1990),* Murdy (1998),* Pethiyagoda (1991),* Skelton (1993),* Talwar and Jhingran (1992), Watson (1996b),* Watson and Horsthemke (1995)*

Class Actinopterygii
Subclass Neopterygii

Order Perciformes; Suborder Kurtoidei
(per) Family Kurtidae—**nurseryfishes** [kur'-ti-dē]

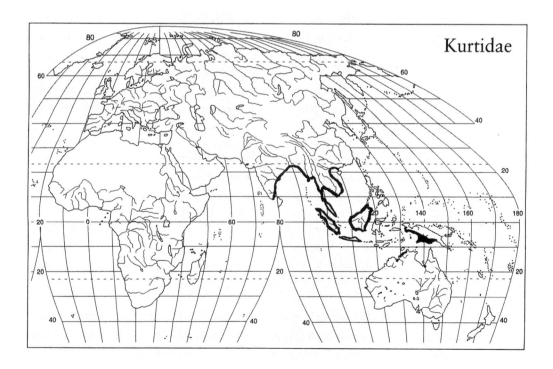

THIS SMALL FAMILY OF TWO SPECIES IN THE GENUS *KURTUS* IS assigned to its own suborder, Kurtoidei. Tominaga (1968) compared *Kurtus*

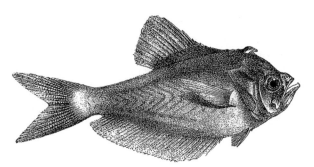

FIGURE 303. *Kurtus indicus* (Day, 1878b, Plate XLII, Fig. 1).

with the Beryciformes and Perciformes and considered it intermediate. Johnson (1993) suggested that it may be closely related to the Apogonidae.

Kurtus indicus (Fig. 303) occurs in coastal areas from the Coromandel coast of southeast India to China, Indonesia, and Borneo (Cantor, 1849; Day, 1878a; Hardenberg, 1936; DeBeaufort, 1951; Kottelat et al., 1993). It occasionally enters estuaries but is mostly found in marine waters. *Kurtus gulliveri* is found in northern Australia and southern New Guinea and is more likely to occur in large, turbid rivers and estuaries. Allen (1991) reported it as far upstream as 830 km from the mouth of the Fly River in New Guinea. In Australia, *Kurtus* has been recorded from the Ord River in Western Australia and from the Saxby River, a tributary of the Flinders River that empties into the Gulf of Carpentaria, Queensland (Allen, 1989). It also occurs from the Daly to the East Alligator Rivers in the Northern Territory. Although their range is wide, nurseryfish appear to be patchy in their geographical distribution and are found most often in small schools in brackish lower reaches of large rivers (Merrick and Schmida, 1984).

Kurtus has a compressed, oblong body with small cycloid scales and a rudimentary anterior lateral line. The opercular bones are very thin. Four spines are present at the angle of the preopercle. The dorsal fin is single with a reduced anterior spinous part followed by soft rays. The back is elevated into a hump. The mouth is large with villiform jaw teeth. The anal fin is long with two anal spines. *Kurtus indicus* has 31 or 32 soft anal rays, and *K. gulliveri* has 44–47 soft anal rays. The pelvic fin has one spine and 5 rays. The caudal fin is deeply forked. Expanded ribs completely enclose the posterior portion of the swim bladder and partially enclose the anterior portion.

The most distinctive feature of this genus is the presence in males of an occipital hook directed anteriorly and downward nearly forming a closed ring. This hook is formed from the supraoccipital and modified dorsal spines (Fig. 304). Females lack a hook. Young males show only a slight protuberance that eventually enlarges as the fish grows. Spawning has not been observed (Breder and Rosen, 1966), but somehow the eggs become

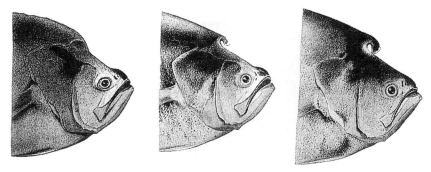

FIGURE 304. *Kurtus gulliveri* males showing hook development (Weber, 1913, Plate 12, Figs. 4–6).

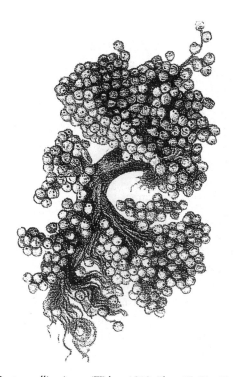

FIGURE 305. *Kurtus gulliveri* eggs (Weber, 1913, Plate 12, Fig. 2).

attached to the male's hook by a twisted cord of egg membranes (Fig. 305), and the male carries the eggs around like a bunch of grapes. A cluster of eggs rests on the left and right side of the male's head where they develop until hatching (Fig. 306). For this reason, these fish are known as nursery-fish. Virtually nothing is known about this bizarre fish and its unique parental care system, which was first reported by Weber (1910). DeBeaufort (1914) described the skeletal anatomy of the hook. *Kurtus indicus* has a similar hook, but Hardenberg (1936) reported that he examined thousands of individuals and never found a specimen with eggs. Weber

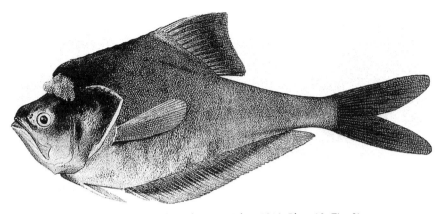

FIGURE 306. *Kurtus gulliveri* male with eggs (Weber, 1913, Plate 13, Fig. 3)

(1910, 1913) illustrated *K. gulliveri* carrying an egg mass, and Lake (1971) reported a male with eggs from fresh water in the Saxby River at Taldora, Queensland, hundreds of kilometers inland. Northern Territory Fisheries staff have observed males carrying eggs 44.5 km upstream from the mouth of the Daly River in water 0.4–0.7 ppt from August 18 to August 26, 1983, and from June 28 to July 4, 1986 (H. Larson, personal communication).

Because these fish live in warm, slow-moving, turbid rivers, this form of parental care may be an adaptation to a low-oxygen environment. Males with eggs can move into surface waters where oxygen levels are high and where siltation can be avoided (Lake, 1978). Weber (1910) speculated that this form of parental care protected the eggs from being swept away by floods. Balon (1975) referred to this system as forehead brooding. Blumer (1979) reviewed male parental care in bony fishes and noted that there are species in 10 families in which the male alone orally broods or carries the eggs externally. He believed that oral brooding and external egg carrying are probably derived from site-guarding ancestors. He further speculated that protection of the eggs may have developed as an incidental effect of males remaining and defending a site from conspecific males. In these cases, males may be expected to have a relatively high probably of genetic relatedness to their mate's offspring. The behavior that enhances mating success could lead to parental behavior.

Blumer (1979) further pointed out that external fertilization is the rule in all cases except one among the 61 families of bony fishes in which males give parental care, alone or biparentally. Females that abandon their eggs to male care may be free to feed more readily than caregiving females and thus reach a greater size. This, in turn, could result in increased egg production, higher quality of egg yolk, or more frequent oviposition. Males carrying eggs, however, are not tied to a nest and are also free to pursue further mating while simultaneously caring for eggs. The eggs of *Kurtus* have been described by Guitel (1913) and appear similar to apogonid eggs (Mooi,

1990). Apogonids are oral brooders. If egg carrying did evolve from oral brooding, this may support Johnson's (1993) position that nurseryfish are related to apogonids.

The diet of nurseryfish is composed of small fishes and prawns (Roberts, 1978; Allen, 1991). They are not considered to be strong swimmers. *Kurtus gulliveri* may reach a total length of 60 cm (Merrick and Schmida, 1984). Grant (1982) wrote that *K. gulliveri* was an esteemed table fish and considered by some to be superior to barramundi, *Lates calcarifer*, Australia's most sought-after food fish. Northern Territory prawn fishers call it the "breakfast fish"; it is part of the banana prawn bycatch in the Fog Bay area west of Darwin (H. Larson, personal communication). Virtually nothing is known about the behavior, ecology, spawning, incubation, parental care, or development of *Kurtus*. This lack of information is due to its small numbers, patchy distribution, and remote habitat that it shares with the dangerous saltwater crocodile, *Crocodylus porosus*.

Map references: Allen (1989, 1991),* Cantor (1849), Day (1878a), DeBeaufort (1951), Hardenberg (1936), Kottelat *et al.* (1993)

Class Actinopterygii
Subclass Neopterygii

Order Perciformes, Suborder Acanthuroidei
(per) Family Scatophagidae—scats   [ska-tō-fāj'-i-dē]

THE SUBORDER ACANTHUROIDEI CONTAINS SIX FAMILIES, including surgeonfishes and their relatives. These fishes are primarily tropical marine herbivores that live in shallow coral reef areas. The phylogeny of this group has been studied by Tyler *et al.* (1989) and Guiasu and Winterbottom (1993) based on osteology and by Winterbottom (1993b) based on muscle morphology; also see Winterbottom and McLennan (1993) regarding the evolution and biogeography of acanthuroids. The Scatophagidae is the sister group to four derived acanthuroid families.

Scats range from the east African coast through India and southeastern Asia, China, Taiwan, the Philippines, and the Indo-West Pacific to the Society Islands (Weber and DeBeaufort, 1936). They are found in marine and brackish waters and often penetrate coastal rivers. There are two genera and about four species, one of which, *Scatophagus tetracanthus*, can breed in fresh water.

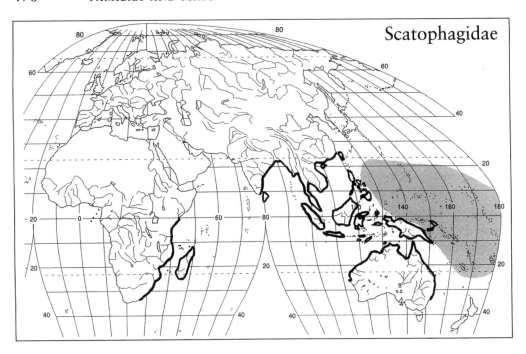

The deep, compressed, disc-shaped body of scats resembles that of marine butterfly fishes (Chaetodontidae); however, scats have four anal fin spines, whereas chaetodontids have three. There is a deep notch in the dorsal fin, and the sharp anterodorsal prong of the first dorsal fin pterygiophore sometimes protrudes through the skin as a procumbent spiny process. There are 16 branched rays in the caudal fin instead of the more common 17 of perch-like fishes. The small mouth is not protractile. Small, ctenoid scales cover the head and body and extend onto the soft dorsal and anal fins. A pelvic axillary process is present. The lateral line is distinct and follows the curve of the back.

It is thought that some scats spawn near coral reefs, and that the young move to the fresh water of river mouths where they grow until they return to the sea to spawn. They pass through an armored tholichthys-like larval stage which later regresses. The name Scatophagidae means "feces eater." Talwar and Jhingran (1992) wrote that scats collect in large numbers near sewage outfalls of towns. Scats usually feed on detritus. They are food fishes in some parts of their range.

There are two species of *Scatophagus. Scatophagus argus* (Fig. 307) ranges from India to the Society Islands and is common along the southeast coast of Australia, northward to New Guinea (Merrick and Schmida, 1984; Allen, 1991). In Australia, members of this family are called butterfishes. Juveniles enter rivers within a few kilometers of the sea. Small specimens are covered with black spots and vertical bands. Their coloration is variable throughout their range. Young scats make beautiful aquarium animals, but

FIGURE 307. *Scatophagus argus* (Day, 1878b, Plate XXIX, Fig. 3).

the colors and patterns fade as the fish grow. Maximum TL is 35 cm. There are venom glands at the base of the fin spines, and care must be taken in handling both juveniles and adults in order to avoid a painful wound (Cameron and Endean, 1970; Halstead, 1978).

Scatophagus tetracanthus occurs in rivers and lagoons of east Africa from Kenya to South Africa and is very common in rivers along the east and west coast of Madagascar. It can survive and reproduce in fresh waters without going to the sea (Arnoult, 1986b).

Selenotoca multifasciata occurs along the coast from Shark Bay in Western Australia to central New South Wales, decreasing near Sydney. It is also present in New Guinea (Kuiter, 1993). It too is kept as an aquarium animal and is venomous. *Selenotoca papuensis* occurs in New Guinea.

Map references: Allen (1991), Arnoult (1986b), Herre (1953), Kottelat *et al.* (1993), Kuiter (1993), Talwar and Jhingran (1992), Weber and DeBeaufort (1936)

Class Actinopterygii
Subclass Neopterygii

Order Perciformes, Suborder Anabantoidei
(1st) Family Luciocephalidae—**pikehead** [lū'-shi-ō-se-fal'-i-dē]

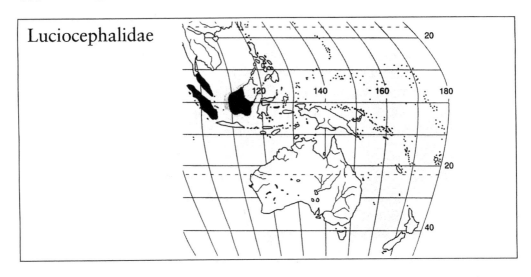

Luciocephalidae

THE SUBORDER ANABANTOIDEI CONSISTS OF 5 FRESHWATER
families with about 80 species from Africa and southern Asia (Lauder and
Liem, 1983). Most of these fishes are known as gouramies, and they are
very important in the aquarium trade. Anabantoid fishes have paired
suprabranchial organs that function as accessory breathing structures.
The suprabranchial organs are formed from the expansion of the
epibranchial (upper part) of the first gill arch that is covered with
highly vascularized epithelium. This structure is very convoluted and is
called a labyrinth organ. Its complex folding greatly increases the surface
area for oxygen absorption. The labyrinth organ fills most of the opercular
cavity and is positioned behind and above the gill chamber and extends
into the suprabranchial space; hence the name suprabranchial organ.
The gills of these fishes are considerably reduced, probably due to lack
of space in the opercular chamber. Anabantoids must gulp air at the
surface since gill respiration alone is inadequate to prevent suffocation.
An air bubble is taken from the surface and held in the labyrinth organs
as oxygen is extracted over the vascularized surface. The remains of the
air bubble are ejected through the gill openings. This adaptation
allows anabantoid fishes to live in warm waters with very little dissolved
oxygen. See Liem (1963, 1980, 1987), Lauder and Liem (1983), and
Graham (1997) for details of the structure and function of labyrinth
organs.

Anabantoids depend on bubbles of air for purposes other than respiration. While in the labyrinth organs the air bubbles touch a tympanic-like
membranous window in the cranium and transmit vibrations to the inner
ears (Alexander, 1967; Lauder and Liem, 1983). This amplifies their hearing ability. Male anabantoid fishes also eject long-lasting, mucus-coated air
bubbles from their mouth to form a bubble nest that floats on the surface.
Most pairs embrace under the bubble nest, and the eggs are extruded by the

female. The eggs drift upward but are usually picked up by the male and placed in the nest by mouth. The male guards and tends the nest (Breder and Rosen, 1966).

Another character of anabantoid fishes is a posteriorly divided, physoclystic swim bladder that extends into the tail region. Lauder and Liem (1983) reported four derived features that unite the five anabantoid families (Luciocephalidae, Anabantidae, Helostomatidae, Belontiidae, and Osphronemidae) in a monophyletic suborder. They did not consider the Channidae to be closely related to the anabantoids.

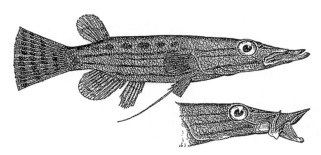

FIGURE 308. *Luciocephalus pulcher* (Hornaday, 1885, opposite p. 387).

The Luciocephalidae consists of one species, *Luciocephalus pulcher* (Fig. 308), from fresh waters of the Malay Peninsula, Sumatra, Borneo, Banka (Bangka), and Billiton (Belitung) (Weber and DeBeaufort, 1922; Roberts, 1989a). Banka and Billiton are small islands off the southeast coast of Sumatra. There may be a second species from Borneo (W. Burgess, personal communication).

Luciocephalus has spineless dorsal and anal fins set far posteriorly. The anal fin is notched. The pelvic fins are the only fins with a spine. The first pelvic ray is extended as a thread-like filament. The tail is rounded, and a lateral line is present. Scales are ctenoid. A gular bone is present in the throat similar to the gular plate in some primitive fishes (coelacanth, bowfin, bichir, and some elopomorphs). Its function is uncertain (Liem, 1967a), but it may be related to mouthbrooding. As many as 91 eggs, 3 or 4 mm in diameter, have been counted in the mouth of a 109-mm SL male *L. pulcher* (Kottelat et al., 1993).

The suprabranchial organ of *Luciocephalus* is much less elaborate than that of other anabantoids. It has less surface area and is poorly vascularized. It may be more important for sound detection than for respiration (Pinter, 1986). Maximum TL of these fishes is about 18 cm.

As the common name implies, the head is *Esox*-like with a highly protractile mouth. In fact, Lauder and Liem (1981) remarked that *Luciocephalus* has the most protrusible jaws of any teleost. The premaxillaries can extend one-third the head length, but prey is captured by a sudden open-mouth lunge rather than inhalation. It feeds on aquatic insects, crustaceans, and small fishes.

Britz (1994), based on osteological studies, and Britz *et al.* (1995), based on egg surface structure and reproductive behavior, cast doubt on the hypothesis embodied in the current classification that *Luciocephalus* is the sister group of the remaining anabantoids. They suggested it is more closely related to a group within Liem's (1963) family Belontiidae, namely, *Parasphaerichthys*, *Sphaerichthys*, and more distantly, *Ctenops*.

Map references: Roberts (1989a), Sterba (1966),* Weber and DeBeaufort (1922)

Class Actinopterygii
Subclass Neopterygii

Order Perciformes, Suborder Anabantoidei
(1st) Family Anabantidae—**climbing gouramies**
(an'-a-ban'-ti-dē)

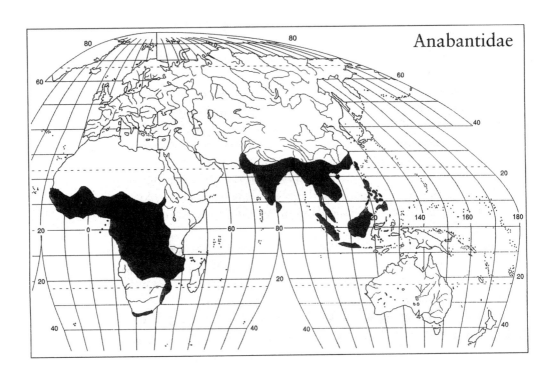

Anabantidae

ANABANTIDS OCCUR IN FRESH WATERS OF AFRICA, PAKISTAN and India to Thailand, southern China, Indonesia, and the Philippines (Talwar and Jhingran, 1992; Weber and DeBeaufort, 1922). They are the most generalized group within the Anabantoidei (Liem, 1963). There are 4 described genera and about 30 species. *Ctenopoma*, an African genus with about 26 species, was recently subdivided by Norris (1995), who erected the genus *Microctenopoma* to include 12 dwarf species formerly included within *Ctenopoma*. Further revision of *Ctenopoma*, which will remove about 10 fairly deep-bodied species, is in preparation (S. M. Norris, personal communication). *Sandelia* includes two African species (Gosse, 1986c). Two or more species of *Anabas* are from Pakistan, India, Bangladesh, Sri Lanka, Burma, the Malay Peninsula, Indonesia, Borneo, the Philippines, and Taiwan (Kottelat et al., 1993; Talwar and Jhingran, 1992; Roberts, 1989a).

Anabantids have an accessory air-breathing capability localized in highly vascularized bony convolutions above the gill chambers. This labyrinth organ enables these fishes to live in oxygen-poor habitats. The labyrinth organ, with the help of pelvic and opercular spines and a thick skin with heavy scales, allows terrestrial locomotion, at least in some species such as the oriental *Anabas testudineus*; hence the common name climbing perches, which more properly should be climbing gouramies. The bodies of *Anabas* and *Sandelia* are elongated and more terete. *Ctenopoma* and *Microctenopoma* are deep-bodied and compressed forms. Anabantids have well-developed, long dorsal and anal fins. The paired fins are short, and the pelvic fins lack elongated rays (except in male *Microctenopoma*). The caudal fin is rounded or truncate. The mouth is large, with small conical teeth. The upper jaw is weakly protrusible in most species. *Anabas*, *Ctenopoma*, and *Microctenopoma* have ctenoid scales. *Sandelia* has cycloid scales. All members of this family are carnivorous.

Anabas testudineus (Fig. 309) has a wide range throughout Southeast Asia. It occurs in large streams, ditches, canals, lakes, ponds, and swamps and can reach 23 cm TL. According to Pinter (1986), it has been observed to leave the water and ramble about on land as far as 180 m in one night. Feeding has not been observed during the terrestrial forays (Liem, 1987). The climbing theme originated with Daldorf (1797), who apparently found a live specimen in a tree 5 ft off the ground. Local reports suggested that *Anabas* could actually climb trees; however, Olson et al. (1986) discounted this as beyond its physical capabilities. Smith (1945) described the walking process in detail. Its overland progress is jerky and ungraceful and is accomplished by lateral movements of the caudal fin while the fish is supported by the spread paired fins. It can remain out of water for extended periods provided it is kept moist. *Anabas* is a food fish and is sold alive in the markets of the Philippines and Thailand in wicker baskets. While on display the fish are occasionally sprinkled with water during the day. The distribution of *Anabas* across

FIGURE 309. *Anabas testudineus* as *A. scandens* (Lydekker, 1903, p. 410).

Wallace's line (not shown on the map) is probably the result of people carrying this hardy, air-breathing species from place to place (Myers, 1951). This species has the largest labyrinth organs relative to body mass of any anabantoid fish (Graham, 1997). There is a great deal of morphological variation in *Anabas* throughout its range, which almost certainly signals the presence of more than one species (Dutt and Ramaseshaiah, 1980, 1983).

In southern Africa, *Sandelia bainsii* and *S. capensis* are found in eastern and southern Cape coastal rivers, respectively, far from other members of the family which occur in tropical areas (Skelton, 1993). The surface area of *Sandelia's* labyrinth organs is small, but they live in habitats where exposure to hypoxia is not great (Beadle, 1981).

Ctenopoma is broadly distributed throughout Africa from the Sahara to the Cape region. Their greatest diversity occurs in the rain forests of central Africa near the central basin of the Congo (Zaire) River system (Norris et al., 1988; Norris and Douglas, 1991). *Ctenopoma ocellatum* (Fig. 310) is a typical representative and reaches 14 cm TL in the Congo basin. The genus also occurs in rivers of west Africa (Norris, 1992). *Ctenopoma*, as previously constituted, included three distinct morphologies—"deep-bodied," "shallow-bodied," and "dwarf"—and was considered paraphyletic (Norris and Teugels, 1990; Norris and Douglas, 1991, 1992). The dwarf species were separated into *Microctenopoma* (Norris, 1995), and a new genus will be created for the shallow-bodied species (S. M. Norris, personal communication).

FIGURE 310. *Ctenopoma ocellatum* as *Anabas ocellata* (Boulenger, 1916, Fig. 42).

Two reproductive strategies are present. The most advanced strategy, according to Norris and Douglas (1991), involves sexually dimorphic males that are brightly colored with elaborate fins. These fish are assigned to *Microctenopoma*. The males establish and defend territories within which they construct a bubble nest. The males guard the eggs in the nest until hatching. Fishes with the more primitive reproductive strategy lack sexual dimorphism and parental care behavior. Jackson (1961) provided details of overland movements of *C. multispinis*.

Map references: Darlington (1957), Gosse (1986c), Norris (1992, 1995),* Norris and Douglas (1991, 1992),* Norris *et al.* (1988),* Skelton (1993),* Talwar and Jhingran (1992), Weber and DeBeaufort (1922)

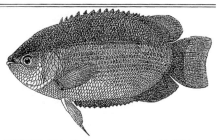

Class Actinopterygii
Subclass Neopterygii

FIGURE 311 *Helostoma temmincki* (Weber and DeBeaufort, 1922, Fig. 88).

Order Perciformes, Suborder Anabantoidei
(1st) Family Helostomatidae—**kissing gourami**
[hē-lō-stō-mat'-i-dē]

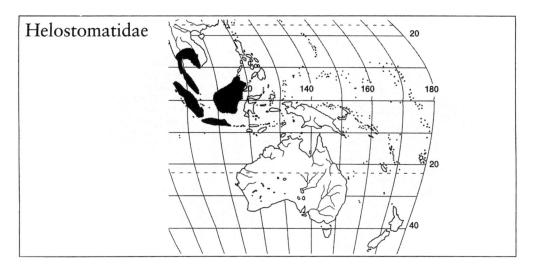

HELOSTOMA TEMMINCKI (FIG. 311) IS THE ONLY SPECIES IN
this family. It occurs in central Thailand, the Malay Peninsula, Sumatra,
Java, and Borneo (Smith, 1945; Roberts, 1989a). The Helostomatidae is
considered to be the sister group to the Anabantidae (Lauder and Liem,
1983).

Helostoma has long dorsal and anal fins. The dorsal fin may
have 16–18 spines and 13–16 soft rays. The anal fin has 13–15 spines
and 17–19 rays. The pelvic fin rays are not produced into elongated
filaments as in some other gouramies such as Trichogaster. There are
two incomplete lateral lines. The lower one begins below the end of
the upper one. The compressed body is covered with ctenoid scales,
whereas the top of the head has cycloid scales. A labyrinth organ is
present, and Helostoma can breathe atmospheric air (Liem, 1987). The
mouth is small and protractile. Teeth are absent from the jaws, vomer, and
pharynx.

Helostoma temmincki is called the kissing gourami because two individuals may extend their broad-lipped mouths and touch. This interesting
behavior may represent aggression between males and has earned a spot for
this species in the tropical fish hobbyist's menagerie. Kissing gouramies feed
on algae and plankton. Their enlarged lips are covered with moveable,
horny teeth that enable them to scrape algae off rocks. Their gill rakers are
very closely spaced and serve as a highly efficient plankton filter. Helostoma
may be the most highly specialized filter-feeding freshwater fish in Asia
(Roberts, 1989a). Liem (1967b) described the functional anatomy of the
head of this species.

Kissing gouramies tend to inhabit slow-moving waters with abundant
vegetation. It is an important food fish, and the maximum TL is about
30 cm (Smith, 1945). They produce a large number of floating eggs
without a nest or parental care (Breeder and Rosen, 1966).

Nonindigenous populations have been reported from two localities in Florida as a result of either an aquarium release or a fish farm escape (Fuller et al., 1999).

Map references: Roberts (1989a), Smith (1945), Weber and DeBeaufort (1922)

Class Actinopterygii
Subclass Neopterygii

Order Perciformes; Suborder Anabantoidei
(1st) Family Belontiidae—**gouramies**
(be-lōn-tī'-i-dē) [be-lōn-tē'-i-dē]

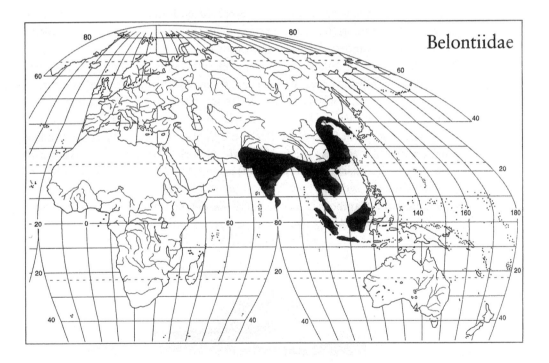

THE BELONTIIDAE IS THE LARGEST AND MOST DIVERSE FAMILY of anabantoids, with 12 genera and about 47 species. They occur in fresh waters of Pakistan, India, China, Korea to the Malay archipelago, Indonesia, and Borneo (Liem, 1963, 1965; Roberts, 1989a; Kottelat et al., 1993).

Some belontiids are deep-bodied and compressed, but most are elongate and cylindrical. They have a small mouth with protusible, toothed jaws. Teeth are absent from the prevomer and palatines. The lateral line is usually vestigial or absent. The anal fin base is usually much longer than the dorsal fin base. The pelvic fins are inserted below the base of the pectoral fins, and the first ray of the pelvics in some species is elongated and functions as a tactile organ. These pelvic filaments may be directed forward to sense another individual or to gather information about the fish's surroundings. The caudal fin is usually rounded.

Three subfamilies were established by Liem (1963). The Belontiinae consists of a single, disjunctly distributed genus, *Belontia*, with two species. *Belontia signata* occurs in Sri Lanka (Pethiyagoda, 1991), whereas *B. hasselti* inhabits southern Thailand and the Malay Peninsula, Sumatra, Java, and Borneo (Roberts, 1989a). This subfamily has two partial lateral lines, the lower one beginning below the end of the upper one. This arrangement may be considered a single complete lateral line that rises and falls in two step-like gradations (Talwar and Jhingran, 1992). The first ray of the pelvic fin is split and elongated into two filaments. The dorsal fin base is longer than the anal fin base, unlike that in other genera in the family. The caudal fin of adults is filamented with males having longer filaments, thus earning the common name "combtail gouramies." These colorful fishes are important in the aquarium trade.

The subfamily Macropodinae includes the Siamese fighting fishes and the paradisefishes. These fishes are extremely colorful and are very popular with tropical fish hobbyists. The Macropodinae has a vestigial lateral line. The dorsal fin base is shorter than the anal fin base. This subfamily has 7 genera and about 32 species. *Betta*, the fighting fishes, is the largest genus with about 20 species, some of which are oral brooders and others are bubble nest builders. Roberts (1989a) suggested that generic division based on brood care would result in an unnatural classification. *Betta* occurs in Thailand, Cambodia, Laos, Vietnam, Malaysia, Sumatra, Java, and Borneo. Smith (1945) provided a fascinating, extended discussion of *Betta splendens* (Fig. 312) life history and the extraordinary pugnacity of males

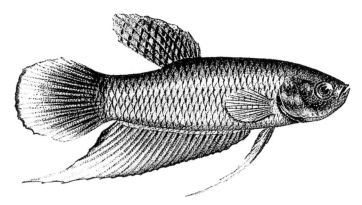

FIGURE 312. *Betta splendens* as *B. pugnax* (Waite, 1904b, Plate XXXVIII, Fig. 1).

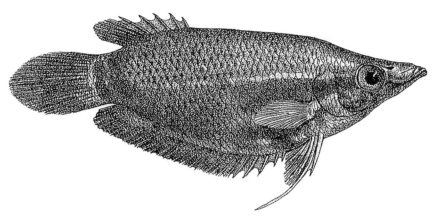

FIGURE 313. *Ctenops nobilis* as *Osphronemus nobilis* (Day, 1878b, Plate LXXVIII, Fig. 5).

whose fighting behavior is utilized for entertainment and gambling in Thailand. *Ctenops nobilis* (Fig. 313) is a paradisefish endemic to India (Talwar and Jhingran, 1992), and *Malpulutta kretseri* is a Sri Lankan endemic (Pethiyagoda, 1991). *Macropodus* is a genus of three species of paradisefish from India, the Malay Peninsula, Vietnam, China, and Korea. Other genera in this subfamily are *Parosphromenus*, *Pseudosphromenus*, and *Trichopsis*.

The subfamily Trichogastrinae is composed of *Colisa*, *Parasphaerichthys*, *Sphaerichthys*, and *Trichogaster*, with about 13 species. Four species of *Colisa* occur in Pakistan, India, Bangladesh, Burma, and Thailand (Talwar and Jhingran, 1992). *Colisa fasciatus* (Fig. 314) reaches 12 cm TL and is utilized as food and as an aquarium animal. Dill (1977)

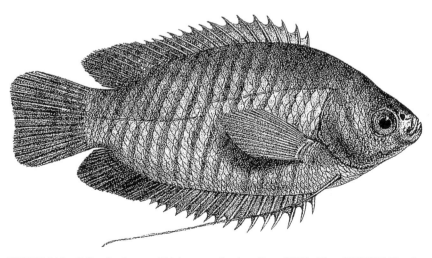

FIGURE 314. *Colisa fasciatus* as *Trichogaster fasciata* (Day, 1878b, Plate LXXVIII, Fig. 6).

reported that *Colisa* ejected water droplets at terrestrial insects in a fashion similar to that of archerfishes (*Toxotes*). *Parasphaerichthys ocellatus* is endemic to Burma. *Sphaerichthys* consists of 4 species from the Malay Peninsula, Sumatra, and Borneo (Roberts, 1989a). *Trichogaster* is a Southeast Asian genus with four species that have extremely long pelvic fin rays. They make graceful aquarium animals. *Trichogaster pectoralis* may reach 24 cm TL and 0.5 kg. It has been introduced into other tropical areas such as New Guinea as a food fish in pond culture (Glucksman, 1976). Because of the air-breathing capabilities of their labyrinth organs, *Trichogaster* and other belontiids are frequently transported alive from place to place throughout, and even beyond, their natural range. For example, *T. trichopterus* has been introduced across Wallace's line into the Philippines and Bali (Roberts, 1989a). At least 11 species of anabantoid fishes have been reported from fresh waters in the United States, mostly from Florida (Courtenay et al., 1984; Fuller et al., 1999). These are escapees from the aquarium trade. See Lever (1996) for information on introductions throughout the world.

Interesting courtship, nest building, flowing fins, and gorgeous coloration have made belontiids one of the most popular aquarium families. Britz (1994) and Britz *et al.* (1995) suggested that *Luciocephalus* is closely related to some belontiids.

Map references: Darlington (1957), Grzimek (1974),* Kottelat (1993), Nichols (1943), Roberts (1989a), Smith (1945), Talwar and Jhingran (1992)

Class Actinopterygii
Subclass Neopterygii

Order   Perciformes; Suborder Anabantoidei
(1st) Family Osphronemidae—**giant gouramies**
(os'-frō-nem'-i-dē) [os-frō-nē'-mi-dē]

ONE GENUS, *OSPHRONEMUS*, WITH FOUR SPECIES, MAKES UP the family Osphronemidae, which occurs naturally in fresh waters of Sumatra, Java, and Borneo (Roberts, 1992b). It has been widely introduced elsewhere. The Osphronemidae is considered to be the sister group to the Belontiidae (Lauder and Liem, 1993).

Giant gouramies have a deep, oval body that is strongly compressed. The lateral line is complete and continuous. Scales are ctenoid. The mouth is small and moderately protractile. Teeth are present on the jaws but

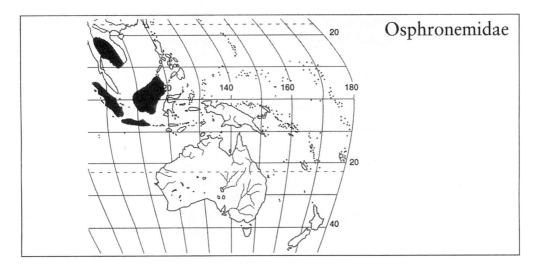

Osphronemidae

absent from the prevomer and palatines. The dorsal fin base is shorter than the anal fin base. The pelvic fin has one spine and five soft rays, with the first ray being produced into a filament that reaches beyond the caudal fin. The remaining pelvic soft rays are vestigial.

The native range of *Osphronemus goramy* (Fig. 315) was probably Sumatra, Java, and Borneo (Roberts, 1989a), and possibly Thailand and the Malay Peninsula (Roberts, 1992b). Its preferred habitat includes swamps, streams, and rivers, but it easily adapts to pond culture. This anabantoid species is so hardy and has been transported by humans to such an extent that it is difficult to determine its exact original native range (Smith, 1945). There are early reports from Thailand, China, and India, but these are almost certainly introductions (Smith, 1945; Talwar and Jhingran, 1992). It has also been naturalized in the Philippines, Sri Lanka, Madagascar, Colombia, Papua New Guinea, Mauritius, and New

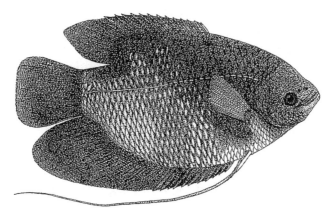

FIGURE 315. *Osphronemus goramy* (Weber and DeBeaufort, 1922, Fig. 89).

Caledonia (Lever, 1996). This species reaches 60 cm TL and 10 kg and is widely cultured in tropical countries as a food fish. It is considered one of the most desirable table fishes in several Asian countries. Its air-breathing ability allows the giant gouramy to remain alive and fresh while it is marketed.

Giant gouramies are primarily herbivorous but will also consume insects and other small invertebrates. Eggs are placed in a nest constructed of plant material within aquatic vegetation in shallow water. The parents guard the eggs and young. Males can be distinguished from females by their more pointed dorsal and anal fins. Juveniles are kept as aquarium specimens. Older adults develop a rather unlovely swollen head region. The original scientific name of the giant gourami, dating back to 1777, was O. olfax, which reflected the idea that the labyrinth organ was an olfactory organ rather than an accessory respiratory structure (Roberts, 1989a).

Roberts (1992b) described two new species, O. laticlavius and O. septemfasciatus, from Borneo and O. exodon with odd dentition from the Mekong River (Roberts, 1994).

Map references: Roberts (1989a, 1992b, 1994)

Class Actinopterygii
Subclass Neopterygii

Order Perciformes; Suborder Channoidei
(1st) Family Channidae—snakeheads   [kan'-ni-dē]

GREENWOOD ET AL. (1966) RECOGNIZED THE CHANNIDS IN their own order, Channiformes, that they placed between the Gasterosteiformes and the Synbranchiformes. Current thinking places this family as a suborder of the Perciformes (Nelson, 1994). An older name for the family is Ophiocephalidae. One should be careful not to confuse the Channidae with the marine gonorynchiform milkfish (*Chanos chanos*) of the family Chanidae or with the Asiatic glassfishes, Chandidae (= Ambassidae).

Snakeheads occur in fresh waters of tropical Africa and southern Asia. There are about 18 species of *Channa* (= *Ophiocephalus*) in Asia (Reddy, 1978; Roberts, 1989a) and 3 species of *Parachanna* in Africa (Teugels and Daget, 1984; Teugels et al., 1986; Teugels, 1992). *Parachanna obscura* is

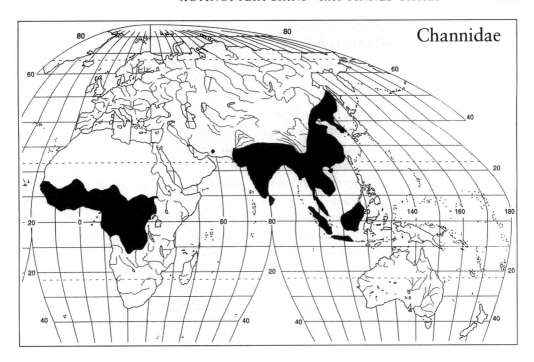

Channidae

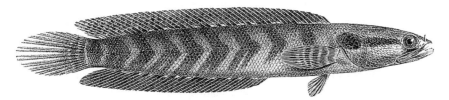

FIGURE 316. *Parachanna africana* as *Ophiocephalus africanus* (Steindachner, 1879a, Plate III, Fig. 2).

the most widely distributed African species, occurring in the White Nile (Sterba, 1966) and in west Africa from the Senegal River to the Chad system and into the Congo basin. It can grow to 455 mm TL. *Parachanna africana* (Fig. 316) from west Africa reaches 320 mm TL. A specimen of *Ophicephalus punctatus* reported by Smith (1950) from Delagoa Bay, southern Africa, is considered to be an import from Asia (Teugels et al., 1986). *Channa* occurs from the Indian subcontinent to Southeast Asia and China. *Channa striata* occurs in Sulawesi and the Philippines, probably via introductions (Darlington, 1957).

Unlike anabantoids, snakeheads are elongate and cylindrical with no spines in their fins. The dorsal and anal fins are long. The mouth is large with a protruding lower jaw. Teeth are present on the jaws and palate. The eyes are dorsolateral and positioned on the anterior half of the head. The anterior nostrils are tubular. The caudal fin is rounded. The cycloid or ctenoid scales are small except on the head, where large scales resemble cephalic plates of snakes. The large head scales, large mouth, and eye position give the fish a snake-like appearance; hence the common name. The swim bladder extends the length of the body (Talwar and Jhingran, 1992). The young of some *Channa* such as *C. orientalis* from Sri Lanka often have a large ocellus on the last five dorsal fin rays. This gives them a bowfin-like (*Amia*) appearance. Pethiyagoda (1991) suggested that the ocellus is a secondary sexual character of females. Individuals of this species lack pelvic fins (Dewitt, 1960). Roberts (1989a) described a dramatic color change between juvenile and adult members of several species of *Channa*.

A labyrinth organ is present, but it is not as complicated as in the anabantoids. Paired suprabranchial chambers are located behind and above the gills in the dorsal and medial recesses of the skull. These chambers are bounded laterally by the opercular bones and lined with respiratory epithelium. An interior labyrinth increases surface area. Ishimatsu and Itazawa (1981) and Liem (1984) explained the structure and function of the air-breathing organs of *Channa*.

Snakeheads can live for several days out of the water if they are kept moist. They are popular food fishes because of their excellent tasting flesh with few bones and because they can be sold alive at the marketplace. Air-breathing ability makes it possible to transport live fishes great distances. There is a description from China in 1822 that slices of living snakeheads were sold until the fish expired (Graham, 1997). Smith (1945) observed that *C. striata* was the most common food fish in Thailand during the 1930s. This species has been naturalized in Indonesia, Madagascar, Papua New Guinea, and many islands of Oceania such as Fiji, Hawaii, Mauritius, and New Caledonia (Lever, 1996). Snakeheads are voracious ambush predators and will eat native fishes, frogs, and snakes when introduced into ponds (Talwar and Jhingran, 1992). Some species of *Channa* are capable of wiggling overland (Kottelat et al., 1993) or of burrowing into soft mud to avoid a drought (Smith, 1945; Graham, 1997). Fishers uncover buried *C. striata* by cutting away the mud in layers during the dry season (Smith, 1945).

Many snakeheads are bubble nesters, and they may be monogamous (Breder and Rosen, 1966). *Channa* constructs a nest in a swampy, slow-flowing area among dense vegetation. The fertilized eggs float up into the nest and are guarded by both parents until the young are about 50 mm TL. The young often form a dense school guarded by the parents (Kottelat et al., 1993). Smith (1945) reported that *C. micropeltes* (Fig. 317) would savagely attack swimmers or fishers that approach a nest. This species is the largest snakehead and can reach about 1 m in length and a weight of more than 20 kg. *Channa micropeltes* has a reputation as a sporting fish among anglers in Thailand. It apparently leaps like a salmon and is difficult to land

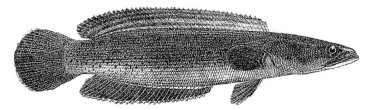

FIGURE 317. *Channa micropeltes* as *Ophiocephalus micropeltes* (Day, 1878b, Plate LXXVII, Fig. 4).

(Smith, 1945). See Ng and Lim (1990) for a review of the natural history of snakeheads.

Map references: Berg (1948/1949), Darlington (1957), Nichols (1943), Roberts (1989a), Talwar and Jhingran (1992), Teugels (1992),* Teugels *et al.* (1986)

Class **Actinopterygii**
Subclass **Neopterygii**

Order **Pleuronectiformes; Suborder Pleuronectoidei**
(per) Family **Achiridae—american soles** [ā-kī'-ri-dē]

PLEURONECTIFORMES IS THE ORDER OF FLATFISHES. THESE distinctive, highly derived fishes, such as flounder, sole, halibut, plaice, and turbot, are very important commercially. After hatching, flatfish begin life as a bilaterally symmetrical larva that swims upright like a normal fish. Early in development, usually between 10 and 25 mm, one eye migates to the other side of the skull. The bones of the skull are not completely ossified at this point. This migration usually takes 1–5 days. This is, of course, very disruptive to the skull bones, muscles, blood vessels, and nerves of the head. The optic nerve may cross itself twice (Policansky, 1982a,b; Ahlstrom et al., 1984). Likewise, the nasal organ also migrates from the blind side to the eyed side. The pectoral and pelvic fins on the blind side are reduced. The end product of this metamorphosis is a benthic, nonbilaterally symmetrical adult. If both eyes end up on the right side the flatfish is considered dextral or right-eyed. The blind and usually colorless left side rests on the substrate. If the eyes are on the left side, the flatfish is sinistral or left-eyed. The blind right side rests on the bottom. Most species are either dextral or sinistral, but some species have individuals of both types. This character is under genetic control. The eyed side is cryptically pigmented and flatfish are

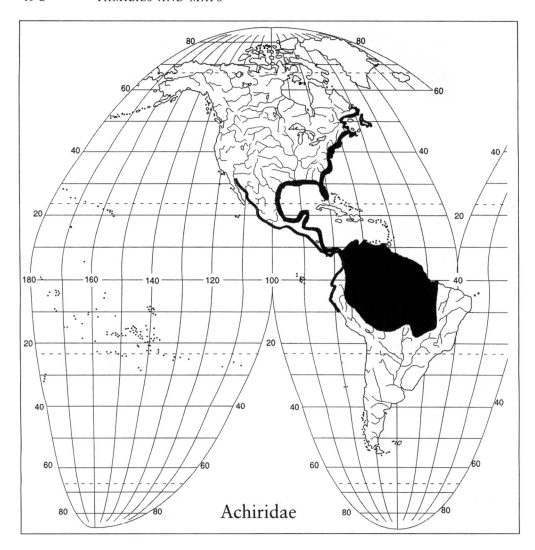

Achiridae

famous for the ability to change their color and pattern to blend in with the bottom coloration. Flatfish can swim close to the bottom by undulating their body.

Flatfish are highly compressed and have long dorsal and anal fins. The dorsal fin base usually overlaps the head. The blind side tends to be flat, whereas the eyed side is more convex. The eyes protrude above the head and may be visible when the flatfish is buried in the substrate. The swim bladder is absent, and the body cavity is small. The skin may be covered with tubercles or cycloid or ctenoid scales. Flatfishes are carnivorous and overwhelmingly marine. Most live on the continental shelves on soft bottoms. Nelson (1994) recognized about 570 species in 11 families, 3 of which have freshwater representatives. Pleuronectiform classification is based on Ahlstrom *et al.* (1984), Hensley and Ahlstrom (1984), and Chapleau (1993). The evolutionary history of flatfishes is not clearly understood. They are considered to be monophyletic, but their sister group remains unknown (Chapleau, 1993).

The Achiridae consists of 9 genera and about 28 species of soles distributed throughout the Americas (Nelson, 1994). Some species enter fresh water. American soles are dextral, and the caudal fin is free from the dorsal and anal fins. The right pelvic fin is attached to the anal fin. Chapleau and Keast (1988) considered the American soles to be a separate family from the eastern Atlantic, Mediterranean, and Indo-Pacific soles (Soleidae). In their phylogeny the Achiridae is the sister group of the Soleidae plus the Cynoglossidae. These three families were placed in the suborder Soleoidei in past classifications. Robins *et al.* (1991) and Nelson (1994) suggested that they be united into one family as three subfamilies.

The hogchoker, *Trinectes maculatus*, enters North American fresh waters from Massachusetts to Panama and ascends coastal rivers as far upstream as the fall line as a normal part of its life cycle (Burgess, 1980e). There is a spring downstream spawning migration. Larval hogchokers move upstream after hatching. Dovel *et al.* (1969) reported on its life history. Its common name is said to derive from the idea that its very rough scales would choke a hog (Bigelow and Schroeder, 1953). Adult size is 80–140 mm TL. *Achirus mazatlanus* may enter fresh water along the Pacific coast from Sonora, Mexico, to Peru (Miller, 1966).

In North America most achirids are marine, such as *A. lineatus* (Fig. 318) that occurs from Florida to Uruguay, but in South America they

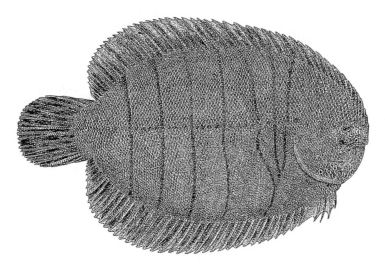

FIGURE 318. *Achirus lineatus* (Goode, 1884, Plate 41).

abound in fresh and littoral waters (Wheeler, 1975). *Achirus achirus* is said to ascend the Amazon River as far as 1000 km (Münzing, 1974). Géry (1969) stated that *Achirus* occurs in the Peruvian Amazon 3200 km from the sea. Other genera include *Apionichthys*, *Baiostoma*, *Catathyridium*, *Gymnachirus*, and *Hypoclinemus*.

Map reference: Allen and Robertson (1994), Burgess (1980e),* Bussing (1998),* Géry (1969), Miller (1966)

Class Actinopterygii
Subclass Neopterygii

Order Pleuronectiformes; Suborder Pleuronectoidei
(per) Family Soleidae—**soles** (sō-lē'-i-dē)

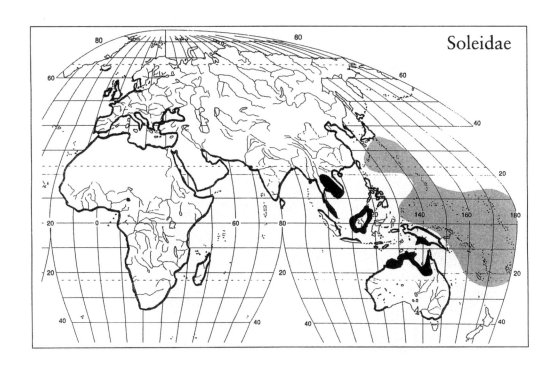

THERE ARE ABOUT 20 GENERA AND 89 SPECIES OF SOLES from the eastern Atlantic, Mediterranean, and Indo-Pacific regions

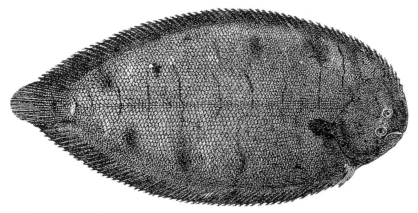

FIGURE 319. *Euryglossa orientalis* as *Synaptura orientalis* (Day 1878b, Plate, Fig. 2).

(Chapleau and Keast, 1988). These fishes are primarily marine but some may enter fresh waters in Africa, Asia, and Australia. The eyes are on the right side (dextral) as in the American soles (Achiridae). The dorsal and anal fins may be free or united with the caudal fin. The pelvic fins are not attached to the anal fin. Chapleau and Keast concluded that the Soleidae is more closely related to the Cynoglossidae than to the Achiridae.

In Africa, *Dagetichthys lakdoensis* is known only from one locality in Cameroon about 1300 km from the coast (Desoutter, 1986, 1992; Chapleau and Desoutter, 1996). It is a small sole that reaches about 40 mm TL. *Achiroides* occurs in fresh waters of the Malay peninsula, Sumatra, Java, and Borneo (Roberts, 1989a).

Aseraggodes klunzingeri is found in central and southern New Guinea and northern Australia as far upstream as 350 km (Allen, 1989, 1991; Merrick and Schmida, 1984). Two species of *Brachirus* also inhabit rivers in the Gulf of Carpentaria drainage system of northern Australia (Allen, 1989). *Synaptura villosa* is found in central and southern New Guinea and has been taken 900 km up the Fly River (Allen, 1991). The species mentioned previously reach about 10 cm TL.

Euryglossa (= *Synaptura*) *orientalis* is known from coastal fresh waters of Thailand (Smith, 1945) and throughout estuaries of southern Asia (Talwar and Jhingran, 1992) (Fig. 319).

Other genera include *Aesopia, Heteromycteris, Liachirus, Microchirus, Monochirus, Pardachirus, Pegusa, Solea,* and *Zebrias.*

Map references: Allen (1989, 1991),* Allen and Swainston (1988), Desoutter (1986b, 1992), Herre (1953), Kottelat *et al.* (1993), Kuiter (1993), Masuda *et al.* (1984), Quéro *et al.* (1986a),* Roberts (1989a), Talwar and Jhingran (1992)

Class Actinopterygii
Subclass Neopterygii

Order Pleuronectiformes; Suborder Pleuronectoidei
(per) Family Cynoglossidae—tonguefishes   (sin'-ō-glos'-i-dē)

THE CYNOGLOSSIDAE IS A MARINE FAMILY OF 3 GENERA AND about 110 species which occurs on the Pacific and Atlantic coasts of the Americas, in the eastern Atlantic, and in the Indo-West Pacific (Menon, 1977; Munroe, 1990, 1991; Munroe et al., 1991). Most species are found between 40°N and 40°S (Chapleau, 1988). Six species are known from fresh waters (Menon, 1977; Roberts, 1989a).

Tonguefishes, often called tongue-soles, have their eyes on the left side (sinistral). The eyes are tiny and so close together that they may actually touch. The dorsal and anal fins are confluent with the pointed caudal fin. The resulting elongate teardrop shape resembles a tongue; hence the common name. Pectoral fins are absent, and the mouth is asymmetrical. Most species are less than 30 cm TL. Chapleau (1988, 1993) considered the family monophyletic with two subfamilies: Symphurinae and Cynoglossinae.

The 57 species of *Symphurus* in the subfamily Symphurinae are mostly deep-sea fishes that occur throughout the family range (Munroe, 1990, 1991; Munroe et al., 1991). In this subfamily the pelvic fins are not connected to the anal fin, there is no lateral line, and the mouth is anterior and straight. The only species reported from fresh water is *S. orientalis* from Peking, China (Menon, 1977). *Symphurus* is the only tonguefish genus in the Americas (Menon, 1977).

The subfamily Cynoglossinae is characterized by pelvic fins connected to the anal fin, a lateral line on the eyed side, and an inferior mouth. There are two genera: *Cynoglossus* (50 species) and *Paraplagusia* (3 species). *Cynoglossus punticeps* (Fig. 320) from the Indo-Australian region has two

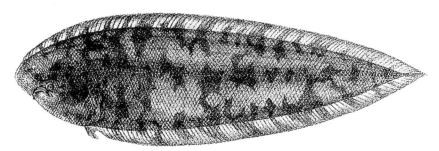

FIGURE 320. *Cynoglossus punticeps* as *C. brachyrhynchus* (Day, 1878b, Plate XCVI, Fig. 4).

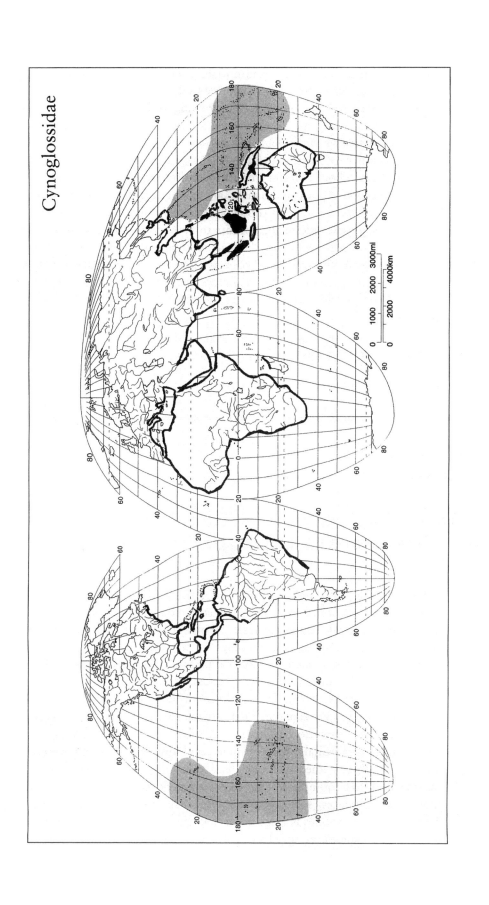

Cynoglossidae

lateral lines on the ocular side and no lateral lines on the blind side. This subfamily occurs from the eastern Atlantic to the western Pacific. Most are marine or estuary, shallow-water, burrowing animals. Menon (1977) reviewed *Cynoglossus* and listed 5 species that are known from fresh waters. *Cynoglossus heterolepis* occurs in central and southern New Guinea and northern Australia (Allen, 1991). It has been recorded from the upper Fly River, about 900 km from the sea (Roberts, 1978). *Cynoglossus feldmanni* occurs in Borneo, Sumatra, and Cambodia. *Cynoglossus waandersi* and *C. kapuasensis* are known from the Kapuas River of western Borneo. *Cynoglossus microlepis* appears to be restricted to fresh waters in Thailand, Cambodia, Vietman, Borneo, and Sumatra. Roberts (1989a) suggested that all 5 freshwater species are known only from fresh waters and do not occur in marine habitats. *Paraplagusia* differs from *Cynoglossus* by the presence of fringed lips in the former. No freshwater forms have been identified.

Map references: Allen (1991),* Kottelat *et al.* (1993), Menon (1977),* Quéro *et al.* (1986b),* Roberts (1989a), Springer (1982)

Class Actinopterygii
Subclass Neopterygii

Order Tetraodontiformes; Suborder Tetraodontoidei
(per) Family Tetraodontidae—**puffers**   (tet'-ra-ō-don'-ti-dē)

THIS ORDER OF BIZARRELY SHAPED FISHES IS THE MOST derived of the Acanthopterygians. Various tetraodontiforms may be globular (puffers), oval (triggerfishes), triangular (boxfishes), or highly compressed (ocean sunfish). They range in size from 22 mm TL and 30 g (*Rudarius excelsus*, a filefish, Monacanthidae) to 3 m TL and 1000 kg (*Mola mola*, the ocean sunfish, Molidae) (Lauder and Liem, 1983; Matsuura and Tyler, 1994). Their unhydrodynamic bodies make for slow swimmers, but they are capable of delicate maneuvering. They show a great deal of reductive evolution, lacking many skeletal elements including parietals, nasals, infraorbitals, anal fin spines, and lower ribs. The gill openings are very small, as is the mouth, which contains a few enlarged teeth or a massive beak-like toothplate. They feed on a variety of stationary or slow-moving invertebrates, such as sponges, jellyfish, corals, and sea urchins. The skin of some species is extremely thick and may be covered with scales modified into spines, armored plates, or ossicles. Some of them have the

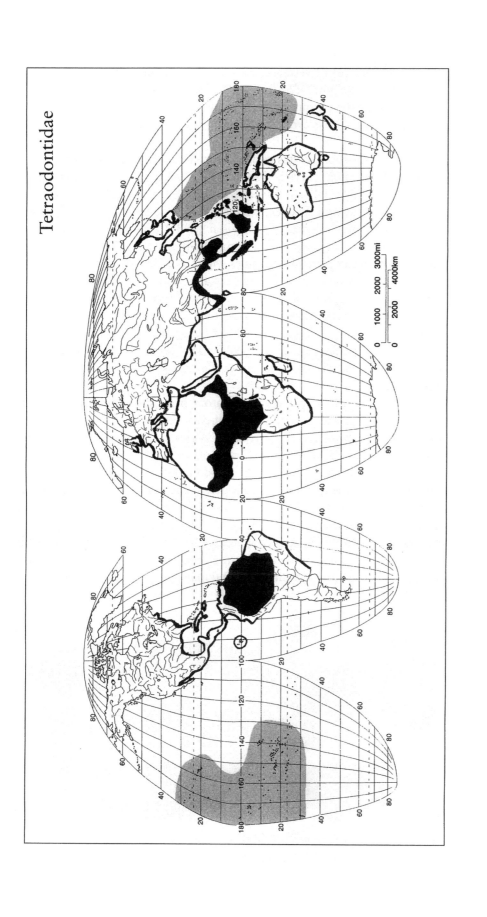

Tetraodontidae

lowest number (16) of vertebrae of any fishes. Brainerd and Patek (1998) proposed that the low vertebral number and its resulting body stiffness reduced the escape swimming performance of ancestral tetraodontiformes and thereby drove the evolution of mechanical defenses common in this group.

The classification of this order is based on myology (Winterbottom, 1974) and osteology (Tyler, 1980). Tyler's monograph contains many exquisite osteological illustrations. Leis's (1984) study of larvae provided an alternate classification. Lauder and Liem (1983) considered the tetraodontiforms to be monophyletic, and Rosen (1984) presented evidence that zeiforms are the sister group of tetraodontiforms. There are 9 marine families within the Tetraodontiformes with about 339 species.

The worldwide puffer family, Tetraodontidae, includes 19 genera with about 121 species (Nelson, 1994), and about 20 species occur in fresh water in South America, Africa, Southeast Asia, and New Guinea. About 12 of these are found only in fresh water (Dekkers, 1975; Roberts, 1982c, 1986c, 1989a; Su et al., 1986). The Tetraodontidae is the only family in the order Tetraodontiformes that has some species restricted to fresh water.

The "teeth" of puffers are fused into a beak-like dental plate with a median suture, giving the appearance of two teeth on each jaw (as shown by *Tetrodon lineatus* from large rivers of Africa; Fig. 321); hence the name Tetraodontidae as opposed to the Triodontidae (three fused teeth) and the Diodontidae (two fused teeth). The Diodontidae and Tetraodontidae are sister groups. These two families considered together are the sister group to the Triodontidae (Lauder and Liem, 1983). Puffers can make considerable noise by grinding their teeth together, and their biting power is significant. Puffers, both marine and freshwater, have been implicated in unprovoked biting attacks on humans. Su *et al.* (1986) reviewed several cases, including attacks in which a girl lost three toes and males suffered genital mutilation.

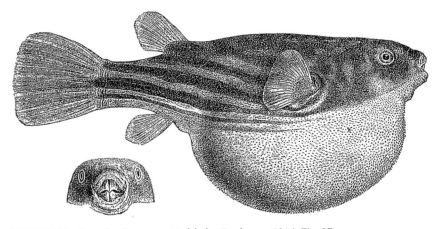

FIGURE 321. *Tetrodon lineatus* as *T. fahaka* (Boulenger, 1916, Fig. 97).

The body of puffers is naked or covered with short prickles. There are no fin spines. The gill openings are slit-like and immediately anterior to the pectoral fin base. Pelvic fins and ribs are absent. Puffers are not strong swimmers. They propel themselves by fluttering the dorsal and anal fins which originate far posteriorly.

The common name "puffer" is due to the remarkable inflation ability of this family. This is a defense mechanism. When a puffer is frightened or disturbed it swallows water into a ventral diverticulum of the stomach. This causes the body to distend into a globe shape. The inflated puffer floats and bobs along in the water, thus making it difficult for a predator to swallow something that resembles a prickly basketball. If removed from the water, an uninflated puffer can inflate its stomach with air. Eventually, when the threat has passed, the swallowed water or air is released and the fish resumes its normal shape.

If that is not weird enough, the organs of puffers, especially the skin, blood, and internal organs such as liver, intestines, and gonads, contain a deadly neurotoxic poison known as tetrodotoxin or tetraodotoxin. In 1774, Captain Cook nearly died of puffer poisoning during his second voyage when he ate a small amount of liver and roe in New Caledonia. The white flesh of some puffers is reputed to be delicious, and in Japan licensed chefs learn to prepare it in a way that avoids contamination with poisonous tissues. It is said that fugu (puffer flesh from any of several species of *Takifugu*) provides a tingly narcotic high probably due to the presence of a small dose of tetrodotoxin in the white meat. Cooking does not inactivate the toxin. One must pay a high price for a fugu meal. Each year, some people pay with their life. The fatality rate of those affected with puffer poisoning is about 60%, with death occurring within 24 hr of ingestion. As a neurotoxin, tetrodotoxin kills by paralysis of the muscles of respiration. First-aid before the onset of paralysis should include the use of an emetic. After paralysis, mouth-to-mouth respiration may be necessary to sustain life until medical help can be obtained. Halstead (1978) provided a resumé of clinical characteristics in all their gory details as well as information on toxicology, pharmacology, and chemistry. Su *et al.* (1986) included an account of the symptoms preceding death and the pathological conditions resulting from consumption of a west Australian puffer. Smith (1945) reported that *Chelonodon* (= *Tetraodon*) *fluviatilis* (Fig. 322) from rivers in Thailand was very poisonous to ducks and other domestic animals as well as to humans. Davis (1983, 1985, 1988a) suggested that tetrodotoxin may be the source of Haitian zombie behavior. This idea has been both challenged (Booth, 1988), and defended (Davis 1988b).

There are two subfamilies: Tetraodontinae and Canthigastrinae. *Canthigaster* is the only genus in the latter subfamily. It contains about 26 species that occur in the tropical Indo-Pacific from South Africa and the Red Sea to Central America. Only 1 species occurs in the Atlantic (Randal and Cea Egaña, 1989).

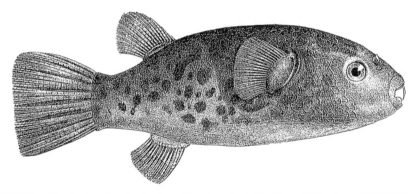

FIGURE 322. *Chelonodon fluviatilis* as *Tetrodon fluviatilis* (Day, 1878b, Plate CLXXXIII, Fig. 1).

Only the Tetraodontinae has freshwater representatives, and these species tend to be much smaller than their marine relatives (Smith, 1945). Large marine puffers can reach 1 m TL. Freshwater puffers, especially juveniles, are often kept by tropical fish hobbyists. *Colomesus asellus* occurs in the Guianas and the Orinoco and Amazon drainages of South America to the Peruvian headwaters along the eastern slope of the Andes 4800 km from the sea as well as in the West Indies (Tyler, 1964). Its most southernly record is from the Rio Araguaia, Brazil, at about 15°S. Six species of *Tetraodon* are recorded from African fresh waters, including the Nile, Chad, Niger, Volta, Gambia, Geba, Senegal, and Congo basins and Lake Tanganyika (Roberts, 1986; Lévêque, 1992).

Talwar and Jhingran (1992) reported two species of *Chelonodon* and two species of *Tetraodon* from fresh waters of the Indian subcontinent. *Xenopterus naritus* (Fig. 323) is apparently restricted to fresh waters of Burma, Thailand, Malay Peninsula, Vietman, and China.

Roberts (1982c, 1989a) recorded two species of *Tetraodon* from Laos, Kampuchea, Thailand, Malay Peninsula, Sumatra, Java, Borneo, and Billiton. This genus is widely distributed and has many marine species. *Tetraodon* from fresh waters was reviewed by Dekkers (1975); however,

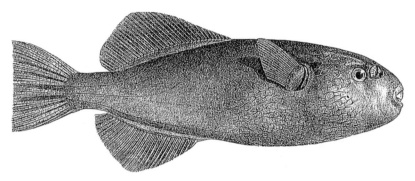

FIGURE 323. *Xenopterus naritus* (Day, 1878b, Plate CLXXXII, Fig. 1).

generic assignments within the family are not always clear (Kottelat et al., 1993). Roberts (1982c, 1989a) reported five species of exclusively freshwater *Chonerhinos* from Kampuchea, Thailand, Malay Peninsula, Sumatra, and Borneo. Allen (1991) listed one species in each of the following genera that sometimes penetrate fresh water in southern New Guinea: *Chelonodon*, *Marilyna*, and *Tetraodon*. There are no freshwater puffers in Australia (Allen, 1989), but the family is represented in estuaries, especially in northern Queensland.

Map references: Allen (1991),* Berg (1948/1949), Dekkers (1975), Kottelat *et al.* (1993), Kuiter (1993), Lévêque (1992),* Masuda *et al.* (1984), Miller (1966), Roberts (1982c,* 1986c, 1989a), Talwar and Jhingran (1992), Tortonese (1986d),* Tyler (1964)

Glossary

The following words are defined as used in this book. They may have different meanings when used in a different context.

adipose fin The fleshy fin on the back behind the dorsal fin as found in trout and catfish.

ammocoete Larval lamprey.

amphidromous Of fish whose migration from fresh water to the sea and vice versa is not for breeding purposes but is a regular part of the life cycle, such as Australian grayling (*Prototroctes*).

anadromous Of fish that ascend rivers to spawn, as do salmons.

anterior The front end.

apomorphic Derived from and differing from an ancestral condition.

autochthonous Native in the sense of having evolved in a given place.

axillary process Bony splint at the base of the pelvic fin.

barbels Fleshy projection near mouth, chin, or snout as in catfish or cod.

benthic Bottom dwelling.

biconcave Concave on both ends.

branchial Pertaining to the gills.

buccal Pertaining to the mouth.

bycatch The accidental species caught in a fisheries.

cartilage Tough, resilient, white skeletal tissue.

catadromous Of fish that descend rivers to spawn in the sea, as do the American and European eels.

caudal peduncle The fleshy tail end of the body between the anal and caudal fins.

circumpolar Surrounding either pole of the earth.

clade A lineage or branch of a cladogram; a monophyletic group of taxa that share a closer common ancestry with one another than with members of any other clade.

compressed Flattened from side to side.

conspecific A term applied to individuals or populations of the same species.

cosmopolitan Worldwide distribution.

countershading Coloration with parts normally in shadow being light or parts normally illuminated being dark.

ctenoid Of scales whose posterior margin has teeth-like projections, as in perciform fishes.

cycloid Of scales whose posterior margin is smooth, as in cypriniform fishes.

deciduous Of scales easily shed, not firmly attached.

demersal Sinking to the bottom as some fish eggs do.

depauperate Including few species, impoverished.

depressed Flattened from top to bottom.

detritus Organic debris.

dextral Right side.

dimorphism Two body forms in the same species.

disjunct Separated.

disseminule That part of an organism which is capable of dispersal.

dorsal Of, on, or near the back.

ecological equivalent Unrelated groups that occupy similar niches in different geographical regions.

ecophenotype A form exhibiting nongenetic adaptations associated with a given habitat or environmental factor.

elver A young eel chiefly found along shores or in estuaries.

endangered Actively threatened with extinction.

endemic Restricted to a particular area.

estivate To pass the unfavorable conditions of summer in a state of torpor.

estuary Wide brackish mouth of a river where tide meets current.

fall line The line joining the waterfalls on many approximately parallel rivers; the fall line marks the point where each river leaves the uplands for the lowlands and thus the limits of its navigability.

family A group of phylogenetically related genera or a single genus forming a taxonomic category ranking between the order and the genus.

fimbriae Slender, fringe-like processes.

forage Food (noun); to feed (verb).

fusiform Cigar-shaped, rounded, broadest in the middle, and tapering at each end.

ganoid scale Hard, glossy, enameled scales, as in gars.

genus, pl. genera A group of phylogenetically related species or a single species forming a taxonomic category ranking between the family and the species.

gill racker Bony projection on anterior edge of gill arch.

Gondwana A single, ancient, immense southern continent composed of Africa, South America, Antarctica, Australia, New Zealand, and peninsular India.

gonopodium Intromittent organ of male poecilids formed from modified anal rays.

gular Behind the chin and between the sides of the lower jaw.

hemibranch Gill with filaments on only one side.

hermaphroditic Having both testes and ovaries in one body.

heterocercal Having vertebral column extend up into larger lobe of caudal fin.

holarctic Northern parts of Old and New World.

ichthyology The branch of zoology that deals with the scientific study of fishes, especially their taxonomy, zoogeography, evolution, and ecology.

keel A sharp ridge on the ventral midline.

kype Curved or hooked lower jaw of a male salmon.

lanceolate Spear shaped.

lateral line Series of openings to sensory canals along the sides of a fish.

lepidophagous Scale-eater.

leptocephalus Larval eel.

mandible Lower jaw.

maxillary Lateral part of upper jaw.

medial Toward the midline of the body.

metamorphosis A change in structure due to development.

molariform Molar-like crushing teeth.

monophyletic Derived from a common ancestor.

monotypic Having only a single representative as a family with one genus or a genus with one species.

naked Without scales.

nape Dorsal surface immediately behind head.

neoteny Sexual maturity attained in an organism retaining juvenile characteristics.

niche The role of an organism in its environment.

nomen dubium Term used to indicate insufficient evidence exists to allow recognition of a nominal (named) species.

notochord A turgid rod of cells lying immediately beneath and parallel to the nerve cord in vertebrate embryos. The vertebral centra are derived from it.

nuptial Referring to courtship or breeding.

odontode Small tooth-derived projections on bones or scales.

operculum Bony gill cover.

opisthocoelous Having vertebrae convex anteriorly and concave posteriorly, as in the gars.

osmoregulation The control of osmotic pressure of body fluids within an organism.

ostariophysan A fish of the superorder Ostariophysi, including the series Anotophysi (order Gonorynchiformes) and the series Otophysi (cypriniforms, characiforms, and siluriforms), having the anterior three to five vertebrae modified as in Weberian ossicles which connect the swim bladder with the inner ear.

oviparous Egg laying.

pedomorphic Retention of juvenile characters of ancestral forms by adults.

peripheral division The designation of fishes in fresh waters that are highly tolerant of salt water, such as salmon, plotosid catfishes, sticklebacks, and gobies; includes marine fishes that enter fresh water.

pharyngeal teeth Tooth-like projections from the pharyngeal gill arches as in the Cyprinidae.

phytoplankton Microscopic algae and plant plankton.

piscivorous Fish eating.

pleisomorphic Primitive; the ancestral state.

poikilothermic Cold-blooded; body temperature approximates that of the environment.

population All the individuals of the same species living in a given area.

posterior The tail end.

primary division The division of strictly freshwater fishes having little salt tolerance, such as bowfins, pikes, most Ostariophysi, sunfishes, and darters.

primitive Early in a given evolutionary sequence; ancestral; pleisomorphic.

protandrous A hermaphrodite that functions first as a male and produces sperm and then transforms to a female and produces eggs.

protogynous A hermaphrodite that functions first as a female and produces eggs and then transforms to a male and produces sperm.

protractile Protrusible.

protrusible Having a mouth that can be extended forward.

pungent Sharp.

pyloric caeca Tubular blind pouches at the junction of stomach and intestine.

ray Usually branched and flexible rod that supports the fin membrane.

relict A survivor that continues to exist after extinction of other members of its group.

rheotaxis Movement toward or away from the current.

secondary division The division of freshwater fishes which have some salt tolerance, such as gars, cyprinodonts, and cichlids.

sensu lato In the broadest sense.

sinistral Left side.

species, pl. species A group of actually or potentially interbreeding natural populations which are reproductively isolated from other such groups.

species flock Evolution of large numbers of distinct but related species from a common ancestor in an isolated area, such as cichlids in African Great Lakes.

specific epithet The second name of the scientific binomial.

spermatophore Small packet of sperm produced by males and inserted into females, as in the Poeciliidae.

spine Sharp, hardened, unbranched fin rays.

spiracle Dorsally located remnant of the first gill slit.

spiral value Spiral fold of mucous membrane projecting into the intestines.

subtropical Nearly tropical; bordering tropical zone.

superior mouth Mouth that opens upward, with lower jaw more anterior than upper jaw.

sympatric Occurring in the same area.

synapomorphic Shared, derived character.

synonymy A term referring to the existence of two or more different scientific names for the same taxon.

syntopic Species or populations that occupy the same macrohabitat.

tactile Relating to the sense of touch.

tapetum lucidum Iridescent pigmented layer of the choroid coat in eyes. It reflects light back onto the retina, thus improving visual sensitivity.

taxonomy The science of classification.

terete Cylindrical and tapering at both ends.

terminal mouth Mouth that opens at anterior end of head with upper and lower jaws equal.

territory Any area defended by an animal.

troglobitic Cave or subterranean dwelling.

tropical. Referring to the region of the earth between the Tropic of Cancer and the Tropic of Capricorn.

tubercle Temporary epidermal projection on head, body, or fins of males of some species which facilitates contact with females during spawning or which is used for defense of territories; common among the Cypriniformes.

vagility Inherent power of movement of individuals or their disseminules.

vascularized Supplied with blood vessels.

venomous Able to inject a toxin by biting or stinging.

ventral Of, on, or near the lower surface.

viviparous Giving birth to live young.

Wallace's line Hypothetical line between Bali and Lombok, Borneo and the Celebes, and the Philippines and the Moluccas which separates the characteristic Asian fauna from the Australian, thereby marking the boundary of the Oriental and Australian biogeographic realms.

Weberian apparatus or **ossicles** Series of three to five modified vertebrae which connect the swim bladder to the inner ear of Ostariophysian fishes.

Weber's line Hypothetical line lying approximately along the Australo-Paupan shelf which separates the islands that have a majority of Oriental animals from those which have a majority of Australian ones; a line of faunal balance sometimes preferred to Wallace's line as the boundary between the Oriental and Australian realms.

Appendix A

Principal Rivers of the World

River	Countries of transit	Outflow	Length Kilometers	Miles	Rank
Alabama	United States	Mobile R.	507	315	161
Albany	Canada	Hudson Bay	985	610	105
Aldan	Russian Federation	Lena R.	2414	1500	34
Allegheny	United States	Ohio R.	523	325	158
Amazon	Peru–Brazil	Atlantic O.	6276	3900	3
Amu Darya	Russian Federation	Aral Sea	2253	1400	38
Amur	China–Russian Federation	Tatar Strait	4345	2700	10
Angara	Russian Federation	Yensiei R.	1852	1151	57
Araguaia	Brazil	Tocantins R.	1770	1100	59
Arkansas	United States	Mississippi R.	2334	1450	36
Athabasca	Canada	Lake Athabasca	1231	756	84
Back	Canada	Arctic O.	974	605	106
Big Black	United States	Mississippi R.	531	330	156
Bighorn	United States	Yellowstone R.	540	336	154
Black	United States	White R.	451	280	175
Brahmaputra	China–India–Bangladesh	Bay of Bengal	2769	1700	29
Brazos	United States	Gulf of Mexico	1400	870	77
Bug	Russian Federation	Dnieper Estuary	853	530	115
Bug	Russian Federation–Poland	Vistula R.	724	450	128
Canadian	United States	Arkansas R	1458	906	73
Cedar	United States	Iowa R.	529	329	157
Cheyenne	United States	Missouri R.	805	500	119
Chiangjiang (see Yanzi)					
Churchill	Canada	Hudson Bay	1609	1000	64
Cimarron	United States	Arkansas R.	966	600	108
Colorado	United States–Mexico	Gulf of California	2334	1450	37
Colorado (Texas)	United States	Gulf of Mexico	1352	840	78

(continues)

Principal Rivers of the World (*continued*)

River	Countries of transit	Outflow	Length Kilometers	Miles	Rank
Columbia	Canada–United States	Pacific O.	1955	1215	47
Congo	Republic of Congo	Atlantic O.	4828	3000	5
Coosa	United States	Alabama R.	460	286	173
Cumberland	United States	Ohio R.	1106	687	94
Dakota	United States	Missouri R.	1143	710	89
Danube	West Germany–Austria–Chech Republic–Hungary–Yugoslavia–Rumania–Bulgaria–Russian Federation	Black Sea	2776	1725	27
Darling	Australia	Murray R.	1867	1160	56
Delaware	United States	Delaware Bay	476	296	169
Deschutes	United States	Columbia R.	402	250	185
Des Moines	United States	Mississippi R.	853	530	116
Dnieper	Russian Federation	Black Sea	2253	1400	39
Dniester	Russian Federation	Black Sea	1408	875	76
Don	Russian Federation	Sea of Azov	1931	1200	50
Donets	Russian Federation	Don R.	1046	650	98
Drava	Austria–Yugoslavia–Hungary	Danube R.	724	450	129
Dvina, Northern	Russian Federation	White Sea	756	470	125
Dvina, Western	Russian Federation	Baltic Sea	1030	640	100
Ebro	Spain	Mediterranean Sea	756	470	126
Elbe	Czech Republic–Germany	North Sea	1167	725	88
Euphrates	Turkey–Syria–Iraq	Persian Gulf	1736	1700	28
Flint	United States	Apalachicola R.	426	265	180
Fly	Papua New Guina	Coral Sea	1127	700	90
Fraser	Canada	Pacific O.	1118	695	92
Gambia	Guinea–Senegal–Gambia	Atlantic O.	805	500	120
Ganges	India	Bay of Bengal	2494	1550	33
Garonne	France	Gironde R.	563	350	150
Gila	United States	Colorado R.	1014	630	101
Godavari	India	Bay of Bengal	1448	900	74
Grand	United States	Lake Michigan	418	260	181
Green	United States	Colorado R.	1175	730	87
Green	United States	Ohio R.	579	360	147
Hsi (see Xijiang)					
Huanghe (see Yellow)					
Hudson	United States	New York Bay	492	306	164
Humboldt	United States	Humboldt Lake	467	290	171
Illionia	United States	Mississippi R.	439	273	178
Indus	Pakistan	Arabian Sea	3058	1900	21
Iowa	United States	Mississippi R.	468	291	170
Irrawaddy	Burma (Myanmar)	Bay of Bengal	2012	1250	45
Irtysh	Russian Federation	Ob R.	2961	1840	22
James	United States	Chesapeake Bay	547	340	152
Japura	Colombia–Brazil	Amazon R.	2816	1750	25
John Day	United States	Columbia R.	452	281	174
Jordan	Lebanon–Israel–Jordan	Dead Sea	322	200	191
Jurua	Peru–Brazil	Amazon R.	1931	1200	51
Kama	Russian Federation	Volga R.	1931	1200	49
Kentucky	United States	Ohio R.	417	259	183
Klamath	United States	Pacific O.	402	250	186
Kolyma	Russian Federation	Arctic O.	1609	1000	65

(continues)

Principal Rivers of the World (continued)

River	Countries of transit	Outflow	Length Kilometers	Miles	Rank
Kootenay	Canada–United States	Columbia R.	644	400	141
Lena	Russian Federation	Arctic O.	4506	2800	6
Little Colorado	United States	Colorado R.	483	300	166
Little Missouri	United States	Missouri R.	901	560	111
Loire	France	Atlantic O.	1006	625	102
Mackenzie	Canada	Beaufort Sea	4064	2525	13
Madeira	Brazil	Amazon R.	3380	2100	18
Magdalena	Columbia	Caribbean Sea	1706	1060	61
Maranon	Peru	Amazon R.	1609	1000	66
Marne	France	Seine R.	523	325	159
Mekong	China–Burma–Thailand– Laos–Kampuchea	South China Sea	4184	2600	11
Meuse	France–Belgium–Netherlands	North Sea	925	575	109
Milk	United States	Missouri R.	1004	624	103
Minnesota	United States	Mississippi R.	534	332	155
Mississippi	United States	Gulf of Mexico	3975	2470	15
Missouri	United States	Mississippi R.	4382	2723	9
Missouri–Mississippi	United States	Gulf of Mexico	6418	3988	2
Mureşul	Rumania–Hungary	Tisza R.	644	400	142
Murray	Australia	Indian O.	1931	1200	52
Narmada	India	Arabian Sea	1287	800	80
Neches	United States	Sabine Lake	451	280	176
Negro, Rio	Argentina	Atlantic O.	1127	700	91
Negro, Rio	Colombia–Brazil	Amazon R.	2253	1400	40
Nelson	Canada	Hudson Bay	644	400	143
Neosho	United States	Arkansas R.	740	460	127
Neuse	United States	Pamlico Sound	418	260	182
New	United States	Kanawha R.	410	255	84
Niger	Guinea–Mali–Niger–Benin Nigeria	Gulf of Guinea	4184	2600	12
Nile	Uganda–Sudan–Egypt	Mediterranean Sea	6437	4000	1
Niobrara	United States	Missouri R.	694	431	132
North Platte	United States	Platte R.	995	618	104
Nueces	United States	Corpus Christi Bay	544	338	153
Ob	Russian Federation	Gulf of Ob	4023	2500	14
Oder	Poland–Germany	Baltic Sea	885	550	112
Ohio	United States	Mississippi R.	1579	981	69
Oka	Russian Federation	Volga R.	1529	950	71
Orange	Lesotho–South Africa–Namibia	Atlantic O.	2092	1300	42
Orinoco	Venezuela–Colombia	Atlantic O.	2575	1600	32
Osage	United States	Missouri R.	805	500	121
Ottawa	Canada	Saint Lawrence R.	1102	685	95
Ouachita	United States	Red R.	974	605	107
Owyhee	United States	Snake R.	402	250	187
Paraguay	Brazil–Paraguay–Argentina	Paraná R.	2414	1500	35
Paraná	Brazil–Paraguay–Argentina	Río la Plata	3943	2450	16
Parnaíba	Brazil	Atlantic O.	1448	900	75
Peace	Canada	Slave R.	1690	1050	62
Pearl	United States	Gulf of Mexico	781	485	122
Pechora	Russian Federation	Arctic O.	1770	1100	60
Pecos	United States	Rio Grande	1183	735	86
Pee Dee	United States	Atlantic O.	700	435	131

(continues)

Principal Rivers of the World *(continued)*

River	Countries of transit	Outflow	Length Kilometers	Miles	Rank
Pilcomayo	Bolivia–Argentina–Paraguay	Paraguay R.	1609	1000	60
Plata, Río de la	Argentina–Uruguay	Atlantic O.	298	185	192
Platte	United States	Rio Grande	1183	735	86
Po	Italy	Adriatic Sea	673	418	138
Potomac	United States	Chesapeake Bay	462	287	172
Powder	United States	Yellowstone R.	604	375	146
Purus	Brazil	Amazon R.	3219	2000	19
Red	United States	Mississippi R.	1931	1200	53
Red (of the North)	United States–Canada	Lake Winnipeg	877	545	113
Republican	United States	Kansas R.	679	422	136
Rhine	Switzerland–Germany–France–Netherlands	North Sea	1320	820	79
Rhone	Switzerland–France	Mediterranean Sea	811	504	118
Rio Grande	United States–Mexico	Gulf of Mexico	2897	1800	23
Roanoke	United States	Albemarle Sound	612	380	145
Rock	United States	Mississippi R.	483	300	167
Roosevelt, Río	Brazil	Madeíra R.	1529	950	72
Sabine	United States	Gulf of Mexico	649	403	140
Sacramento	United States	San Francisco Bay	615	382	144
Saguenay	Canada	Saint Lawrence R.	201	125	193
Saint Francis	United States	Mississippi R.	684	425	134
Saint John	United States–Canada	Bay of Fundy	673	418	139
Saint Johns	United States	Atlantic O.	444	276	177
Saint Lawrence	United States–Canada	Gulf of Saint Lawrence	1223	760	85
Saint Maurice	Canada	Saint Lawrence R.	523	325	160
Salado, Río	Argentina	Paraná R.	1931	1200	54
Salmon	United States	Snake R.	676	420	137
Salween	China–Burma	Bay of Bengal	2816	1750	26
San Joaquin	United States	San Francisco Bay	563	350	151
San Juan	United States	Colorado R.	579	360	148
São Francisco	Brazil	Atlantic O.	2897	1800	24
Saskatchewan	Canada	Lake Winnipeg	1939	1205	48
Savannah	United States	Atlantic O.	505	314	162
Seine	France	English Channel	772	480	123
Senegal	Mali–Mauritania–Senegal	Atlantic O.	1609	1000	68
Shannon	Ireland	Atlantic O.	386	240	189
Si *(see Xiiang)*					
Smoky Hill	United States	Kansas R.	869	540	114
Snake	United States	Columbia R.	1670	1038	63
South Platte	United States	Platte R.	682	424	135
Sungari	China	Amur R.	1287	800	81
Susquehanna	United States	Chesapeake Bay	715	444	130
Syr Darya	Russian Federation	Aral Sea	2092	1300	43
Tagus	Spain–Portugal	Atlantic O.	911	566	110
Tallahatchie	United States	Yazoo R.	484	301	165
Tallapoosa	United States	Alabama R.	431	268	179
Tennessee	United States	Ohio R.	1049	652	97
Thames	United Kingdom	North Sea	336	209	190
Tiber	Italy	Mediterranean Sea	393	244	188
Tigris	Turkey–Syria–Iraq	Euphrates R.	1851	1150	58
Tisza	Hungary–Yugoslavia	Danube R.	1287	800	82

(continues)

Principal Rivers of the World *(continued)*

River	Countries of transit	Outflow	Length Kilometers	Miles	Rank
Tobol	Russian Federation	Irtysh R.	1287	800	83
Tocantins	Brazil	Para R.	2736	1700	30
Tombigbee	United States	Mobile R.	845	525	117
Trinity	United States	Galveston Bay	579	360	149
Ucayali	Peru	Amazon R.	1931	1200	55
Ural	Russian Federation	Caspian Sea	2235	1400	41
Uruguay	Brazil–Argentina–Uruguay	Río de la Plata	1579	981	70
Vistula	Poland	Baltic Sea	1046	650	99
Volga	Russian Federation	Caspian Sea	3742	2325	17
Wabash	United States	Ohio R.	764	475	124
Weser	Germany	North Sea	483	300	168
White	United States	Mississippi R.	1110	690	93
Wisconsin	United States	Mississippi R.	692	430	133
Xijiang (Hsi)	China	South China Sea	2012	1250	46
Xingu	Brazil	Amazon R.	2092	1300	44
Yangzi (Chiangjiang)	China	East China Sea	5150	3200	4
Yellow (Huanghe)	China	Gulf of Chihli (Bohaiwan)	4506	2800	8
Yellowstone	United States	Missouri R.	1080	671	96
Yenisei	Russian Federation	Arctic O.	4506	2800	7
Yukon	Canada–United States	Bering Sea	3219	2000	20
Zambezi	Angola–Zambia–southern Zimbabwe–Mozambique	Indian O.	2655	1650	31

Appendix B

Principal Lakes of the World

Lake	Countries	Locality	Area in Square kilometers	Square miles	Rank
Albert	Congo–Uganda	East Africa	5343	2064	32
Aral Sea	Russian Federation	Western Turkestan	67,741	26,166	4
Athabasca	Canada	Northeast Alberta and, ˌnorthwest Saskatchewan	7767	3000	24
Baikal	Russian Federation	Southern Siberia	34,173	13,200	7
Balaton	Hungary	Western Hungary	595	230	54
Balkhash	Russian Federation	Kazakhstan	18420	7115	16
Bangweulu	Zambia	Northeastern Zambia	4323	1670	41
Caspian Sea	Iran–Russian Federation	Southwest Asia	437,521	169,000	1
Cayuga	United States	Central New York State	171	66	67
Chad	Chad–Niger–Nigeria	Northwest central Africa	25,889	10,000	12
Champlain	Canada–United States	Between New York and Vermont	1553	600	50
Como	Italy	Lombardy (Italian Lake District)	145	56	68
Constance, Lake (Bodensee)	Austria–Switzerland– Germany	Between southwest Germany and northeast Switzerland	536	207	58
Dead Sea	Israel–Jordan	Near eastern end of Mediterranean	958	370	52
Dongtinghu (Tung-T'ing)	China	Hunan Province	3754	1450	47
Dubawnt	Canada	Southwestern Keewatin District	4282	1654	42
Erie	Canada–United States	Great Lakes, between Ontario and Ohio	25,733	9940	13
Eyre	Australia	Northeastern South Australia	8880	3430	21
Gairdner	Australia	Southern South Australia	3883	1500	44

(continues)

Principal Lakes of the World *(continued)*

Lake	Countries	Locality	Area in Square kilometers	Square miles	Rank
Garda	Italy	Eastern Lombardy (Italian Lake District)	370	143	60
Geneva, Lake (Lake Leman)	France–Switzerland	Eastern France and southwest Switzerland	582	225	56
Great Bear	Canada	Central Mackenzie District	33,138	12,800	8
Great Salt	United States	Northwestern Utah	5954	2300	27
Great Slave	Canada	Southern Mackenzie District	28,923	11,172	10
Helmand	Iran	Eastern Iran	5178	2000	34
Huron	Canada–United States	Great Lakes between Ontario and Michigan	59,570	23,010	5
Issyk-Kul	Russian Federation	Northeastern Kirghizia	5825	2250	29
Kariba	Zimbabwe	South central Africa	5307	2050	33
Khanka	China–Russian Federation	Between Manchuria and maritime Siberia	4401	1700	39
Kyoga	Uganda	South central Uganda	2589	1000	48
Ladoga	Russian Federation	Northwestern Russian Federation	18,122	7000	17
Lake of the Woods	Canada–United States	Minnesota–southwestern Manitoba–southwestern Ontario	3844	1485	45
Leopold II	Congo	Western Congo	4401	1700	40
Maggiore	Italy–Switzerland	Lombardy (Italian Lake District)	215	83	63
Manitoba	Canada	Southern Manitoba	4704	1817	37
Maracaibo	Venezuela	Along coast of northwest Venezuela	16,310	6300	18
Mead	United States	Northwest Arizona and southeast Nevada	588	227	55
Michigan	United States	Great Lakes between Michigan and Wisconsin	57,991	22,400	6
Murray	Papua New Guinea	Papua	777	300	53
Mweru	Congo–Zambia	Central Africa	4194	1620	43
Nettiling	Canada	Baffin Island	5064	1956	36
Neuchatel, Lake of	Switzerland	Western Switzerland	220	85	62
Nicaragua	Nicaragua	Southwest Nicaragua	7922	3060	23
Nipigon	Canada	Southern Ontario	4479	1730	38
Nyasa	Mozambique–Malawi–Tanzania	Southeastern Africa	28,478	11,000	11
Okeechobee	United States	Southern Florida	1890	730	49
Onega	Russian Federation	Northwestern Russian Federation	9745	3764	19
Oneida	United States	Central New York State	207	80	64
Ontario	Canada–United States	Great Lakes between Ontario and New York	19,520	7540	15
Pontchartrain	United States	Southeast Louisiana	1553	600	51
Qinghai (Tsinghai, Koko Nor)	China	Northeastern Qinghai Province	5954	2300	28
Reindeer	Canada	Northern part of Manitoba–Saskatchewan boundary	6327	2444	25
Rudolf	Kenya–Ethiopia–Sudan	East Africa	9061	3500	20

(continues)

Principal Lakes of the World (*continued*)

Lake	Countries	Locality	Area in Square kilometers	Square miles	Rank
Seneca	United States	Western New York State	173	67	66
Superior	Canada–United States	Great Lakes, between Ontario and Michigan	82,378	31,820	2
Tahoe	United States	On California–Nevada boundary	518	20	59
Tanganyika	Burundi–Congo–Zambia–Tanzania	Central Africa	32,879	12,700	9
Titicaca	Bolivia–Peru	Altiplano, Andes Mts.	8284	3200	22
Torrens	Australia	Eastern South Australia	6231	2400	26
Tsing Hai (Koko Nor). See Qinghai					
Tung-T'ing. See Dongtinghu					
Urmia	Iran	Northwestern Iran	5178	2000	35
Van	Turkey	Eastern Turkey in Asia	3764	1454	46
Vanern	Sweden	Southern Sweden	5543	2141	30
Victoria	Kenya–Tanzania–Uganda	East Africa	69,454	26,828	3
Winnebago	United States	East central Wisconsin	557	215	57
Winnipeg	Canada	Southern Manitoba	24,077	9300	14
Winnipegosis	Canada	Southwestern Manitoba	5400	2086	31
Winnipesaukee	United States	Central New Hampshire	184	71	65
Yellowstone	United States	Yellowstone National Park, northwestern Wyoming	362	140	61
Zurich, Lake of	Switzerland	Northern Switzerland	88	34	69

Bibliography

Aarn, Ivantsoff, W, and Hansen, B. (1997). East coast sympatry. *Fishes Sahul* **11**(2), 507–519.

Aarn, Ivantsoff, W., and Kottelat, M. (1988). Phylogenetic analysis of Telmatherinidae (Teleostei: Atherinomorpha) with description of *Marosatherina*, a new genus from Sulawesi. *Ichthyol. Explorations Freshwaters* **9**(3), 311–323.

Ahlstrom, E. H., Amaoka, K., Hensley, D. A., Moser, H. G., and Sumida, B. Y. (1984). Pleuronectiformes: Development. In *Ontogeny and Systematics of Fishes* (H. G. Moser *et al.*, Eds.), Special Publication No. 1, pp. 640–670. American Society of Ichthyologists and Herpetologists, Lawrence, KS.

Albert, J. S. (2000). Species diversity and phylogenetic systematics of American knifefishes (Gymnotiformes, Teleostei), Miscellaneous Publications of the Museum of Zoology 190, 1–129. University of Michigan, Ann Arbor.

Albert, J. S., and Campos-da-Paz, R. (1998). Phylogenetic systematics of Gymnotiformes with diagnoses of 58 clades: A review of available data. In *Phylogeny and Classification of Neotropical Fishes* (L. R. Malabarba, R. E. Reis, R. P. Vari, Z. M. S. Lucena, and C. A. S. Lucena, Eds.), pp. 419–446. EDIPUCRS, Porto Alegre, Brazil.

Albert, J. S., de Campos Fernandes-Matioli, F. M., and de Almeida-Toledo, L. F. (1999). New species of *Gymnotus* (Gymnotiformes, Teleostei) from southeastern Brazil: Toward the deconstruction of *Gymnotus carapo*. *Copeia* **1992**(2), 410–421.

Albert, J. S., Lannoo, M. J., and Yuri, T. (1998). Testing hypotheses of neural evolution in gymnotiform electric fishes using phylogenetic character data. *Evolution* **52**(6), 1760–1780.

Albert, J. S., and Miller, R. R. (1995). *Gymnotus maculosus*, a new species of electric fish (Chordata: Teleostei: Gymnotoidei) from Middle America, with a key to species of *Gymnotus*. *Proc. Biol. Soc. Washington* **108**(4), 662–678.

Alexander, R. McN. (1964). The structure of the Weberian apparatus in the Siluri. *Proc. Zool. Soc. London* **142**(3), 419–440.

Alexander, R. McN. (1965). Structure and function in the catfish. *J. Zool.* **148**, 88–152.

Alexander, R. McN. (1967). *Functional Design in Fishes*. Hutchinson Library, London.

Allen, G. R. (1973). Nature's sharpshooters. *Tropical Fish Hobbyist* **21**, 18–27.

Allen, G. R. (1978). A review of the archerfishes (family Toxotidae). *Rec. Western Austr. Museum* **6**(4), 355–378.

Allen, G. R. (1980). A generic classification of the rainbowfishes (family Melanotaeniidae). *Rec. Western Austr. Museum* **8**(3), 449–490.

Allen, G. R. (1989). *Freshwater Fishes of Australia*. T. F. H., Neptune City, NJ.

Allen, G. R. (1991). *Field Guide to the Freshwater Fishes of New Guinea*, Christensen Research Institute Publication No. 9. Madang, Papua New Guinea.

Allen, G. R. (1993). *Variichthys*, a replacement name for the terapontid *Varia* and the first record of *V. lacustris* from Australia. *Rec. Western Austr. Museum* **16**(3), 459–460.

523

Allen, G. R. (1994). Silversides and their allies. In *Encyclopedia of Fishes* (J. P. Paxton and W. N. Eschmeyer, Eds.), pp. 153–156. Academic Press, San Diego.

Allen, G. R. (1995). *Rainbowfishes in Nature and in the Aquarium.* Tetra Verlad, Melle, Germany.

Allen, G. R. (1996a). Two new species of rainbow-fishes (*Melanotaenia*: Melanotaeniidae), from the Kikori River system, Papua New Guinea. *Rev. Francaise Aquariologie* 23(1/2), 9–16.

Allen, G. R. (1996b). Freshwater fishes of Irian Jaya. In *Proceedings of the First International Conference on Eastern Indonesian–Australian Vertebrate Fauna* (D. J. Kitchener and A. Suyanto, Eds.), pp. 15–21. Manado, Indonesia.

Allen, G. R. (1996c). Rainbowfishes. In *Freshwater Fishes of South-Eastern Australia* (R. M. McDowall, Ed.), pp. 134–140. Reed, Sydney.

Allen, G. R. (1997). A new species of rainbowfish (*Melanotaenia*: Melanotaeniidae), from the Lakekamu Basin, Papua New Guinea. *Rev. Francaise Aquariologie* 24(1/2), 37–42.

Allen, G. R. (1998a). A review of the marine catfish genus *Paraplotosus* (Plotosidae) with the description of a new species from north-western Australia. *Raffles Bull. Zool.* 46(1), 123–134.

Allen, G. R. (1998b). A new genus and species of rain-bowfish (Melanotaeniidae) from fresh waters of Irian Iaya, Indonesia. *Rev. Francaise Aquariologie* 25(1/2), 11–16.

Allen, G. R., and Berra, T. M. (1989). Life history aspects of the west Australian salamanderfish *Lepidogalaxias salamandroides* Mees. *Rec. Western Austr. Museum* 14(3), 253–267.

Allen, G. R., and Burgess, W. E. (1990). A review of the glassfishes (Chandidae) of Australia and New Guinea. *Rec. Western Austr. Museum Suppl. No.* 34, 139–206.

Allen, G. R., and Cross, N. J. (1982). *Rainbowfishes of Australia and Papua New Guinea.* T. F. H., Neptune City, NJ.

Allen, G. R., and Feinberg, M. N. (1998). Descriptions of four new species of freshwater catfishes (Plotosidae) from Australia. *Aqua J. Ichthyol. Aquat. Biol.* 3(1), 9–18.

Allen, G. R., and Renyaan, S. J. (1996). *Chilatherina pricei,* a new species of rainbowfish (Melanotaeniidae) from Irian Jaya. *Rev. Francaise Aquariologie* 23(1/2), 5–8.

Allen, G. R., and Renyaan, S. J. (1998). Three new species of rainbowfishes (Melanotaeniidae) from Irian Jaya, Indonesia. *Aqua J. Ichthyol. Aquat. Biol.* 3(2), 69–80.

Allen, G. R., and Robertson, D. R. (1994). *Fishes of the Tropical Eastern Pacific.* Univ. of Hawaii Press, Honolulu.

Allen, G. R., and Swainston, R. (1988). *The Marine Fishes of North-Western Australia.* Western Australian Museum, Perth.

Allen, G. R., Ivantsoff, W., Shepard, M. A., and Renyaan, S. J. (1998). *Pseudomugil pellucidus* (Pisces: Pseudomugilidae), a newly discovered blue-eye from Timika-Tembagapura region, Irian Jaya. *Aqua J. Ichthyol. Aquat. Biol.* 3(1), 1–8.

Allen, K. R. (1951). *The Horokiwi Stream,* Bulletin No. 10. New Zealand Marine Department of Fisheries.

Allibone, R. M., Crowl, T. A., Holmes, J. M., King, T. M., McDowall, R. M., Townsend, C. R., and Wallis, G. P. (1996). Isozyme analysis of *Galaxias* species (Teleostei: Galaxiidae) from the Taieri River, South Island, New Zealand: A species complex revealed. *Biol. J. Linnean Soc.* 57, 107–127.

Alves-Gomes, J. A. (1998). The phylogenetic position of the South American electric fish genera *Sternopygus* and *Archolaemus* (Ostariophysi: Gymnotiformes) according to 12S and 16S mitochondrial DNA sequences. In *Phylogeny and Classification of Neotropical Fishes* (L. R. Malabarba, R. E. Reis, R. P. Vari, Z. M. S. Lucena, and C. A. S. Lucena, Eds.), pp. 447–460. EDIPUCRS, Porto Alegre, Brazil.

Alves-Gomes, J. A. (1999). Systematic biology of gymnotiform and mormyriform electric fishes: Phylogenetic relationships, molecular clocks and rates of evolution in the mitochondrial rRNA genes. *J. Exp. Biol.* 202(10), 1167–1183.

Alves-Gomes, J., and Hopkins, C. D. (1997). Molecular insights into the phylogeny of mormyriform fishes and the evolution of their electric organs. *Brain Behav. Evol.* 49, 324–351.

Alves-Gomes, J. A., Ortí, G., Haygood, M., Heiligenberg, W., and Meyer, A. (1995). Phylogenetic analysis of the South American electric fishes (Order Gymnotiformes) and the evolution of their electric system: A synthesis based on morphology, electrophysiology, and mitrochondrial sequence data. *Mol. Biol. Evol.* 12, 298–318.

Anderson, M. E. (1998). A Late Cretaceous (Maastrichtian) galaxiid fish from South Africa. J. L. B. Smith Institute of Ichthyology Special Publication No. 60, pp. 1–12.

Anderson, W. D., III, and Collette, B. B. (1991). Revision of the freshwater viviparous halfbeaks of the genus *Hemirhamphodon* (Teleostei: Hemiramphidae). *Ichthyol. Exploration Freshwaters* 2(2), 151–176.

Andrews, A. P. (1996). Family Bovichtidae. In *Freshwater Fishes of South-Eastern Australia* (R. M. McDowall, Ed.), pp. 198–199. Reed, Sydney.

Annandale, N. (1918). Fish and fisheries of the Inle Lake. *Rec. Indian Museum* 14, 33–64.

Anonymous (1994, December 15). Withdraw of proposed rule for endangered status and critical habitat for the Alabama sturgeon. *Fed. Regist.* 59(240), 64784–64809.

Applegate, V. C. (1950). Natural history of the sea lamprey, *Petromyzon marinus*, in Michigan. Special Scientific Report. U.S. Wildlife Service. *Fisheries* 55, 1–237.

Araujo-Lima, C., and Goulding, M. (1997). *So Fruitful a Fish: Ecology, Conservation, and Aquaculture of the Amazon's Tambaqui.* Columbia Univ. Press, New York.

Armbruster, J. W. (1998a). Modifications of the digestive tract for holding air in loricariid and scoloplacid catfishes. *Copeia* 1998(3), 663–675.

Armbruster, J. W. (1998b). Phylogenetic relationships of the suckermouth armored catfish of the *Rhinelepis* group (Loricariidae: Hypostominae). *Copeia* 1998(3), 620–636.

Arnoult, J. (1986a). Gasterosteidae. In *Check-list of the Freshwater Fishes of Africa* (J. Daget, J.-P. Gosse, and D. F. E. Thys van den Audenaerde, Eds.), p. 280. CLOFFA II, ORSTOM, Paris.

Arnoult, J. (1986b). Scatophagidae. In *Check-list of the Freshwater Fishes of Africa* (J. Daget, J.-P. Gosse, and D. F. E. Thys van den Audenaerde, Eds.), p. 314. CLOFFA II, ORSTOM, Paris.

Arratia, G. F. (1981). Géneros de peces de aguas continentales de Chile. Museo Nacional de Historia Natural (Santiago), Publicación Ocasional No. 34, pp. 1–108.

Arratia, G. F. (1982). A review of freshwater percoids from South America (Pisces, Osteichthyes, Perciformes, Percichthyidae, and Perciliidae). *Abhandlungen Senckenbergischen Naturforschended Gesellschaft* 540, 1–52.

Arratia, G. (1983). Peces de la region sureste de los Andes y sus probables relaciones biogeograficas actuales. *Deserta* 7, 48–107.

Arratia, G. (1987). Description of the primitive family Diplomystidae (Siluriformes, Teleostei, Pisces): Morphology, taxonomy and phylogenetic implications. *Bonner Zool. Monogr.* 24, 1–120.

Arratia, G. (1990). The South American Trichomycterinae (Teleostei: Siluriformes), a problematic group. In *Vertebrates in the Tropics* (G. Peters and R. Hutterer, Eds.), pp. 395–403. Museum Alexander Koenig, Bonn.

Arratia, G. (1992). Development and variation of the suspensorium of primitive catfishes (Teleostei: Ostariophysi) and their phylogenetic relationships. *Bonner Zool. Monogr.* 32, 1–149.

Arratia, G. (1998). Basal teleosts and teleostean phylogeny: Response to C. Patterson. *Copeia* 1998(4), 1109–1113.

Arratia, G. (1999). The monophyly of Teleostei and stem-group teleosts. Consensus and disagreements. In *Mesozoic Fishes 2—Systematics and Fossil Record* (G. Arratia and H.-P. Schultz, Eds.), pp. 265–334. Verlag Dr. Friedrich Pfeil, Munich.

Arratia, G., and Huaquin, L. (1995). Morphology of the lateral line system and of the skin of diplomystid and certain primitive loricarioid catfishes and systematic and ecological considerations. *Bonner Zool. Monogr.* 36, 1–110.

Arratia, G., and Menu-Marque, S. (1981). Revision of the freshwater catfishes of the genus *Hatcheria* (Siluriformes, Trichomycteridae) with commentaries on ecology and biogeography. *Zool. Anz.* 207(1/2), 88–111.

Arratia, G., Chang, A., Menu-Marque, S., and Rojas, G. (1978). About *Bullockia* gen. nov., *Trichomycterus mendozensis* n. sp., and revision of the family Trichomycteridae (Pisces, Siluriformes). *Stud. Neotropical Fauna Environ.* 13, 157–194.

Arratia, G., Peñafort, B., and Menu-Marque, S. (1985). Peces de la región sureste de los Andes y sus probables relaciones biogeográficas actuales. *Deserta* 7, 48–108.

Arthington, A. H., and Lloyd, L. N. (1989). Introduced poeciliids in Australia and New Zealand. In *Ecology and Evolution of Livebearing Fishes* (G. K. Meffe and F. F. Snelson, Jr., Eds.), pp. 333–348. Prentice-Hall, Englewood Cliffs, NJ.

Audley-Charles, M. G. (1981). Geogical history of the region of Wallace's Line. In *Wallace's Line and Plate Tectonics* (T. C. Whitmore, Ed.), pp. 24–35. Clarendon, Oxford.

Avise, J. C., and Selander, R. K. (1972). Evolutionary genetics of cave-dwelling fishes of the genus *Astyanax. Evolution* 26, 1–19.

Avise, J. C., and Smith, M. H. (1974). Biochemical genetics of sunfish. II. Genic similarity between hybridizing species. *Am. Nat.* 108, 458–472.

Avise, J. C., and Smith, M. H. (1977). Gene frequency comparisons between sunfish (Centrarchidae) populations at various stages of evolutionary divergence. *Syst. Zool.* 26, 319–335.

Avise, J. C., Straney, D. O., and Smith, M. H. (1977). Biochemical genetics of Sunfish. IV. Relationships of gentrarchid genera. *Copeia* 1977(2), 250–258.

Avise, J. C., Helfman, G. S., Saunders, N. C., and Hales, L. S. (1986). Mitochondrial DNA differentiation in North Atlantic eels: Population genetic consequences of an unusual life history pattern. *Proc. Natl. Acad. Sci. USA* 83, 4350–4354.

Axelrod, H. R., and Burgess, W. E. (1982, April). Loaches of the world. *Tropical Fish Hobbyist*, 32–44.

Azpelicueta, M. M. (1994a). Three East-Andean species of *Diplomystes* (Siluriformes: Diplomystidae). *Ichthyol. Exploration Freshwater* 5(3), 223–240.

Azpelicueta, M. M. (1994b). Los diplomístidos en Argentian (Siluriformes, Diplomystidae). Fauna de Agua Dulce de la República Argentina. PROFADU-CONICET, La Plata.

Babiker, M. M. (1984). Development of dependence on aerial respiration in *Polypterus senegalus* (Cuvier). *Hydrobiologia* 110, 351–363.

Bailey, R. M. (1938). A systematic revision of the centrarchid fishes, with a discussion of their distribution, variations, and probable interrelationships. Ph.D. dissertation, University of Michigan, Ann Arbor.

Bailey, R. M. (1971). Pisces (zoology). In *McGraw-Hill Encyclopedia of Science and Technology*, 3rd ed., Vol.10, pp. 281–282. McGraw-Hill, New York.

Bailey, R. M. (1980). Comments on the classification and nomenclature of lampreys—An alternate view. *Can. J. Fish. Aquat. Sci.* 37, 1626–1629.

Bailey, R. M. (1982). Reply [to Vladykov and Kott, 1982]. *Can. J. Fish. Aquat. Sci.* 39, 1217–1220.

Bailey, R. M., and Baskin, J. N. (1976). *Scoloplax dicra*, a new armored catfish from the Bolivian Amazon. *Occasional Papers Museum Zool. Univ. Michigan* 674, 1–14.

Bailey, R. M., and Etnier, D. A. (1988). Comments on the subgenera of darters (Percidae) with descriptions of two new species of *Etheostoma (Ulocentra)* from southeastern United States. *Miscellaneous Publ. Museum Zool. Univ. Michigan* 175, 1–48.

Bailey, R. M., and Gosline, W. A. (1955). Variation and systematic significance of vertebral counts in the American fishes of the family Percidae. *Miscellaneous Publ. Museum Zool. Univ. Michigan* 93, 5–44.

Bailey, R. M., and Stewart, D. J. (1984). Bagrid catfishes from Lake Tanganyika, with a key and descriptions of new taxa. *Miscellaneous Publ. Museum Zool. Univ. Michigan* 168, 1–41.

Bailey, R. M., Winn, H. E., and Smith, C. L. (1954). Fishes from the Escambia River, Alabama and Florida, with ecologic and taxonomic notes. *Proc. Acad. Natural Sci. Philadelphia* 106, 109–164.

Bailey, R. M., Lachner, E. A., Lindsey, C. C., Robins, C. R., Roedel, P. M., Scott, W. B., and Woods, L. P. (1960). *A List of Common and Scientific Names of Fishes from the United States and Canada*, 2nd ed., Special Publication No. 2. American Fisheries Society, Bethesda, MD

Bailey, R. M., Fitch, J. E., Herald, E. S., Lachner, E. A., Lindsey, C. C., Robins, C. R., and Scott, W. B. (1970). *A List of Common and Scientific Names of Fishes from the United States and Canada*, 3rd ed., Special Publication No. 6. American Fisheries Society, Bethesda, MD.

Baldridge, H. D. (1974). Shark attack: A program of data reduction and analysis. Contributions of Mote Marine Laborator, Vol. 1, No. 2. (Reprinted as *Shark Attack*, Berkley, New York)

Baldwin, C. C., and Johnson, G. D. (1999). *Paxton concilians*: A new genus and species of pseudamine apogonid (Teleostei: Percoidei) from northwestern Australia: The sister group of the enigmatic *Gymnapogon*. *Copeia* 1999(4), 1050–1071.

Ball, I. R. (1975). Nature and formulation of biogeographical hypotheses. *Syst. Zool.* 24(4), 407–430.

Balon, E. K. (1975). Reproductive guilds of fishes: A proposal and definition. *J. Fish. Res. Board Can.* 32, 821–864.

Balon, E. K. (Ed.) (1980). *Charrs, Salmonid Fishes of the Genus Salvelinus*. Junk, The Hague.

Baltz, D. M. (1984). Life history variation in female surfperches (Perciformes, Embiotocidae). *Environ. Biol. Fishes* 10, 159–171.

Balushkin, A. V. (1993). Classification, phylogenetic relationships, and origins of the families of the suborder Notothenioidei (Perciformes). *J. Ichthyol.* 32(7), 90–110.

Banarescu, P. (1990). *Zoogeography of Fresh Waters. Vol. 1. General Distribution and Dispersal of Freshwater Animals*. AULA-Verlag, Wiesbaden, Germany.

Banarescu, P. (1992). *Zoogeography of Fresh Waters. Vol. 2. Distribution and Dispersal of Freshwater Animals in North America and Eurasia*. AULA-Verlag, Wiesbaden, Germany.

Banarescu, P. (1995). *Zoogeography of Fresh Waters. Vol. 3. Distribution and Dispersal of Freshwater Animals in Africa, Pacific Areas and South America*. AULA-Verlag, Wiesbaden, Germany.

Banarescu, P., and Coad, B. W. (1991). Cyprinids of Eurasia. In *Cyprinid Fishes: Systematics, Biology and Exploitation* (I. J. Winfield and J. S. Nelson, Eds.), pp. 127–155. Chapman & Hall, London.

Banister, K. E. (1970). The anatomy and taxonomy of *Indostomus paradoxus* Prashad and Mukerji. *Bull. Br. Museum Nat. History Zool.* 19, 179–209.

Banister, K. (1986). Gasterosteidae. In *Fishes of the North-Eastern Atlantic and the Mediterranean* (P. J. P. Whitehead, M.-L. Bauchot, J.-C. Hureau, J. Nielsen, and E. Tortonese, Eds.), Vol. 2, pp. 640–643. UNESCO, Paris.

Barbour, C. D. (1973). A biogeographical history of *Chirostoma* (Pisces: Atherinidae): A species flock from the Mexican plateau. *Copeia* **1973**(3), 533–556.

Barila, T. Y. (1980). *Stizostedion*. In *Atlas of North American Freshwater Fishes* (D. S. Lee *et al.*, Eds.), pp. 745–748. North Carolina State Museum of Natural History, Raleigh.

Barlow, C. G., and Lisle, A. (1987). Biology of the Nile perch *Lates niloticus* (Pisces: Centropomidae) with reference to its proposed role as a sport fish in Australia. *Biol. Conserv.* **39**, 269–289.

Barlow, G. W. (2000). *The Cichlid Fishes: Nature's Grand Experiment in Evolution*. Perseus, Cambridge, MA.

Barlow, G. W., Liem, K. F., and Wickler, W. (1968). Badidae, a new fish family—Behavioural, osteological, and developmental evidence. *J. Zool.* **156**, 415–447.

Barthem, R., and Goulding, M. (1997). *The Catfish Connection*. Columbia Univ. Press, New York.

Bartholomew, J. G., Clarke, W. E., and Grimshaw, P. H. (1911). *Atlas of Zoogeography*. Bartholomew, Edinburgh, UK.

Baskin, J. N. (1973). Structure and relationships of the Trichomycteridae. Ph.D. thesis, City University of New York.

Bauchot, M.-L. (1986). Anguillidae. In *Fishes of the North-Eastern Atlantic and the Mediterranean* (P. J. P. Whitehead, M.-L. Bauchot, J.-C. Hureau, J. Nielsen, and E. Tortonese, Eds.), Vol. 2, pp. 535–536. UNESCO, Paris.

Bauchot, M.-L. (1992). Moronidae. In *Faune des Poissons d'eaux Douces et Saumâtres de l'Afrique de l'Ouest* (C. Lévêque, D. Paugy, and G. G. Teugels, Eds.), Vol. 2, pp. 668–670. ORSTOM, Paris.

Beadle, L. C. (1981). *The Inland Waters of Tropical Africa: An Introduction to Tropical Limnology*, 2nd ed. Longman, London.

Beamish, R. J., Merriles, M. J., and Crossman, E. J. (1971). Karyotypes and DNA values for members of the suborder Esocoidei (Osteichthyes: Salmoniformes). *Chromosoma* **34**, 436–447.

Beaufort, L. F. de (1951). *Zoogeography of the Land and Inland Waters*. Sidgwick & Jackson, London.

Becker, G. C. (1983). *Fishes of Wisconsin*. Univ. of Wisconsin Press, Madison.

Beebe, W. (1945). Vertebrate fauna of a tropical dry season mud-hole. *Zoologica* **30**, 81–87.

Begle, D. P. (1991). Relationships of the osmeroid fishes and the use of reductive characters in phylogenetic analysis. *Syst. Zool.* **40**(1), 33–53.

Begle, D. P. (1992). Monophyly and relationships of the argentinoid fishes. *Copeia* **1992**(2), 350–366.

Behnke, R. (1972). The systematics of salmonid fishes of recently glaciated lakes. *J. Fish. Res. Board Can.* **29**, 639–671.

Behnke, R. (1986). Salmonidae. In *Check-list of the Freshwater Fishes of Africa* (J. Daget, J.-P. Gosse, and D. F. E. Thys van den Audenaerde, Eds.), p. 125. CLOFFA I, ORSTOM, Paris.

Behnke, R. J. (1992). *Native trout of western North America*. *Am. Fish. Soc. Monogr.* **6**.

Belbenoit, P., Moller, P., Serrier, J., and Push, S. (1979). Ethological observations on the electric organ discharge behavior of electric catfish *Malopterurus electricus* (Pisces). *Behav. Ecol. Sociobiol.* **4**, 321–330.

Bell, J. D., Berra, T. M., Jackson, P. D., Last, P. R., and Sloane, R. D. (1980). Recent records of the Australian grayling *Prototroctes maraena* Günther (Pisces: Prototroctidae) with notes on its distribution. *Austr. Zoologist* **20**(3), 419–431.

Bell, M. A., and Foster, S. A. (Eds.) (1994). *The Evolutionary Biology of the Three-Spine Stickleback*. Oxford Univ. Press, Oxford.

Bemis, W., Burggren, W. W., and Kemp, N. E. (Eds.) (1987). *The Biology and Evolution of Lungfish*. A. R. Liss, New York.

Bemis, W. E., and Grande, L. (1999). Development of the median fins of the North American paddlefish (*Polyodon spathula*), and a reevaluation of the lateral fin-fold hypothesis. In *Mesozoic Fishes 2—Systematics and Fossil Record* (G. Arratia and H.-P. Schultze, Eds.), pp. 41–68. Verlag Dr. Friedrich Pfeil, Munich.

Bemis, W. E., and Kynard, B. (1997). Sturgeon rivers: An introduction to acipenseriform biogeography and life history. *Environ. Biol. Fishes* **48**, 167–183.

Bemis, W. E., Findeis, E. K., and Grande, L. (1997). An overview of Acipenseriformes. *Environ. Biol. Fishes* **48**, 25–71.

Bennett, G. W. (1971). *Management of Lakes and Ponds*, 2nd ed. Van Nostrand Reinhold, New York.

Bennett, M. V. L. (1971a). Electric organs. In *Fish Physiology* (W. S. Hoar and D. J. Randall, Eds.), Vol. 5, pp. 347–491. Academic Press, New York.

Bennett, M. V. L. (1971b). Electroreception. In *Fish Physiology* (W. S. Hoar and D. J. Randall, Eds.), Vol. 5, pp. 493–574. Academic Press, New York.

Bennett, W. A., Currie, R. J., Wagner, P. F., and Beitinger, T. L. (1997). Cold tolerance and potential overwintering of red-bellied piranha *Pygocentrus natterei* in the United States. *Trans. Am. Fish. Soc.* **126**(5), 841–849.

Ben-Tuvia, A. (1986a). Teraponidae. In *Fishes of the North-Eastern Atlantic and the Mediterranean* (P. J. P. Whitehead, M.-L. Bauchot, J.-C. Hureau, J. Nielsen, and E. Tortonese, Eds.), Vol. 2, pp. 797–799. UNESCO, Paris.

Ben-Tuvia, A. (1986b). Mugilidae. In *Fishes of the North-Eastern Atlantic and the Mediterranean* (P. J. P. Whitehead, M.-L. Bauchot, J.-C. Hureau, J. Nielsen, and E. Tortonese, Eds.), Vol. 3, pp. 1197–1204. UNESCO, Paris.

Berg, L. S. (1948–1949). *Freshwater Fishes of the USSR and Adjacent Countries*, 4th ed., 3 vols. Israel Program for Scientific Translation, Jerusalem.

Berggren, W. A., and Hollister, C. D. (1974). Paleography, paleobiology, and the history of circulation in the Atlantic Ocean. In *Studies in Paleo-Oceanography* (W. Hay, Ed.), Special Publication No. 20, pp. 126–186. Society of Economic Paleontology and Mineralogy.

Bernardi, G. (1997). Molecular phylogeny of the Fundulidae (Teleostei, Cyprinodontiformes) based on the cytochrome b gene. In *Molecular Systematics of Fishes* (T. D. Kocher and C. A. Stepien, Eds.), pp. 189–197. Academic Press, San Diego.

Bernardi, G., and Powers, D. A. (1995). Phylogenetic relationships among nine species from the genus *Fundulus* (Cyprinodontiformes, Fundulidae) inferred from sequences of the cytochrome *b* gene. *Copeia* **1995**(2), 469–471.

Berra, T. M. (1973). A home range study of *Galaxias bongbong* in Australia. *Copeia* **1973**(2), 363–366.

Berra, T. M. (1974). The trout cod, *Maccullochella macquariensis*, a rare freshwater fish of eastern Australia. *Biol. Conserv.* **6**(1), 53–56.

Berra, T. M. (1981). *An Atlas of Distribution of the Freshwater Fish Families of the World*. Univ. of Nebraska Press, Lincoln.

Berra, T. M. (1982). Life history of the Australian grayling, *Prototroctes maraena* (Salmoniformes: Prototroctidae) in the Tambo River, Victoria. *Copeia* **1982**(4), 795–805.

Berra, T. M. (1984). Reproductive anatomy of the Australian grayling, *Prototroctes maraena* Gunther. *J. Fish Biol.* **25**, 241–251.

Berra, T. M. (1987). Speculations on the evolution of life history tactics of the Australian grayling. *Am. Fish. Soc. Symp.* **1**, 519–530.

Berra, T. M. (1989). *Scleropages leichardti* Günther (Osteoglossiformes): The case of the missing H. *Bull. Austr. Soc. Limnol.* **12**, 15–19.

Berra, T. M. (1990). *Evolution and the Myth of Creationism*. Stanford Univ. Press, Stanford, CA.

Berra, T. M. (1997a). Some 20th century fish discoveries. *Environ. Biol. Fishes* **50**, 1–12.

Berra, T. M. (1997b). William Bartram (1739–1823) and North American ichthyology, In *Collection Building in Ichthyology and Herpetology* (T. W. Pietsch and W. D. Anderson, Jr., Eds.), Special Publication No. 3, pp. 439–446. American

Society of Ichthyologists and Herpetologists, Lawrence, KS.

Berra, T. M., and Allen, G. R. (1989a). Burrowing, emergence, behavior, and functional morphology of the Australian salamanderfish, *Lepidogalaxias salamandroides. Fisheries* **14**(2), 2–10.

Berra, T. M., and Allen, G. R. (1989b). Clarification of the differences between *Galaxiella nigrostriata* (Shipway, 1953) and *Galaxiella munda* McDowall, 1978 (Pisces: Galaxiidae) from Western Australia. *Rec. Western Austr. Museum* **14**(3), 293–297.

Berra, T. M., and Allen, G. R. (1991). Population structure and development of *Lepidogalaxias salamandroides* (Pisces: Salmoniformes) from Western Australia. *Copeia* **1991**(3), 845–850.

Berra, T. M., and Allen, G. R. (1995). Inability of salamanderfish, *Lepidogalaxias salamandroides*, to tolerate hypoxic waters. *Rec. Western Austr. Museum* **17**, 117.

Berra, T. M., and Barbour, C. D. (1998). Is the Chilean *Galaxias globiceps* (Teleostei: Galaxiidae) extant or extinct? *Ichthyol. Exploration Freshwaters* **9**(3), 273–278.

Berra, T. M., and Berra, R. M. (1977). A temporal and geographical analysis of new teleost names proposed at 25 year intervals from 1869–1970. *Copeia* **1977**(4), 640–647.

Berra, T. M., and Cadwallader, P. L. (1983). Age and growth of Australian grayling, *Prototroctes maraena* Günther (Salmoniformes: Prototroctidae), in the Tambo River, Victoria. *Austr. J. Mar. Freshwater Res.* **34**, 451–460.

Berra, T. M., and Gunning, G. E. (1970). Repopulation of experimentally decimated sections of streams by longear sunfish, *Lepomis megalotis megalotis* (Rafinesque). *Trans. Am. Fish. Soc.* **99**(4), 776–781.

Berra, T. M., and Gunning, G. E. (1972). Seasonal movement and home range of the longear sunfish, *Lepomis megalotis* (Rafinesque) in Louisiana. *Am. Midland Nat.* **88**(2), 368–375.

Berra, T. M., and Pusey, B. J. (1997). Threatened fishes of the world: *Lepidogalaxias salamandroides* Mees, 1961 (Lepidogalaxiidae). *Environ. Biol. Fishes* **50**, 201–202.

Berra, T. M., and Ruiz, V. H. (1994). Rediscovery of *Galaxias globiceps* Eigenmann from southern Chile. *Trans. Am. Fish. Soc.* **123**(4), 595–600.

Berra, T. M., and Weatherley, A. H. (1972). A systematic study of the Australian freshwater serranid fish genus *Maccullochella. Copeia* **1972**(1), 53–64.

Berra, T. M., Campbell, A., and Jackson, P. D. (1987). Diet of the Australian grayling, *Prototroctes maraena* Günther (Salmoniformes: Prototroctidae), with notes on the occurrence of a trematode parasite and black peritoneum. *Austr. J. Mar. Freshwater Res.* 38, 661–669.

Berra, T. M., Crowley, L. E. L. M., Ivantsoff, W., and Fuerst, P. A. (1996). *Galaxias maculatus*: An explanation of its biogeography. *Mar. Freshwater Res.* 47, 845–849.

Berra, T. M., Feltes, R. M., and Ruiz, V. H. (1995). *Brachygalaxias gothei* from south-central Chile, a synonym of *B. bullocki* (Osteichthys: Galaxiidae). *Ichthyol. Exploration Freshwaters* 6(3), 227–234.

Berra, T. M., Moore, R., and Reynolds, L. F. (1975). The freshwater fishes of the Laloki River system of New Guinea. *Copeia* 1975(2), 316–326.

Berra, T. M., Sever, D. M., and Allen, G. R. (1989). Gross and histological morphology of the swim bladder and lack of accessory respiratory structures in *Lepidogalaxias salamandroides*, an aestivating fish from Western Australia. *Copeia* 1989(4), 850–856.

Berra, T. M., Smith, J. F., and Morrison, J. D. (1982). Probable identification of the cucumber odor of the Australian grayling *Prototroctes maraena*. *Trans. Am. Fish. Soc.* 111(1), 78–82.

Bertin, L. (1956). *Eels: A Biological Study*. Cleaver-Hume Press, London.

Bertin, L., and Arambourg, C. (1958). Ichthyogeographie. In *Traité de Zoologie 13* (P. P. Grasse, Ed.), Vol. 3, pp. 1944–1966.

Beumer, J. P. (1979a). Reproductive cycles of two Australian freshwater fishes: The spangled perch, *Therapon unicolor* Günther, 1859, and the east Queensland rainbowfish, *Nematocentris splendida* Peters, 1866. *J. Fish Biol.* 15, 111–134.

Beumer, J. P. (1979b). Temperature and salinity tolerance of the spangled perch, *Terapon unicolor* Günther, 1859, and the east Queensland rainbowfish *Nematocentris splendida* Peters, 1866. *Proc. R. Soc. Queensland* 90, 85–91.

Bianco, P. G., and Miller, R. R. (1989). First record of *Valencia letournexui* (Sauvage, 1880) in Peloponnese (Greece) and remarks on the Mediterranean family Valenciidae (Cyprinodontiformes). *Cybium* 13(4), 385–387.

Bigelow, H. B., and Schroeder, W. C. (1953). Fishes of the Gulf of Maine. *Fish. Bull.* 53(74), 1–577.

Bigorne, R. (1990a). Mormyridae. In *Faune des Poissons d'eaux Douce et Saumâtres de l'Afrique de l'Ouest* (C. Lévêque, D. Paugy, and G. G. Teugels, Eds.), Vol. 1, pp. 122–184. ORSTOM, Paris.

Bigorne, R. (1990b). Gymnarchidae. In *Faune des Poissons d'eaux Douce et Saumâtres de l'Afrique de l'Ouest* (C. Lévêque, D. Paugy, and G. G. Teugels, Eds.), Vol. 1, pp. 185–186. ORSTOM, Paris.

Birdsong, R. S., and Yerger, R. W. (1967). A natural population of hybrid sunfishes: *Lepomis marcochirus* × *Chaenobryttus gulosus*. *Copeia* 1967(1), 62–71.

Birdsong, R. S., Murdy, E. O., and Pezold, F. L. (1988). A study of the vertebral column and median fin osteology in gobioid fishes with comments on gobioid relationships. *Bull. Mar. Sci.* 42(2), 174–214.

Birkhead, W. S. (1972). Toxicity of stings of ariid and ictalurid catfishes. *Copeia* 1972(4), 790–807.

Birkhead, W. S. (1980). *Astyanyx mexicanus* (Filippi) Mexican tetra. In *Atlas of North American Freshwater Fishes* (D. S. Lee *et al.*, Eds.), p. 139. North Carolina State Museum of Natural History, Raleigh.

Birstein, V. J., Waldman, J. R., and Bemis, W. E. (Eds.) (1997). *Sturgeon Biodiversity and Conservation.* Kluwer, Dordrecht.

Blake, B. F. (1977). Food and feeding habits of mormyrid fishes of Lake Kainji, Nigeria, with special reference to seasonal variations and interspecific differences. *J. Fish Biol.* 11, 315–328.

Blouw, D. M., and Hagen, D. W. (1984). The adaptive significance of dorsal spine variation in the fourspine stickleback, *Apeltes quadracus*. I. Geographic variation in spine number. *Can. J. Zool.* 62, 1329–1339.

Blouw, D. M., and Hagen, D. W. (1990). Breeding ecology and evidence of reproductive isolation of a widespread stickleback fish (Gasterosteidae) in Nova Scotia, Canada. *Biol. J. Linnean Soc.* 39, 195–217.

Blumer, L. S. (1979). Male parental care in the bony fishes. *Q. Rev. Biol.* 54, 149–161.

Boden, G., Teugels, G. G., and Hopkins, C. D. (1997). A systematic revision of the large-scaled *Marcusenius* with description of a new species from Cameroon (Teleostei; Osteoglossomorpha; Mormyridae). *J. Nat. History* 31, 1645–1682.

Böhlke, J. E., and Rohde, F. C. (1980). *Elassoma*. In *Atlas of North American Freshwater Fishes* (D. S. Lee *et al.*, Eds.), pp. 584–586. North Carolina State Museum of Natural History, Raleigh.

Boltz, J., and Stauffer, J. R., Jr. (1986). Branchial brooding in the pirate perch, *Aphredoderus sayanus* (Gilliams). *Copeia* 1984(4), 1030–1031.

Boltz, J. M., and Stauffer, J. R., Jr. (1993). Systematics of *Aphredoderus sayanus* (Teleostei: Aphredoderidae). *Copeia* 1993(1), 81–98.

Bond, C. E. (1996). *Biology of Fishes*, 2nd ed. Saunders, Fort Worth, TX.

Bone, Q., Marshall, N. B., and Blaxter, J. H. S. (1995). *Biology of Fishes*, 2nd ed. Blackie, London.

Booth, W. (1988). Voodoo science. *Science* **240**, 274–277.

Bornbusch, A. H. (1991a). Monophyly of the catfish family Siluridae (Teleostei: Siluriformes), with a critique of previous hypotheses of the family's relationships. *Zool. J. Linnean Soc.* **101**, 105–120.

Bornbusch, A. H. (1991b). Redescription and reclassification of the silurid catfish *Apodoglanis furnessi* Fowler (Siluriformes: Siluridae), with diagnosis of three intrafamilial silurid subgroups. *Copeia* **1991**(4), 1070–1084.

Borror, D. J. (1960). *Dictionary of Word Roots and Combining Forms.* Mayfield, Palo Alto, CA.

Boughton, D. A., Collette, B. B., and McCune, A. R. (1991). Heterochrony in jaw morphology of needlefishes (Teleostei: Belonidae). *Syst. Zool.* **40**(3), 329–354.

Boulenger, G. A. (1909–1916). *Catalogue of the Fresh-Water Fishes of Africa in the British Museum (Natural History)*, 4 vols: Vol. 1, 1909; Vol. 2, 1911; Vol. 3, 1915; Vol. 4, 1916. British Museum (Natural History), London.

Brainerd, E. L., and Patek, S. N. (1998). Vertebral column morphology, C-start curvature, and the evolution of mechanical defenses in tetraodontiform fishes. *Copeia* **1988**(4), 971–984.

Branson, B. A., and Moore, G. A. (1962). The lateralis components of the acoustico-lateralis system in the sunfish family Centrarchidae. *Copeia* **1962**(1), 1–108.

Bratton, B. O., and Kramer, B. (1989). Patterns of the electric organ discharge during courtship and spawning in the mormyrid fish, *Pollymirus isidori. Behav. Ecol. Sociobiol.* **24**, 349–368.

Breder, C. M., Jr., and Rosen, D. E. (1966). *Modes of Reproduction in Fishes.* Natural History Press, Garden City, NY.

Brewster, B. (1986). A review of the genus *Hydrocynus* Cuvier, 1819 (Teleostei: Characiformes). *Bull. Br. Museum Nat. History (Zool.)* **50**(3), 163–206.

Brichard, P. (1989). *Pierre Brichard's Book of Cichlids and All the Other Fishes of Lake Tanganyika.* T. F. H., Neptune City, NJ.

Briggs, J. C. (1974). Operation of zoogeographic barriers. *Syst. Zool.* **23**, 248–256.

Briggs, J. C. (1979). Ostariophysan zoogeography: An alternative hypothesis. *Copeia* **1979**(1), 111–118.

Briggs, J. C. (1984). Freshwater fishes and biogeography of Central America and the Antilles. *Syst. Zool.* **33**, 428–435.

Briggs, J. C. (1986). Introduction to the zoogeography of North American fishes. In *The Zoogeography of North American Freshwater Fishes* (C. H. Hocutt and E. O. Wiley, Eds.), pp. 1–16. Wiley, New York.

Briggs, J. C. (1987). Antitropical distribution and evolution in the Indo-West Pacific Ocean. *Syst. Zool.* **36**, 237–247.

Briggs, J. C. (1995). *Global Biogeography.* Elsevier, Amsterdam.

Brill, J. S., Jr. (1982, January). Observations on the unique reproductive behavior of *Fundulus lima* Vaillant. A killifish from Baja California. *Freshwater Mar. Aquarium*, 1–15, 74–79, 82–87.

Britz, R. (1994). Ontogenetic features of *Luciocephalus* (Perciformes, Anabantoidei) with a revised hypothesis of anabantoid intrarelationships. *Zool. J. Linnean Soc.* **112**, 491–508.

Britz, R. (1997). Egg surface structure and larval cement glands in nandid and badid fishes with remarks on phylogeny and biogeography. *Am. Museum Novitates* **3195**, 1–17.

Britz, R., and Kottelat, M. (1999a). *Sundasalanx mekongensis*, a new species of clupeiform fish from the Mekong basin (Teleostei: Sundasalangidae). *Ichthyol. Exploration Freshwaters* **10**(4), 337–344.

Britz, R., and Kottelat, M. (1999b). Two new species of gasterosteiform fishes of the genus *Indostomus* (Teleostei: Indostomidae). *Ichthyol. Exploration Freshwaters* **10**(4), 327–336.

Britz, R., Kokoscha, M., and Riehl, R. (1995). The anabantoid genera *Ctenops, Luciocephalus, Parasphaerichthys,* and *Spaerichthys* (Teleostei: Perciformes) as a monophyletic group: Evidence from egg surface structure and reproductive behavior. *Jpn J. Ichthyol.* **42**(1), 71–79.

Brooks, D. R., Thorson, T. B., and Mayes, M. A. (1981). Freshwater stingrays (Potamotrygonidae) and their helminth parasites: Testing hypotheses of evolution and coevolution. In *Advances in Cladistics, Proceedings of the First Meeting of the Willi Hennig Society* (V. A. Funk and D. R. Brooks, Eds.), pp. 147–175. New York Botanical Garden, New York.

Brooks, J. L. (1950). Speciation in ancient lakes. *Q. Rev. Biol.* **25**, 131–176.

Broughton, R. E., and Gold, J. R. (2000). Phylogenetic relationships in the North American cyprinid genus *Cyprinella* (Actinopterygii: Cyprinidae) based on sequences of the mitochondrial ND2 and ND4L genes. *Copeia* **2000**(1), 1–10.

Brown, B. A., and Ferraris, C. J., Jr. (1988). Comparative osteology of the Asian catfish family Chacidae, with the description of a new species from Burma. *Am. Museum Novitates* **2907**, 1–16.

Brown, J. H., and Lomolino, M. V. (1998). *Biogeography*, 2nd ed. Sinauer, Sunderland, MA.

Brundin, L. (1975). Circum-Antarctic distribution patterns and continental drift. *Mém. Muséum d'Histoire Nat.* **88**, 19–27.

Bruner, J. C. (1991). Comments on the genus *Amyzon* (family Catostomidae). *J. Paleontol.* 65(4), 678–686.

Buckup, P. A. (1991). The Characidiinae: A phylgenetic study of the South American darters and their relationship with other characiform fishes. Ph.D. thesis, University of Michigan, Ann Arbor.

Buckup, P. A. (1993a). The monophyly of the Characidiinae, a Neotropical group of characiform fishes (Teleostei: Ostariophysi). *Zool. J. Linnean Soc.* 108, 225–245.

Buckup, P. A. (1993b). Review of the characidiin fishes (Teleostei: Characiformes), with descriptions of four new genera and ten new species. *Ichthyol. Exploration Freshwaters* 4(2), 97–154.

Buckup, P. A. (1993c). Phylogenetic interrelationships and reductive evolution in neotropical characidiin fishes (Characiformes, Ostariophysi). *Cladistics* 9, 305–341.

Buckup, P. A. (1998). Relationships of the Characidiinae and phylogeny of characiform fishes (Teleostei: Ostariophysi). In *Phylogeny and Classification of Neotropical Fishes* (L. R. Malabarba, R. E. Reis, R. P. Vari, Z. M. S. Lucena, and C. A. S. Lucena, Eds.), pp. 123–144. EDIPUCRS, Porto Alegre, Brazil.

Buckup, P. A., and Hahn, L. (2000). *Charadidium vestigipinne*: A new species of Characidiinae (Teleostei, Characiformes) from southern Brazil. *Copeia* 2000(1), 150–155.

Buckup, P. A, and Reis, R. E. (1997). Characidiin genus *Characidium* (Teleostei, Characiformes) in southern Brazil, with description of three new species. *Copeia* 1997(3), 531–548.

Bullock, T. H., and Heiligenberg, W. (Eds.) (1986). *Electroception*. Wiley, New York.

Burgess, G. H. (1980a). *Anchoa mitchilli* (Valenciennes) bay anchovy. In *Atlas of North American Freshwater Fishes* (D. S. Lee *et al.*, Eds.), p. 73. North Carolina State Museum of Natural History, Raleigh.

Burgess, G. H. (1980b). *Alosa* sp. In *Atlas of North American Freshwater Fishes* (D. S. Lee *et al.*, Eds.), pp. 61–68, 70. North Carolina State Museum of Natural History, Raleigh.

Burgess, G. H. (1980c). *Strongylura* marina (Walbaum). In *Atlas of North American Freshwater Fishes* (D. S. Lee *et al.*, Eds.), p. 489. North Carolina State Museum of Natural History, Raleigh.

Burgess, G. H. (1980d). *Centropomus*. In *Atlas of North American Freshwater Fishes* (D. S. Lee *et al.*, Eds.), pp. 569–572. North Carolina State Museum of Natural History, Raleigh.

Burgess, G. H. (1980e). *Trinectes maculatus* (Bloch and Schneider). In *Atlas of North American Freshwater Fishes* (D. S. Lee *et al.*, Eds.), p. 831. North Carolina State Museum of Natural History, Raleigh.

Burgess, G. H. (1983a). *Atractosteus tristoechus* (Bloch and Schneider). In *Atlas of North American Freshwater Fishes* (D. S. Lee, S. P. Platania, and G. H. Burgess, Eds.), Occasional Papers of the North Carolina Biological Survey 1983–1986, 1983 Supplement, p. 7.

Burgess, G. H. (1983b). *Lucifuga*. In *Atlas of North American Freshwater Fishes* (D. S. Lee, S. P. Platania, and G. H. Burgess, Eds.), Occasional Papers of the North Carolina Biological Survey 1983–1986, 1983 Supplement, pp. 8–10.

Burgess, G. H., and Gilbert, C. R. (1980). *Amia calva* Linneas bowfin. In *Atlas of North American Freshwater Fishes* (D. S. Lee *et al.*, Eds.), pp. 53–54. North Carolina State Museum of Natural History, Raleigh.

Burgess, G. H., and Ross, S. W. (1980). *Carcharhinus leucas* (Valenciennes) bull shark. In *Atlas of North American Freshwater Fishes* (D. S. Lee *et al.*, Eds.), p. 36. North Carolina State Museum of Natural History, Raleigh.

Burgess, W. E. (1989). *An Atlas of Freshwater and Marine Catfishes—A Preliminary Survey of the Siluriformes*. T. F. H., Neptune City, NJ.

Burgess, W. E., and Finley, L. (1996). An atlas of freshwater and marine catfishes: Update. *Tropical Fish Hobbyist* 45(2, No. 488), 163–174.

Burgess, W. E., and Walls, J. G. (1993). *Cichlasoma*: The next step. *Tropical Fish Hobbyist* 41(5), 80–82.

Burggren, W. W., and Bemis, W. E. (1992). Metabolism and ram ventilation in juvenile paddlefish *Polyodon spathula* (Chondrostei: Polyodontidae). *Physiol. Zool.* 65, 515–539.

Burns, J. R., Weitzman, S. H., Grier, H. J., and Menezes, N. A. (1995). Internal fertilization, testis and sperm morphology in Glandulocaudine fishes (Teleostei: Characidae: Glandulocaudinae). *J. Morphol.* 224, 131–145.

Burns, J. R., Weitzman, S. H., and Malabarba, L. R. (1997). Insemination in eight species of cheirodontine fishes (Teleostei: Characidae: Cheirodontinae). *Copeia* 1997(2), 433–438.

Burns, J. R., Weitzman, S. H., Lange, K. R., and Malabarba, L. R. (1999). Sperm ultrastructure in characid fishes (Teleostei: Ostariophysi). In *Phylogeny and Classification of Neotropical Fishes* (L. R. Malabarba, R. E. Reis, R. P. Vari, Z. M. S. Lucena, and C. A. S. Lucena, Eds.), pp. 236–244. EDIPUCRS, Porto Alegre, Brazil.

Burr, B. M. (1980). *Polyodon spathula* (Walbaum), paddlefish. In *Atlas of North American Freshwater Fishes* (D. S. Lee, et al., Eds.), pp. 45–46. North Carolina State Museum of Natural History, Raleigh.

Bussing, W. A. (1998). *Peces de las Aguas Continentales de Costa Rica* [Frshwater Fishes of Costa Rica] [Editorial]. Univ. of Costa Rica, San José.

Buth, D. G. (1979). Duplicate gene expression in tetraploid fishes of the tribe Moxostomatini (Cypriniformes: Catostomidae). *Comp. Biochem. Physiol. B* 63, 7–12.

Cadwallader, P. L. (1981). Past and present distributions and translocations of Macquarie perch *Macquaria australasica* (Pisces: Percichthyidae), with particular reference to Victoria. *Proc. R. Soc. Victoria* 93, 23–30.

Cadwallader, P. L., and Backhouse, G. N. (1983). *A Guide to the Freshwater Fish of Victoria.* Victorian Government Printing Office, Melbourne.

Camerini, J. R. (1993). Evolution, biogeography, and maps: An early history of Wallace's Line. *Isis* 84, 700–727.

Cameron, A. M., and Endean, R. (1970). Venom glands in scatophagid fish. *Toxicon* 8, 171–178.

Campos, H., Arratia, G., and Cuevas, C. (1997). Karyotypes of the most primitive catfishes (Teleostei: Siluriformes: Diplomystidae). *J. Zool. Syst. Evol. Res.* 35, 113–119.

Campos-da-Paz, R. (1995). Revision of the South American freshwater fish genus *Sternarchorhamphus* Eigenmann, 1905 (Ostariophysi: Gymnotiformes: Apteronotidae), with notes on its relationships. *Proc. Biol. Soc. Washington* 108(1), 29–44.

Campos-da-Paz, R. (1996). Redescription of the Central American electric fish *Gymnotus cylindricus* (Ostariophysi: Gymnotiformes: Gymnotidae), with comments on character ambiguity within the ostariophysan clade. *J. Zool. London* 240, 371–382.

Campos-da-Paz, R. (1999). New species of *Megadontognathus* from the Amazon basin, with phylogenetic and taxonomic discussions on the genus (Gymnotiformes: Apteronotidae). *Copeia* 1999(4), 1041–1049.

Campos-da-Paz, R., and Albert, J. S. (1998). The gymnotiform "eels" of Tropical America: A history of classification and phylogeny of the South American electric knifefishes (Teleostei: Ostariophysi: Siluriphysi). In *Phylogeny and Classification of Neotropical Fishes* (L. R. Malabarba, R. E. Reis, R. P. Vari, Z. M. S. Lucena, and C. A. S. Lucena, Eds.), pp. 401–417. EDIPUCRS, Porto Alegre, Brazil.

Campos-da-Paz, R., and Costa, W. J. E. M. (1996). *Gymnotus bahianus* sp. nov., a new gymnotid fish from eastern Brazil (Teleostei: Ostariophysi: Gymnotiformes), with evidence for the monophyly of the genus. *Copeia* 1996(4), 937–944.

Cantor, T. E. (1849). Catalogue of Malayan fishes. *J. Asiatic Soc. Bengal* 18(2), 983–1443. (1966 reprint, A. Asher, Amsterdam)

Carlander, K. D. (1977). *Handbook of Freshwater Fishery Biology*, Vol. 2. Iowa State Univ. Press, Ames.

Carlson, D. M., and Bonislawsky, P. S. (1981). The paddlefish (*Polyodon spathula*) fisheries of the midwestern United States. *Fisheries* 6(2), 17–27.

Carter, G. S., and Beadle, L. C. (1930). Notes on the habits and development of *Lepidosiren paradoxa. J. Linnean Soc. Zool.* 37, 197–203.

Casciotta, J., and Arratia, G. (1993a). Tertiary cichlid fishes from Argentina and reassessment of the phylogeny of New World cichlids (Perciformes: Labroidei). *Kaupia Darmstädter Beiträge Naturgeschichte* 2, 195–240.

Casciotta, J., and Arratia, G. (1993b). Jaws and teeth of American cichlids (Pisces: Labroidei). *J. Morphol.* 217, 1–36.

Cashner, R. C. (1980). *Ambloplites.* In *Atlas of North American Freshwater Fishes* (D. S. Lee *et al.*, Eds.), pp. 578–581. North Carolina State Museum of Natural History, Raleigh.

Cashner, R. C., and Jenkins, R. E. (1982). Systematics of the Roanoke bass, *Ambloplites cavifrons. Copeia* 1982(3), 581–594.

Cashner, R. C., and Suttkus, R. D. (1977). *Ambloplites constellatus*, a new species of rock bass from the Ozark Upland of Arkansas and Missouri with a review of western rock bass populations. *Am. Midland Nat.* 98, 147–161.

Cashner, R. C., Burr, B. M., and Rogers, J. S. (1989). Geographic variation of the mud sunfish, *Acantharchus pomotis* (family Centrarchidae). *Copeia* 1989(1), 129–141.

Cashner, R. C., Rogers, J. S., and Grady, J. M. (1998). *Fundulus bifax*, a new species of the subgenus *Xenisma* from the Tallapoosa and Coosa River systems of Alabama and Georgia. *Copeia* 1988(3), 674–683.

Castex, M. N. (1967). Fresh water venomous rays. In *Animal Toxins* (F. E. Russel and P. R. Saunders, Eds.), pp. 167–176. Pergamon, Oxford.

Castro, J. I. (1983). *The Sharks of North American Waters.* Texas A & M Univ. Press, College Station.

Cavender, T. M. (1969). An Oligocene mudminnow (family Umbridae) from Oregon with remarks on relationships within the Esocoidei, Occasional Papers of the Museum of Zoology, University of Michigan, No. 660.

Cavender, T. M. (1986). Review of the fossil history of North American freshwater fishes. In *The Zoogeography of North American Freshwater Fishes* (C. H. Hocut and E. O. Wiley, Eds.), pp. 699–724. Wiley, New York.

Cavender, T. M. (1991). The fossil record of the Cyprinidae. In *Cyprinid Fishes: Systematics,*

Biology, and Exploitation (I. J. Winfield and J. S. Nelson, Eds.), pp. 34–54. Chapman & Hall, London.

Cavender, T. M., and Coburn, M. M. (1992). Phylogenetic relationships of North American Cyprinidae. In *Systematics, Historical Ecology, and North American Freshwater Fishes* (R. L. Mayden, Ed.), pp. 293–327. Stanford Univ. Press, Stanford, CA.

Cederholm, C. J., Kunze, M. D., Murota, T., and Sibatani, A. (1999). Pacific salmon carcasses: Essential contributions of nutrients and energy for aquatic and terrestrial ecosystems. *Fisheries* 24(10), 6–15.

Chao, N. L. (1986a). A synopsis on zoogeography of the Sciaenidae. In *Indo-Pacific Fish Biology: Proceedings of the Second International Conference on Indo-Pacific Fishes* (T. Uyeno, R. Arai, T. Taniuchi, and K. Matsuura, Eds.), pp. 570–589. Ichthyological Society of Japan, Tokyo.

Chao, N. L. (1986b). Sciaenidae. In *Fishes of the North-Eastern Atlantic and the Mediterranean* (P. J. P. Whitehead, M.-L. Bauchot, J.-C. Hureau, J. Nielsen, and E. Tortonese, Eds.), Vol. 2, pp. 865–874. UNESCO, Paris.

Chapleau, F. (1988). Comparative osteology and intergeneric relationships of the tongue soles (Pisces; Pleuronectiformes; Cynoglossidae). *Can. J. Zool.* 66, 1214–1232.

Chapleau, F. (1993). Pleuronectiform relationships: A cladistic reassessment. *Bull. Mar. Sci.* 52(1), 515–540.

Chapleau, F., and Desoutter, M. (1996). Position phylogenetique de *Dagetichthys lakdoensis* (Pleuronectiformes). *Cybium* 20, 103–106.

Chapleau, F., and Keast, A. (1988). A phylogenetic reassessment of the monophyletic status of the family Soleidae, with comments on the suborder Soleoidei (Pisces; Pleuronectiformes). *Can. J. Zool.* 66, 2797–2810.

Chapman, L. J., and Chapman, C. A. (1998). Hypoxia tolerance of the mormyrid *Petrocephalus catostoma*: Implications for persistence in swamp refugia. *Copeia* 1998(3), 762–768.

Chardon, M. (1967). Réflexions sur la dispersion des Ostariophysi à la lumière de recherches morphologiques nouvelles. *Ann. Soc. R. Zool. Belgium* 97(3), 175–186.

Chardon, M. (1968). Anatomie comparée de l'appareil de Weber, et des structures connexes chez les Siluriformes. *Ann. Musee R. l'Afrique Central Sci. Zool.* 169, 1–277.

Charlebois, P. M., Marsden, J. E., Goettel, R. G., Wolfe, R. K., Jude, D. J., and Rudnicka, S. (1997). The round goby, *Neogobius melanostomus* (Pallas), a review of European and North

American literature, Special Publication No. 20. Illinois Natural History Survey.

Chen, I.-S., Hsu, C.-H., Hui, C.-F., Shao, K.-T., Miller, P. J., and Fang, L.-S. (1998). Sequence length and variation in the mitochondrial control region of two freshwater gobiid fishes, belonging to *Rhinogobius* (Teleostei: Gobioidei). *J. Fish Biol.* 53, 179–191.

Chen, X. (1994). Phylogenetic studies of the amblycipitid catfishes (Teleostei, Siluriformes) with species accounts. Ph.D. dissertation, Duke University, Durham, NC.

Chen, X., and Arratia, G. (1996). Breeding tubercles of *Phoxinus* (Teleostei: Cyprinidae): Morphology, distribution, and phylogenetic implications. *J. Morphol.* 228, 127–144.

Chen, X., and Lundberg, J. G. (1995). *Xiurenbagrus*, a new genus of amblycipitid catfishes (Teleostei: Siluriformes), and phylogenetic relationships among the genera of Amblycipitidae. *Copeia* 1995(4), 780–800.

Chereshnev, I. A. (1990). Ichthyofauna composition and features of freshwater fish distribution in northeastern USSR. *J. Ichthyol.* 30, 110–121.

Chernoff, B. (1986a). Phylogenetic relationships and reclassification of menidiine silverside fishes with emphasis in the tribe Membradini. *Proc. Acad. Nat. Sci. Philadelphia* 138(1), 189–249.

Chernoff, B. (1986b). Systematics of American atherinid fishes of the genus *Atherinella* I. The subgenus *Atherinella*. *Proc. Acad. Nat. Sci. Philadelphia* 138(1), 86–188.

Christensen, P. (1982). The distribution of *Lepidogalaxias salamandroides* and other small fresh-water fishes in the lower south-west of Western Australia. *J. R. Soc. Western Austr.* 65, 131–141.

Clausen, H. S. (1959). Denticipitidae, a new family of primitive isospondylous teleosts from West African freshwaters. *Vidensk. Meddr. Dansk. Naturh. Foren.* 121, 141–151.

Clay, W. M. (1975). *The Fishes of Kentucky*. Kentucky Department of Fish and Wildlife Resources, Frankfort.

Clements, J. (1988). *Salmon at the Antipodes*. Published by the author, Ballarat, Victoria.

Cloutier, R., and Ahlberg, P. E. (1996). Morphology, characters, and the interrelationships of basal sarcopterygians. In *Interrelationships of Fishes* (M. L. J. Stiassny, L. R. Parenti, and G. D. Johnson, Eds.), pp. 445–479. Academic Press, San Diego.

Clutton-Brock, T. H. (1991). *The Evolution of Parental Care*. Princeton Univ. Press, Princeton, NJ.

Coad, B. W. (1981a). *Glyptothorax silviae*, a new species of sisorid catfish from southwestern Iran. *Jpn J. Ichthyol.* 27(4), 291–295.

Coad, B. W. (1981b). A bibliography of the stickle-backs. *Syllogeus* 35, 1–142. (National Museum of Canada)

Coad, B. W. (1982). A new genus and species of cichlid endemic to southern Iran. *Copeia* 1982(1), 28–37.

Coad, B. W., and Delmastro, G. B. (1985). Notes on a sisorid catfish from the Black Sea drainage of Turkey. *Cybium* 9(3), 221–224.

Coates, C. W. (1937). Slowly the lungfish gives up its secrets. *Bull. New York Zool. Soc.* 40, 25–34.

Coates, D. (1993). Fish ecology and management of the Sepik-Ramy, New Guinea, a large contemporary tropical river basin. *Environ. Biol. Fishes* 38, 345–368.

Coburn, M. M., and Cavender, T. M. (1992). Interrelationships of North American cyprinid fishes. In *Systematics, Historical Ecology, and North American Freshwater Fishes* (R. L. Mayden, Ed.), pp. 328–373. Stanford Univ. Press, Stanford, CA.

Coburn, M. M., and Gaglione, J. I. (1992). A comparataive study of percid scales (Teleostei: Perciformes). *Copeia* 1992(4), 986–1001.

Cohen, D. M. (1970). How many recent fishes are there? *Proc. California Acad. Sci.* 38, 341–346.

Cohen, D. M. (1984). Gadiformes: An overview. In *Ontogeny and Systematics of Fishes* (H. G. Moser *et al.*, Eds.), Special Publication No. 1, pp. 259–265. American Society of Ichthyologists and Herpetologists, Lawrence, KS.

Cohen, D. M., and McCosker, J. E. (1998). A new species of bythitid fish, genus *Lucifuga*, from the Galápagos Islands. *Bull. Mar. Sci.* 63(1), 179–187.

Cohen, D. M., and Nielsen, J. G. (1978). Guide to the identification of genera of the fish order Ophidiiformes with a tentative classification of the order, NOAA Technical Report, National Marine Fisheries Service Circular 417.

Cohen, D. M., and Robins, C. R. (1970). A new ophidioid fish (genus *Lucifuga*) from a limestone sink, New Providence Island, Bahamas. *Proc. Biol. Soc. Washington* 83(11), 133–144.

Cohen, D. M., Inada, T., Iwamoto, T., and Scialabba, N. (1990). *FAO Species Catalogue. Vol. 10. Gadiform Fishes of the World (Order Gadiformes)*, FAO Fisheries Synopsis No. 125. Food and Agriculture Organization of the United Nations, Rome.

Colbert, E. H. (1973). *Wandering Land and Animals.* Dutton, New York.

Colby, P. J., McNicol, R. E., and Ryder, R. A. (1979). Synopsis of biological data on the Walleye *Stizostedion v. vetreum* (Mitchill 1818), FAO Fisheries Synopsis 119. Food and Agriculture Organization of the United Nations, Rome.

Collette, B. B. (1963). The subfamilies, tribes, and genera of the Percidae (Teleostei). *Copeia* 1963(4), 615–623.

Collette, B. B. (1966a). A review of the venomous toadfishes, subfamily Thalassophryninae. *Copeia* 1966(4), 846–864.

Collette, B. B. (1966b). *Belonion*: a new genus of fresh-water needlefishes from South America. *Am. Museum Novitates* 2274, 1–22.

Collette, B. B. (1968). *Strongylura timucu* (Walbaum): A valid species of western Atlantic needlefish. *Copeia* 1968(1), 189–192.

Collette, B. B. (1973). *Daector quadrizonatus*, a valid species of freshwater venomous toadfish from the Rio Truandó, Columbia with notes on additional material of other species of *Daector*. *Copeia* 1973(2), 355–357.

Collette, B. B. (1974a). *Strongylura hubbsi*, a new species of freshwater needlefish from the Usumacinta Province of Guatemala and México. *Copeia* 1974(3), 611–619.

Collette, B. B. (1974b). *Potamorrhaphis petersi*, a new species of freshwater needlefish (Belonidae) from the Upper Orinoco and Rio Negro. *Proc. Biol. Soc. Washington* 87(5), 31–40.

Collette, B. B. (1974c). South American freshwater needlefishes (Belonidae) of the genus *Pseudotylosurus*. *Zool. Mededelingen* 48(16), 169–186.

Collette, B. B. (1974d). The garfishes (Hemiramphidae) of Australia and New Zealand. *Rec. Austr. Museum* 29(2), 11–105.

Collette, B. B. (1974e). Geographic variation in the Central Pacific halfbeak, *Hyporhamphus acutus* (Günther). *Pacific Sci.* 28(2), 111–122.

Collette, B. B. (1976). Indo-west Pacific halfbeaks (Hemiramphidae) of the genus *Rhychorhamphus* with descriptions of two new species. *Bull. Mar. Sci.* 26(1), 72–98.

Collette, B. B. (1977). Epidermal breeding tubercles and bony contact organs in fishes. *Symp. Zool. Soc. London* 39, 225–268.

Collette, B. B. (1981). Rediscovery of *Hypporhamphus xanthopterus*, a halfbeak endemic to Vembanad Lake, Kerala, southern India. *Matsya* 7, 29–40.

Collette, B. B. (1982a). South American freshwater needlefishes of the genus *Potamorrhaphis* (Beloniformes: Belonidae). *Proc. Biol. Soc. Washington* 95(4), 714–747.

Collette, B. B. (1982b). Two new species of freshwater halfbeaks (Pisces: Hemiramphidae) of the genus *Zenarchopterus* from New Guinea. *Copeia* 1982(2), 265–276.

Collette, B. B. (1985). *Zenarchopterus ornithocephala*, a new species of freshwater halfbeak

(Pisces: Hemiramphidae) from the Vogelkop Peninsula of New Guinea. *Proc. Biol. Soc. Washington* **98**(1), 107–111.

Collette, B. B. (1986). Hemiramphidae. In *Check-list of the Freshwater Fishes of Africa* (J. Daget, J.-P. Gosse, and D. F. E. Thys van den Audenaerde, Eds.), pp. 163–164. CLOFFA II, ORSTOM, Paris.

Collette, B. B. (1995a). *Tondanichthys kottelati*, a new genus and species of freshwater halfbeak (Teleostei: Hemiramphidae) from Sulawesi. *Ichthyol. Exploration Freshwater* **6**(2), 171–174.

Collette, B. B. (1995b). *Potamobatrachus trispinosus*, a new freshwater toadfish (Batrachoididae) from the Rio Tocantins, Brazil. *Ichthyol. Exploration Freshwaters* **6**(4), 333–336.

Collette, B. B., and Banarescu, P. (1977). Systematics and zoogeography of the fishes of the family Percidae. *J. Fish. Res. Board Can.* **34**(10), 1450–1463.

Collette, B. B., and Berry, F. H. (1965). Recent studies on the needlefishes (Belonidae): An evaluation. *Copeia* **1965**(3), 386–392.

Collette, B. B., and Knapp, L. W. (1967). Catalog of type specimens of the darters (Pisces, Etheostomatini). *Proc. U.S. Natl. Museum* **119**, 1–88.

Collette, B. B., and Parin, N. V. (1986a). Belonidae. In *Fishes of the North-Eastern Atlantic and the Mediterranean* (P. J. P. Whitehead, M.-L. Bauchot, J.-C. Hureau, J. Nielsen, and E. Tortonese, Eds.), Vol. 2, pp. 604–619. UNESCO, Paris.

Collette, B. B., and Parin, N. V. (1986b). Belonidae.In *Fishes of the North-Eastern Atlantic and the Mediterranean* (P. J. P. Whitehead, M.-L. Bauchot, J.-C. Hureau, J. Nielsen, and E. Tortonese, Eds.), Vol. 2, pp. 620–622. UNESCO, Paris.

Collette, B. B., and Parin, N. V. (1994). Flyingfishes and their allies. In *Encyclopedia of Fishes* (J. P. Paxton and W. N. Eschmeyer, Eds.), pp. 144–147. Academic Press, San Diego.

Collette, B. B., and Russo, J. L. (1981). A revision of the scaly toadfishes, genus *Batrachoides*, with descriptions of two new species from eastern Atlantic. *Bull. Mar. Sci.* **31**(2), 197–233.

Collette, B. B., and Su, J. (1986). The halfbeaks (Pisces, Beloniformes, Hemiramphidae) of the Far East. *Proc. Acad. Nat. Sci. Philadelphia* **138**(1), 250–301.

Collette, B. B., Ali, M. A., Hokanson, K., Nagiec, M., Smirnov, S. A., Thorpe, J. E., Weatherley, A. H., and Willemsen, J. (1977). Biology of the percids. *J. Fish. Res. Board Can.* **34**(10), 1890–1899.

Collette, B. B., McGowen, G. E., Parin, N. V., and Mito, S. (1984). Beloniformes: Development and relationships. In *Ontogeny and Systematics of Fishes* (H. G. Moser *et al.*, Eds.), Special Publication No. 1, pp. 335–354. American Society of Ichthyologists and Herpetologists, Lawrence, KS.

Collin, H. B., and Collin, S. P. (1996). The fine structure of the cornea of the salamanderfish, *Lepidogalaxias salamandroides* (Lepidogalaxiidae, Teleostei). *Cornea* **15**(4), 414–426.

Compagno, L. J. V. (1984). *FAO Species Catalogue. Vol. 4. Sharks of the World. Part 2. Carcharhiniformes*, FAO Fisheries Synopsis No. 125, pp. 251–655. Food and Agriculture Organization of the United Nations, Rome.

Compagno. L. J. V. (1999a). Systematics and body form. In *Shark, Skates, and Rays: The Biology of Elasmobranch Fishes* (W. C. Hamlet, Ed.), pp. 1–42. Johns Hopkins Univ. Press, Baltimore.

Compagno. L. J. V. (1999b). Checklist of living elasmobranchs. In *Shark, Skates, and Rays: The Biology of Elasmobranch Fishes* (W. C. Hamlet, Ed.), pp. 471–498. Johns Hopkins Univ. Press, Baltimore.

Compagno, L. J. V., and Roberts, T. R. (1982). Freshwater stingrays (Dasyatidae) of Southern Asia and New Guinea, with description of a new species of *Himantura* and reports of unidentified species. *Environ. Biol. Fishes* **7**(4), 321–339.

Compagno, L. J. V., and Roberts, T. R. (1984). Marine and freshwater stingrays (Dasyatidae) of West Africa, with description of a new species. *Proc. California Acad. Sci.* **43**, 283–300.

Conant, E. B. (1987). Bibliography of lungfishes, 1811–1985. In *The Biology and Evolution of Lungfishes* (W. E. Bemis, W. W. Burggren, and N. E. Kemp, Eds.), pp. 305–373. A. R. Liss, New York.

Conniff, R. (1999). Relax, its only a piranha. *Smithsonian* **30**(4), 42–50.

Constantz, G. D. (1989). Reproductive biology of poeciliid fishes. In *Ecology and Evolution of Livebearing Fishes* (G. K. Meffe and F. F. Snelson, Jr., Eds.), pp. 33–50. Prentice Hall, Englewood Cliffs, NJ.

Cooper, J. E. (1980). Amblyopsidae. In *Atlas of North American Freshwater Fishes* (D. S. Lee *et al.*, Eds.), pp. 478–483. North Carolina State Museum of Natural History, Raleigh.

Cooper, J. E., and Kuehne, R. A. (1974). *Speoplatyrhinus poulsoni*, a new genus and species of subterranean fish from Alabama. *Copeia* **1974**(2), 486–493.

Cooper, J. E., and Longley, G. (1980). *Satan eurystomus* Hubbs and Bailey & *Trogloglanis pattersoni* Eigenmann. In *Atlas of North American*

Freshwater Fishes (D. S. Lee *et al.*, Eds.), pp. 473–474. North Carolina State Museum of Natural History, Raleigh.

Copeland, J. W., and Grey, D. L. (Eds.) (1987). *Management of Wild and Cultured Sea Bass/Barramundi (Lates calcarifer)*. Australian Centre for International Agricultural Research, Canberra.

Copley, H. (1941). A short account on the fresh water fishes of Kenya. *J. East African Nat. History Soc.* 16, 1–24.

Costa, W. J. E. M. (1991). Systematics and distribution of the neotropical annual fish genus *Plesiolebias* (Cyprinodontiformes: Rivulidae), with description of a new species. *Ichthyol. Exploration Freshwater* 1(4), 369–378.

Costa, W. J. E. M. (1996). Relationships, monophyly and three new species of neotropical miniature poeciliid genus *Fluviphylax* (Cyprinodontiformes: Cyprinodontoidei). *Ichthyol. Exploration Freshwaters* 7(2), 111–130.

Costa, W. J. E. M. (1998a). Phylogeny and classification of Rivulidae revisited: Origin and evolution of annualism and miniaturization in rivulid fishes (Cyprinodontiformes: Aplocheiloidei). *J. Comp. Biol.* 3(1), 33–92.

Costa, W. J. E. M. (1998b). Phylogeny and classification of the Cyprinodontiformes (Euteleostei: Atherinomorpha): A reappraisal. In *Phylogeny and Classification of Neotropical Fishes* (L. R. Malabarba, R. E. Reis, R. P. Vari, Z. M. S. Lucena, and C. A. S. Lucena, Eds.), pp. 537–560. EDIPUCRS, Porto Alegre, Brazil.

Costa, W. J. E. M., and Brasil, G. C. (1991a). Three new species of *Cynolebias* (Cyprinodontiformes: Rivulidae) from the São Francisco basin, Brazil. *Ichthyol. Exploration Freshwater* 2(1), 55–62.

Costa, W. J. E. M., and Brasil, G. C. (1991b). Description of a new species of *Rivulus* (Cyprinodontiformes: Rivulidae) from the coastal plains of eastern Brazil. *Ichthyol. Exploration Freshwater* 1(4), 379–383.

Costa, W. J. E. M., and Campos-da-Paz, R. (1992). Description d'une nouvelle espèce de poisson électrique du genre néotropical *Hypopomus* (SiluriformesL Gymnotoidei: Hypopomidae) du sud-est de Brésil. *Rev. Fr. Aquariol.* 18(4), 117–120.

Costa, W. J. E. M., and Le Bail, P.-Y. (1999). *Fluviphylax palikur*: A new poeciliid from the Rio Oiapoque basin, northern Brazil (Cyprinodontiformes: Cyprinodontoidei), with comments on miniaturization in *Fluviphylax* and other neotropical freshwater fishes. *Copeia* 1999(4), 1027–1034.

Courtenay, W. R., Jr. (1978). Additional range expansion in Florida of the introduced walking catfish. *Environ. Conserv.* 5, 273–276.

Courtenay, W. R., Jr., and Meffe, G. K. (1989). Small fishes in strange places: A review of introduced poeciliids. In *Ecology and Evolution of Livebearing Fishes* (G. K. Meffe and F. F. Snelson, Jr., Eds.), pp. 319–331. Prentice Hall, Englewood Cliffs, NJ.

Courtenay, W. R., Jr., and Stauffer, J. R., Jr. (Eds.) (1984). *Distribution, Biology, and Management of Exotic Fishes*. Johns Hopkins Univ. Press, Baltimore.

Courtenay, W. R., Jr., and Stauffer, J. R., Jr. (1990). The introduced fish problem and the aquarium fish industry. *J. World Aquaculture Soc.* 21, 145–159.

Courtenay, W. R., Jr., Hensley, D. A., Taylor, J. N., and McCann, J. A. (1984). Distribution of exotic fishes in the continental United States. In *Distribution, Biology, and Management of Exotic Fishes* (W. R. Courtenay, Jr., and J. R. Stauffer, Jr., Eds.), pp. 41–77. Johns Hopkins Univ. Press, Baltimore.

Cox, C. B. (1974). Vertebrate paleodistributional paterns and continental drift. *J. Biogeogr.* 1, 75–94.

Cox, C. B., and Moore, P. D. (1993). *Biogeography*, 5th ed. Blackwell, Oxford.

Crabtree, C. B. (1987). Allozyme evidence for the phylogenetic relationships within the silverside subfamily Atherinopsinae. *Copeia* 1987(4), 860–867.

Cracraft, J. (1974). Continental drift and vertebrate distribution. *Annu. Rev. Ecol. Syst.* 5, 215–261.

Cracraft, J. (1975). Mesozoic dispersal of terrestrial faunas around the southern end of the world. *Mémoires Muséum Natl. d'Histoire Nat.* 88, 29–52.

Craig, J. F. (1987). *The Biology of Perch and Related Fish*. Croom Helm, Kent, UK.

Craig, J. F. (Ed.) (1996). *Pike: Biology and Exploitation*. Chapman & Hall, London.

Craw, R. C. (1979). Generalized tracks and dispersal in biogeography: A response to R. M. McDowall. *Syst. Zool.* 28(1), 99–107.

Crawford, J. D. (1997). Hearing and acoustic communication in mormyrid electric fishes. *Mar. Freshwater Behav. Physiol.* 29, 65–86.

Crawford, J. D., and Hopkins, C. D. (1989). Detection of a previously unrecognized mormyrid fish (*Mormyrus subundulatus*) by electric discharge characters. *Cybium* 13, 319–326.

Crawford, R. H. (1974). Structure of an air-breathing organ and the swim bladder in the Alaskan blackfish, *Dallia pectoralis* Bean. *Can. J. Zool.* 52, 1221–1225.

Cressey, R. F., and Collette, B. B. (1971). Copepod and needlefishes: A study in host–parasite relationships. *Fish. Bull.* 68(3), 347–432.

Croizat, L., Nelson, G. J., and Rosen, D. E. (1974). Centers of origin and related concepts. *Syst. Zool.* **23**(2), 265–287.

Cross, F. B. (1967). *Handbook of Fishes of Kansas,* Publication No. 45. Museum of Natural History, University of Kansas, Lawrence.

Crossman, E. J. (1966). A taxonomic study of *Esox americanus* and its subspecies in eastern North America. *Copeia* **1966**(1), 1–20.

Crossman, E. J. (1978). Taxonomy and distribution of North American esocids. In *Selected Coolwater Fishes of North America* (R. L. Kendall, Ed.), Special Publication No. 11, pp. 13–26. American Fisheries Society, Washington, DC.

Crossman, E. J. (1980). *Esox.* In *Atlas of North American Freshwater Fishes* (D. S. Lee *et al.,* Eds.), pp. 131–138. North Carolina State Museum of Natural History, Raleigh.

Crossman, E. J., and Casselman, J. M. (1987). *An Annotated Bibliography of the Pike, Esox lucius (Osteichthyes: Salmoniformes).* Royal Ontario Museum, Toronto.

Crossman, E. J., and Ráb, P. (1996). Chromosome-banding study of the Alaska blackfish, *Dallia pectoralis* (Euteleostei: Esocae), with implications for karyotype evolution and relationship of esocoid fishes. *Can. J. Zool.* **74**, 147–156.

Crowley, L. E. L. M., and Ivantsoff, W. (1991). Genetic similarity among populations of rainbowfishes (Pisces: Melanotaeniidae) from Atherton Tableland, Northern Queensland. *Ichthyol. Exploration Freshwaters* **2**(2), 129–137.

Cunningham, J. T. (Ed.) (1912). *Reptiles, Amphibia, Fishes and Lower Chordata.* Methuen, London.

Cunningham, J. T., and Reid, D. M. (1933). Pelvic filaments of *Lepidosiren. Nature* **131**, 913.

Curran, D. J. (1989). Phylogenetic relationships among the catfish genera of the family Auchenipteridae (Teleostei: Siluroidea). *Copeia* **1989**(2), 408–419.

Daget, J. (1984a). Osteoglossidae. In *Check-list of the Freshwater Fishes of Africa* (J. Daget, J.-P. Gosse, and D. F. E. Thys van den Audenaerde, Eds.), pp. 57–58. CLOFFA I, ORSTOM, Paris.

Daget, J. (1984b). Notopteridae. In *Check-list of the Freshwater Fishes of Africa* (J. Daget, J.-P. Gosse, and D. F. E. Thys van den Audenaerde, Eds.), pp. 61–62. CLOFFA I, ORSTOM, Paris.

Daget, J. (1984c). Denticipitidae. In *Check-list of the Freshwater Fishes of Africa* (J. Daget, J.-P. Gosse, and D. F. E. Thys van den Audenaerde, Eds.), p. 40. CLOFFA I, ORSTOM, Paris.

Daget, J. (1984d). Citharinidae. In *Check-list of the Freshwater Fishes of Africa* (J. Daget, J.-P. Gosse, and D. F. E. Thys van den Audenaerde, Eds.), pp. 212–216. CLOFFA I, ORSTOM, Paris.

Daget, J. (1986a). Centropomidae. In *Check-list of the Freshwater Fishes of Africa* (J. Daget, J.-P. Gosse, and D. F. E. Thys van den Audenaerde, Eds.), pp. 293–296. CLOFFA II, ORSTOM, Paris.

Daget, J. (1992a). Ariidae. In *Faune des Poissons d'eaux Douces et Saumâtres de l'Afrique de l'Ouest* (C. Lévêque, D. Paugy, and G. G. Teugels, Eds.), Vol. 2, pp. 564–568. ORSTOM, Paris.

Daget, J. (1992b). Synbranchidae. In *Faune des Poissons d'eaux Douces et Saumâtres de l'Afrique de l'Ouest* (C. Lévêque, D. Paugy, and G. G. Teugels, Eds.), Vol. 2, pp. 659–661. ORSTOM, Paris.

Daget, J. (1992c). Polymenidae. In *Faune des Poissons d'eaux Douces et Saumâtres de l'Afrique de l'Ouest* (C. Lévêque, D. Paugy, and G. G. Teugels, Eds.), Vol. 2, pp. 792–795. ORSTOM, Paris.

Daget, J., and Gosse, J.-P. (1984). Distichodontidae. In *Check-list of the Freshwater Fishes of Africa* (J. Daget, J.-P. Gosse, and D. F. E. Thys van den Audenaerde, Eds.), pp. 184–211. CLOFFA I, ORSTOM, Paris.

Daget, J., and Njock, J. C. (1986). Polynemidae. In *Check-list of the Freshwater Fishes of Africa* (J. Daget, J.-P. Gosse, and D. F. E. Thys van den Audenaerde, Eds.), pp. 352–354. CLOFFA II, ORSTOM, Paris.

Daget, J., and Smith, C. L. (1986). Serranidae. In *Check-list of the Freshwater Fishes of Africa* (J. Daget, J.-P. Gosse, and D. F. E. Thys van den Audenaerde, Eds.), pp. 299–303. CLOFFA II, ORSTOM, Paris.

Daget, J., Gosse, J.-P., Teugels, G. G., and Thys van den Audenaerde, D. F. E. (Eds.) (1991). *Check-list of the Freshwater Fishes of Africa.* CLOFFA IV, ORSTOM, Paris.

Daldorf, L. (1797). Natural history of *Perca scandens. Trans. Linnean Soc.* **3**, 62–63.

Darlington, P. J., Jr. (1957). *Zoogeography.* Wiley, New York.

Davis, T. L. O. (1977a). Age determination and growth of the freshwater catfish, *Tandanus tandanus* Mitchell, in the Gwydir River, Australia. *Austr. J. Mar. Freshwater Res.* **28**(2), 119–137.

Davis, T. L. O. (1977b). Reproductive biology of the freshwater catfish, *Tandanus tandanus* Mitchell, in the Gwydir River, Australia. I. Structure of the gonads. *Austr. J. Mar. Freshwater Res.* **28**(2), 139–158.

Davis, T. L. O. (1977c). Reproductive biology of the freshwater catfish, *Tandanus tandanus* Mitchell, in the Gwydir River, Australia. II. Gonadal cycle and fecundity. *Austr. J. Mar. Freshwater Res.* **28**(2), 159–169.

Davis, T. L. O. (1977d). Food habits of the freshwater catfish, *Tandanus tandanus* Mitchell, in the Gwydir River, Australia, and effects associated with impoundment of this river by the Copleton Dam. *Austr. J. Mar. Freshwater Res.* 28(4), 455–465.

Davis, W. (1983). The ethnobiology of the Haitian zombi. *J. Ethnopharmacol.* 9, 85.

Davis, W. (1985). *The Serpent and the Rainbow.* Simon & Schuster, New York.

Davis, W. (1988a). *Passage of Darkness.* Univ. of North Carolina Press, Chapel Hill.

Davis, W. (1988b). Zombification. *Science* 240, 1715–1716.

Davis, W. P., Taylor, D. S., and Turner, B. J. (1995). Does the autecology of the mangrove rivulus fish (*Rivulus marmoratus*) reflect a paradigm for mangrove ecosystem sensitivity? *Bull. Mar. Sci.* 57(1), 208–214.

Dawes, J. (1991). *Livebearing Fishes.* Blandford, London.

Dawley, R. M., and Bogart, J. P. (Eds.) (1989). Evolution and ecology of unisexual vertebrates. *New York State Museum Bull.* 466, 1–302.

Daws, G., and Fujita, M. (1999). *Archipelago: The Islands of Indonesia.* Univ. of California Press, Berkeley.

Dawson, C. E. (1981). Review of the Indo-Pacific doryrhamphine pipefish genus *Doryichthys. Jpn J. Ichthyol.* 28, 1–18.

Dawson, C. E. (1982). Fishes of the Western North Atlantic. Family Sygnathidae. The pipefishes. *Sears Foundation Mar. Res. Memoir* 1(8), 1–172.

Dawson, C. E. (1985). *Indo-Pacific Pipefishes (Red Sea to the Americas).* Gulf Coast Research Laboratory, Ocean Springs, MS.

Dawson, C. E. (1986). Syngnathidae. In *Fishes of the North-eastern Atlantic and the Mediterranean* (P. J. P. Whitehead, M.-L. Bauchot, J.-C. Hureau, J. Nielsen, and E. Tortonese, Eds.), Vol. 2, pp. 628–639. UNESCO, Paris.

Day, F. (1878b). *The Fishes of India,* Vol. 1. Bernard Quaritch, London. (Reprint 1971, Today and Tomorrow's Book Agency, New Delhi)

Day, F. (1878b). *The Fishes of India; Being a Natural History of the Fishes Known to Inhabit the Seas and Fresh Waters of India, Burma, and Ceylon,* Vol. 2. Bernard Quaritch, London. (198 plates)

DeBeaufort, L. F. (1914). Die anatomie und systematische stellung des genus *Kurtis* Bloch. *Gegenbaurs Morphol. Jahrbuch* 48, 391–410.

DeBeaufort, L. F. (1951). *The Fishes of the Indo-Australian Archipelago. IX Percomorphi (Concluded, Blennoidea).* Brill, Leiden.

Dekkers, W. J. (1975). Review of the Asiatic freshwater puffers of the genus *Tetraodon* Linnaeus, 1758

(Pisces, Tetraodontiformes, Tetraodontidae). *Bijdrage Dierkunde* 45(1), 87–142.

DeMartini, E. E. (1969). A correlative study of the ecology and comparative feeding mechanism morphology of the Embiotocidae (surfperches) as evidence of the family's adaptive radiation into available ecological niches. *Wassman J. Biol.* 27(2), 177–247.

Desoutter, M. (1986a). Monodactylidae. In *Check-list of the Freshwater Fishes of Africa* (J. Daget, J.-P. Gosse, and D. F. E. Thys van den Audenaerde, Eds.), pp. 338–339. CLOFFA II, ORSTOM, Paris.

Desoutter, M. (1986b). Soleidae. In *Check-list of the Freshwater Fishes of Africa* (J. Daget, J.-P. Gosse, and D. F. E. Thys van den Audenaerde, Eds.), pp. 430–431. CLOFFA II, ORSTOM, Paris.

Desoutter, M. (1992). Solidae. In *Faune des Poissons d'eaux Douces et Saumâtres de l'Afrique de l'Ouest* (C. Lévêque, D. Paugy, and G. G. Teugels, Eds.), Vol. 2, pp. 860–865. ORSTOM, Paris.

De Vos, L. (1986). Schilbeidae. In *Check-list of the Freshwater Fishes of Africa* (J. Daget, J.-P. Gosse, and D. F. E. Thys van den Audenaerde, Eds.), pp. 36–53. CLOFFA II, ORSTOM, Paris.

De Vos, L. (1992). Schilbeidae. In *Faune des Poissons d'eaux Douces et Saumâtres de l'Afrique de l'Ouest* (C. Lévêque, D. Paugy, and G. G. Teugels, Eds.), Vol. 2, pp. 432–449. ORSTOM, Paris.

DeWitt, H. (1960). A contribution to the ichthyology of Nepal. *Stanford Ichthyological Bull.* 7(4), 63–88.

DeWitt, H. H. (1969). A second species of the family Cottidae from the New Zealand region. *Copeia* 1969(1), 30–34.

Dietz, R. S., and Holden, J. C. (1970). Reconstruction of Pangaea: Breakup and dispersion of continents, Permian to present. *J. Geophys. Res.* 75(26), 4939–4956

Dill, L. M. (1977). Refraction and the spitting behavior of the archerfish (*Toxotes chatareus*). *Behav. Ecol. Sociobiol.* 2, 169–184.

Dillard, J. G., Graham, L. K., and Russell, T. R. (Eds.) (1986). The paddlefish: Status, management and propagation, Special Publication No. 7. North Central Division of the American Fisheries Society, Washington, DC.

Dingerkus, G. (1987). Shark distribution. In *Sharks* (J. Stevens, Ed.), pp. 36–50. Facts on File, New York.

Dingerkus, G. (1995). Relationships of potamotrygonin stingrays (Chondrichthyes: Batiformes: Myliobatidae). *J. Aquariculture Aquat. Sci.* 7, 32–37.

Dingerkus, G., and Seret, B. (1992). *Rhyacichthys guilberti,* a new species of loach goby from northeastern New Caledonia (Teleostei: Rhyacichthyidae). *Tropical Fish Hobbyist* **40**(10), 174–176.

Dominey, W. J. (1988). Mating tactics in bluegill sunfish. *Assoc. Southeast Biologists Bull.* **35**, 61.

Donnelly, B. G. (1973). Aspects of behaviour in the catfish, *Clarias gariepinus* (Pisces: Clariidae), during periods of habitat desiccation. *Arnoldia Rhodesia* **6**(9), 1–8.

Douglas, J. W., Gooley, G. J., and Ingram, B. A. (1994). *Trout Cod, Maccullochella macquariensis (Cuvier)* (*Pisces: Percichthyidae), Resource Handbook and Research and Recovery Plan.* Victorian Department of Conservation and Natural Resources, Snobs Creek, Victoria.

Douglas, J. W., Gooley, G. J., Ingram, B. A., Murray, N. D., and Brown, L. D. (1995). Natural hybridization between Murray cod, *Maccullochella peelii peelii* (Mitchell), and trout cod, *Maccullochella macquariensis* (Cuvier) (Percichthyidae), in the Murray River, Australia. *Mar. Freshwater Res.* **46**, 729–734.

Dovel, W. L., Mihursky, J. A., and McErlean, A. J. (1969). Life history aspects of the hogchoker, *Trinectes maculatus*, in the Patuxent River estuary, Maryland. *Chesapeake Sci.* **10**(2), 104–119.

Dunn, J. R. (1989). A provisional phylogeny of gadid fishes based on adult and early life-history characters. In *Papers on the Systematics of Gadiform Fishes* (D. M. Cohen, Ed.), Science Series No. 32, pp. 209–235. Natural History Museum of Los Angles County, Los Angeles.

Dutt, S., and Ramaseshaiah, M. (1980). Chromosome number in *Anabas testudineus* (Bloch, 1975) [sic] and *A. oligolepis* Bleeker, 1855 (Osteichthys: Anabantidae). *Matsya* **6**, 71–74.

Dutt, S., and Ramaseshaiah, M. (1983). Taxonomic and biometric studies on *Anabas testudineus* (Bloch) and *A. oligolepis* Bleeker (Osteichthys: Anabantidae). *Proc. Indian Natl. Sci. Acad.* **49B**(4), 317–326.

Dyer, B. S. (1997). Phylogenetic revision of Atherinopsinae (Teleostei, Atherinopsidae), with comments on the systematics of the South American freshwater fish genus *Basilichthys* Girard, Publication No. 185, pp. 1–64. University of Michigan, Museum of Zoology, Ann Arbor.

Dyer, B. S. (1998). Phylogenetic systematics and historical biogeography of the Neotropical silverside family Atherinopsidae (Teleostei: Atheriniformes). In *Phylogeny and Classification of Neotropical Fishes* (L. R. Malabarba, R. E. Reis, R. P. Vari, Z. M. S. Lucena, and C. A. S. Lucena, Eds.), pp. 519–536. EDIPUCRS, Porto Alegre, Brazil.

Dyer, B. S., and Chernoff, B. (1996). Phylogenetic relationships among atheriniform fishes (Teleostei: Atherinomorpha). *Zool. J. Linnean Soc.* **117**(1), 1–69.

Eastman, J. T. (1991). Evolution and diversification of Antarctic notothenioid fishes. *Am. Zoologist* **31**, 93–109.

Eastman, J. T. (1993). *Antarctic Fish Biology.* Academic Press, San Diego.

Eastman, J. T., and Clarke, A. (1998). A comparison of adaptive radiations of Antarctic fish with those of non-Antarctic fish. In *Fishes of Antarctic: A Biological Overview* (G. diPrisco, E. Pisano, and A. Clarke, Eds.), pp. 3–26. Springer-Verlag, New York.

Eastman, J. T., and DeVries, A. L. (1986). Antarctic fishes. *Sci. Am.* **254**(11), 106–114.

Eastman, J. T., and DeVries, A. L. (1997). Morphology of the digestive system of Antarctic nototheniid fishes. *Polar Biol.* **17**, 1–13.

Eccles, D. H., and Trewavas, E. (1989). *Malawian Cichlid Fishes: The Classification of Some Haplochromine Genera.* Lake Fish Movies, Herten, Germany.

Echelle, A. A., and Echelle, A. F. (1984). Evolutionary genetics of a "species flock": Atherinid fishes on the Mesa Central of Mexico. In *Evolution of Fish Species Flocks* (A. A. Echelle and I. Kornfield, Eds.), pp. 93–110. Univ. of Maine Press, Orono.

Echelle, A. A., and Kornfield, I. (1984). *Evolution of Fish Species Flocks.* Univ. of Maine Press, Orono.

Echelle, A. A., Echelle, A. F., and Middaugh, D. P. (1989). Evolutionary biology of the *Menidia clarkhubbsi* complex of unisexual fishes (Atherinidae): Origins, clonal diversity, and more of reproduction. In *Evolution and Evology of Unisexual Vertebrates* (R. M. Dawley and J. P. Bogart, Eds.), Bulletin No. 466, pp. 144–152. New York State Museum, Albany.

Eigenmann, C. H. (1909a). The fresh-water fishes of Patagonia and an examination of the Archiplata–Archhelenis theory. *Rep. Princeton Univ. Expedition Patagonia 1896–1899* **3**(pt. 3), 225–374.

Eigenmann, C. H. (1909b). *Cave Vertebrates of America*, Publication No. 104. Carnegie Institute of Washington, Washington, DC.

Eigenmann, C. H. (1912). The freshwater fishes of British Guiana, including a study of the ecological grouping of species and the relation of the fauna of the plateau to that of the lowlands. *Memoirs Carnegie Museum* **5**, 1–578.

Eigenmann, C. H. (1915). XV. The Serrasalminae and Mylinae. *Ann. Carnegie Museum* **9**(3/4), 226–272.

Eigenmann, C. H. (1917–1921). The American Characidae. *Mem. Museum Comp. Zool.* **43**(Pt. 1, 1917; Pt. 2, 1918; Pt. 3, 1921).

Eigenmann, C. H. (1918). The Pygidiidae, a family of South American catfishes. *Memoirs Carnegie Museum* **7**(5), 259–398.

Eigenmann, C. H. (1921). The nature and the origin of the fishes of the Pacific slope of Ecuador, Peru, and Chili. *Proc. Am. Philos. Soc.* **60**, 503–523.

Eigenmann, C. H. (1922). The fishes of western South America. Part I. The fresh-water fishes of north-western South America, including Colombia, Panama, and the Pacific slopes of Ecuador and Peru, together with an appendix upon the fishes of the Rio Meta in Columbia. *Mem. Carnegie Museum* **9**(1), 1–346.

Eigenmann, C. H. (1925). A review of the Doradidae, a family of South American Nematognathi, or catfishes. *Trans. Am. Philos. Soc.* **22**, 280–365.

Eigenmann, C. H. (1928). The fresh-water fishes of Chile. *Memoir Natl. Acad. Sci.* **22**, 1–63.

Eigenmann, C. H., and Allen, W. R. (1942). *Fishes of Western South America.* Univ. of Kentucky, Lexington.

Eigenmann, C. H., and Eigenmann, R. S. (1891). A catalogue of the fresh-water fishes of South America. *Proc. U.S. Natl. Museum* **14**, 1–81.

Eigenmann, C. H., and Ward, D. P. (1905). The Gymnotidae. *Proc. Washington Acad. Sci.* **7**, 159–188.

Ellis, M. M. (1913). The Gymnotid eels of Tropical America. *Memoirs Carnegia Museum* **6**(3), 109–195.

Ellis, R. (1976). *The Book of Sharks.* Grosset & Dunlap, New York.

Epifanio, J. M., Koppelman, J. B., Nedbal, M. A., and Philipp, D. P. (1996). Geographic variation of paddlefish allozymes and mitochondrial DNA. *Trans. Am. Fish. Soc.* **125**(4), 546–561.

Eschmeyer, W. N. (Ed.) (1998). *Catalog of Fishes*, Vol. 3. California Academy of Sciences, San Francisco.

Eschmeyer, W. N., and Herald, E. S. (1983). *A Field Guide to Pacific Coast Fishes of North America.* Houghton Mifflin, Boston.

Etnier, D. A., and Starnes, W. C. (1993). *The Fishes of Tennessee.* Univ. of Tennessee Press, Knoxville.

Farber, J., and Rahn, H. (1970). Gas exchange between air and water and the ventilation pattern in the electric eel. *Respir. Physiol.* **9**, 151–161.

Farber, J. E., and Stepian, C. A. (1997). The utility of mitochondrial DNA control region sequence for analyzing phylogenetic relationships among populations, species, and genera of the Percidae. In *Molecular Systematics of Fishes* (T. D. Kocher and C. A. Stepien, Eds.), pp. 129–143. Academic Press, San Diego.

Farias, I. P., Schneider, H., and Sampaio, I. (1998). Molecular phylogeny of Neotropical cichlids:

The relationships of cichlasomines and heroines. In *Phylogeny and Classification of Neotropical Fishes* (L. R. Malabarba, R. E. Reis, R. P. Vari, Z. M. S. Lucena, and C. A. S. Lucena, Eds.), pp. 499–508. EDIPUCRS, Porto Alegre, Brazil.

Farias, I. P., Ortí, G., Campaio, I., Schneider, H., and Meyer, A. (1999). Mitochondrial DNA phylogeny of the family Cichlidae: Monophyly and fast molecular evolution of the Neotropical assemblage. *J. Mol. Evol.* **48**, 703–711.

Fedorov, V. V. (1986). Cottidae. In *Fishes of the North-Eastern Atlantic and the Mediterranean* (P. J. P. Whitehead, M.-L. Bauchot, J.-C. Hureau, J. Nielsen, and E. Tortonese, Eds.), Vol. 3, pp. 1243–1260. UNESCO, Paris.

Feltes, R. M. (1986). A systematic revision of the Polynemidae (Pisces). Ph.D. thesis, The Ohio State University, Columbus.

Feltes, R. M. (1991). Revision of the polynemid fish genus *Filimanus*, with the description of two new species. *Copeia* **1991**(2), 302–322.

Feltes, R. M. (1993). *Parapolynemus*, a new genus for the polynemid fish previously known as *Polynemus verekeri. Copeia* **1993**(1), 207–215.

Fernandes, C. C. (1998). Sex-related morphological variation in two species of apteronotid fishes (Gymnotiformes) from the Amazon River Basin. *Copeia* **1998**(3), 730–735.

Fernandez, J. M., and Weitzman, S. H. (1987). A new species of *Nannostomus* (Teleostei: Lebiasinidae) from near Puerto Ayacucho, Rio Orinoco drainage, Venezuela. *Proc. Biol. Soc. Washington* **100**(1), 164–172.

Ferraris, C. J., Jr. (1988a). The Auchenipteridae: Putative monophyly and systematics, with a classification of the neotropical doradoid catfishes (Ostariophysi: Siluriformes). Ph.D. thesis, City University of New York, New York.

Ferraris, C. J., Jr. (1988b). Relationships of the neotropical catfish genus *Nemuroglanis*, with a description of a new species (Osteichthys: Siluriformes: Pimelodidae). *Proc. Biol. Soc. Washington* **101**(3), 509–516.

Ferraris, C. J., Jr. (1991a). *Catfish in the Aquarium.* Tetra Press, Morris Plains, NJ.

Ferraris, C. J., Jr. (1991b). On the type species of *Bunocephalus* (Siluriformes: Aspredinidae). *Copeia* **1991**(1), 224–225.

Ferraris, C. J., Jr. (1996). *Denticetopsis*, a new genus of South American whale catfish (Siluriformes: Cetopsidae, Cetopsinae), with two new species. *Proc. California Acad. Sci.* **49**(6), 161–170.

Ferraris, C. J., Jr., and Brown, B. A. (1991). A new species of *Pseudocetopsis* from the Río Negro drainage of Venezuela (Siluriformes: Cetopsidae). *Copeia* **1991**(1), 161–165.

Ferraris, C. J., Jr., and Vari, R. P. (1999). The South American catfish genus *Auchenipterus*

Valenciennes, 1840 (Ostariophysi: Siluriformes: Auchenipteridae): Monophyly and relationships, with a revisionary study. *Zool. J. Linnean Soc.* 126, 387–450.

Findeis, E. K. (1997). Osteology and phylogenetic interrelationships of sturgeons (Acipenseridae). *Environ. Biol. Fishes* 48, 73–126.

Fine, M. L. (1997). Endocrinology of sound production in fishes. *Mar. Freshwater Behav. Physiol.* 29, 23–45.

Fine, M. L., Friel, J. P., McElroy, D., King, C. B., Loesser, K. E., and Newton, S. (1997a). Pectoral spine locking and sound production in the channel catfish *Ictalurus punctatus*. *Copeia* 1997(4), 777–790.

Fine, M. L., McElroy, D., Rafl, J., King, C. B., Loesser, K. E., and Newton, S. (1997b). Lateralization of pectoral stridulation sound production in the channel catfish. *Physiol. Behav.* 60(3), 753–757.

Fink, S. V., and Fink, W. L. (1981). Interrelationships of the ostariophysan fishes (Teleostei). *J. Linnean Soc. Zool.* 72(4), 297–353.

Fink, S. V., and Fink, W. L. (1996). Interrelationships of ostariophysan fishes (Teleostei). In *Interrelationships of Fishes* (M. L. J. Stiassny, L. R. Parenti, and G. D. Johnson, Eds.), pp. 209–249. Academic Press, San Diego.

Fink, W. L. (1984). Basal euteleosts: Relationships. In *Ontogeny and Systematics of Fishes* (H. G. Moser *et al.*, Eds.), Special Publication No. 1, pp. 202–206. American Society of Ichthyologists and Herpetologists, Lawrence, KS.

Fink, W. L. (1989). Ontogeny and phylogeny of shape change and diet in the South American fishes called pirahnas. *Ontogenèse Évolution Geobios Mémoire Spécial* 12, 167–172.

Fink, W. L., and Weitzman, S. H. (1982). Relationships of the stomiiform fishes (Teleostei), with a description of *Diplophos*. *Bull. Museum Comp. Zool.* 150(2), 31–93.

Fishman, A. P., Pack, A. I., Delaney, R. G., and Galante, R. J. (1987). Estivation in *Protopterus*. In *The Biology and Evolution of Lungfishes* (W. E. Bemis, W. W. Burggren, and N. E. Kemp, Eds.), pp. 237–248. A. R. Liss, New York.

Fitzsimons, J. M. (1972). A revision of two genera of goodeid fishes (Cyprinodontiformes, Osteichthyes) from the Mexican Plateau. *Copeia* 1972(4), 728–756.

Fitzsimons, J. M. (1976). Ethological isolating mechanisms in goodeid fishes of the genus *Xenotoca* (Cyprinodontiformes, Osteichthyes). *Bull. Southern California Acad. Sci.* 75, 84–99.

Fletcher, D. E. (1992). Male ontogeny and size-related variation in mass allocation of bluenose shiners (*Pteronotropis welaka*). *Copeia* 1992(2), 479–486.

Forey, P. L., Littlewood, D. T. J., Richie, P., and Meyer, A. (1996). Interrelationships of elopomorph fishes. In *Interrelationships of Fishes* (M. L. J. Stiassny, L. R. Parenti, and G. D. Johnson, Eds.), pp. 175–191. Academic Press, San Diego.

Foster, N. R. (1969). Field and aquarium observations on *Chriopeoides pengelleyi* Fowler, 1939, on the Jamaican killifish. *J. Am. Killifish Assoc.* 6(1), 17–25.

Fowler, H. W. (1934). Zoological results of the third de Schauensee Siamese expedition, Part I.—Fishes. *Proc. Acad. Nat. Sci. Philadelphia* 86, 67–163.

Foxon, G. E. H. (1933a). Pelvic fins of the *Lepidosiren*. *Nature* 131, 732–733.

Foxon, G. E. H. (1933b). Pelvic filaments of the *Lepidosiren*. *Nature* 131, 913–914.

Fraser, T. H. (1972). Comparative osteology of the shallow water cardinal fishes (Perciformes: Apogonidae) with reference to the systematics and evolution of the family. *Ichthyol. Bull. J. L. B. Smith Inst. Ichthyol.* 34, 1–105.

Fraser, T. H., and Lachner, E. A. (1985). A revision of the cardinalfish subgenera *Pristiapogon* and *Zoramia* (genus *Apogon*) of the Indo-Pacific region (Teleostei: Apogonidae). *Smithsonian Contrib. Zool.* 412, 1–46.

Fraser, T. H., and Struhsaker, P. J. (1991). A new genus and species of cardinalfish (Apogonidae) from the Indo-west Pacific, with a key to Apogonine genera. *Copeia* 1991(3), 718–722.

Fraser-Brunner, A. (1950). A revision of the fishes of the family Gasteropelecidae. *Ann. Magazine Nat. History* 3(12th ser.), 959–970.

Fraser-Brunner, A. (1955). A synopsis of the centropomid fishes of the subfamily Chandinae, with descriptions of a new genus and two new species. *Bull. Raffles Museum* 25, 185–213.

Fremling, C. R. (1978). *Biology and Functional Anatomy of the Freshwater Drum Aplodinotus grunniens Rafinesque.* Nasco, Fort Atkinson, WI.

Fremling, C. R. (1980). *Aplodinotus grunniens* Rafinesque. In *Atlas of North American Freshwater Fishes* (D. S. Lee *et al.*, Eds.), p. 756. North Carolina State Museum of Natural History, Raleigh.

Friel, J. P. (1994). A phylogenetic study of the Neotropical banjo catfishes (Teleostei: Siluriformes: Aspredinidae). Ph.D. thesis, Duke University, Durham, NC.

Fryer, G., and Iles, T. D. (1972). *The Cichlid Fishes of the Great Lakes of Africa.* T. F. H., Neptune City, NJ.

Fuller, P. L., Nico, L. G., and Williams, J. D. (1999). *Nonindigenous Fishes Introduced into Inland Waters of the United States,*, Special Publication No. 27. American Fisheries Society, Bethesda, MD.

Fumihito, A. (1989). Morphological comparison of the Mekong giant catfish, *Pangasianodon gigas*, with other pangasiid species. *Jpn. J. Ichthyol.* 36(1), 113–119.

Gardiner, B. G. (1966). *A Catalogue of Canadian Fossil Fishes*, No. 68. Royal Ontario Museum, Univeristy of Toronto, Toronto.

Gardiner, B. G. (1984). Sturgeons as living fossils. In *Living Fossils* (N. Eldredge and S. M. Stanley, Eds.), pp. 148–152. Springer-Verlag, New York.

Gardiner, B. G., and Schaeffer, B. (1989). Interrelationships of lower actinopterygian fishes. *Zool. J. Linnean Soc.* 97, 135–187.

Gardiner, B. G., Maisey, J. G., and Littlewood, D. T. J. (1996). Interrelationships of basal neopterygians. In *Interrelationships of Fishes* (M. L. J. Stiassny, L. R. Parenti, and G. D. Johnson, Eds.), pp. 117–146. Academic Press, San Diego.

Garman, S. (1913). The Plagiostomia (sharks, skates, and rays). *Mem. Museum Comp. Zool. Harvard College* 36, i–xiii, 1–515.

Garrick, J. A. F., and Schultz, L. P. (1963). A guide to the kinds of potentially dangerous sharks. In *Sharks and Survival* (P. W. Gilbert, Ed.), pp. 3–60. Heath, Boston.

Gee, J. H. (1980). Respiratory patterns and antipredator responses in the central mudminnow *Umbra limi*, a continuous, faculative, air-breathing fish. *Can. J. Zool.* 58, 819–827.

Gee, J. H. (1981). Coordination of respiratory and hydrostatic functions of the swimbladder in the central mudminnow, *Umbra limi. J. Exp. Biol.* 92, 37–52.

Géry, J. (1969). The fresh-water fishes of South America. In *Biogeography and Ecology in South America* (E. J. Fittkau, J. Illies, H. Klinge, G. H. Schwabe, and H. Sioli, Eds.), Vol. 2, pp. 828–848. Junk, The Hague.

Géry, J. (1977). *Characoids of the World.* T. F. H., Neptune City, NJ.

Géry, J., and Römer, U. (1997). *Tucanoichthys tucano* gen. n. sp., a new miniature characid fish (Teleostei: Characiformes: Characidae) from the Rio Uaupès basin in Brazil. *Aqua J. Ichthyol. Aquat. Biol.* 2(4), 65–72.

Ghedotti, M. J. (1998). Phylogeny and classification of the Anablepidae (Teleostei: Cyprinodontiformes). In *Phylogeny and Classification of Neotropical Fishes* (L. R. Malabarba, R. E. Reis, R. P. Vari, Z. M. S. Lucena, and C. A. S. Lucena, Eds.), pp. 561–582. EDIPUCRS, Porto Alegre, Brazil.

Ghedotti, M. J., and Weitzman, S. H. (1996). A new species of *Jenynsia* (Cyprinodontiformes: Anablepidae) from Brazil with comments on the composition and taxonomy of the genus, Occasional Paper No. 179, pp. 1–25.

Natural History Museum, The University of Kansas,Lawrence.

Gilbert, C. R. (1976). Composition and derivation of the North American freshwater fish fauna. *Florida Scientist* 39(2), 104–111.

Gilbert, C. R. (1980a). *Hiodon alosoides* (Rafinesque) Goldeneye and *Hiodon tergisus* Lesueur Mooneye. In *Atlas of North American Freshwater Fishes* (D. S. Lee *et al.*, Eds.), pp. 74–75. North Carolina State Museum of Natural History, Raleigh.

Gilbert, C. R. (1980b). *Umbra limi* (Kirkland) and *Umbra pygmaea*. In *Atlas of North American Freshwater Fishes* (D. S. Lee *et al.*, Eds.), pp. 129–130. North Carolina State Museum of Natural History, Raleigh.

Gilbert, C. R., and Burgess, G. H. (1980). *Rivulus marmoratus* Poey. In *Atlas of North American Freshwater Fishes* (D. S. Lee *et al.*, Eds.), p. 536. North Carolina State Museum of Natural History, Raleigh.

Gilbert, C. R., and Lee, D. S. (1980). *Menidia beryllina* (Cope) Tidewater silverside. In *Atlas of North American Freshwater Fishes* (D. S. Lee *et al.*, Eds.), p. 558. North Carolina State Museum of Natural History, Raleigh.

Gill, H. S., and Morgan, D. L. (1999). Larval development of the salamanderfish, *Lepidogalaxias salamandroided* Mees (Lepidogalaxiidae). *Copeia* 1999(1), 219–224.

Gill, H. S., and Neira, F. J. (1994). Larval descriptions of three galaxiid fishes endemic to south-western Australia: *Galaxias occidentalis, Galaxiella munda* and *Galaxiella nigrostriata* (Salmoniformes: Galaxiidae). *Austr. J. Mar. Freshwater Res.* 45, 1307–1317.

Gill, H. S., and Neira, F. J. (1998). Galaxiidae (Galaxiinae): Southern galaxias. In *Larvae of Temperate Australian Fishes* (F. J. Neira, A. G. Miskiewicz, and T. Trnski, Eds.), pp. 70–77. Univ. of Western Australia Press, Perth.

Gilmore, R. G., Donohoe, C. J., and Cooke, D. W. (1983). Observations on the distribution and biology of east-central Florida populations of the common snook, *Centropomus undecimalis* (Bloch). *Florida Scientist* 46(3/4), 313–336.

Girard, C. F. (1858). Fishes. In *General Report upon Zoology of the Several Pacific Railroad Routes, 1857*, pp. i–xiv, 1–400. (Part of U.S. Senate Document No. 78, 33rd Congress, 2nd session) More plates are in Suckley 1860. (Also published separately under the title of Fishes of North America, observed on a survey for a railroad route from the Mississippi river to the Pacific Ocean. Washington, 1858. 21 pls.)

Girard, C. F. (1859). Ichthyology. In *United States and Mexican Boundary Survey, under the Order of Lieut. Col. W. H. Emory, Major First Cavalry and United States Commissioner*, Vol. 2(Pt. 2), pp. 1–85; Fishes, Plates 1–41. Government Printing Office, Washington, DC.

Glucksman, J., West, G., and Berra, T. M. (1976). The introduced fishes of Papua New Guinea with special reference to *Tilapia mossambica*. *Biol. Conserv.* 9, 37–44.

Goldschmidt, T. (1996). *Darwin's Dreampond: Drama in Lake Victoria*. MIT Press, Cambridge, MA.

Goldschmidt, T., Witte, F., and Wanink, J. (1993). Cascading effects of the introduced Nile perch on the detritivorous/phytoplanktivorous species in the sublittoral areas of Lake Victoria. *Conserv. Biol.* 7(3), 686–700.

Gommon, M. F., Glover, J. C. M., and Kuiter, R. H. (1994). *The Fishes of Australia's South Coast*. State Print, Adelaide, Australia.

Gon, O., and Heemstra, P. C. (Eds.) (1990). *Fishes of the Southern Ocean*. J. L. B. Smith Institute of Ichthyology, Grahamstown, South Africa.

Goode, G. B. (1884). *The Fisheries and Fishing Industries of the United States*. U.S. Commission of Fish and Fisheries, Government Printing Office, Washington, DC.

Gosline, W. A. (1940). A revision of the neotropical catfishes of the family Callichthyidae. *Stanford Ichthyological Bull.* 2(1), 1–29.

Gosline, W. A. (1945). Catálogo dos nematognatos de água-doce da América do Sul e Central. *Boletin Museu Nacional. Rio de Janiero Zool.* 33, 1–138.

Gosline, W. A. (1947). Contributions to the classification of the loricariid catfishes. *Arquivos Museu Nacional (Brasil)* 41, 79–134.

Gosline, W. A. (1963). Considerations regarding the relationships of the Percopsiform, Cyprinodontiform, and Gadiform fishes, Occasional Paper No. 629, pp. 1–38. Museum of Zoology, University of Michigan, Ann Arbor..

Gosline, W. A. (1966). The limits of the fish family Serranidae with notes on other lower percoids. *Proc. California Acad. Sci.* 33, 91–112.

Gosline, W. A. (1968). The suborders of perciform fishes. *Proc. U.S. Natl. Museum* 124, 1–78.

Gosline, W. A. (1971). *Functional Morphology and Classification of Teleostean Fishes*. Univ. Press of Hawaii, Honolulu.

Gosline, W. A. (1983). The relationships of the mastacembelid and synbranchid fishes. *Jpn. J. Ichthyol.* 29(4), 323–328.

Gosline, W. A., and Brock, V. E. (1966). *Handbook of Hawaiian Fishes*. Univ. of Hawaii Press, Honolulu.

Gosse, J.-P. (1984a). Protoperidae. In *Check-list of the Freshwater Fishes of Africa* (J. Daget, J.-P. Gosse, and D. F. E. Thys van den Audenaerde, Eds.), pp. 8–17. CLOFFA I, ORSTOM, Paris.

Gosse, J.-P. (1984b). Polypteridae. In *Check-list of the Freshwater Fishes of Africa* (J. Daget, J.-P. Gosse, and D. F. E. Thys van den Audenaerde, Eds.), pp. 18–29. CLOFFA I, ORSTOM, Paris.

Gosse, J.-P. (1984c). Pantodontidae. In *Check-list of the Freshwater Fishes of Africa* (J. Daget, J.-P. Gosse, and D. F. E. Thys van den Audenaerde, Eds.), pp. 59–60. CLOFFA I, ORSTOM, Paris.

Gosse, J.-P. (1984d). Mormyridae. In *Check-list of the Freshwater Fishes of Africa* (J. Daget, J.-P. Gosse, and D. F. E. Thys van den Audenaerde, Eds.), pp. 63–122. CLOFFA I, ORSTOM, Paris.

Gosse, J.-P. (1984e). Gymnarchidae. In *Check-list of the Freshwater Fishes of Africa* (J. Daget, J.-P. Gosse, and D. F. E. Thys van den Audenaerde, Eds.), pp.123–124. CLOFFA I, ORSTOM, Paris.

Gosse, J.-P. (1986a). Malapteruridae. In *Check-list of the Freshwater Fishes of Africa* (J. Daget, J.-P. Gosse, and D. F. E. Thys van den Audenaerde, Eds.), pp. 102–104. CLOFFA II, ORSTOM, Paris.

Gosse, J.-P. (1986b). Mochokidae. In *Check-list of the Freshwater Fishes of Africa* (J. Daget, J.-P. Gosse, and D. F. E. Thys van den Audenaerde, Eds.),pp. 105–152. CLOFFA II, ORSTOM, Paris.

Gosse, J.-P. (1986c). Anabantidae. In *Check-list of the Freshwater Fishes of Africa* (J. Daget, J.-P. Gosse, and D. F. E. Thys van den Audenaerde, Eds.), pp. 402–414. CLOFFA II, ORSTOM, Paris.

Gosse, J.-P. (1988). Révision systématique de deux espèces du genre *Polypterus* (Pisces, Polypteridae). *Cybium* 12(3), 239–245.

Gosse, J.-P. (1990a). Polypteridae. In *Faune des Poissons d'eaux Douce et Saumâtres de l'Afrique de l'Ouest* (C. Lévêque, D. Paugy, and G. G. Teugels, Eds.), Vol. 1, pp. 79–87. ORSTOM, Paris.

Gosse, J.-P. (1990b). Citharinidae. In *Faune des Poissons d'eaux Douce et Saumâtres de l'Afrique de l'Ouest* (C. Lévêque, D. Paugy, and G. G. Teugels, Eds.), Vol. 1, pp. 261–268. ORSTOM, Paris.

Gosse, J.-P., and Coenen, E. J. (1990). Distichodontidae. In *Faune des Poissons d'eaux Douce et Saumâtres de l'Afrique de l'Ouest* (C. Lévêque, D. Paugy, and G. G. Teugels, Eds.), Vol. 1, pp. 237–260. ORSTOM, Paris.

Goulding, M. (1980). *The Fishes and the Forest*. Univ. of California Press, Berkeley.

Gourene, G., and Teugels, G. G. (1990). Clupeidae. In *Faune des Poissons d'eaux Douce et Saumâtres de l'Afrique de l'Ouest* (C. Lévêque, D. Paugy, and G. G. Teugels, Eds.), Vol. 1, pp. 98–113. ORSTOM, Paris.

Gradwell, N. (1971). Observations on jet propulsion in banjo catfishes. *Can. J. Zool.* 49, 1611–1612.

Grady, J. M., and LeGrande, W. H. (1992). Phylogenetic relationships, modes of speciation, and historical biogeography of the madtom catfishes, genus *Notutus* Rafinesque (Siluriformes: Ictaluridae). In *Systematics, Historical Ecology, and North American Freshwater Fishes* (R. L. Mayden, Ed.), pp. 747–777. Stanford Univ. Press, Stanford, CA.

Graham, J. B. (1972). Aerial vision in amphibious fishes. *Fauna* 3, 14–23.

Graham, J. B. (1997). *Air-Breathing Fishes*. Academic Press, San Diego.

Graham, K. (1997). Contemporary status of the North American paddlefish, *Polyodon spathula*. *Environ. Biol. Fishes* 48, 279–289.

Grande, L. (1979). *Eohiodon falcatus*, a new species of hiodontid (Pisces) from the Late Early Eocene Green River formation of Wyoming. *J. Paleontol.* 53(1), 103–111.

Grande, L. (1984). Paleontology of the Green River formation, with a review of the fish fauna, Bulletin No. 16. Geological Survey of Wyoming.

Grande, L. (1985). Recent and fossil clupeomorph fishes with materials for revision of the subgroups of clupeoids. *Bull. Am. Museum Nat. History* 181(2), 231–372.

Grande, L. (1987). Redescription of +*Hypsidoris farsonensis* (Teleostei: Siluriformes), with a reassessment of its phylogenetic relationships. *J. Vertebr. Paleontol.* 7(1), 24–54.

Grande, L. (1990). Vicariance biogeography. In *Palaeobiology: A Systhesis* (D. E. G. Briggs and P. Crowther, Eds.), pp. 448–451. Blackwell, Oxford.

Grande, L. (1996). Using the extant *Amia calva* to test the monophyly of Mesozoic groups of fishes. In *Mesozoic Fishes* (G. Arratia and G. Viohl, Eds.), pp. 181–189. Verlag/Dr. Friedrich Pfeil, Munich.

Grande, L. (1999). The first *Esox* (Esocidae: Teleostei) from the Eocene Green River formation, and a brief review of esocid fishes. *J. Vertebr. Paleontol.* 19(2), 271–292.

Grande, L., and Bemis, W. E. (1991). Osteology and phylogenetic relationships of fossil and recent paddlefish (Polyodontidae) with comments on the interrelationships of Acipenseriformes. *J. Vertebr. Paleontol. Memoir* 1, 1–121. [Supplement to *J. Vertebr. Paleontol.* 11(1)]

Grande, L., and Bemis, W. E. (1996). Interrelationships of Acipenseriformes, with comments on "Chondrostei." In *Interrelationships of Fishes* (M. L. J. Stiassny, L. R. Parenti, and G. D. Johnson, Eds.), pp. 85–115. Academic Press, San Diego.

Grande, L., and Bemis, W. E. (1998). A comprehensive phylogenetic study of amiid fishes (Amiidae) based on comparative skeletal anatomy. An empirical search for interconnected patterns of natural history. *Soc. Vertebr. Paleontol. Memoir* 4, 1–690. [Supplement to *J. Vertebr. Paleontol.* 18(1)]

Grande, L., and Bemis, W. E. (1999). Historical biogeography and historical paleoecology of Amiidae and other halecomorph fishes. In *Mesozoic Fishes 2—Systematics and Fossil Record* (G. Arratia and H.-P. Schultze, Eds.), pp. 413–424. Verlag/Dr. Friedrich Pfeil, Munich.

Grande, L., and de Pinna, M. (1998). Description of a second species of the catfish †*Hypsidoris* and a reevaluation of the genus and the family †Hypsidoridae. *J. Vertebr. Paleontol.* 18(3), 451–474.

Grant, E. M. (1982). *Guide to Fishes* [of Queensland]. Department of Harbours and Marine, Brisbane, Queensland, Australia.

Greenfield, D. W. (1997). *Allenbatrachus*, a new genus of Indo-Pacific toadfish (Batrachoididae). *Pacific Sci.* 51(3), 306–313.

Greenfield, D. W., and Thomerson, J. E. (1997). *Fishes of the Continental Waters of Belize*. Univ. Press of Florida, Gainesville.

Greenwood, P. H. (1960). Fossil denticipitid fishes from east Africa. *Bull. Br. Museum Nat. History Geol.* 5, 1–11.

Greenwood, P. H. (1961). A revision of the genus *Dinotopterus* (Pisces, Clariidae) with notes on the comparative anatomy of the suprabranchiae organs in the Clariidae. *Bull. Br. Museum Nat. History Zool.* 7(4), 215–241.

Greenwood, P. H. (1963). The swimbladder in African Notopteridae (Pisces) and its bearing on the taxonomy of the family. *Bull. Br. Museum* 11, 379–412.

Greenwood, P. H. (1968a). The osteology and relationships of the Denticipitidae, a family of clupeomorph fishes. *Bull. Br. Museum Nat. History Zool.* 16, 213–273.

Greenwood, P. H. (1968b). Notes on the visceral anatomy of *Denticeps clupeoides* Clausen, 1959, a west African clupeomorph fish. *Rev. Zool. Bot. Africa* 77, 1–10.

Greenwood, P. H. (1974a). The cichlid fishes of Lake Victoria, East Africa: The biology and evolution of a species flock. *Bull. Br. Museum Nat. History Zool. Suppl.* 6, 1–134.

Greenwood, P. H. (1974b). Review of Cenozoic freshwater fish faunas in Africa. *Ann. Geologic Surv. Egypt* 4, 211–231.

Greenwood, P. H. (1976). A review of the family Centropomidae (Pisces, Perciformes). *Bull. Br. Museum Nat. History Zool.* 29(1), 1–81.

Greenwood, P. H. (1981). *The Haplochromine Fishes of the East African Lakes*. Cornell Univ. Press, Ithaca, NY.

Greenwood, P. H. (1983). The zoogeography of African freshwater fishes: Bioaccountancy or biogeography? In *The Emergence of the Biosphere* (R. W. Sims, J. H. Price, and P. E. S. Whalley, Eds.), pp. 179–199. Academic Press, London.

Greenwood, P. H. (1984a). *Polypterus* and *Erpetoichthys*: Anachronistic osteichthyans. In *Living Fossils* (N. Eldredge and S. M. Stanley, Eds.), pp. 143–147. Springer-Verlag, New York.

Greenwood, P. H. (1984b). *Denticeps clupeoides* Clausen 1959: The static clupeomorph. In *Living Fossils* (N. Eldredge and S. M. Stanley, Eds.), pp. 140–142. Springer-Verlag, New York.

Greenwood, P. H. (1984c). What is a species flock? In *Evolution of Fish Species Flocks* (A. A. Echelle and I. Kornfield, Eds.), pp. 13–19. Univ. of Maine Press, Orono.

Greenwood, P. H. (1987a). The natural history of African lungfishes. In *The Biology and Evolution of Lungfishes* (W. E. Bemis, W. W. Burggren, and N. E. Kemp, Eds.), pp. 163–179. A. R. Liss, New York.

Greenwood, P. H. (1987b). The genera of pelmatochromine fishes (Teleostei, Cichlidae). A phylogenetic review. *Bull. Br. Museum Nat. History Zool.* 53, 139–203.

Greenwood, P. H. (1994a). Bonytongues and their allies. In *Encyclopedia of Fishes* (J. P. Paxton and W. N. Eschmeyer, Eds.), pp. 80–84. Academic Press, San Diego.

Greenwood, P. H. (1994b). Cichlids. In *Encyclopedia of Fishes* (J. P. Paxton and W. N. Eschmeyer, Eds.), pp. 203–206. Academic Press, San Diego.

Greenwood, P. H., and Stiassny, M. L. J. (1998). Cichlids. In *Encyclopedia of Fishes* (J. P. Paxton and W. N. Eschmeyer, Eds.), 2nd ed., pp. 200–204. Academic Press, San Diego.

Greenwood, P. H., and Thomson, K. S. (1960). The pectoral anatomy of *Pantodon buchholzi* (a freshwater flying fish) and the related Osteoglossidae. *Proc. Zool. Soc. London* 135, 283–301.

Greenwood, P. H., Rosen, D. E., Weitzman, S. H., and Myers, G. S. (1966). Phyletic studies of teleostean fishes, with a provisional classification of living forms. *Bull. Am. Museum Nat. History* 131(4), 339–456.

Grey, D. L. (1987). An overview of *Lates calcarifer* in Australia and Asia. In *Management of Wild and Cultured Sea Bass/Barramundi (Lates calcarifer)* (J. W. Copland and D. L. Grey, Eds.), Proceedings No. 20, pp. 15–21. ACIAP, Canberra, Australia.

Grier, H. J. (1984). Testis structure and formation of spermatophores in the atherinomorph teleost *Hoaraichthys setnai*. *Copeia* 1984(4), 833–839.

Grier, H. J., and Collette, B. B. (1987). Unique spermatozeugmata in testes of halfbeaks of the genus

Zenarchopterus (Teleostei: Hemiramphidae). *Copeia* 1987(2), 300–311.

Grier, H. J., Moody, D. P., and Cowell, B. C. (1990). Internal fertilization and sperm morphology in the brook silverside, *Labidesthes sicculus* (Cope). *Copeia* 1990(1), 221–226.

Grier, H. J., and Parenti, L. R. (1999). Systematics and gonadal morphology of the Everglades pygmy sunfish. In *Abstracts of the American Society of Ichthyologists and Herpetologists*, p. 117. Pennsylvania State University, College Station.

Griffith, R. W. (1974). Environmental and salinity tolerance in the genus *Fundulus*. *Copeia* 1974(2), 319–331.

Groot, C., and Margolis, L. (Eds.) (1991). *Pacific Salmon Life Histories*. Univ. of British Columbus Press, Vancouver.

Gross, M. R. (1984). Sunfish, salmon, and the evolution of alternative reproductive strategies and tactics in fishes. In *Fish Reproduction: Strategies and Tactics* (G. Potts and R. Wootton, Eds.), pp. 55–75. Academic Press, New York.

Groves, P. (1998). Leafy sea dragons. *Sci. Am.* 279(6), 84–89.

Grundfest, H. (1960). Electric fishes. *Sci. Am.* 203(4), 115–124.

Grzimek, B. (Ed.) (1974). *Grzimek's Animal Life Encyclopedia. Vol. 5. Fishes II: Amphibians*. Van Nostrand–Reinhold, New York.

Gudger, E. W. (1930). *The Candiru—The Only Vertebrate Parasite of Man*. Hoeber, New York.

Guiasu, R. C., and Winterbottom, R. (1993). Osteological evidence for the phylogeny of recent genera of surgeonfishes (Percomorpha, Acanthuridae). *Copeia* 1993(2), 300–312.

Guitel, F. (1913). L'Appareil fixateur de l'oeuf du *Kurtus gulliveri*. *Arch. Zool. Exp. Gén.* 52, 1–11.

Haas, R. (1976a). Sexual selection in *Nothobranchius guentheri* (Pisces, Cyprinodontidae). *Evolution* 20, 614–622.

Haas, R. (1976b). Behavioral biology of the annual killifish, *Nothobranchius guentheri*. *Copeia* 1976(1), 80–91.

Hackney, P. A., and Holbrook, J. A. (1978). *Sauger, Walleye, and Yellow Perch in the Southeastern United States*, Special Publication No. 11, pp. 74–81. American Fisheries Society, Bethesda, MD.

Haglund, T. R., Buth, D. W., and Blouw, D. M. (1990). Allozyme variation and the recognition of the "white" stickleback." *Bioch. Syst. Ecol.* 18, 559–563.

Haglund, T. R., Buth, D. G., and Lawson, R. (1992). Allozyme variation and phylogenetic relationships of Asian, North American, and European populations of ninespine stickleback, *Pungitius pungitius*. In *Systematics, Historical Ecology,* and North American Freshwater Fishes (R. L. Mayden, Ed.), pp. 438–452. Stanford Univ. Press, Stanford, CA.

Haig, J. (1950). Studies on the classification of the catfishes of the Oriental and Palaearctic family Siluridae. Rec. Indian Museum 48, 59–116.

Hall, G. E. (Ed.) (1986). Managing Muskies: A treatise on the Biology and Propagation of Muskellunge in North America, Special Publication No. 15. American Fisheries Society, Bethesda, MD.

Halstead, B. W. (1978). Poisonous and Venomous Marine Animals of the World, Rev. ed. Darwin Press, Princeton, NJ.

Hardenberg, J. D. F. (1936). On a collection of fishes from the estuary of the lower and middle course of the River Kapuas (W. Borneo). Treubia Buitzenzorg 15(3), 225–254.

Hardisty, M. W. and I. C. Potter (Eds.) (1971). The Biology of Lampreys. Academic Press, London and NY.

Harlan, J. R., and Speaker, E. B. (1969). Iowa Fish and Fishing, 4th ed. Iowa Conservation Commission, Des Moines.

Harland, W. B., Cox, A. V., Llewellyn, P. G., Pickton, C. A. G., Smith, A. G., and Walters, R. (1982). A Geologic Time Scale. Cambridge Univ. Press, Cambridge, UK.

Harold, A. S., and Vari, R. P. (1994). Systematics of the Trans-Andean species of Creagrutus (Ostariophysi: Characiformes: Characidae). Smithsonian Contrib. Zool. 551, 1–31.

Harrington, R. W., Jr. (1961). Oviparous hermaphroditic fish with internal self-fertilization. Science 134, 1749–1750.

Harrington, R. W., Jr. (1971). How ecological and genetic factors interact to determine when self-fertilizing hermaphrodites of Rivulus marmoratus change into functional secondary males, with a reappraisal of the modes of intersexuality among fishes. Copeia 1971(3), 389–432.

Harrington, R. W., Jr., and Kallman, K. D. (1968). The homozygosity of clones of the self-fertilizing hermaphroditic fish Rivulus marmoratus (Cyprinodontidae: Atheriniformes). Am. Nat. 102, 337–343.

Harris, J. H., and Rowland, S. J. (1996). Family Percichthyidae: Australian freshwater cods and basses. In Freshwater Fishes of South-Eastern Australia (R. M. McDowall, Ed.), pp. 150–163. Reed, Sydney.

Harrison, I. J. (1989). Specialization of the gobioid palatoquadrate complex and its rlevance to gobioid systematics. J. Nat. History 23, 325–353.

Harrison, I. J., and Howes, G. J. (1991). The pharyngobranchial organ of mugilid fishes; Its structure, variability, ontogeny, possible function and taxonomic utility. Bull. Br. Museum Nat. History Zool. 57(2), 111–132.

Harrison, I. J., and Miller, P. J. (1992a). Eleotridae. In Faune des Poissons d'eaux Douces et Saumâtres de l'Afrique de l'Ouest (C. Lévêque, D. Paugy, and G. G. Teugels, Eds.), Vol. 2, pp. 822–836. ORSTOM, Paris.

Harrison, I. J., and Miller, P. J. (1992b). Gobiidae. In Faune des Poissons d'eaux Douces et Saumâtres de l'Afrique de l'Ouest (C. Lévêque, D. Paugy, and G. G. Teugels, Eds.), Vol. 2, pp. 798–821. ORSTOM, Paris.

Harry, R. R. (1953). A contribution to the classification of the African catfishes of the family Amphiliidae, with description of collections from Cameroon. Rev. Zool. Zotanique Africaines 47(3/4), 177–232.

Hasler, A. D. (1966). Underwater Guideposts. Univ. of Wisconsin Press, Madison.

Hastings, R. W., and Yerger, R. W. (1971). Ecology and life history of the diamond killifish, Adinia xenica (Jordan and Gilbert). Am. Midland Nat. 86(2), 276–291.

Heiligenberg, W. (1993). Electrosensation. In The Physiology of Fishes (D. H. Evans, Ed.), pp. 137–160. CRC Press, Boca Raton, Fl.

Helfman, G. S., Collete, B. B., and Facey, D. E. (1997). The Diversity of Fishes. Blackwell, Malden, MD.

Hensley, D. A., and Ahlstrom, E. H. (1984). Pleuronectiformes: Relationships. In Ontogeny and Systematics of Fishes (H. G. Moser et al., Eds.), Special Publication No. 1, pp. 670–687. American Society of Ichthyologists and Herpetologists, Lawrence, KS.

Hensley, D. A., and Courtenay, W. R., Jr. (1980a). Misgurnus anguillicaudatus (Cantor) Oriental weatherfish. In Atlas of North American Freshwater Fishes (D. S. Lee et al., Eds.), p. 436. North Carolina State Museum of Natural History, Raleigh.

Hensley, D. A., and Courtenay, W. R., Jr. (1980b). Hypostomus spp. Armored catfish. In Atlas of North American Freshwater Fishes (D. S. Lee et al., Eds.), p. 477. North Carolina State Museum of Natural History, Raleigh.

Herald, E. S. (1959). From pipefish to seahorse—A study of phylogenetic relationships. Proc. California Acad. Sci. 29(13), 465–473.

Herald, E. S. (1962). Living Fishes of the World. Doubleday, Garden City, NY.

Herre, A. W. (1927). Gobies of the Philippines and the China Sea, Monograph No. 23. Philippines Bureau of Science.

Herre, A. W. (1940). New species of fishes from the Malay Peninsula and Borneo. Bull. Raffles Museum Singapore 16, 5–26.

Herre, A. W. (1949). A case of poisoning by a stinging catfish in the Philippines. Copeia 1949(3), 222.

Herre, A. W. (1953). Check list of Philippine fishes, Research Report No. 20. Fish and Wildlife Association, Washington, DC.

Higuchi, H. (1992). A phylogeny of the South American thorny catfishes (Osteichthys; Siluriformes, Doradidae). Ph.D. thesis, Harvard University, Cambridge, MA.

Hilton, E. J., and Bemis, W. E. (1999). Skeletal variation in shortnose sturgeon (*Acipenser brevirostrum*) from the Connecticut River: Implications for comparative osteological studies of fossil and living fishes. In *Mesozoic Fishes 2— Systematics and Fossil Record* (G. Arratia and H.-P. Schultze, Eds.), pp. 69–94. Verlag Dr. Friedrich Pfeil, Munich.

Hixon, M. A. (1980). Comparative interactions between California reef fishes of the genus *Embiotoca*. *Ecology* 61, 918–931.

Hocutt, C. H., and Wiley, E. O. (Eds.) (1986). *The Zoogeography of North American Freshwater Fishes*. Wiley, New York.

Hoedeman, J. J. (1974). *Naturalist's Guide to Fresh-Water Aquarium Fish*. Sterling, New York.

Hoese, D. F. (1984). Gobioidei: Relationships. In *Ontogeny and Systematics of Fishes* (H. G. Moser *et al.*, Eds.), Special Publication No. 1, pp. 588–591. American Society of Ichthyologists and Herpetologists, Lawrence, KS.

Hoese, D. F. (1994). Gobies. In *Encyclopedia of Fishes* (J. P. Paxton and W. N. Eschmeyer, Eds.), pp. 220–224. Academic Press, San Diego.

Hoese, D. F., and Gill, A. C. (1993). Phylogenetic relationships of eleotridid fishes (Perciformes: Gobioidei). *Bull. Mar. Sci.* 52(1), 415–440.

Hoover, J. J., George, S. G., and Killgore, K. J. (2000). Rostrum size of paddlefish (*Polydon spathula*) (Acipenseriformes: Polyodontidae) from the Mississippi Delta. *Copeia* 2000(1), 288–290.

Hopkins, C. D. (1974). Electric communication in fish. *Am. Scientist* 62(4), 426–437.

Hopkins, C. D. (1981). On the diversity of electric signals in a community of mormyrid electric fishes in west Africa. *Am. Zool.* 21, 211–222.

Hopkins, C. D. (1986). Behavior of Mormyridae. In *Electroreception* (T. H. Bullock and W. Heiligenberg, Eds.). Wiley, New York.

Hopkins, C. D. (1988). Neuroethology of electric communication. *Annu. Rev. Neurosci.* 11, 497.

Hopkins, C. D. (1991). *Hypopomus pinnicaudatus* (Hypopomidae), a new species of gymnotiform fish from French Guiana. *Copeia* 1991(1), 151–161.

Hopkins, C. D. (1999). Design features for electric communication. *J. Exp. Biol.* 202(10), 1217–1228.

Hopkins, C. D., Comfort, N. C., Bastiant, J., and Bass, A. (1990). Functional analysis of sexual dimorphism in an electric fish, *Hypopomus pinnicaudatus*. *Br. Behav. Evol.* 35, 350–367.

Hora, S. L. (1933). Siluroid fishes of India, Burma, and Ceylon, I. Loachlike fishes of the genus *Amblyceps* Blyth. *Rec. Indian Museum* 35, 607–621.

Hora, S. L. (1935). Physiology, bionomics, and evolution of air breathing fishes of India. *Trans. Natl. Instit. Sci. India* 1, 1–16.

Hora, S. L. (1936). Siluroid fishes of India, Burma and Ceylon. 2. Fishes of the genus *Akysis* Bleeker. 3. Fishes of the genus *Olyra* McClelland. 4. On the use of the generic name *Wallago* Bleeker. 5. Fishes of the genus *Heteropneustes* Müller. *Rec. Indian Museum* 38(2), 199–209.

Hora, S. L., and Silas, E. G. (1952). Evolution and distribution of glyptosternoid fishes of the family Sisoridae (Order Siluroidea). *Proc. Natl. Inst. Sci. India* 18(4), 309–322.

Hori, M. (1993). Frequency-dependent natural selection in the handed-ness of scale-eating cichlid fish. *Science* 260, 216–219.

Horn, M. H. (1972). The amount of space available for marine and freshwater fishes. *Fish. Bull.* 70, 1295–1297.

Horn, M. H., and Riggs, C. D. (1973). Effects of temperature and light on the rate of air breathing of the bowfin, *Amia calva*. *Copeia* 1973, 653–657.

Horn, M. H., Martin, K. L. M., and Chotkowski, M. A. (Eds.) (1999). *Intertidal Fishes*. Academic Press, San Diego.

Hornaday, W. T. (1885). *Two Years in the Jungle: The Experiences of a Hunter and Naturalist in India, Ceylon, the Malay Peninsula and Borneo*. Scribner's, New York.

Hortle, M. E., and White, R. W. G. (1980). Diet of *Pseudaphritis urvillii* (Cuvier and Valenciennes) (Pisces: Bovichthyidae) from south-eastern Australia. *Austr. J. Mar. Freshwater Res.* 31, 533–539.

Howden, H. F. (1974). Problems in interpreting dispersal of terrestrial organisms related to continental drift. *Biotropica* 6, 1–6.

Howes, G. J. (1976). The cranial musculature and taxonomy of characoid fishes of the tribes Cynodontini and Characini. *Bull. Br. Museum Nat. History Zool.* 29(4), 203–248.

Howes, G. J. (1983a). Problems in catfish anatomy and phylogeny exemplified by the neotropical Hypophthalmidae (Teleostei: Siluroidei). *Bull. Br. Museum Nat. History Zool.* 45(1), 1–39.

Howes, G. J. (1983b). The cranial muscles of loricarioid catfishes, their homologies and value as taxonomic characters (Teleostei: Siluroidei).

Bull. Br. Museum Nat. History Zool. **45**(6), 309–345.

Howes, G. J. (1985). The phylogenetic relationships of the electric catfish family Malapteruridae (Teleostei: Siluroidei). *J. Nat. History* **19**, 37–67.

Howes, G. J. (1989). Phylogenetic relationships of macrouroid and gadoid fishes based on cranial myology and arthrology. In *Papers on the Systematics of Gadiform Fishes* (D. M. Cohen, Ed.), Science Series 32, pp. 113–128. Natural History Museum of Los Angles County, Los Angeles..

Howes, G. J. (1991a). Systematics and biogeography: An overview. In *Cyprinid Fishes: Systematics, Biology, and Exploitation* (I. J. Winfield and J. S. Nelson, Eds.), pp. 1–33. Chapman & Hall, London.

Howes, G. J. (1991b). Anatomy, phylogeny and taxonomy of the gadoid fish genus *Macruronus* Günther, 1873, with a revised hypothesis of gadoid phylogeny. *Bull. Br. Museum Nat. History Zool.* **51**(1), 77–110.

Howes, G. J., and Fumihito, A. (1991). Cranial anatomy and phylogeny of the south-east Asian catfish genus *Belodontichthys. Bull. Br. Museum Nat. History Zool.* **57**(2), 133–160.

Howes, G. J., and Sanford, C. P. J. (1987). The phylogenetic position of the Plecoglossidae (Teleostei, Salmoniformes), with comments on the Osmeridae and Osmeroidei. In *Proceedings of the Fifth Congress of European Ichthyology, Stockholm (1985)* (S. O. Kullander and B. Fernholm, Eds.), pp. 17–30.

Hubbs, C. L. (1938). Fishes from the caves of Yucatan. Carnegie Institute of Washington Publication No. 491, pp. 261–295.

Hubbs, C. L. (1941). A new family of fishes. *J. Bombay Nat. History Soc.* **42**, 446–447.

Hubbs, C. L. (1950). Studies of cyprinodont fishes. XX. A new subfamily from Guatemala, with ctenoid scales and a unilateral pectoral clasper, Publication No. 78, pp. 1–28. Museum of Zoology, University of Michigan, Ann Arbor.

Hubbs, C. L. (1955). Hybridization between fish species in nature. *Syst. Zool.* **4**, 1–20.

Hubbs, C. L., and Bailey, R. M. (1947). Blind catfishes from artesian waters of Texas. *Occasional Papers Museum Zool. Univ. Michigan* **499**, 1–15.

Hubbs, C. L., and Potter, I. C. (1971). Distribution, phylogeny and taxonomy. In *The Biology of Lampreys* (M. W. Hardisty and I. C. Potter, Eds.), Vol. 1, pp. 1–65. Academic Press, London.

Hubbs, C. L., and Turner, C. L. (1939). Studies of the fishes of the Order Cyprinodontes. XVI. A revision of the Goodeidae, Publication No. 42, pp.

1–80. Museum of Zoology, University of Michigan, Ann Arbor.

Hubbs, C. L., Hubbs, L. C., and Johnson, R. E. (1943). Hybridization in nature between species of catostomid fishes. In *Contribution of the Laboratory of Vertebrate Biology*, No. 22, pp. 1–69. University of Michigan, Ann Arbor.

Hubbs, C. L., Miller, R. R., and Hubbs, L. C. (1974). Hydrographic history and relict fishes of the north-central Great Basin. *Mem. California Acad. Sci.* **7**, 1–259.

Huber, J. H. (1992). *Review of Rivulus Ecobiogeography Relationships*. Cybium, Société Française d'Ichthyologie.

Hureau, J.-C. (1986). Polynemidae. In *Fishes of the North-Eastern Atlantic and the Mediterranean* (P. J. P. Whitehead, M.-L. Bauchot, J.-C. Hureau, J. Nielsen, and E. Tortonese, Eds.), Vol. 3, pp. 1205–1206. UNESCO, Paris.

Hutchings, J. A., and Morris, D. W. (1985). The influence of phylogeny, size, and behaviour on patterns of covariatin in salmonid life histories. *Oikos* **45**, 118–124.

Hutchings, J. A., and Myers, R. A. (1994). What can be learned from the collapse of a renewable resource? Atlantic cod, *Gadus morhua*, of Newfoundland and Labrador. *Can. J. Fish. Aquat. Sci.* **51**, 2126–2146.

Ibara, R. M., Penny, L. T., Ebeling, A. W., Van Dykhuizen, G., and Cailliet, G. (1983). The mating call of the plainfin midshipman fish, *Porichthys notatus*. In *Predators and Prey in Fishes* (D. L. Noakes, D. G. Linquist, G. S. Helfman, and J. A. Ward, Eds.). Junk, The Hague.

Idyll, C. P. (1969, May). Grunion, the fish that spawns on land. *Natl. Geographic* **135**, 714–723.

Iles, R. B. (1960). External sexual differences and their significance in *Mormyrus kannume* Forskål, 1775. *Nature* **188**, 516.

Inger, R. F., and Kong, C. P. (1962). The fresh-water fishes of north Borneo. *Fieldiana Zool.* **45**, 1–268.

Ingram, B. A., and Douglas, J. W. (1995). Threatened fishes of the world: *Maccullochella macquariensis* (Cuvier, 1829) (Percichthyidae). *Environ. Biol. Fishes* **43**, 38.

Ingram, B. A., and Rimmer, M. A. (1992). Induced breeding and larval rearing of the endangered Australian freshwater fish trout cod, *Maccullochella macquariensis* (Cuvier) (Percichthyidae). *Aquaculture Fish. Management* **24**, 7–17.

Ingram, B. A., Barlow, C. G., Burchmore, J. J., Gooley, G. J., Rowland, S. J., and Sanger, A. C. (1990). Threatened native freshwater fishes in Australia—Some case histories. *J. Fish Biol.* **37A**, 175–182.

International Game Fish Association (IGFA) (1993). *World Record Game Fishes*. IGFA, Pompano Beach, FL.

Isbrücker, I. J. H. (1980). Classification and catalogue of the mailed Loricariidae (Pisces, Siluriformes), No. 22. Verslangen en Technische Gegevens, Institut voor Taxonomische Zoölogie, Universiteit van Amsterdam, Amsterdam.

Isbrücker, I. J. H., and Nijssen, H. (1982). New data on *Metaloricaria paucidens* from French Guinana and Surinom (Pisces, Siluriformes, Loricariidae). *Bijdragen Dierkunde* 52(2), 155–168.

Isbrücker, I. J. H., and Nijssen, H. (1988). Review of the South American characiform fish genus *Chilodus*, with description of a new species, *C. gracilis* (Pisces, Characiformes, Chilodontidae). *Beaufortia* 38(3), 47–56.

Isbrücker, I. J. H., and Nijssen, H. (1989). Diagnose dreier neuer harnischwelsgattungen mit fünf neuen arten aus Brasilien. *Aquarien Terrarien. Z.* 42(9), 541–547.

Isbrücker, I. J. H., and Nijssen, H. (1991). *Hypancistrus zebra*, a new genus and species of uniquely pigmented ancistrine loricariid fish from the Rio Xingu, Brazil (Pisces: Siluriformes: Loricariidae). *Ichthyol. Exploration Freshwaters* 1(4), 345–350.

Isbrücker, I. J. H., and Nijssen, H. (1992). Sexualdimorphismus bei hamischwelsen (Loricariidae). *Aquarien Terrarienzeitschrift. Sonderheft Harnischwelse*, 19–33.

Ishimatsu, A., and Itazawa, Y. (1981). Ventilation of the air-breathing organ in the snakehead *Channa argus. Jpn. J. Ichthyol.* 28, 276–282.

Ivantsoff, W., and Aarn (1999). Detection of predation on Australian native fishes by *Gambusia holbrooki. Mar. Freshwater Res.* 50, 467–468.

Ivantsoff, W., and Allen, G. R. (1984). Two new species of *Pseudomugil* (Pisces: Melanotaeniidae) from Irian Jaya and New Guinea. *Austr. Zool.* 21, 479–489.

Ivantsoff, W., and Crowley, L. E. L. M. (1991). Review of the Australian silverside fishes of the genus *Atherinomorus* (Atherinidae). *Austr. J. Mar. Freshwater Res.* 42, 479–505.

Ivantsoff, W., and Crowley, L. E. L. M. (1996). Silversides or hardyheads. In *Freshwater Fishes of South-Eastern Australia* (R. M. McDowall, Ed.), pp. 123–133. Reed, Sydney.

Ivantsoff, W., Crowley, L. E. L. M., and Allen, G. R. (1987). Descriptions of three new species and one subspecies of freshwater hardyhead (Pisces: Atherinidae: *Craterocephalus*) from Australia. *Rec. Western Austr. Museum* 13(2), 171–188.

Ivantsoff, W., Unmack, P., Saeed, B., and Crowley, L. E. L. M. (1991). A redfinned blue-eye, a new species and genus of the family Pseudomugilidae from central western Queensland. *Fishes Sahul (J. Austr. New Guinea Fishes Assoc.)* 6(4), 277–282.

Ivantsoff, W., Aarn, Shepherd, M., and Allen, G. R. (1996). *Pseudomugil reticulatus* (Pisces: Pseudomugilidae), a review of the species originally described from a single specimen of Vogelkop Peninsula, Irian Jaya, with further evaluation of the systematics of Atherinoidea. *Aqua J. Ichthyol. Aquat. Biol.* 2(4), 1–12.

Ivoylov, A. A. (1986). Classification and nomenclature of tilapias of the tribe Tilapinii (Cichlidae). New commercial fishes in warm waters of the USSR. *J. Ichthyol.* 26(3), 97–109.

Iwamaatsu, T., Imaki, A., Kawamoto, A., and Inden, A. (1982). On *Oryzias javanicus* collected from Jakarta, Singapore and West Kalimantan. *Medaka* 1, 5–6.

Iwata, A., Jeon, S.-R., Mizuno, N., and Choi, K.-C. (1985). A revision of the eleotrid goby genus *Odontobutis* in Japan, Korea, and China. *Jpn. J. Ichthyol.* 31, 373–388.

Jackson, P. B. N. (1961). *The Fishes of Northern Rhodesia*. Government Printer, Lusaka, Rhodesia.

Jackson, P. D. (1998). Australian threatened fishes—1998 supplement. *Austr. Soc. Fish Biol. Newslett.* 28(2), 30–37.

Jackson, P. D., Koehn, J. D., Lintermans, M., and Sanger, A. C. (1996). Family Gadopsidae: Freshwater blackfishes. In *Freshwater Fishes of South-Eastern Australia* (R. M. McDowall, Ed.), pp. 186–190. Reed, Sydney.

Jamieson, I. G., Blouw, D. M., and Colgan, P. W. (1992). Parental care as a constraint on male mating success in fishes: A comparative study of threespine and white sticklebacks. *Can. J. Zool.* 70, 956–962.

Janssens, P. A., and Cohen, P. P. (1968). Nitrogen metabolism in African lungfish. *Comp. Biochem. Physiol.* 24, 879–886.

Janvier, P., and Lund, R. (1983). *Hardistiella montanensis* n. gen. et sp. (Petromyzontidae) from the Lower Carboniferous of Montana, with remarks on the affinities of the lampreys. *J. Vertebr. Paleontol.* 2(4), 407–413.

Jayaram, K. C. (1956). Taxonomic status of the Chinese catfish family Cranoglanididae, Myers, 1931. *Proc. Natl. Inst. Sci. India* 21B(6), 256–263.

Jayaram, K. C. (1976). Contributions to the study of bagrid fishes. 13. Interrelationships of Indo-African catfishes of the family Bagridae. *Matsya* 2, 47–53.

Jayaram, K. C. (1977). Zoogeography of Indian freshwater fishes. *Proc. Indian Acad. Sci.* 86B, 265–274.

Jayaram, K. C. (1981). *The Freshwater Fishes of India, Pakistan, Bangladesh, Burma and Sri Lanka—A Handbook.* Zoological Survey of India, Calcutta.

Jayaram, K. C., and Majumdar, N. (1964). Siluroid fishes of India, Burma and Ceylon. 15. Fishes of the genus *Chaca* Gray, 1831. *Proc. Zool. Soc. Calcutta* **17**(2), 177–181.

Jenkins, R. E., and Burkhead, N. M. (1994). *Freshwater Fishes of Virginia.* American Fisheries Society, Bethesda, MD.

Jenkins, R. E., and Cashner, R. C. (1983). Records and distributional relationships of the Roanoke bass, *Ambloplites cavifroms,* in the Roanoke River drainage, Virginia. *Ohio J. Sci.* **83**, 146–155.

Jenkins, R. E., Haxo, W. H., and Burkhead, N. M. (1994). Perches: Family Percidae. In *Freshwater Fishes of Virginia* (R. E. Jenkins and N. M. Burkhead, Eds.), pp. 755–889. American Fisheries Society, Bethesda, MD.

Jensen, N. H. (1976). Reproduction of the bull shark, *Carcharinus leucas,* in the Lake Nicaragua-Rio San Juan system. In *Investigations of the Ichthyofauna of Nicaraguan Lakes* (T. B. Thorson, Ed.), pp. 539–559. School of Life Sciences, University of Nebraska, Lincoln.

Jiajian, Z. (1990). The Cyprinidae fossils from middle Miocene of Shanwang Basin. *Vertebr. Palasiat.* **28**(2), 95–127.

Johansen, K., and Lenfant, C. (1967). Respiratory function in the South American lungfish, *Lepidosiren paradoxa* (Fitz.). *J. Exp. Biol.* **46**, 205–218.

Johnels, A. G., and Svensson, G. S. O. (1954). On the biology of *Protopterus annectens* Owen. *Arkiv. Zool.* **7**, 131–158.

Johnson, G. D. (1984). Percoidei: Development and relationships. In *Ontogeny and Systematics of Fishes* (H. G. Moser *et al.*, Eds.), Special Publication No. 1, pp. 464–498. American Society of Ichthyologists and Herpetologists, Lawrence, KS.

Johnson, G. D. (1992). Monophyly of the euteleostean clades—Neoteleostei, Eurypterygii, and Ctenosquamata. *Copeia* **1992**(1), 8–25.

Johnson, G. D. (1993). Percomorph phylogeny: Progress and problems. *Bull. Mar. Sci.* **52**(1), 3–28.

Johnson, G. D., and Gill, A. C. (1994). Perches and their allies. In *Encyclopedia of Fishes* (J. P. Paxton and W. N. Eschmeyer, Eds.), pp. 181–196. Academic Press, San Diego.

Johnson, G. D., and Patterson, C. (1993). Percomorph phylogeny: A survey of acanthomorphs and a new proposal. *Bull. Mar. Sci.* **52**(1), 554–626.

Johnson, G. D., and Patterson, C. (1996). Relationships of lower Euteleostean fishes. In *Interrelationships of Fishes* (M. L. J. Stiassny, L. R. Parenti, and G. D. Johnson, Eds.), pp. 251–331. Academic Press, San Diego.

Johnson, G. D., and Springer, V. G. (1997). *Elassoma: Another Look.* Abstracts of the 77th annual meeting, p. 176. American Society of Ichthyologists and Herpetologists, Univ. Washington, Seattle.

Johnson, R. D. O. (1912). Notes on the habits of a climbing catfish (*Arges marmoratus*) from the Republic of Colombia. *Ann. N. Y. Acad. Sci.* **22**, 327–333.

Johnson, T. C., Scholz, C. A., Talbot, M. R., Kelts, K., Ricketts, R. D., Ngobi, G., Beunig, K., Ssemmanda, I., McGill, J. W. (1996). Late Pleistocene desiccation of Lake Victoria and rapid evolution of cichlid fishes. *Science* **273**, 1091–1093.

Johnston, C. E., and Knight, C. L. (1999). Life-history traits of the bluenose shiner, *Pteronoptopis welaka* (Cypriniformes: Cyprinidae). *Copeia* **1999**(1), 200–205.

Johnston, C. E., and Page, L. M. (1992). The evolution of complex reproductive strategies in North American minnows (Cyprinidae). In *Systematics, Historical Ecology, and North American Freshwater Fishes* (R. L. Mayden, Ed.), pp. 600–621. Stanford Univ. Press, Stanford, CA.

Jones, W. J., and Quattro, J. M. (1999). Phylogenetic affinities of pygmy sunfishes (*Elassoma*) inferred from mitochondrial DNA sequences. *Copeia* **1992**(2), 470–474.

Jordan, D. S., and Evermann, B. W. (1896). Fishes of North and Middle America. *Bull. U.S. Natl. Museum* **47**(1–4).

Jordan, D. S., and Evermann, B. W. (1900). The fishes of North and Middle America. *Bull. U.S. Natl. Museum* **47**(4, Pts. 1/2).

Jordan, D. S., and Evermann, B. W. (1905). *American Food and Game Fishes.* Doubleday, Page, New York.

Jordan, D. S., and Metz, C. W. (1913). A catalog of the fishes known from the waters of Korea. *Mem. Carnegie Museum* **11**(1), 1–65, Plates 1–10.

Jordan, D. S., and Richardson, R. E. (1909). A catalog of the fishes of the island of Formosa, or Taiwan, based on the collections of Dr. Hans Sauter. *Mem. Carnegie Museum* **4**(4), 159–204, Plates LXIII–LXXIV.

Joss, J. M. P., Rajasekar, S., Raj-Prasad, R. A., and Ruitenberg, K. (1997). Developmental endocrinology of the dipnoan, *Neoceratodus forsteri. Am. Zool.* **37**, 461–469.

Joubert, C. S. W. (1975). The food and feeding habits of *Mormyrops deliciosus* (Leach, 1818) and

Mormyrus longirostris Peters, 1852 (Pisces: Mormyridae) in Lake Kariba. *Kariba Stud.* 5, 68–85. (Trustees of the National Museums and Monuments of Rhodesia)

Jubb, R. A. (1981). *Nothobranchius*. T. F. H., Neptune City, NJ.

Jude, D. L., Reider, R. H., and Smith, G. R. (1992). Establishment of Gobiidae in the Great Lakes basin. *Can. J. Fish. Aquat. Sci.* 49, 416–421.

Kailola, P. (1989). Fork-tailed catfishes family Ariidae. In *Freshwater Fishes of Australia* (G. R. Allen, Ed.), pp. 47–55. T. F. H., Neptune City, NJ.

Kallman, K. D., and Harrington, R. W., Jr. (1964). Evidence for the existence of homozygous clones in the self-fertilizing hermaphroditic teleost *Rivulus marmoratus* (Poey). *Biol. Bull.* 126, 101–114.

Karplus, I. (1979). The tactile communication between *Cryptocentrus steinitzi* (Pisces, Gobiidae) and *Alpheus purpurilenticularis* (Crustacea, Alpheiidae). *Z. Tierpsychol.* 49, 173–196.

Kaufman, L. (1992). Catastrophic change in species-rich freshwater ecosystems: The lessons of Lake Victoria. *BioScience* 42(11), 846–858.

Kaufman, L. S., and Liem, K. F. (1982). Fishes of the suborder Labroidei (Pisces: Perciformes): Phylogeny, ecology, and evolutionary significance. *Breviora* 472, 1–19.

Keenleyside, M. H. A. (1991). *Cichlid Fishes: Behaviour, Ecology, and Evolution*. Chapman & Hall, London.

Kelly, W. E., and Atz, J. W. (1964). A pygidiid catfish that can suck blood from goldfish. *Copeia* 1964(4), 702–704.

Kemp, A. (1981). Rearing of embryos and larvae of the Australian lungfish, *Neoceratodus forsteri*, under laboratory conditions. *Copeia* 1981, 776–784.

Kemp, A. (1982). The embryological development of the Queensland lungfish, *Neoceratodus forsteri* (Krefft). *Mem. Queensland Museum* 20(3), 553–597.

Kemp, A. (1987). The biology of the Australian lungfish, *Neoceratodus forsteri* (Krefft 1870). In *The Biology and Evolution of Lungfishes* (W. E. Bemis, W. W. Burggren, and N. E. Kemp, Eds.), pp. 181–198. A. R. Liss, New York.

Kemp, A., and Molnar, R. E. (1981). *Neoceratodus forsteri* from the Lower Cretaecous of New South Wales, Australia. *J. Paleontol.* 55(1), 211–217.

Kenmuir, D. H. S. (1973). The ecology of the tigerfish, *Hydrocynus vittatus* Castlenau, in Lake Kariba. *Occasional Papers Natl. Museums Monuments Rhodesia B* 5(3), 115–170.

Kennedy, M., and Fitzmaurice, P. (1972). The biology of the bass *Dicentrarchus labrax* in Irish waters. *J. Mar. Biol. Assoc. UK* 52(3), 557–597.

Kerr, J. G. (1900). The external features in the development of *Lepidosiren paradoxa*, Fitz. *Philos. Trans. R. Soc. London B* 19, 299–330.

Kershaw, D. R. (1970). The cranial osteology of the "butterfly fish," *Pantodon buchholzi* Peters. *Zool. J. Linnean Soc.* 49, 5–19.

Khalaf, K. T. (1961). *The Marine and Fresh Water Fishes of Iraq*. Ar-Rabitta Press, Baghdad.

Kirby, R. F., Thompson, K. W., and Hubbs, C. (1977). Karyotypic similarities between the Mexican and blind tetras. *Copeia* 1977(3), 578–580.

Klinger, S., Magnuson, J. J., and Gallepp, G. W. (1982). Survival mechanisms of the central mudminnow (*Umbra limi*), fathead minnow (*Pimephales promelas*) and brook stickleback (*Culaea inconstans*) for low oxygen in winter. *Environ. Biol. Fishes* 7, 113–120.

Kner, R. (1859). Zur Familie der Characinen. III. Folge der Ichthyologischen Beiträge. *Denkschriften Akad. Wissenschaften Wien.* 17, 137–182, Plates 1–9.

Knöppel, H. A. (1970). Food of Central Amazonia fishes. *Amazoniana* 2(3), 257–352.

Kocher, T. D., Conroy, J. A., McKaye, K. R., Stauffer, J. R., Jr. (1993). Similar morphologies of cichlid fish in lakes Tanganyika and Malawi are due to convergence. *Mol. Phylogenet. Evol.* 2, 158–165.

Konings, A. (1990). *A. Konings's Book of Cichlids and All the Other Fishes of Lake Malawi*. T. F. H., Neptune City, NJ.

Kornfield, I., and Parker, A. (1997). Molecular systematics of a rapidly evolving species flock: The mbuna of Lake Malawi and the search for phylogenetic signals. In *Molecular Systematics of Fishes* (T. D. Kocher and C. A. Stepien, Eds.), pp. 25–37. Academic Press, San Diego.

Kottelat, M. (1982a). Notes d'ichtyologie asiatique. *Bull. Muséum Natl. d'Histoire Nat. Ser. A* 4, 523–529.

Kottelat, M. (1982b). A new noemacheilinae loach from Thailand and Burma. *Jpn. J. Ichthyol.* 29(2), 169–172.

Kottelat, M. (1983). A new species of *Erethistes* (Osteichthys: Siluriformes: Sisoridae) from Thailand and Burma. *Hydrobiologia* 107, 71–74.

Kottelat, M. (1987). Nomenclatorial status of the fish names created by J. C. van Hasselt (1823) and of some cobitoid genera. *Jpn. J. Ichthyol.* 33(4), 368–375.

Kottelat, M. (1988). Indian and Indochinese species of *Balitoria* (Osteichthyes: Cypriniformes) with descriptions of two new species and comments on the family-group names Balitoridae and

Homalopteridae. *Rev. Suisse Zool.* 95(2), 487–504.

Kottelat, M. (1989). Zoogeography of the fishes from Indochinese inland waters with an annotated check-list. *Bull. Zoöl. Museum (Univ. Amsterdam)* 12(1), 1–54.

Kottelat, M. (1990a). *Indochinese Nemacheilines, a Revision of Nemacheiline Loaches (Pisces: Cypriniformes) of Thailand, Burma, Laos, Cambodia and southern Viet Nam.* Verlag Dr. Friedrich Pfeil, Munich.

Kottelat, M. (1990b). Sailfin silversides (Pisces: Telmatherinidae) of Lakes Towuti, Mahalona and Wawontoa (Sulawesi, Indonesia) with descriptions of two new genera and two new species. *Ichthyol. Exploration Freshwater* 1(3), 227–246.

Kottelat, M. (1990c). Synopsis of the endangered buntingis (Osteichthys: Adrianichthyidae and Oryziidae) of Lake Poso, Central Sulawesi, Indonesia, with a new reproductive guild and descriptions of three new species. *Ichthyol. Exploration Freshwaters* 1(1), 49–67.

Kottelat, M. (1991a). Sailfin silversides (Pisces: Telmatherinidae) of Lake Matano, Sulawesi, Indonesia with descriptions of six new species. *Ichthyol. Exploration Freshwater* 1(4), 321–344.

Kottelat, M. (1991b). Notes on the taxonomy and distribution of some western Indonesian freshwater fishes, with diagnoses of a new genus and six new species (Pisces: Cyprinidae, Belontiidae, and Chaudhuriidae). *Ichthyol. Exploration Freshwaters* 2(3), 273–287.

Kottelat, M. (1997). European freshwater fishes. *Biologia* 52(Suppl. 5), 1–271.

Kottelat, M. (1998). Fishes of the Nam Theun and Xe Bangfai basins (Laos), with diagnoses of twenty new species (Cyprinidae, Balitoridae, Cobitidae, Coiidae, and Odontobutidae). *Ichthyol. Exploration Freshwaters* 9(1), 1–128.

Kottelat, M., and Chu, X.-L. (1988). Revision of *Yunnanilus* with descriptions of a miniature species flock and six new species from China (Cypriniformes: Homalopteridae). *Environ. Biol. Fishes* 23, 65–93.

Kottelat, M., and Lim, K. P. (1994). Diagnoses of two new genera and three new species of earthworm eels from the Malay Peninsula and Borneo (Teleostei: Chaudhuriidae). *Ichthyol. Exploration Freshwaters* 5(2), 181–190.

Kottelat, M., and Lim, K. P. (1999). Mating behavior of *Zenarchopterus gilli* and *Zenarchopterus buffonis* and function of the modified dorsal and anal fin rays in some species of *Zenarchopterus* (Teleostei: Hemiramphidae). *Copeia* 1999(4), 1097–1101.

Kottelat, M., and Ng, H. H. (1999). *Belodontichthys truncatus*, a new species of silurid catfish from

Indochina (Teleostei: Siluridae). *Ichthyol. Exploration Freshwaters* 10(4), 387–391.

Kottelat, M., and Vidthayanon, C. (1993). *Boraras micros*, a new genus and species of minute freshwater fish from Thailand (Teleostei: Cyprinidae). *Ichthyol. Exploration Freshwaters* 4(2), 161–176.

Kottelat, M., Whitten, A. J., Kartikasari, S. N., and Wirjoatmodjo, S. (1993). *Freshwater Fishes of Western Indonesia and Sulawesi.* Periplus Editions (HK), Singapore.

Koumans, F. P. (1953). *The Fishes of the Indo-Australian Archipelago. Vol. X. Gobioidea.* Brill, Leiden.

Kozhov, M. (1963). *Lake Baikal and Its Life. Monographie Biologicae 11.* Junk, The Hague.

Kramer, B. (1974). Electric organ discharge interaction during interspecific agonistic behabior in freely swimming mormyrid fish. *J. Comp. Physiol.* 93, 203–235.

Kramer, B. (1978). Spontaneous discharge rythms and social signalling in the weakly electric fish *Pollimyrus isidori* (Curvier and Valenciennes) (Mormyridae, Teleostei). *Behav. Ecol. Sociobiol.* 4, 61–74.

Kramer, B. (Ed.) (1996). *Electroception and Communication in Fishes.* Fischer, Stuttgart.

Krekorian. C. O'N. (1976). Field observations in Guyana on the reproductive biology of the spraying characid, *Copeina arnoldi* Regan. *Am. Midland Nat.* 96(1), 88–97.

Krupp, F. (1983). Freshwater fishes of Saudi Arabia and adjacent regions of the Arabian Peninsula. *Fauna Saudi Arabia* 5, 568–636.

Kuehne, R. A., and Barbour, R. W. (1983). *The American Darters.* Univ. Press of Kentucky, Lexington.

Kuiter, R. H. (1993). *Coastal Fishes of South-Eastern Australia.* Crawford House, Bathurst, NSW, Australia.

Kuiter, R. H., Humphries, P. A., and Arthington, A. H. (1996). Family Nannopercidae: Pygmy perches. In *Freshwater Fishes of South-Eastern Australia* (R. M. McDowall, Ed.), pp. 168–175. Reed, Sydney.

Kulkarni, C. V. (1940). On the systematic position, structural modification, bionomics, and development of a remarkable new family of cyprinodont fishes from the province of Bombay. *Rec. Indian Museum* 42(2), 379–423.

Kullander, S. O. (1983). *A Revision of the South American Cichlid Genus Cichlasoma (Teleostei: Cichlidae).* Naturhistoriska Riksmuseet, Stockholm.

Kullander, S. O. (1996). *Heroina isonycterina*, a new genus and species of cichlid fish from Western Amazonia, with comments on cichlasomine systematics. *Ichthyol. Exploration Freshwaters* 7(2), 149–172.

Kullander, S. O. (1998). A phylogeny and classification of the South American Cichlidae (Teleostei: Perciformes). In *Phylogeny and Classification of Neotropical Fishes* (L. R. Malabarba, R. E. Reis, R. P. Vari, Z. M. S. Lucena, and C. A. S. Lucena, Eds.), pp. 461–498. EDIPUCRS, Porto Alegre, Brazil.

Kullander, S. O., and Hartel, K. E. (1997). The systematic status of cichlid genera described by Louis Agassiz in 1859: *Amphilophus, Baiodon, Hypsophrys,* and *Parachromis* (Teleostei: Cichlidae). *Ichthyol. Exploration Freshwaters* 7(3/4), 193–202.

Kweon, H.-S., Park, E.-H., and Peters, N. (1998). Spermatozoon ultrastructure in the internally self-fertilizing hermaphroditic teleost, *Rivulus marmoratus* (Cyprinodontiformes, Rivulidae). *Copeia* 1998(4), 1101–1106.

Ladich, F. (1997). Agonistic behavior and significance of sounds in vocalizing fish. *Mar. Freshwater Behav. Physiol.* 29, 87–108.

Lagler, K. F., Bardach, J. E., Miller, R. R., and Passino, D. R. M. (1977). *Ichthyology*, 2nd ed. Wiley, New York.

Lake, J. S. (1967). Rearing experiments with five species of Australian freshwater fishes I. Inducement to spawning. *Austr. J. Mar. Freshwater Res.* 18, 137–153.

Lake, J. S. (1971). *Freshwater Fishes and Rivers of Australia.* Nelson, Melbourne.

Lake, J. S. (1978). *Australian Freshwater Fishes.* Nelson, Melbourne.

Lake, J. S., and Midgley, S. H. (1970). Reproduction of freshwater Ariidae in Australia. *Austr. J. Sci.* 32(11), 441.

Langeani, F. (1998). Phylogenetic study of the Hemiodontidae (Ostariophysi: Characiformes). In *Phylogeny and Classification of Neotropical Fishes* (L. R. Malabarba, R. E. Reis, R. P. Vari, Z. M. S. Lucena, and C. A. S. Lucena, Eds.), pp. 145–160. EDIPUCRS, Porto Alegre, Brazil.

Langeani, F. (1999). New species of *Hemiodus* (Ostariophysi, Characiformes, Hemiodontidae) from the Rio Tocantins, Brazil, with comments on color patterns and tooth shapes within the species and genus. *Copeia* 1999(3), 718–722.

Larimore, R. W. (1957). Ecological life history of the warmouth (Centrarchidae). *Bull. Illinois Nat. History Surv.* 27(1), 1–83.

Larson, H. K. (1995). A review of the Australian endemic gobiid fish genus *Chlamydogobius*, with description of five new species. *The Beagle Rec. Museums Art Galleries Northern Territory* 12, 19–51.

Larson, H. K., and Hoese, D. F. (1996a). Gudgeons. In *Freshwater Fishes of South-Eastern Australia* (R. M. McDowall, Ed.), pp. 200–219. Reed, Sydney.

Larson, H. K., and Hoese, D. F. (1996b). Gobies. In *Freshwater Fishes of South-Eastern Australia* (R. M. McDowall, Ed.), pp. 220–228. Reed, Sydney.

Larson, H. K., and Kottelat, M. (1992). A new species of *Mugilogobius* (Pisces: Gobiidae) from Lake Matano, central Sulawesi, Indonesia. *Ichthyol. Exploration Freshwater* 3(3), 225–234.

Larson, H. K., and Martin, K. C. (1989). *Freshwater Fishes of the Northern Territory*, Handbook Series No. 1. Northern Territory Museum of Arts and Sciences, Darwin, NT, Australia.

Last, P. R., and Stevens, J. D. (1994). *Sharks and Rays of Australia.* CSIRO, Melbourne.

Lauder, G. V., and Liem, K. F. (1981). Prey capture by *Luciocephalus pulcher*: Implications for models of jaw protrusion in teleost fishes. *Environ. Biol. Fishes* 6, 257–268.

Lauder, G. V., and Liem, K. F. (1983). The evolution and interrelationships of the actinopterygian fishes. *Bull. Museum Comp. Zool.* 150, 95–197.

Laur, D. R., and Ebeling, A. E. (1983). Predator–prey relationships in surfperches. *Environ. Biol. Fishes* 8, 217–229.

Lavoué, S., Bigorne, R., Lecointre, G., and Agnése, J. F. (2000). Phylogenetic relationships of mormyrid electric fishes (Mormyridae: Teleostei) inferred from cytochrome b sequences. *Mol. Phylogenet. Evol.* 14(1), 1–10.

Lecointre, G., and Nelson, G. (1996). Clupeomorpha, sister-group of Ostariophysi. In *Interrelationships of Fishes* (M. L. J. Stiassny, L. R. Parenti, and G. D. Johnson, Eds.), pp. 193–207. Academic Press, San Diego.

Lecointre, G., Bonillo, C., Ozouf-Costaz, C. and Hureau, J.-C. (1997). Molecular evidence for the origin of Antarctic fishes: Paraphyly of the Bovichthidae and no indication for the monophyly of the Notothenioidei (Teleostei). *Polar Biol.* 18, 193–208.

Lee, D. S. (1980). *Labidesthes sicculus* (Cope) Brook silversides. In *Atlas of North American Freshwater Fishes* (D. S. Lee *et al.,* Ed.), p. 557. North Carolina State Museum of Natural History, Raleigh.

Lee, D. S., Gilbert, C. R., Hocutt, C. H., Jenkins, R. E., McAllister, D. E., and Stauffer, J. R., Jr. (Eds.) (1980). *Atlas of North American Freshwater Fishes.* North Carolina State Museum of Natural History, Raleigh.

Lee, D. S., Platania, S. P., and Burgess, G. H. (1983). *Atlas of North American Freshwater Fishes,* Occasional Papers of the North Carolina Biological Survey 1983–1986, 1983 Supplement.

Legendre, M., Teugels, G. G., Cauty, C., and Jalabert, B. (1992). A comparative study on morphology, growth rate and reproduction of *Clarias gariepinus* (Burchell, 1822), *Heterobranchus longifilis* Valenciennes, 1840, and their reciprocal hybrids (Pisces, Clariidae). *J. Fish Biol.* **40**, 59–79.

LeGrande, W. H. (1981). Chromosomal evolution in North American catfishes (Siluriformes: Ictaluridae) with particular emphasis on the madtoms, *Noturus. Copeia* **1981**(1), 33–52.

Leis, J. M. (1984). Tetradontiformes: Relationships. In *Ontogeny and Systematics of Fishes* (H. G. Moser et al., Eds.), Special Publication No. 1, pp. 459–463. American Society of Ichthyologists and Herpetologists, Lawrence, KS.

Lenfant, C., Johansen, K., and Hanson, D. (1970). Bimodal gas exchange and ventilation–perfusion relationships in lower vertebrates. *Proc. Fed. Am. Soc. Exp. Biol.* **29**, 1124–1129.

Lenglet, G. (1973). Contributions à l'étude de l'anatomie viscérale des Kneriidae. *Ann. Soc. R. Zool. Belgium* **103**(2/3), 239–270.

Lévêque, C. (1990a). Protopteridae. In *Faune des Poissons d'eaux Douce et Saumâtres de l'Afrique de l'Ouest* (C. Lévêque, D. Paugy, and G. G. Teugels, Eds.), Vol. 1, pp. 76–78. ORSTOM, Paris.

Lévêque, C. (1990b). Phractolaemidae. In *Faune des Poissons d'eaux Douce et Saumâtres de l'Afrique de l'Ouest* (C. Lévêque, D. Paugy, and G. G. Teugels, Eds.), Vol. 1, pp. 189–191. ORSTOM, Paris.

Lévêque, C. (1990c). Relict tropical fish fauna in Central Sahara. *Ichthyol. Exploration Freshwaters* **1**, 39–48.

Lévêque, C. (1992). Tetraodontidae. In *Faune des Poissons d'eaux Douce et Saumâtres de l'Afrique de l'Ouest* (C. Lévêque, D. Paugy, and G. G. Teugels (Eds.), Vol. 2, pp. 868–870. ORSTOM, Paris.

Lévêque, C., Paugy, D., and Teugels, G. G. (Eds.) (1990). *Faune des Poissons d'eaux Douces et Saumâtres de l'Afrique de l'Ouest*, 2 vols. ORSTOM/MRAC, Tervuren/Paris.

Lever, C. (1996). *Naturalized Fishes of the World.* Academic Press, San Diego.

Li, G.-Q., and Wilson, M. V. H. (1996). Phylogeny of Osteoglossomorpha. In *Interrelationships of Fishes* (M. L. J. Stiassny, L. R. Parenti, and G. D. Johnson, Eds.), pp. 163–174. Academic Press, San Diego.

Li, G.-Q., and Wilson, M. V. H. (1999). Early divergence of Hiodontiformes *sensu stricto* in East Asia and phylogeny of some Late Mesozoic teleosts from China. In *Mesozoic Fishes 2— Systematics and the Fossil Record* (G. Arratia and H.-P. Schultze, Eds.), pp 369–384. Verlag Dr. Friedrich Pfiel, Munich.

Li, G-Q., Wilson, M. V. H., and Grande, L. (1993). Review of *Eohiodon* (Teleostei:

Osteoglossomorpha) from Western America, with a phylogenetic reassessment of Hiodontidae. *J. Paleontol.* **71**(6), 1109–1124.

Liem, K. F. (1963). The comparative osteology and phylogeny of the Anabantoidei (Teleostei, Pisces). *Illinois Biol. Monogr.* **30**, 1–149.

Liem, K. F. (1965). The status of the anabantoid fish genera *Ctenops* and *Trichopsis. Copeia* **1965**(2), 206–213.

Liem, K. F. (1967a). A morphological study of *Luciocephalus pulcher*, with notes on gular elements in other recent teleosts. *J. Morphol.* **121**, 103–133.

Liem, K. F. (1967b). Functional morphology of the head of the anabantoid teleost fish *Helostoma temmincki. J. Morphol.* **121**, 135–158.

Liem, K. F. (1968). Geographical and taxonomic variation in the pattern of natural sex reversal in the teleost fish order Synbranchiformes. *J. Zool.* (*London*) **156**, 225–238.

Liem, K. F. (1970). Comparative functional anatomy of the Nandidae (Pisces: Teleostei). *Fieldiana Zool.* **56**, 1–166.

Liem, K. F. (1973). Evolutionary strategies and morphological innovations: Cichlid pharyngeal jaws. *Syst. Zool.* **22**(4), 425–441.

Liem, K. F. (1980). Air ventilation in advanced teleosts: Biomechanical and evolutionary aspects. In *Environmental Physiology of Fishes* (M. A. Ali, Ed.), pp. 57–91. Plenum, New York.

Liem, K. F. (1981). Larvae of air-breathing fishes as countercurrent flow devices in hypoxic environments. *Science* **211**, 1177–1179.

Liem, K. F. (1984). The muscular basis of aquatic and aerial ventilation in the air-breathing teleost fish *Channa. J. Exp. Biol.* **113**, 1–18.

Liem, K. F. (1986). The pharyngeal jaw apparatus of the Embiotocidae (Teleostei): A functional and evolutionary perspective. *Copeia* **1986**(2), 311–323.

Liem, K. F. (1987). Functional design of the air ventilation apparatus and overland excursions by teleosts. *Fieldiana Zool.* **37**, 1–29.

Liem, K. F. (1994). Swampeels. In *Encyclopedia of Fishes* (J. P. Paxton and W. N. Eschmeyer, Eds.), pp. 173–174. Academic Press, San Diego.

Liem, K. F., Eclancher, B., and Fink, W. L. (1984). Aerial respiration in the banded knife fish *Gymnotus carapo* (Teleostei: Gymnotoidei). *Physiol. Zool.* **57**(1), 185–195.

Limbaugh, C. (1961). Cleaning symbiosis. *Sci. Am.* **205**, 42–49.

Lindquist, D. G. (1980a). *Dormitator maculatus* (Bloch). In *Atlas of North American Freshwater Fishes* (D. S. Lee et al., Eds.), p. 781. North Carolina State Museum of Natural History, Raleigh.

Lindquist, D. G. (1980b). Gobies. In *Atlas of North American Freshwater Fishes* (D. S. Lee *et al.*, Eds.), pp. 785–798. North Carolina State Museum of Natural History, Raleigh.

Linnaeus, C. (1758). *Systema Naturae: Regnum Animale*, 10th ed. (reprint, 1894). Engelmann, Leipzig.

Llewellyn, L. C. (1973). Spawning, development and temperature tolerance of the spangled perch *Madigania unicolor* (Günther) from inland waters in Australia. *Austr. J. Mar. Freshwater Res.* 24(1), 73–94.

Longley, G., and Karnei, H., Jr. (1979). Status of *Trogloglanis pattersoni* Eigenmann, the toothless blindcat and status of *Satan eurystomus* Hubbs and Bailey, the widemouth blindcat, Endangered Species Report No. 5. U.S. Fish and Wildlife Service, Albuquerque, NM.

López, H. L., Menni, R. C., and Miquelarena, A. M. (1987). Lista de los peces de agua dulce de la Argentina. *Biol. Acuatica* 12, 1–50.

Losey, G. S., Jr. (1987). Cleaning symbiosis. *Symbiosis* 4, 229–258.

Lovejoy, N. R. (1996). Systematics of myliobatoid elasmobranch: With emphasis on the phylogeny and biogeography of Neotropical freshwater stingrays (Potamotrygonidae). *Zool. J. Linnean Soc.* 117, 207–257.

Lovejoy, N. R. (1997). Stingrays, parasites, and Neotropical biogeography: A closer look at Brooks *et al.*'s hypotheses concerning the origin of Neotropical freshwater rays (Potamotrygonidae). *Syst. Biol.* 46, 218–230.

Lovejoy, N. R., Bermingham, E., and Martin, A. P. (1998). Marine incursion into South America. *Nature* 396, 421–422.

Lowe-McConnell, R. H. (1964). The fishes of the Rupununi savanna district of British Guiana. Pt. 1. Groupings of fish species and effects of seasonal cycles on the fish. *J. Linnean Soc. Zool.* b 103–144.

Lowe-McConnell, R. H. (1987). *Ecological Studies in Tropical Fish Communities*. Cambridge Univ. Press, Cambridge, UK.

Lu, L. (1994). A new paddlefish from the Upper Jurassic of northeast China. *Vertebrata Pal Asiatica* 32(2), 32.

Lucena, C. A. S., and Menezes, N. A. (1998). A phylogenetic analysis of *Roestes* Günther and *Gilbertolus* Eigenmann, with a hypothesis on the relationships of the Cynodontidae and Acestrorhynchidae (Teleostei: Ostariophysi: Characiformes). In *Phylogeny and Classification of Neotropical Fishes* (L. R. Malabarba, R. E. Reis, R. P. Vari, Z. M. S. Lucena, and C. A. S. Lucena, Eds.), pp. 261–278. EDIPUCRS, Porto Alegre, Brazil.

Lundberg, J. G. (1975a). The fossil catfishes of North America. *Univ. Michigan Museum Paleontol. Papers Paleontol.* 11, 1–51.

Lundberg, J. G. (1975b). Homologies of the upper shoulder girdle and temporal region bones in catfishes (Order Siluriformes) with comments on the skull of the Helogeneidae. *Copeia* 1975(1), 66–74.

Lundberg, J. G. (1982). The comparative anatomy of the toothless blindcat, *Trogloglanis pattersoni* Eigenmann, with a phylogenetic analysis of the ictalurid catfishes. *Miscellaneous Publications Museum Zool. Univ. Michigan* 163, 1–85.

Lundberg, J. G. (1992). The phylogeny of ictalurid catfishes: A synthesis of recent work. In *Systematics, Historical Ecology, and North American Freshwater Fishes* (R. L. Mayden, Ed.), pp. 392–420. Stanford Univ. Press, Stanford, CA.

Lundberg, J. G. (1993). African-South American freshwater fish clades and continental drift: Problems with a paradigm. In *The Biotic Relationships between Africa and South America* (P. Goldblatt, Ed.), pp. 156–199. Yale Univ. Press, New Haven, CT.

Lundberg, J. G., and Baskin, J. N. (1969). The caudal skeleton of the catfish, order Siluriformes. *Am. Museum Novitates* 2398.

Lundberg, J. G., and Mago-Leccia, F. (1986). A review of *Rhabdolichops* (Gymnotiformes, Sternopygidae), a genus of South American freshwater fishes, with descriptions of four new species. *Proc. Acad. Nat. Sci. Phildelphia* 138, 53–85.

Lundberg, J. G., and McDade, L. A. (1986). On the South American catfish *Brachyrhamdia imitator* Myers (Siluriformes, Pimelodidae), with phylogenetic evidence for a large intrafamilial lineage. *Acad. Nat. Sci. Philadelphia Notulae Naturae* 463, 1–24.

Lundberg, J. G., and Rapp Py-Daniel, L. (1994). *Bathycetopsis oliveirai*, Gen. et sp. nov., a blind and depigmented catfish (Siluriformes: Cetopsidae) from the Brazilian Amazon. *Copeia* 1994(2), 381–390.

Lundberg, J. G., Machado-Allison, A., and Kay, R. F. (1986). Miocene characid fishes from Columbia: Evolutionary stasis and extirpation. *Science* 234, 208–209.

Lundberg, J. G., Linares, O. J., Antonio, M. E., and Nass, P. (1988). *Phractocephalus hemiliopterus* (Pimelodidae, Siluriformes) from the Upper Miocene Urumaco formation, Venezuela: A further case of evolutionary stasis and local extinction among South American fishes. *J. Vertebr. Paleontol.* 8(2), 131–138.

Lundberg, J. G., Bornbush, A. H., and Mago-Leccia, F. (1991a). *Gladioglanis conquistador* n. sp.

from Ecuador with diagnoses of the subfamilies Rhamdiinae Bleeker and Pseudopimelodinae n. subf. (Siluriformes: Pimelodidae). *Copeia* **1991**(1), 190–209.

Lundberg, J. G., Mago-Leccia, F., and Nass, P. (1991b). *Exallodontus aguanai*: A new genus and species of Pimelodidae (Pisces: Siluriformes) from deep river channels of South America, and delimitations of the subfamily Pimelodinae. *Proc. Biol. Soc. Washington* **104**(4), 840–869.

Lundberg, J. G., Fernandes, C. C., Albert, J. S., and Garcia, M. (1996). *Magosternarchus*, a new genus with two new species of electric fishes (Gymnotiformes: Apteronotidae) from the Amazon River Basin, South America. *Copeia* **1996**(3), 657–670.

Lundberg, J. G., Kottelat, M., Smith, G. R., Stiassny, M. L. J., and Gill, A. C. (2000). So many fishes, so little time: An overview of recent ichthyological discovery in continental waters. *Ann. Missouri Bot. Garden* **87**, 26–62.

Lydekker, R. (ca. 1903). *The New Natural History*, Vol. 5. Merril & Baker, New York.

Lythgoe, J., and Lythgoe, G. (1992). *Fishes of the Sea: The North Atlantic and Mediterranean*. MIT Press, Cambridge, MA.

Mabee, P. M. (1988). Supraneural and predorsal bones in fishes: Development and homologies. *Copeia* **1988**(4), 827–838.

Machado, F. A., and Sazima, I. (1983). Compartamento alimentar do peixe hematófago *Branchioica bertonii* (Siluriformes, Trichomycteridae). *Ciência Cultura* **35**, 344–348.

MacDonald, C. M. (1978). Morphological and biochemical systematics of Australian freshwater and estuarine percichthyid fishes. *Austr. J. Mar. Freshwater Res.* **29**(5), 667–698.

Magnuson, J. L., and Smith, L. L. (1963). Some phases of the life history of the trout-perch. *Ecology* **44**, 83–95.

Magnuson, J. J., Keller, J. W., Beckel, A. L., and Gallepp, G. W. (1983). Breathing gas mixtures different from air: An adaptation for survival under ice of a facultative air-breathing fish. *Science* **220**, 312–314.

Magnuson, J. J., Beckel, A. L., Mills, K., and Brandt, S. B. (1985). Surviving winter hypoxia: Behavioral adaptations of fishes in a northern Wisconsin winterkill lake. *Environ. Biol. Fishes* **14**, 241–250.

Mago-Leccia, F. (1970). Estudos preliminares sobre la ecologia de los peces de los llanos de Venezuela. *Acta Biol. Venezuelica* **7**(1), 71–102.

Mago-Leccia, F. (1978). Los peces de la familia Sternopygidae de Venezuela. *Acta Cientifica Venezolana* **29**(1), 1–89.

Mago-Leccia, F. (1994). Electric fishes of the continental waters of America. *Biblioteca Acad. Ciencias Fisicas Matematicas Nat. Venezuela* **29**, 1–225.

Mago-Leccia, F., and Zaret, T. M. (1978). The taxonomic status of *Rhabdolichops troscheli* (Kaup, 1856) and speculations of gymnotiform evolution. *Environ. Biol. Fishes* **3**(4), 379–384.

Mago-Leccia, F., Lundberg, J. G., and Baskin, J. N. (1985). Systematics of the South American freshwater fish genus *Adontosternarchus* (Gymnotiformes, Apteronotidae). *Contrib. Sci. Nat. History Museum Los Angeles County* **358**, 1–19.

Maisey, J. G. (1996). *Discovering Fossil Fishes*. Holt, New York.

Maitland, P. S. (1977). *Freshwater Fishes of Britain and Europe*. Hamlyn, London.

Malabarba, L. R. (1998). Monophyly of the Cheirodontinae, characters and major clades (Ostariophysi: Characidae). In *Phylogeny and Classification of Neotropical Fishes* (L. R. Malabarba, R. E. Reis, R. P. Vari, Z. M. S. Lucena, and C. A. S. Lucena, Eds.), pp. 193–233. EDIPUCRS, Porto Alegre, Brazil.

Malabarba, L. R., Reis, R. E., Vari, R. P., Lucena, Z. M. S., and Lucena, C. A. S. (Eds.) (1998). *Phylogeny and Classification of Neotropical Fishes*. EDIPUCRS, Porto Alegre, Brazil.

Mansueti, A. J. (1963). Some changes in morphology during ontogeny in the pirateperch, *Aphredoderus s. sayanus*. *Copeia* **1963**(3), 546–557.

Mansueti, A. J., and Hardy, J. D., Jr. (1967). *Development of Fishes of the Chesapeake Bay Region*. Natural Resources Institute, Univ. of Maryland, College Park.

Marcus, J. M., and McCune, A. R. (1999). Ontogeny and phylogeny in the northern swordtail clade of *Xiphophorus*. *Syst. Biol.* **48**(3), 491–522.

Markle, D. F. (1989). Aspects of character homology and phylogeny of the Gadiformes. In *Papers on the Systematics of Gadiform Fishes* (D. M. Cohen, Ed.), Science Series No. 32, pp. 59–88. Natural History Museum of Los Angles County, Los Angeles.

Marquet, G., and Mary, N. (1999). Comments on some New Caledonian freshwater fishes of economical and biogeographical interest. In *Proceedings of the 5th Indo-Pacific Conference (Nouméa, 3–8 November 1997)* (B. Séret and J.-Y. Sire, Eds.), pp. 29–39. Société Française d'Ichthyologie and Institut de Resherche pour le Développement, Paris.

Marshall, C. R. (1987). A list of fossil and extant dipnoans. In *The Biology and Evolution of Lungfish* (W. E. Bemis, W. W. Burggren, and

N. E. Kemp, Eds.), pp. 15–23. A. R. Liss, New York.

Martin, A. P., and Bermingham, E. (1998). Systematics and evolution of lower Central American cichlids inferred from analysis of cytochrome b gene sequences. *Mol. Phylogenet. Evol.* **9**, 192–203.

Martin, F. D., and Hubbs, C. (1973). Observations on the development of pirate perch, *Aphredoderus sayanus* (Pisces: Aphredoderidae) with comments on yolk circulation patterns as a possible taxonomic tool. *Copeia* **1973**(2), 377–379.

Martin, K. L. M., Berra, T. M., and Allen, G. R. (1993). Cutaneous aerial respiration during forced emergence in the Australian salamander-fish, *Lepidogalaxias salamandroides. Copeia* **1993**(3), 875–879.

Masters, C. O. (1968). The most dreaded fish in the Amazon River. *Carolina Tips* **31**(2), 5–6.

Masuda, H., Amaoka, K., Araga, C., Uyeno, T., and Yoshino, T. (Eds.) (1984). *The Fishes of the Japanese Archipelago.* Tokai Univ. Press, Tokyo.

Matsuura, K., and Tyler, J. C. (1994). Triggerfishes and their allies. In *Encyclopedia of Fishes* (J. P. Paxton and W. N. Eschmeyer, Eds.), pp. 229–233. Academic Press, San Diego.

Matthews, O. (1998, June 22). Bye, beluga. Later, Sevruga. *Newsweek*, 42.

Matthews, W. J. (1998). *Patterns in Freshwater Fish Ecology.* Chapman & Hall, New York.

Maugé, A. L. (1986a). Ambassidae. In *Check-list of the Freshwater Fishes of Africa* (J. Daget, J.-P. Gosse, and D. F. E. Thys van den Audenaerde, Eds.), pp. 297–298. CLOFFA II, ORSTOM, Paris.

Maugé, A. L. (1986b). Kuhliidae. In *Check-list of the Freshwater Fishes of Africa* (J. Daget, J.-P. Gosse, and D. F. E. Thys van den Audenaerde, Eds.), pp. 306–307. CLOFFA II, ORSTOM, Paris.

Maugé, A. L. (1986c). Eleotridae. In *Check-list of the Freshwater Fishes of Africa* (J. Daget, J.-P. Gosse, and D. F. E. Thys van den Audenaerde, Eds.), pp. 389–398. CLOFFA II, ORSTOM, Paris.

Maugé, A. L. (1986d). Gobiidae. In *Check-list of the Freshwater Fishes of Africa* (J. Daget, J.-P. Gosse, and D. F. E. Thys van den Audenaerde, Eds.), pp. 358–388. CLOFFA II, ORSTOM, Paris.

Mayden, R. L. (1989). Phylogenetic studies of North American minnows, with emphasis on the genus *Cyprinella* (Teleostei: Cypriniformes), Miscellaneous Publication No. 80. University of Kansas Museum of Natural History, Lawrence.

Mayden, R. L. (1991). Cyprinids of the New World. In *Cyprinid Fishes: Systematics, Biology, and Exploitation* (I. J. Winfield and J. S. Nelson, Eds.), pp. 240–263. Chapman & Hall, London.

Mayden, R. L. (Ed.) (1992). *Systematics, Historical Ecology, & North American Freshwater Fishes.* Stanford Univ. Press, Stanford, CA.

Mayden, R. L. (1993). *Elassoma alabamae*, a new species of pygmy sunfish endemic to the Tennessee River drainage of Alabama (Teleostei: Elassomatidae). *Bull. Alabama Museum Nat. History* **16**, 1–14.

Mayden, R. L., and Kuhajda, B. R. (1996). Systematics, taxonomy, and conservation status of endangered Alabama sturgeon, *Scaphirhynchus suttkusi* Williams and Clemmer (Actinopterygii, Acipenseridae). *Copeia* **1996**(2), 241–273.

Mayden, R. L., Burr, B. M., Page, L. M., and Miller, R. R. (1992). The native freshwater fishes of North America. In *Systematics, Historical Ecology, and North American Freshwater Fishes* (R. L. Mayden, Ed.), pp. 827–863. Stanford Univ. Press, Stanford, CA.

Mayr, E. (1944). Wallace's Line in light of recent zoogeographic studies. *Q. Rev. Biol.* **19**, 1–14.

Mayr, E. (Ed.) (1952). The problem of land connections across the south Atlantic with special reference to the Mesozoic. *Bull. Am. Museum Nat. History* **99**, 79–258.

McAllister, D. E. (1963). A revision of the smelt family. *Bull. Natl. Museum Canada* **191**, 1–53.

McAllister, D. (1984). Osmeridae. In *Fishes of the North-Eastern Atlantic and the Mediterranean* (P. J. P. Whitehead, M.-L. Bauchot, J.-C. Hureau, J. Nielsen, and E. Tortonese, Eds.), Vol. 1, pp. 399–402. UNESCO, Paris.

McCosker, J. E. (1989). Freshwater eels (Family Anguillidae) in California: Current conditions and future scenarios. *California Fish Game* **75**(1), 4–10.

McCoy, F. (1884). *Prodromus of the Zoology of Victoria. Vol. 1, Decade IX.* Melbourne.

McCulloch, A. R. (1917). Studies in Australian fishes No. 4. *Rec. Austr. Museum* **11**(7), 163–188, Plates xxix–xxxi.

McCulloch, A. R. (1920). Studies in Australian fishes No. 6. *Rec. Austr. Museum* **13**(2), 41–71, Plates x–xiv.

McCulloch, A. R., and Phillips, W. J. (1923). Notes on New Zealand fishes. *Rec. Austr. Museum* **14**(1), 18–22, Plates i–iii.

McDowall, R. M. (1967). Some points of confusion in galaxiid nomenclature. *Copeia* **1967**(4), 841–843.

McDowall, R. M. (1968). The status of *Nesogalaxias neocaledonicus* (Weber and de Beaufort) (Pisces, Galaxiidae). *Breviora* **286**, 1–8.

McDowall, R. M. (1969). Relationships of galaxioid fishes with further discussion of salmoniform classification. *Copeia* **1969**(4), 796–824.

McDowall, R. M. (1970). The galaxiid fishes of New Zealand. *Bull. Museum Comp. Zool.* **139**, 341–432.

McDowall, R. M. (1971a). Fishes of the family Aplochitonidae. *J. R. Soc. New Zealand* **1**(1), 31–52.

McDowall, R. M. (1971b). The galaxiid fishes of South America. *Zool. J. Linnean Soc.* **50**(1), 33–73.

McDowall, R. M. (1973a). The status of the South African galaxiid (Pisces: Galaxiidae). *Ann. Cape Provincial Museum (Nat. History)* **9**(5), 91–101.

McDowall, R. M. (1973b). *Galaxias indicus* Day, 1888—A *nomen dubium*. *J. R. Soc. New Zealand* **3**(2), 191–192.

McDowall, R. M. (1973c). Relationships and taxonomy of the New Zealand torrent fish, *Cheimarrichthys fosteri* Haast (Pisces: Mugiloididae). *J. R. Soc. New Zealand* **3**(2), 199–217.

McDowall, R. M. (1976a). Fishes of the family Prototroctidae (Salmoniformes). *Austr. J. Mar. Freshwater Res.* **27**(4), 641–659.

McDowall, R. M. (1976b). Notes on some *Galaxias* fossils from the Pliocene of New Zealand. *J. R. Soc. New Zealand* **6**, 17–22.

McDowall, R. M. (1978). Generalized tracks and dispersal in biogeography. *Syst. Zool.* **27**(1), 88–104.

McDowall, R. M. (1979). Fishes of the family Retropinnidae (Pisces: Salmoniformes): A taxonomic revision and synopsis. *J. R. Soc. New Zealand* **9**(1), 85–121.

McDowall, R. M. (1984a). Southern hemisphere freshwater salmoniforms: Development and relationships. In *Ontogeny and Systematics of Fishes* (H. G. Moser *et al.*, Eds.), Special Publication No. 1, pp. 150–153. American Society of Ichthyologists and Herpetologists, Lawrence, KS.

McDowall, R. M. (1984b). *The New Zealand Whitebait Book*. Reed, Wellington, New Zealand.

McDowall, R. M. (1988). *Diadromy in Fishes.* Timber Press/Croom Helm, Portland, OR/London.

McDowall, R. M. (1990). *New Zealand Freshwater Fishes: A Natural History and Guide.* Heinemann Reed, Auckland, New Zealand.

McDowall, R. M. (Ed.) (1996). *Freshwater Fishes of South-Eastern Australia.* Reed, Sydney.

McDowall, R. M. (1997a). Two further new species of *Galaxias* (Teleostei: Galaxiidae) from the Taieri River, southern New Zealand. *J. R. Soc. New Zealand* **27**(2), 199–217.

McDowall, R. M. (1997b). Affinities, generic classification and biogeography of the Australian and New Zealand Mudfishes (Salmoniformes: Galaxiidae). *Rec. Austr. Museum* **49**, 121–137.

McDowall, R. M. (1997c). An accessory lateral line in some New Zealand and Australian galaxiids (Teleostei: Galaxiidae). *Ecol. Freshwater Fish* **6**, 217–224.

McDowall, R. M. (1998a). Phylogenetic relationships and ecomorphological divergence in sympatric and allopatric species of *Paragalaxias* (Teleostei: Galaxiidae) in high elevation Tasmanian lakes. *Environ. Biol. Fishes* **53**, 235–257.

McDowall, R. M. (1998b). Driven by diadromy: Its role in the historical and ecological biogeography of the New Zealand freshwater fish fauna. *Italian J. Zool.* **65**(Suppl.), 73–85.

McDowall, R. M. (1999). Caudal skeleton in *Galaxias* and allied genera (Teleostei: Galaxiidae). *Copeia* **1999**(4), 932–939.

McDowall, R.M. (2000). Biogeography of the New Zealand Torrentfish, *Cheimarrichthys fosteri* (Teleostei: Pinguipedidae): A distribution driven mostly by ecology and behavior. *Environ. Biol. Fishes* **58**, 119–131.

McDowall, R. M., and Chadderton, W. L. (1999). *Galaxias gollumoides* (Teleostei: Galaxiidae), a new fish species from Stewart Island, with notes on other non-migratory freshwater fishes present on the island. *J. R. Soc. New Zealand* **29**(1), 77–88.

McDowall, R. M., and Frankenberg, R. S. (1981). The galaxiid fishes of Australia. *Rec. Austr. Museum* **33**(10), 443–605.

McDowall, R. M., and Nakaya, K. (1988). Morphological divergence in the two species of *Aplochiton* Jenyns (Salmoniformes: Aplochitonidae): A generalist and a specialist. *Copeia* **1988**(1), 233–236.

McDowall, R. M., and Pole, M. (1977). A large galaxiid fossil (Teleostei) from the Miocene of Central Otago, New Zealand. *J. R. Soc. New Zealand* **27**(2), 193–198.

McDowall, R. M., and Pusey, B. J. (1983). *Lepidogalaxias salamandroides* Mees—A redescription, with natural history notes. *Rec. Western Austr. Museum* **11**, 11–23.

McDowall, R. M., and Wallis, G. P. (1996). Description and redescription of *Galaxias* species (Teleostei: Galaxiidae) from Otago and Southland. *J. R. Soc. New Zealand* **26**(3), 401–427.

McDowall, R. M., Robertson, D. A., and Saito, R. (1975). Occurrence of galaxiid larvae and juveniles in the sea. *New Zealand J. Mar. Freshwater Res.* **9**, 1–9.

McDowall, R. M., Clark, B. M., Wright, G. J., and Northcote, T. G. (1993). *Trans*-2-*cis*-6-nonadienal: The cause of cucumber odor in osmerid and retropinnid smelts. *Trans. Am. Fish. Soc.* **122**(1), 144–147.

McDowall, R. M., Mitchell, C. P., and Brothers, E. B. (1994). Age at migration from the sea of juvenile *Galaxias* in New Zealand (Pisces: Galaxiidae). *Bull. Mar. Sci.* **54**, 385–402.

McDowall, R. M., Jellyman, D. J., and Dijkstra, L. H. (1998). Arrival of an Australian anguillid eel in New Zealand: An example of transoceanic dispersal. *Environ. Biol. Fishes* **51**, 1–6.

McEachran, J. D., and Capapé, C. (1984). Dasyatidae. In *Fishes of the North-Eastern Atlantic and the Mediterranean* (P. J. P. Whitehead, M.-L. Bauchot, J.-C. Hureau, J. Nielsen, and E. Tortonese, Eds.), Vol. 1, pp. 197–202. UNESCO, Paris.

McEachran, J. D., Dunn, K. A., and Miyake, T. (1996). Interrelationships of the batoid fishes (Chondrichthyes: Batoidea). In *Interrelationships of Fishes* (M. L. J. Stiassny, L. R. Parenti, and G. D. Johnson, Eds.), pp. 63–84. Academic Press, San Diego.

McKay, R. J. (1984). Introductions of exotic fishes in Australia. In *Distribution, Biology, and Management of Exotic Fishes* (W. R. Courtenay, Jr., and J. R. Stauffer, Jr., Eds.), pp. 177–199. Johns Hopkins Univ. Press, Baltimore.

McKenzie, D. J., and Randall, D. J. (1990). Does *Amia calva* estivate? *Fish Physiol. Biochem.* **8**, 147–158.

McLennan, D. A. (1993). Phylogenetic relationships in the Gasterosteidae: An updated tree based on behavioral characters with a discussion of homoplasy. *Copeia* **1993**(2), 318–326.

Meek, S. E. (1904). The fresh-water fishes of Mexico north of the Isthmus of Tehuantepec. *Field Columbian Museum Zool. Ser. 5*, v–xiii, 1–252, Plates 1–17.

Meek, S. E., and Hildebrand, S. F. (1916). The fishes of the fresh waters of Panama. Publication No. 191, Zoological Series 10(15), pp. 217–374, Plates VI–XXXII. Field Museum of Natural History.

Mees, G. F. (1961). Description of a new fish of the family Galaxiidae from Western Australia. *J. R. Soc. Western Austr.* **44**, 33–38.

Mees, G. F. (1987). The members of the subfamily Aspredininae, family Aspredinidae in Surinam (Pisces, Negatognathi). *Proc. Koninkklijke. Nederlandse Akademie voor Wetenschappen.* C **90**(2), 173–192.

Mees, G. F. (1988). The genera of the subfamily Bunocephalinae (Pisces, Negatognathi, Aspredinidae). *Proc. Koninkklijke. Nederlandse Akademie voor Wetenschappen.* C **91**(1), 85–102.

Mees, G. F. (1989). Notes on the genus *Dysichthys*, subfamily Bunocephalinae (Pisces, Negatognathi). *Proc. Koninkklijke. Nederlandse Akademie voor Wetenschappen.* C **92**(2), 189–205.

Mees, G. F., and Kailola, P. J. (1977). The freshwater Therapontidae of New Guinea. *Zool. Verhand.* **153**, 1–89.

Meffe, G. K., and Snelson, F. F., Jr. (Eds.) (1989). *Ecology and Evolution of Livebearing Fishes.* Prentice Hall, Englewood Cliffs, NJ.

Megrey, B. A. (1980). *Dorosoma cepedianum* (Lesueur) gizzard shad. In *Atlas of North American Freshwater Fishes* (D. S. Lee et al., Eds.), p. 69. North Carolina State Museum of Natural History, Raleigh.

Meisner, A. D., and Collette, B. B. (1998). A new species of viviparous halfbeak, *Dermogenys bispina* (Teleostei: Hemiramphidae) from Sabah (North Borneo). *Raffles Bull. Zool.* **46**(2), 373–380.

Meisner, A. D., and Collette, B. B. (1999). Generic relationships of the internally-fertilized southeast Asian halfbeaks (Hemiramphidae: Zenarchopterinae). In *Proceedings of the 5th Indo-Pacific Conference (Nouméa, 3–8 November 1997)* (B. Séret and J.-Y. Sire, Eds.), pp. 69–76. Société Française d'Ichthyologie and Institut de Resherche pour le Développement, Paris.

Meldrim, J. W. (1980). *Novumbra hubbsi* Schultz Olympic mudminnow. In *Atlas of North American Freshwater Fishes* (D. S. Lee et al., Eds.), p. 128. North Carolina State Museum of Natural History, Raleigh.

Menezes, N. A. (1969). The food of *Brycon* and three closely related genera of the tribe Acestrorhynchini. *Papeis Avulsos Zool.* **22**, 217–223.

Menezes, N. A., and Géry, J. (1983). Seven new ascestrorhynchin characid species (Osteichthyes, Ostariophysi, Characiformes) with comments on the systematics of the group. *Rev. Suisse Zool.* **5**(5), 563–592.

Menezes, N. A., and Lucena, C. A. S. (1998). Revision of the subfamily Roestinae (Ostariophysi: Characiformes: Cynodontidae). *Ichthyol. Exploration Freshwaters* **9**(3), 279–291.

Menezes, N. A., and Weitzman, S. H. (1990). Two new species of *Mimagoniates* (Teleostei: Characidae: Glandulocaudinae), their phylogeny and biogeography and a key to the glandulocaudin fishes of Brazil and Paraguay. *Proc. Biol. Soc. Washington* **103**(2), 380–426.

Menni, R. C., and Gómez, S. E. (1995). On the habitat and isolation of *Gymnocharacinus bergi* (Osteichthyes: Characidae). *Environ. Biol. Fishes* **42**, 15–23.

Menon, A. G. K. (1977). A systematic monograph of the tongue soles of the genus *Cynoglossus* Hamilton–Buchanan (Pisces: Cynoglossidae). *Smithsonian Contrib. Zool.* 238, 1–129.

Merrick, J. R. (1982). Pond culture of the spotted barramundi, *Scleropages leichardti* (Pisces: Osteoglossidae). *Aquaculture* 29, 171–176.

Merrick, J. R. (1996). Family Terapontidae. Freshwater grunters or perches. In *Freshwater Fishes of South-Eastern Australia* (R. M. McDowall, Ed.), pp. 164–167. Reed, Sydney.

Merrick, J. R., and Schmida, G. E. (1984). *Australian Freshwater Fishes*. J. R. Merrick School of Biological Sciences, Macquarie University, North Ryde, N. S. W, Australia.

Merron, G. S., Holden, K. K., and Bruton, M. N. (1990). The reproductive biology of early development of the African pike *Hepsetus odoe*, in the Okavango delta, Botswana. *Environ. Biol. Fishes* 28, 215–235.

Mettee, M. F. (1974). A study on the reproductive behavior, embryology, and larval development of the pygmy sunfishes in the genus *Elassoma*. Ph.D. dissertation, University of Alabama, Tascaloosa.

Mettee, M. F., O'Neil, P. E., and Pierson, J. M. (1996). *Fishes of Alabama and the Mobil Basin*. Oxmoor House, Birmingham, AL.

Meyer, A. (1993). Phylogenetic relationships and evolutionary processes in East African cichlid fishes. *Trends Ecol. Evol.* 8(8), 279–284.

Meyer, A. (1995). Molecular evidence on the origin of tetrapods and the relationships of the coelacanth. *Trends Ecol. Evol.* 10, 111–116.

Meyer, A., Kocher, T. D., Basasibwaki, P., and Wilson, A. C. (1990). Monophyletic origin of Lake Victoria cichlid fishes suggested by mitochondrial DNA sequences. *Nature* 347, 550–553.

Midgley, S. H. (1968). A study of Nile perch in Africa and its suitability for Australian conditions, Fellowship Report No. 3. Winston Churchill Memorial Trust, Australia.

Midgley, S. H., Midgley, M., and Rowland, S. J. (1991). Fishes of the Bulloo-Bacannia drainage division. *Mem. Queensland Museum* 30(3), 505–508.

Miller, P. J. (1973). The osteology and adaptive features of *Rhyacichthys aspro* (Teleostei: Gobioidei) and the higher classification of gobioid fishes. *J. Zool. (London)* 171, 397–434.

Miller, P. J. (1986). Gobiidae. In *Fishes of the North-Eastern Atlantic and the Mediterranean* (P. J. P. Whitehead, M.-L. Bauchot, J.-C. Hureau, J. Nielsen, and E. Tortonese, Eds.), Vol. 3, pp. 1019–1085. UNESCO, Paris.

Miller, P. J. (1987). Affinities, origin and adaptive features of the Australian desert goby, *Chlamydogobius eremius* (Zietz, 1896)

(Teleostei: Gobiidae). *J. Nat. History* 21, 687–705.

Miller, P. J. (1990). The endurance of endemism: The Mediterranean freshwater gobies and their prospects for survival. *J. Fish Biol.* 37(Suppl. A), 145–156.

Miller, R. G. (1993). *History and Atlas of the Fishes of the Antarctic Ocean*. Foresta Institute for Ocean and Mountain Studies, Carson City, NV.

Miller, R. R. (1948). The cyprinodont fishes of the Death Valley system of eastern California and southwestern Nevada, Publication No. 68, pp. 1–155. Museum of Zoology, University of Michigan, Ann Arbor.

Miller, R. R. (1955a). A systematic review of the Middle American fishes of the genus *Profundulus*, Publication No. 92, pp. 1–74. Museum of Zoology, University of Michigan, Ann Arbor.

Miller, R. R. (1955b). An annotated list of the American cyprinodontid fishes of the genus *Fundulus* with the description of *Fundulus persimilis* from Yucatan, Occasional Papers No. 568, pp. 1–25. Museum of Zoology, University of Michigan, Ann Arbor.

Miller, R. R. (1956). A new genus and species of cyprinodontid fish from San Luis Potosi, Mexico, with remarks on the subfamily Cyprinodontinae, Occasional Papers No. 581, pp. 1–17. Museum of Zoology, University of Michigan, Ann Arbor.

Miller, R. R. (1958). Origin and affinities of the freshwater fish fauna of western North America. In *Zoogeography* (C. L. Hubbs, Ed.), pp. 187–222. Arno Press, New York.

Miller, R. R. (1961). Speciation rates in some freshwater fishes of western North America. In *Vertebrate Speciation* (F. Blair, Ed.), pp. 537–560. Univ. of Texas Press, Austin.

Miller, R. R. (1966). Geographical distribution of Central American freshwater fishes. *Copeia* 1966(4), 773–802.

Miller, R. R. (1979). Ecology, habits and relationships of the Middle American cautro ojos, *Anableps dowi* (Pisces: Anablepidae). *Copeia* 1979(1), 82–91.

Miller, R. R., and Carr, A. (1974). Systematics and distribution of some freshwater fishes from Honduras and Nicaragua. *Copeia* 1974(1), 120–125.

Miller, R. R., and Fitzsimons, J. M. (1971). *Ameca splendens*, a new genus and species of goodeid fish from western México, with remarks on the classification of the Goodeidae. *Copeia* 1971(1), 1–17.

Miller, R. R., and Hubbs, C. L. (1969). Systematics of *Gasterosteus aculeatus*, with particular reference to intergradation and introgression along the

Pacific coast of North America: A commentary on a recent contribution. *Copeia* **1969**(1), 52–69.

Miller, R. R., and Smith, G. R. (1981). The distribution and evolution of *Chasmistes* (Pisces: Catostomidae) in western North America, Occasional Papers No. 696, pp. 1–46. Museum of Zoology, University of Michigan, Ann Arbor.

Miller, R. R., and Smith, M. L. (1986). Origin and geography of fishes of central Mexico. In *The Zoogeography of North American Freshwater Fishes* (C. H. Hocult and E. O. Wiley, Eds.), pp. 487–517. Wiley, New York.

Miller, R. R., and Waters, V. (1972). A new genus of cyprinodontid fish from Nuevo Leon, Mexico, Contributions to Science No. 233, pp. 1–13. Natural History Museum of Los Angeles County, Los Angeles.

Mims, S. D., Dingerkus, G., and Claster, L. S. (1994). Chinese paddlefish research: A model for U.S.–China aquatic conservation. *Diversity* **10**(4), 23–25.

Minckley, W. L., and Deacon, J. E. (Eds.) (1991). *Battle against Extinction*. Univ. of Arizona Press, Tucson.

Minckley, W. L., and DeMarais, B. D. (2000). Taxonomy of chubs (Teleostei, Cyprinidae, genus *Gila*) in the American southwest with comments on conservation. *Copeia* **2000**(1), 251–256.

Mitchell, R. W., Russell, W. H., and Elliot, W. R. (1977). Mexican eyeless characin fishes, genus *Astyanax*: Environment, distribution, and evolution, Special Publication No. 12. Texas Tech University Museum, Lubbock

Mo, T. (1991). *Anatomy, Relationships and Systematics of the Bagridae (Teleostei: Siluroidei) with a Hypothesis of Siluroid Phylogeny*. Koeltz, Koenigstein, Germany.

Mok, E. Y. M., and Munro, A. D. (1991). Observations on the food and feeding adaptations of four species of small pelagic teleosts in streams of Sungei Buloh mangal, Singapore. *Reffles Bull. Zool.* **39**, 235–257.

Mok, H.-K. (1981). The posterior cardinal veins and kidneys of fishes, with notes on their phylogenetic significance. *Jpn. J. Ichthyol.* **27**(4), 281–290.

Moller, P. (1995). *Electric Fishes: History and Behavior*, Fish and Fisheries Series No. 17. Chapman & Hall, London.

Moller, P., and Serrier, J. (1986). Species recognition in mormyrid weakly electric fish. *Anim. Behav.* **34**, 332–339.

Moller, P., Serrier, J., and Bowling, D. (1989). Electric organ discharge displays during social encounter in the weakly electric fish *Brienomyrus niger* L. (Mormyridae). *Ethology* **82**, 177–191.

Mong, S. J., and Berra, T. M. (1979). The effects of increasing dosages of X-radiation on the chromosomes of the central mudminnow, *Umbra limi* (Kirtland) (Salmoniformes: Umbridae). *J. Fish Biol.* **14**, 523–527.

Monkolprasit, S., and Roberts, T. R. (1990). *Himantura chaophraya*, a new giant freshwater stingray from Thailand. *Jpn. J. Ichthyol.* **37**(3), 203–208.

Montoya-Burgos, J.-I., Muller, S., Weber, C., and Pawlowski, J. (1998). Phylogenetic relationships of the Loricariidae (Siluriformes) based on mitochondrial rRNA gene sequences. In *Phylogeny and Classification of Neotropical Fishes* (L. R. Malabarba, R. E. Reis, R. P. Vari, Z. M. S. Lucena, and C. A. S. Lucena, Eds.), pp. 363–374. EDIPUCRS, Porto Alegre, Brazil.

Mooi, R. D. (1990). Egg surface morphology of Pseudochromoids (Perciformes: Percoidei), with comments on its phylogenetic implications. *Copeia* **1990**(2), 455–475.

Moriarty, C. (1978). *Eels: A Natural and Unnatural History*. David & Charles, Newton Abbot, UK.

Morrow, J. E. (1980). *The Freshwater Fishes of Alaska*. Alaska Northwest, Anchorage, Alaska.

Moyle, P. B. (1976). *Inland fishes of California*. Univ. of California Press, Berkeley.

Moyle, P. B. (1977a). In defense of sculpins. *Fisheries* **2**(1), 20–23.

Moyle, P. B. (1977b). Are coarse fish a curse? *Fly Fisherman* **8**(5), 35–39.

Moyle, P. B. (1980). *Archoplites interruptus* (Girard) Sacramento perch. In *Atlas of North American Freshwater Fishes* (D. S. Lee *et al.*, Eds.), p. 582. North Carolina State Museum of Natural History, Raleigh.

Moyle, P. B. (2000). *Inland Fishes of California*, 2nd ed. Univ. of California Press, Berkeley.

Moyle, P. B., and Cech, J. J., Jr. (1996). *Fishes: An introduction to Ichthyology*, 3rd. ed. Prentice Hall, Upper Saddle River, NJ.

Munro, I. S. R. (1967). *The Fishes of New Guinea*. Department of Agriculture, Stocks, and Fisheries, Port Moresby, New Guinea.

Munroe, T. A. (1990). Eastern Atlantic tonguefishes of the *Symphurus plagusia* complex (Cynoglossidae, Pleuronectiformes), with descriptions of two new species. *Fish. Bull.* **89**, 247–287.

Munroe, T. A. (1991). Western Atlantic tonguefishes (*Symphurus*: Cynoglossidae, Pleuronectiformes), with descriptions of two new species. *Bull. Mar. Sci.* **47**(2), 464–515.

Munroe, T. A., Nizinski, M. S., and Mahadeva, M. N. (1991). *Symphurus prolatinaris*, a new species of shallow-water tonguefish (Pleuronectiformes: Cynoglossidae) from the eastern Atlantic. *Proc. Biol. Soc. Washington* **104**(3), 448–458.

Münzing, J. (1974). Flounders. In *Grzimek's Animal Life Encyclopedia* (B. Grzimek, Ed.), Vol. 5, pp. 226–244. Van Nostrand–Reinhold, New York.

Murdy, E. O. (1989). A taxonomic revision and cladistic analysis of the oxudercine gobies (Gobiidae: Oxudercinae). *Rec. Austr. Museum Suppl.* 11, 1–93.

Murdy, E. O. (1998). A review of the gobioid fish genus *Gobioides*. *Ichthyol. Res.* 45(2), 121–133.

Murphy, W. J., Thomerson, J. E., and Collier, G. E. (1999). Phylogeny of the Neotropical killifish family Rivulidae (Cyprinodontiformes, Aplocheiloidei) Inferred from Mitochondrial DNA sequences. *Mol. Phylogenet. Evol.* 13(2), 289–301.

Murray, A. M., and Wilson, M. V. H. (1996). A new Palaeocene genus and species of percopsiform (Teleostei: Paracanthopterygii) from the Paskapoo Formation, Smoky Tower, Alberta. *Can. J. Earth Sci.* 33, 429–438.

Murray, A. M., and Wilson, M. V. H. (1999). Contributions of fossils to the phylogenetic relationships of the percopsiform fishes (Teleostei: Paracanthopterygii): Order restored. In *Mesozoic Fishes 2—Systematics and Fossil Record* (G. Arratia and H.-P. Schultze, Eds.), pp. 397–411. Verlag Dr. Friedrich Pfeil, Munich.

Musyl, M. K., and Keenan, C. P. (1992). Population genetics and zoogeography of Australian freshwater golden perch, *Macquaria ambigua* (Richardson 1845) (Teleostei: Percichthyidae), and electrophoretic identification of a new species from Lake Eyre Basin. *Austr. J. Mar. Freshwater Res.* 43, 1585–1601.

Myers. G. S. (1930). The killifish of San Ignacio and the stickleback of San Ramon, lower California. *Proc. California Acad. Sci. Ser. 4* 19, 95–104.

Myers, G. S. (1931). On the fishes described by Koller from Hainan in 1926 and 1927. *Lignan Sci. J.* 10(2/3), 255–262.

Myers, G. S. (1938). Fresh-water fishes and West Indian zoogeography, Annual Report of the Smithsonian Institution for 1937, pp. 339–364.

Myers, G. S. (1949). Salt-tolerance of fresh-water fish groups in relation to zoogeographical problems. *Bijdragen Dierkunde* 28, 315–322.

Myers, G. S. (1951). Freshwater fishes and East Indian zoogeography. *Stanford Ichthyol. Bull.* 4(1), 11–21.

Myers, G. S. (1956). *Copella*, a new genus of pyrrhulinin characid fishes from the Amazon. *Stanford Ichthyol. Bull.* 7(2), 12–13.

Myers, G. S. (1960). The genera and ecological geography of the South American banjo catfishes, family Aspredinidae. *Stanford Ichthyol. Bull.* 7(4), 132–139.

Myers, G. S. (1966). Derivation of the freshwater fish fauna of Central America. *Copeia* 1966(4), 766–773.

Myers, G. S. (1967). Zoogeographical evidence of the age of the South Atlantic Ocean. *Stud. Tropical Oceanogr.* 5, 614–621.

Myers, G. S. (1979). A freshwater sea horse. *Tropical Fish Hobbyist* 28, 29–34.

Myers, G. S., and de Carvalho, A. L. (1959). A remarkable new genus of anostomin characid fishes from the Upper Rio Xingú in central Brazil. *Copeia* 1959(2), 148–152.

Myers, R. A., Hutchings, J. A., and Barrowman, N. J. (1997). Why do fish stocks collapse? The example of cod in Atlantic Canada. *Ecol. Applications* 7, 91–106.

Naiman, R. J., and Soltz, D. L. (Eds.) (1981). *Fishes in North American Deserts*. Wiley, New York.

Neil, W. T. (1950). An estivating bowfin. *Copeia* 1950, 240.

Neira, F. J., Miskiewicz, A. G., and Trnski, T. (Eds.) (1998). *Larvae of Temperate Australian Fishes*. Univ. of Western Australia Press, Perth.

Nelson, G. J. (1969). Infraorbital bones and their bearing on the phylogeny and geography of Osteoglossomorph fishes. *Am. Museum Novitates* 2394.

Nelson, G. J. (1978). From Candolle to Croizat: Comments on the history of biogeography. *J. History Biol.* 11(2), 269–305.

Nelson, G. (1994). Sardines and their allies. In *Encyclopedia of Fishes* (J. P. Paxton and W. N. Eschmeyer, Eds.), pp. 91–95. Academic Press, San Diego.

Nelson, J. S. (1971). Comparison of pectoral and pelvic skeletons and of some other bones and their phylogenetic implications in the Aulorhynchidae and Gasterosteidae (Pisces). *J. Fish. Res. Board Canada* 28, 427–442.

Nelson, J. S. (1976). *Fishes of the World*. Wiley, New York.

Nelson, J. S. (1984). *Fishes of the World*, 2nd ed. Wiley, New York.

Nelson, J. S. (1990). Redescription of *Antipodocottus elegans* (Scorpaeniformes: Cottidae) from Australia, with comments on the genus. *Copeia* 1990(3), 840–846.

Nelson, J. S. (1994). *Fishes of the World*, 3rd ed. Wiley, New York.

Nelson, J. S., Crossman, E. J., Espinosa-Pérez, H., Gilbert, C. R., Lea, R. N., and Williams, J. D. (1998). Recommend changes in common fish names: Pikeminnow to replace squawfish (*Ptychocheilus* spp.). *Fisheries* 23(9), 37.

Nelson, S. G., and Eldredge, L. G. (1991). Distribution and status of introduced cichlid fishes of the genera *Oreochromis* and *Tilapia* in the islands of the south Pacific and Micronesia. *Asian Fish. Sci.* 4, 11–22.

Netboy, A. (1968). *The Atlantic Salmon.* Houghton Mifflin, Boston.

Netboy, A. (1980). *Salmon: The World's Most Harassed Fish.* Andre Deutsch, London.

Ney, J. J. (1978). A synoptic review of yellow perch and walleye biology. *Am. Fish. Soc. Special Publication* 11, 1–12.

Ng, H. H. (1999a). A review of the southeast Asian catfish genus *Ceratoglanis* (Siluriformes: Siluridae), with the description of a new species from Thailand. *Proc. California Acad. Sci.* 51(9), 385–395.

Ng, H. H. (1999b). *Laides longibarbis,* a valid species of schilbeid catfish from Indochina (Teleostei: Siluriformes). *Ichthyol. Exploration Freshwaters* 10(4), 381–385.

Ng, H. H., and Kottelat, M. (1998a). *Hyalobagrus,* a new genus of miniature bagrid catfish from southeast Asia (Teleostei: Siluriformes). *Ichthyol. Exploration Freshwaters* 9(4), 335–346.

Ng, H. H., and Kottelat, M. (1998b). *Pterocryptis buccata,* a new species of catfish from western Thailand (Teleostei: Siluridae) with epigean and hypogean populations. *Ichthyol. Res.* 45(4), 393–399.

Ng, H. H., and Kottelat, M. (1998c). The catfish genus *Akysis* Bleeker (Teleostei: Akysidae) in Indochina, with descriptions of six new species. *J. Nat. History* 32, 1057–1097.

Ng, H. H., and Kottelat, M. (1999). *Oreoglanis hypsiurus,* a new species of glyptosternine catfish (Teleostei: Siluriformes). *Ichthyol. Exploration Freshwaters* 10(4), 375–380.

Ng, H. H., and Kottelat, M. (2000). *Cranoglanis henrici* (Vaillant, 1893), a valid species of cranoglanidid catfish from Indochina (Teleostei: Cranoglanididae). *Zoosystema,* 22(4): 847–852.

Ng, H. H., and Lim, K. K. P. (1995). A revision of the southeast Asian catfish genus *Parakysis* (Teleostei: Akysidae), with descriptions of two new species. *Ichthyol. Exploration Freshwaters* 6(3), 255–266.

Ng, H. H., and Siebert, D. J. (1998). A revision of the akysid catfish genus *Breitensteinia* Steindachner with descriptions of two new specpcies. *J. Fish Biol.* 53, 645–657.

Ng, P. K. L. (1992). The giant Malayan catfish, *Wallago leerii* Bleeker, 1851, and the identities of *Wallagonia tweediei* Hora & Misra, 1941, and *Wallago maculatus* Inger & Chin, 1959 (Teleostei: Siluridae). *Raffles Bull. Zool.* 40(2), 245–263.

Ng, P. K. L., and Lim, K. K. P. (1990). Snakeheads (Pisces: Channidae): Natural history, biology, and economic importance. In *Essays in Zoology. Papers Commemorating the 40th Anniversary of the Department of Zoology, National University of Singapore* (C.-M. Chou and P. K. L. Ng, Eds.), pp. 127–152. National University of Singapore, Department of Zoology, Singapore.

Ng, P. K. L., and Lim, K. K. P. (1993). The southeast Asian catfish genus *Encheloclarias* (Teleostei: Clariidae), with descriptions of four new species. *Ichthyol. Exploration Freshwaters* 4(1), 21–37.

Nichols, J. T. (1943). The fresh-water fishes of China. In *Natural History of Central Asia,* Vol. 9. American Museum of Natural History, New York.

Nico, L. G., and de Pinna, M. C. C. (1996). Confirmation of *Glanapteryx anguilla* (Siluriformes, Trichomycteridae) in the Orinoco River basin, with notes on the distribution and habitats of the Glanapteryginae. *Ichthyol. Exploration Freshwaters* 7(1), 27–32.

Nijssen, H. (1971). Two new species and one new subspecies of the South American catfish genus *Corydoras* (Pisces, Siluriformes, Callicthyidae). *Beaufortia* 19(250), 89–98.

Nijssen, H., and Isbrücker, I. J. H. (1970). The South American catfish genus *Brochis* Cope, 1872 (Pisces, Siluriformes, Callichthyidae). *Beaufortia* 18(236), 151–168.

Nijssen, H., and Isbrücker, I. J. H. (1979). Chronological enumeration of nominal species and subspecies of *Corydoras* (Pisces, Siluriformes, Callichthyidae). *Bull. Museum Univ. Amsterdam* 6(17), 129–135.

Nijssen, H., and Isbrücker, I. J. H. (1980). A review of the genus *Corydoras* Lacepede, 1803 (Pisces, Siluriformes, Callichthyidae). *Bijdragen Dierkunde* 50(1), 190–220.

Nijssen, H., and Isbrücker, I. J. H. (1983). *Brochis britskii,* a new species of plated catfish from the Upper Rio Paraguai system, Brazil (Pisces, Siluriformes, Callichthyidae). *Bull. Zool. Museum Univ. Amsterdam* 9(20), 177–186.

Nijssen, H., and Isbrücker, I. J. H. (1986). Cinq espèces nouvelles de poissons-chats cuirassés du genre *Coryodras* Lacepède, 1803, du Pérou et de l'Equateur (Pisces, Siluriformes, Callichthyidae). *Rev. Française d'Aquariologie* 12(3), 65–76.

Nijssen, H., and Isbrücker, I. J. H. (1987). *Spectracanthicus murinus,* nouveaux genre et espèce de poisson-chat cuirassé du Rio Tapajós, Est. Pará, Br. e'sil, avec les remarques sur d'autres generes de Loricariidés (Pisces, Siluriformes, Loricariidae). *Rev. Française d'Aquariologie* 13(4), 93–98.

Nilsson, G. E. (1996). Brain and body oxygen requirements of *Gnathonemus petersii,* a fish with an exceptionally large brain. *J. Exp. Biol.* 199, 603–607.

Norman, J. R., and Greenwood, P. H. (1975). *A History of Fishes*, 3rd ed. Wiley, New York.

Normark, B. B., McCune, A. R., and Harris, R. G. (1991). Phylogenetic relationships of neopterygian fishes inferred from mitochondrial DNA sequences. *Mol. Biol. Evol.* **8**, 819–834.

Norris, S. M. (1992). Anabantidae. In *Faune des Poissons d'eaux Douces et Saumâtres de l'Afrique de l'Ouest* (C. Lévêque, D. Paugy, and G. G. Teugels, Eds.), Vol. 2, pp. 837–847. ORSTOM, Paris.

Norris, S. M. (1995). *Microctenopoma uelense* and *M. nigricans*, a new genus and two new species of anabantid fishes from Africa. *Ichthyol. Exploration Freshwater* **6**(4), 357–376.

Norris, S. M., and Douglas, M. E. (1991). A new species of nest building *Ctenopoma* (Teleostei, Anabantidae) from Zaïre, with a redescription of *Ctenopoma lineatum* (Nichols). *Copeia* **1991**(1), 166–178.

Norris, S. M., and Douglas, M. E. (1992). Geographic variation, taxonomic status, and biogeography of two widely distributed African freshwater fishes: *Ctenopoma petherici* and *C. kingsleyae* (Teleostei: Anabantidae). *Copeia* **1992**(3), 709–724.

Norris, S. M., and Minckley, W. L. (1997). Two new species of *Etheostoma* (Osteichthyes: Percidae) from Central Coahuil, Northern Mexica. *Ichthyol. Exploration Freshwaters* **8**(2), 159–176.

Norris, S. M., and Teugels, G. C. (1990). A new species of *Ctenopoma* (Teleostei: Anabantidae) from southeastern Nigeria. *Copeia* **1990**(2), 492–499.

Norris, S. M., Miller, R. J., and Douglas, M. E. (1988). Distribution of *Ctenopoma muriei* and the status of *Ctenopoma ctenotis* (Pisces: Anabantidae). *Copeia* **1988**(2), 487–491.

Novacek, M. J., and Marshall, L. G. (1976). Early biogeographic history of ostariophysan fishes. *Copeia* **1976**, 1–12.

Nursall, J. R. (1981). Behavior and habitat affecting the distribution of five species of sympatric mudskippers in Queensland. *Bull. Mar. Sci.* **31**(3), 730–735.

Oetinger, M. I., and Zorzi, G. D. (Eds.) (1995). The biology of freshwater elasmobranchs. *J. Aquaricultue Aquat. Sci.* **7**, 1–161.

Ogutu-Ohwayo, R. (1993). The effects of predation by Nile perch, *Lates niloticus* L., on the fish of Lake Nabugabo, with suggestions for conservation of endangered endemic cichlids. *Conserv. Biol.* **7**(3), 701–711.

O'Hara, J. J. (1968). Influence of weight and temperature on metabolic rate of sunfish. *Ecology* **49**(1), 159–161.

Olsen, P. E., and McCune, A. R. (1991). Morphology of the *Semionotus elegans* species group from the early Jurassic part of the Newark Supergroup of eastern North America with comments on the family Semionotidae (Neopterygii). *J. Vertebr. Paleontol.* **11**(3), 269–292.

Olson, K. R., Munshi, J. S. D., Ghosh, T. K., and Ojha, J. (1986). Gill microcirculation of the air-breathing climbing perch, *Anabas testudineus* (Bloch): Relationships with the accessory respiratory organs and systemic circulation. *Am. J. Anat.* **176**, 305–320.

Ortega, H., and Vari, R. P. (1986). Annotated checklist of the freshwater fishes of Peru. *Smithsonian Contrib. Zool.* **437**, 1–25.

Ortí, G. (1997). Radiation of characiform fishes: Evidence from mitochondrial and nuclear DNA sequences. In *Molecular Systematics of Fishes* (T. D. Kocher and C. A. Stepien, Eds.), pp. 219–243. Academic Press, San Diego.

Ortí, G., Petry, P., Porto, J. I. R., Jégu, M., and Meyer, A. (1996). Patterns of nucleotide change in mitochondrial ribosomal RNA genes and the phylogeny of piranhas. *J. Mol. Evol.* **42**, 169–182.

Ortubay, S. G., Gómez, S. E., and Cussac, V. E. (1997). Lethal temperatures of a Neotropical fish relic in Patagonia, the scale-less characinid *Gymnocharacinus bergi*. *Environ. Biol. Fishes* **49**, 341–350.

Osmundson, D. B., and Burnham, K. P. (1998). Status and trends of the endangered Colorado squawfish in the Upper Colorado River. *Trans. Am. Fish. Soc.* **127**(6), 957–970.

Osmundson, D. B., Ryel, R. J., Tucker, M. E., Burdick, B. D., Elmblad, W. R., and Chart, T. E. (1998). Dispersal patterns of subadult and adult Colorado squawfish in the Upper Colorado River. *Trans. Am. Fish. Soc.* **127**(6), 943–956.

Ovenden, J. R., White, R. W. G., and Sanger, A. C. (1988). Evolutionary relationships of *Gadopsis* spp. inferred from restriction enzyme analysis of their mitochondrial DNA. *J. Fish Biol.* **32**, 137–148.

Page, L. M. (1974). The subgenera of *Percina* (Percidae: Etheostomatini). *Copeia* **1974**(1), 66–86.

Page, L. M. (1981). The genera and subgenera of darters (Percidae, Etheostomatini). *Occasional Papers Museum Nat. History Univ. Kansas* **90**, 1–69.

Page, L. M. (1983). *Handbook of Darters*. TFH, Neptune City, NJ.

Page, L. M. (1985). Evolution of reproductive behaviors in percid fishes. *Illinois Nat. History Surv. Bull.* **33**(3), 275–295.

Page, L. M., and Burr, B. M. (1991). *A Field Guide to Freshwater Fishes*. Houghton Mifflin, Boston.

Page, L. M., and Johnson, C. E. (1990). Spawning in the creek chubsucker, *Erimyzon oblongus*, with a review of spawning behavior in suckers (Catostomidae). *Environ. Biol. Fishes* **27**, 265–272.

Pardue, G. B. (1993). Life history and ecology of the mud sunfish (*Acantharchus pomotis*). *Copeia* **1993**(2), 533–540.

Parenti, L. R. (1981). A phylogenetic and biogeographic analysis of cyprinodontiform fishes (Teleostei, Atherinomorpha). *Bull. Am. Museum Nat. History* **168**(4), 335–557.

Parenti, L. R. (1984a). On the relationships of phallostethid fishes (Atherinomorpha), with notes on the anatomy of *Phallostethus dunckeri* Regan 1913. *Am. Museum Novitates* **2779**, 1–12.

Parenti, L. R. (1984b). A taxonomic revision of the Andean killifish genus *Orestias* (Cyprinodontiformes, Cyprinodontidae). *Bull. Am. Museum Nat. History* **178**(2), 107–214.

Parenti, L. R. (1984c). Biogeography of the Andean killifish genus *Orestias* with comments on the species flock concept. In *Evolution of Fish Species Flocks* (A. A. Echelle and I. Kornfield, Eds.), pp. 85–92. Univ. of Maine Press, Orono.

Parenti, L. R. (1986a). Bilateral asymmetry in phallostethid fishes (Atherinomorpha) with description of a new species from Sarawak. *Proc. California Acad. Sci.* **44**(10), 225–236.

Parenti, L. R. (1986b). Homology of pelvic fin structures in female phallostethid fishes (Atherinomorpha, Phallostethidae). *Copeia* **1986**(2), 305–310.

Parenti, L. R. (1987). Phylogenetic aspects of tooth and jaw structure of the medaka, *Oryzias latipes*, and other beloniform fishes. *J. Zool. London* **211**, 561–572.

Parenti, L. R. (1989). A phylogenetic revision of the phallostethid fishes (Atherinomorpha, Phallostethidae). *Proc. California Acad. Sci.* **46**(11), 243–277.

Parenti, L. R. (1991). Ocean basins and the biogeography of freshwater fishes. *Austr. Syst. Bot.* **4**, 137–149.

Parenti, L. R. (1993). Relationships of atherinomorph fishes (Teleostei). *Bull. Mar. Sci.* **52**(1), 170–196.

Parenti, L. R. (1994). Killifishes and ricefishes. In *Encyclopedia of Fishes* (J. P. Paxton and W. N. Eschmeyer, Eds.), pp. 148–152. Academic Press, San Diego.

Parenti, L. R. (1996). Phylogenetic systematics and biogeography of phallostethid fishes (Atherinomorpha, Phallostethidae) of northwestern Borneo, with description of a new species. *Copeia* **1996**(3), 703–712.

Parenti, L. R., and Louie, K. D. (1998). *Neostethus djajaorum*, new species, from Sulawesi, Indonesia, the first phallostethid fish (Teleostei: Atherinomorpha) known from east of Wallace's Line. *Raffles Bull. Zool.* **46**(1), 139–150.

Parenti, L. R., and Maciolek, J. A. (1993). New sicydiinae gobies from Ponape and Palau, Micronesia, with comments on systematics of the subfamily Sicydiinae (Teleostei: Gobiidae). *Bull. Mar. Sci.* **53**(3), 945–972.

Parenti, L. R., and Maciolek, J. A. (1996). *Sicyopterus rapa*, new species of sicydiine goby (Teleostei: Gobiidae), from Rapa, French Polynesia. *Bull. Mar. Sci.* **58**(3), 600–667.

Parenti, L. R., and Thomas, K. R. (1998). Pharyngeal jaw morphology and homology in sicydiine gobies (Teleostei: Gobiidae) and allies. *J. Morphol.* **237**, 257–274.

Parker, A. (1997). Combining molecular and morphological data in fish systematics: Examples from the Cyprinodontiformes. In *Molecular Systematics of Fishes* (T. D. Kocher and C. A. Stepien, Eds.), pp. 163–188. Academic Press, San Diego.

Parr, D. (1999). Progress with paddlefish restoration. *Pennsylvania Angler Boater* **68**(1), 13–16.

Patterson, C. (1968). The caudal skeleton in Lower Liassic pholidophoroid fishes. *Bull. Br. Museum Nat. History Geol.* **16**(5), 201–239.

Patterson, C. (1973). Interrelationships of holosteans. In *Interrelationships of Fishes* (P. H. Greenwood, R. S. Miles, and C. Patterson, Eds.), pp. 233–305. Academic Press, London.

Patterson, C. (1981). The development of the North American fish fauna—A problem of historical biogeography. In *The Evolving Biosphere* (P. L. Forey, Ed.), pp. 265–281. British Museum of Natural History/Cambridge Univ. Press, London.

Patterson, C. (1994). Bony fishes. In *Major Features of Vertebrate Evolution* (D. R. Prothero and R. M. Schoch, Eds.), pp. 57–84. Paleontological Society, University of Tennessee, Knoxville.

Patterson, C. (1998). Comments on basal teleosts and teleostean phylogeny, by Gloria Arratia. *Copeia* **1998**(4), 1107–1109.

Patterson, C., and Johnson, G. D. (1995). The intermuscular bones and ligaments of teleostean fishes. *Smithsonian Contrib. Zool.* **559**.

Patterson, C., and Johnson, G. D. (1997). Comments on Begle's "Monophyly and Relationships of Argentinoid Fishes." *Copeia* **1997**(2), 401–409.

Patterson, C., and Longbottom, A. E. (1989). An Eocene amiid fish from Mali, West Africa. *Copeia* **1989**, 827–836.

Patterson, C., and Rosen, D. E. (1989). The Paracanthopterygii revisited: Order and disorder. In *Papers on the Systematics of Gadiform Fishes*

(D. M. Cohen, Ed.), Science Series No. 32, pp. 5–36. Natural History Museum of Los Angeles County, Los Angeles..

Paugy, D. (1984). Characidae. In *Check-list of the Freshwater Fishes of Africa* (J. Daget, J.-P. Gosse, and D. F. E. Thys van den Audenaerde, Eds.), pp. 140–183. CLOFFA I, ORSTOM, Paris.

Paugy, D. (1990a). Osteoglossidae. In *Faune des Poissons d'eaux Douce et Saumâtres de l'Afrique de l'Ouest* (C. Lévêque, D. Paugy, and G. G. Teugels, Eds.), Vol. 1, pp. 112–115. ORSTOM, Paris.

Paugy, D. (1990b). Cromeriidae. In *Faune des Poissons d'eaux Douce et Saumâtres de l'Afrique de l'Ouest* (C. Lévêque, D. Paugy, and G. G. Teugels, Eds.), Vol. 1, pp. 187–188. ORSTOM, Paris.

Paugy, D. (1990c). Hepsetidae. In *Faune des Poissons d'eaux Douce et Saumâtres de l'Afrique de l'Ouest* (C. Lévêque, D. Paugy, and G. G. Teugels, Eds.), Vol. 1, pp. 192–194. ORSTOM, Paris.

Paugy, D. (1990d). Characidae. In *Faune des Poissons d'eaux Douce et Saumâtres de l'Afrique de l'Ouest* (C. Lévêque, D. Paugy, and G. G. Teugels, Eds.), Vol. 1, pp. 195–236. ORSTOM, Paris.

Paugy, D. (1992). Centropomidae. In *Faune des Poissons d'eaux Douces et Saumâtres de l'Afrique de l'Ouest* (C. Lévêque, D. Paugy, and G. G. Teugels, Eds.), Vol. 2, pp. 662–663. ORSTOM, Paris.

Paugy, D., and Guégan, J. F. (1989). Note á propos de trois espèces d'Hydrocynus (Pisces, Characidae) du bassin du Niger, suire de la zehabilitahou de l'espèce *Hydrocynus vittatus* (Castelnau, 1861). *Rev. Hydrobiol. Trop.* **22**, 63–69.

Paugy, D., and Roberts, T. R. (1992). Mochokidae. In *Faune des Poissons d'eaux Douces et Saumâtres de l'Afrique de l'Ouest* (C. Lévêque, D. Paugy, and G. G. Teugels, Eds.), Vol. 2, pp. 500–563. ORSTOM, Paris.

Pelseneer, P. (1904). La "Ligne de Weber": Limite zoologique de l'Asie et de l'Australie. *Bull. Classe Sci. R. Belgique*, 1001–1022.

Peters, N., Jr. (1967). Opercular-und postopercularorgan (occipitalorgan) der gattung Kneria (Kneriidae, Pisces) und ein vergleich mit verwandten strukturen. *Morph. Ökol. Tiere* **59**, 381–435.

Peters, N. (1973). Gonorynchiformes. In *Grzimek's Animal Life Encyclopedia* (B. Grzimek, Ed.), Vol. 4, pp. 269–275. Van Nostrand–Reinhold, New York.

Petersen, J. H., and Ward, D. L. (1999). Development and corroboration of a bioenergetics model for northern pikeminnow feeding on juvenile salmonids in the Columbia River. *Trans. Am. Fish. Soc.* **128**(5), 784–801.

Pethiyagoda, R. (1991a). *Monodactylus kottelati*, ein neues Flossenblatt (Pisces: Monodactylidae) aus Sri Lanka. *Aquar. Terrar. Ztschr.* **44**, 162, 164–167.

Pethiyagoda, R. (1991b). *Freshwater Fishes of Sri Lanka*. Wildlife Heritage Trust of Sri Lanka, Colombo.

Pettit, M. J., and Beitinger, T. L. (1980). Thermal responses of the South American lungfish, *Lepidosiren paradoxa*. *Copeia* **1980**(1), 130–136.

Pettit, M. J., and Beitinger, T. L. (1981). Aerial respiration of the brachiopterygian fish, *Calamoichthys calabricus*. *Comp. Biochem. Physiol. A* **68**, 507–509.

Pettit, M. J., and Beitinger, T. L. (1985). Oxygen acquisition of the reedfish, *Erpetoichthys calabricus*. *J. Exp. Biol.* **114**, 289–306.

Pezold. F. (1993). Evidence for a monophyletic Gobiinae. *Copeia* **1993**(3), 634–643.

Pflieger, W. L. (1975). *The Fishes of Missouri*. Missouri Department of Conservation, Jefferson City.

Phillips, D. P., and Pleyte, K. A. (1991). Nuclear DNA and salmonid phylogenetics. *J. Fish Biol.* **39**(Suppl. A), 259–275.

Phillips, R. B., and Oakley, T. H. (1997). Phylogenetic relationships among the Salmoninae based on nuclear and mitochondrial DNA sequences. In *Molecular Systematics of Fishes* (T. D. Kocher and C. A. Stepien, Eds.), pp. 145–162. Academic Press, San Diego.

Pickette, G. D., and Pawson, M. G. (1994). *Sea Bass: Biology, Exploitation and Conservation*. Chapman & Hall, London.

Pietsch, T. W. (1978). Evolutionary relationships of the sea moths (Teleostei: Pegasidae) with a classification of gasterosteiform families. *Copeia* **1978**(3), 517–529.

Pietsch, T. W. (1989). Phylogenetic relationships of trachinoid fishes of the family Uranoscopidae. *Copeia* **1989**(2), 253–303.

Pietsch, T. W., and Zabetian, C. P. (1990). Osteology and interrelationships of the sand lances (Teleostei: Ammodytidae). *Copeia* **1990**(1), 78–100.

Pinna, M. C. C. de (1988). A new genus of trichomycterid catfish (Siluroidei, Glanapteryginae), with comments on its phylogenetic relationships. *Rev. Suisse Zool.* **95**(1), 113–128.

Pinna, M. C. C. de (1989a). A new Sarcoglanidine catfish, phylogeny of its subfamily, and an appraisal of the phyletic status of the Trichomycterinae (Teleostei, Trichomycteridae). *Am. Museum Novitates* **2950**, 1–39.

Pinna, M. C. C. de (1989b). Redescription of *Glanapteryx anguilla*, with notes on the phylogeny of Glanapteryginae (Siluriformes, Trichomycteridae). *Proc. Acad. Nat. Sci. Philadelphia* 141, 361–374.

Pinna, M. C. C. de (1992a). A new subfamily of Trichomycteridae (Teleostei: Siluriformes), lower loricarioid relationships and a discussion of the impact of a ditional taxa for phylogenetic analysis. *Zool. J. Linnean Soc.* 106, 175–229.

Pinna, M. C. C. de (1992b). *Trichomycterus castroi*, a new species of trichomycterid catfish from the Rio Iguaçu of Southeastern Brazil (Teleostei: Siluriformes). *Ichthyol. Exploration Freshwaters* 3(1), 89–95.

Pinna, M. C. C. de (1996a). Teleostean monophyly. In *Interrelationships of Fishes* (M. L. J. Stiassny, L. R. Parenti, and G. D. Johnson, Eds.), pp. 147–162. Academic Press, San Diego.

Pinna, M. C. C. de (1996b). A phylogenetic analysis of the Asian catfish families Sisoridae, Akysidae, and Amblycipitidae, with a hypothesis on the relationships of the Neotropical Aspredinidae (Teleostei, Ostariophysi). *Fieldiana* 1478, 1–83.

Pinna, M. C. C. de (1998). Phylogenetic relationships of Neotropical Siluriformes (Teleostei: Ostariophysi): Historical overview and synthesis of hypotheses. In *Phylogeny and Classification of Neotropical Fishes* (L. R. Malabarba, R. E. Reis, R. P. Vari, Z. M. S. Lucena, and C. A. S. Lucena, Eds.), pp. 279–330. EDIPUCRS, Porto Alegre, Brazil.

Pinna, M. C. C. de, and Britski, H. A. (1991). *Megalocentor*, a new genus of parasitic catfish from the Amazon basin: the sister group of *Apomatoceros* (Trichomycteridae: Stegophilinae). *Ichthyol. Exploration Freshwaters* 2(2), 113–128.

Pinna, M. C. C. de, and Ferraris, C. (1992). Review of "Anatomy, Relationships and Systematics of the Bagridae (Teleostei: Siluroidei) with a Hypothesis of Siluroid Phylogeny." *Copeia* 1992(4), 1132–1134.

Pinna, M. C. C. de, and Starnes, W. C. (1990). A new genus and species of Sarcoglanidinae from the Rio Mamoré, Amazon Basin, with comments on subfamily phylogeny (Teleostei, Trichomycteridae). *J. Zool. (London)* 222, 75–88.

Pinna, M. C. C. de, and Vari, R. P. (1995). Monophyly and phylogenetic diagnosis of the family Cetopsidae, with synonymization of the Helogenidae (Teleostei: Siluriformes). *Smithsonian Contrib. Zool.* 571.

Pinter, H. (1986). *Labyrinth Fish*. Barron's, Woodbury, NY.

Pitcher, T. J., and Hart, P. J. B. (1995). *The Impact of Species Changes in African Lakes*. Chapman & Hall, New York.

Platania, S. P., and Ross, S. W. (1980). *Arius felis* (Linnaeus) Hardhead catfish. In *Atlas of North American Freshwater Fishes* (D. S. Lee *et al.*, Eds.), p. 476. North Carolina State Museum of Natural History, Raleigh.

Platnick, N. I., and Nelson, G. (1978). A method of analysis for historical biogeography. *Syst. Zool.* 27, 1–16.

Policansky, D. (1982a). Influence of age, size, and temperature on metamorphosis in the starry flounder, *Platichthys stellatus*. *Can. J. Fish. Aquat. Sci.* 39, 514–517.

Policansky, D. (1982b). The asymmetry of flounders. *Sci. Am.* 246(5), 116–122.

Poll, M. (1971). Révision des *Synodontis africains* (famille Mochocidae). *Ann. Museum R. Africa Cent.* 191, 1–497.

Poll, M. (1973). Nombre et distribution geographique des poissons d-eau douce africains. *Bull. Muséum Natl. d'Histoire Nat.* 150(6), 113–128.

Poll, M. (1984a). Kneriidae, Cromeriidae, and Grasseichthyidae. In *Check-list of the Freshwater Fishes of Africa* (J. Daget, J.-P. Gosse, and D. F. E. Thys van den Audenaerde, Eds.), pp. 129–135. CLOFFA I, ORSTOM, Paris.

Poll, M. (1984b). *Parakneria tanzaniae*, espéce nouvelle des chutes de la rivière Kimani, Tanzanie. *Rev. Zool. Africa* 98, 1–8.

Poll, M. (1986). Classification des Cichlidae du lac Tanganika, tribus, genres et espèces. *Acad. R. Belgique Mém. Classe Sci.* 45(2), 1–163.

Poll, M., and Gosse, J.-P. (1969). Révision des Malapteruridae (Pisces, Siluriformes) et description d'une deuxième espèce de silure électrique: *Malapterurus microstoma* sp. n. *Bull. R. Inst. Nat. Sci. Belgium* 45(38), 1–12.

Poll, M., and Gosse, J. P. (1995). Genera des poissons d'eau douce d'Afrique. Académie Royale de Belgique. *Mém. Classe Sci.* 3(9), 1–324.

Poll, M., Teugels, G. G., and Whitehead, P. J. P. (1984). Clupeidae. In *Check-list of the Freshwater Fishes of Africa* (J. Daget, J.-P. Gosse, and D. F. E. Thys van den Audenaerde, Eds.), pp. 41–55. CLOFFA I, ORSTOM, Paris.

Pollard, D. A., Davis, T. L. O., and Llewellyn, L. C. (1996). Eel-tailed catfishes. In *Freshwater Fishes of South-Eastern Australia* (R. M. McDowall, Ed.), pp. 109–113. Reed, Sydney.

Polunin, I. (1972). Who says fish can't climb trees. *Natl. Geographic* 141, 85–91.

Porterfield, J. C., Page, L. M., and Near, T. J. (1999). Phylogenetic relationships among fantail darters (Percidae: *Etheostoma*: *Catonotus*): total Evidence analysis of morphological and molecular data. *Copeia* 1999(3), 551–564.

Poss, S. G., and Miller, R. R. (1983). Taxonomic status of the plains killifish, *Fundulus zebrinus*. *Copeia* **1983**, 55–67.

Potter, I. C. (1980). The Petromyzoniformes with particular reference to paired species. *Can. J. Fish. Aquat. Sci.* **37**(11), 1595–1615.

Potter, I. C. (1986). The distincitve characters of Southern Hemisphere lampreys (Geotriidae and Mordaciidae). In *Indo-Pacific Fish Biology: Proceedings of the Second International Conference on Indo-Pacific Fishes* (T. Uyeno, R. Arai, T. Traniuchi, and K. Matsuura, Eds.), pp. 9–19. Ichthyological Society of Japan, Tokyo.

Potter, I. C. (1996). Family Mordaciidae and Family Geotriidae. In *Freshwater Fishes of South-Eastern Australia* (R. McDowall, Ed.), pp. 32–38. Reed, Sydney.

Poulson, T. L. (1963). Cave adaptation in amblyopsid fishes. *Am. Midland Nat.* **70**, 257–290.

Poulson, T. L. (1986). Evolutionary reduction by neutral mutations: Plausibility arguments and data from amblyopsid fishes and linyphiid spiders. *Natl. Speleol. Soc. Bull.* **47**(2), 109–117.

Poulson, T. L., and White, W. B. (1969). The cave environment. *Science* **165**, 971–981.

Powell, C. M., Johnson, B. D., and Veevers, J. J. (1984). Interaction of Australia and South-east Asia. In *Phanerozoic Earth History of Australia* (J. J. Veevers, Ed.), pp. 38–42. Clarendon, Oxford.

Poyato-Ariza, F. J. (1966). A revision of the ostariophysan fish family Chanidae, with special reference to the Mesozoic forms. *Palaeo Ichthyol.* **6**, 5–52.

Prashad, B., and Mukerji, D. D. (1929). The fish of the Indawygi Lake and the streams of the Myitkina District (Upper Burma). *Rec. Indian Museum* **31**, 161–223.

Preston, J. L. (1978). Communication systems and social interactions in a goby-shrimp symbiosis. *Anim. Behav.* **26**, 791–802.

Prosek, J. (1996). *Trout: An Illustrated History*. Knopf, New York.

Pusey, B. J. (1986). The effect of starvation on oxygen consumption and nitrogen excretion in *Lepidogalaxias salamandroided* (Mees). *J. Comp. Physiol. B* **156**, 701–705.

Pusey, B. J. (1989). Aestivation in the teleost fish *Lepidogalaxias salamandroides* (Mees). *Comp. Biochem. Physiol. A* **92**, 137–138.

Pusey, B. J. (1990). Seasonality, aestivation and the life history of the salamanderfish *Lepidogalaxias salamandroides* (Pisces: Lepidogalaxiidae). *Environ. Biol. Fishes* **29**, 15–26.

Pusey, B. J., and Edward, D. H. D. (1990). Structure of fish assemblages in waters of the southern acid pet flats, south-western Australia. *Austr. J. Mar. Freshwater Res.* **41**, 721–734.

Pusey, B. J., and Stewart, T. (1989). Internal fertilization in *Lepidogalaxias salamandroides* Mees (Pisces: Lepidogalaxiidae). *Zool. J. Linnean Soc.* **97**, 69–79.

Quammen, D. (1966). *The Song of the Dodo: Island Biogeography in an Age of Extinctions*. Scribner's, New York.

Quéro, J.-C., Desoutter, M., and Lagardère, F. (1986a). Soleidae. In *Fishes of the North-Eastern Atlantic and the Mediterranean* (P. J. P. Whitehead, M.-L. Bauchot, J.-C. Hureau, J. Nielsen, and E. Tortonese, Eds.), Vol. 3, pp. 1308–1324. UNESCO, Paris.

Quéro, J.-C., Desoutter, M., and Lagardère, F. (1986b). Soleidae. In *Fishes of the North-Eastern Atlantic and the Mediterranean* (P. J. P. Whitehead, M.-L. Bauchot, J.-C. Hureau, J. Nielsen, and E. Tortonese, Eds.), Vol. 3, pp. 1325–1328. UNESCO, Paris.

Quignard, J.-P., and Pras, A. (1986). Atherinidae. In *Fishes of the North-Eastern Atlantic and the Mediterranean* (P. J. P. Whitehead, M.-L. Bauchot, J.-C. Hureau, J. Nielsen, and E. Tortonese, Eds.), Vol. 3, pp. 1207–1210. UNESCO, Paris.

Raat, A. S. P. (1988). Synopsis of biological data on the northern pike *Esox lucius* Linnaeus, 1758, FAO Fisheries Synopsis No. 30, Rev. 2.. Food and Agriculture Organization of the United Nations, Rome.

Ráb, P., and Crossman, E. J. (1994). Chromosomal NOR phenotypes in North American pike and pickerels, genus *Esox*, with notes on the Umbridae (Euteleostei: Esocae). *Can. J. Zool.* **72**, 1951–1956.

Rahel, F. J. (2000). Homogenization of fish faunas across the United States. *Science* **288**, 854–856.

Rainboth, W. J. (1991). Cyprinids of south east Asia. *In Cyprinid Fishes: Systematics, Biology and Exploitation* (I. J. Winfield and J. S. Nelson, Eds.), pp. 156–210. Chapman & Hall, London.

Rainboth, W. J. (1996). *FAO Species Identification Field Guide for Fishery Purposes. Fishes of the Cambodian Mekong*. Food and Agriculture Organization of the United Nations, Rome.

Randall, J. E., and Cea Egaña, A. (1989). *Canthigaster cyanetron*: A new toby (Teleostei: Tetraodontidae) from Easter Island. *Rev. Fr. Aquariol.* **15**, 93–96.

Rassmussen, A.-S., Janke, A., and Arnason, U. (1998). The mitochondrial DNA molecule of the hagfish (*Myxine glutinosa*) and vertebrate phylogeny. *J. Mol. Evol.* **46**, 382–388.

Rauchenberger, M. (1989a). Annotated species list of the subfamily Poeciliinae. In *Ecology and Evolution of Livebearing Fishes (Poeciliidae)* (G. K. Meffe and F. F. Snelson, Jr., Eds.), pp. 359–367. Prentice Hall, Englewood Cliffs, NJ.

Rauchenberger, M. (1989b). Systematics and biogeography of the genus *Gambusia* (Cyprinodontiformes: Poeciliidae). *Am. Museum Novitates* **2951**, 1–74.

Reash, R. J., and Berra, T. M. (1987). Comparison of fish communities in a clean-water stream and an adjacent polluted stream. *Am. Midland Nat.* **118**(2), 301–322.

Reash, R. J., and Berra, T. M. (1989). Incidence of fin erosion and anomalous fishes in a polluted stream and a nearby clean stream. *Water Air Soil Pollut.* **47**, 47–63.

Reddy, W. P. B. (1978). Studies on the taxonomy of Indian species of the family Channidae (Pisces: Teleostei) and some aspects of the biology of Channa punctata (Bloch, 1793) from Guntur, Andhra Pradesh. *Matsya* **4**, 95–96.

Reighard, J. (1940). The natural history of *Amia calva* Linnaeus. *Mark Anniversary Vol.* **4**, 57–108.

Reis, R. E. (1997). Revision of the neotropical catfish genus *Hoplosternum* (Ostariophysi: Siluriformes: Callichthyidae), with the description of two new genera and three new species. *Ichthyol. Exploration Freshwaters* **7**(4), 299–326.

Reis, R. E. (1998a). Systematics, biogeography, and the fossil record of the Callichthyidae: A review of the available data. In *Phylogeny and Classification of Neotropical Fishes* (L. R. Malabarba, R. E. Reis, R. P. Vari, Z. M. S. Lucena, and C. A. S. Lucena, Eds.), pp. 351–362. EDIPUCRS, Porto Alegre, Brazil.

Reis, R. E. (1998b). Anatomy and phylogenetic analysis of the neotropical callichthyid catfishes (Ostariophysi, Siluriformes). *Zool. J. Linnean Soc.* **124**, 105–168.

Reis, R. E., and Pereira, E. H. L. (1999). *Hemipsilichthys nudulus*, a new, uniquely-plated species of loricariid catfish from the rio Araranguá basin, Brazil (Teleostei: Siluriformes). *Ichthyol. Exploration Freshwaters* **10**(1), 45–51.

Reis, R. E., and Schaefer, S. A. (1992). *Eurycheilus pantherinus* (Siluroidei: Loricariidae), a new genus and species of Hypoptopomatinae from southern Brazil. *Copeia* **1992**(1), 215–223.

Reis, R. E., and Schaefer, S. A. (1998). New Cascudinhos from southern Brazil: Systematics, endemism, and relationships (Siluriformes, Loricariidae, Hypoptopomatinae). *Am. Museum Novitates* **3254**, 1–25.

Reist, J. D. (1987). Comparative morphometry and phenetics of the genera of esocoid fishes (Salmoniformes). *Zool. J. Linnean Soc.* **89**, 275–294.

Reynolds, L. F. (1983). Migration patterns of five fish species in the Murray–Darling River system. *Austr. J. Mar. Freshwater Res.* **34**(6), 857–871.

Rimmer, M. A. (1985a). Reproduction of the fork-tailed catfish *Arius graeffei* Kner and Steindachner (Pisces Ariidae) from the Clarence River, New South Wales. *Austr. J. Mar. Freshwater Res.* **36**(1), 23–32.

Rimmer, M. A. (1985b). Growth, feeding and condition of the fork-tailed catfish *Arius graeffei* Kner and Steindachner (Pisces Ariidae) from the Clarence River, New South Wales. *Austr. J. Mar. Freshwater Res.* **36**(1), 33–39.

Rimmer, M. A. (1985c). Early development and buccal incubation in the fork-tailed catfish *Arius graeffei* Kner and Steindachner (Pisces Ariidae) from the Clarence River, New South Wales. *Austr. J. Mar. Freshwater Res.* **36**(3), 405–411.

Risch, L. (1986). Bagridae. In *Check-list of the Freshwater Fishes of Africa* (J. Daget, J.-P. Gosse, and D. F. E. Thys van den Audenaerde, Eds.), pp. 2–35. CLOFFA II, ORSTOM, Paris.

Risch, L. M. (1992). Bagridae. In *Faune des Poissons d'eaux Douces et Saumâtres de l'Afrique de l'Ouest* (C. Lévêque, D. Paugy, and G. G. Teugels, Eds.), Vol. 2, pp. 395–431. ORSTOM, Paris.

Rivas, L. R. (1986). Systematic review of the perciform fishes of the genus *Centropomus*. *Copeia* **1986**(3), 579–611.

Roberts, C. D. (1993). Comparative morphology of spined scales and their phylogenetic significance in the Teleostei. *Bull. Mar. Sci.* **52**(1), 60–113.

Roberts, T. R. (1969). Osteology and relationships of characoid fishes, particulary the genera *Hepsetus*, *Salminus*, *Hoplias*, *Ctenolucius*, and *Acestrorhynchus*. *Proc. California Acad. Sci. Ser. 4* **36**(15), 391–500.

Roberts, T. R. (1970a). Scale-eating American charocoid fishes with special reference to *Probolodus heterostomus*. *Proc. California Acad. Sci.* **38**, 383–390.

Roberts, T. R. (1970b). Description, osteology and relationships of the Amazonian cyprinodont fish *Fluviphylax pygmaeus* (Myers and Carvalho). *Breviora* **347**.

Roberts, T. R. (1971a). *Micromischodus sugillatus*, a new hemiodontid characin fish from Brazil, and its relationships to the Chilodontidae. *Breviora* **367**, 1–25.

Roberts, T. R. (1971b). Osteology of the Malaysian phallostethoid fish *Ceratostethus bicornis*, with a discussion of the evolution of remarkable structural novelties in its jaws and external genitalia. *Bull. Museum Comp. Zool.* **142**(4), 393–418.

Roberts, T. R. (1971c). The fishes of the Malaysian family Phallostethidae (Atheriniformes). *Breviora* **374**, 1–27.

Roberts, T. R. (1972). Ecology of fishes in the Amazon and Congo basins. *Bull. Museum Comp. Zool.* **143**(2), 117–147.

Roberts, T. R. (1973). Osteology and relationships of the Prochilodontidae, a South American family of characoid fishes. *Bull. Museum Comp. Zool.* 145(4), 213–235.

Roberts, T. R. (1974a). Osteology and classification of the neotropical characoid fishes of the families Hemiodontidae (including Anodontidae) and Parodontidae. *Bull. Museum Comp. Zool.* 146(9), 411–472.

Roberts, T. R. (1974b). Dental polymorphism and systematics in *Saccodon*, a neotropical genus of freshwater fishes (Parodontidae, Characoidei). *J. Zool.* 173(3), 303–321.

Roberts, T. R. (1975). Geographic distribution of African freshwater fishes. *Zool. J. Linnean Soc.* 57, 249–319.

Roberts, T. R. (1978). An ichthyological survey of the Fly River in Papua New Guinea with descriptions of new species. *Smithsonian Contrib. Zool.* 281, 1–72.

Roberts, T. R. (1980). A revision of the Asian mastacembelid fish genus *Macrognathus*. *Copeia* 1980(3), 385–391.

Roberts, T. R. (1981). Sundasalangidae, a new family of minute freshwater salmoniform fishes from Southeast Asia. *Proc. California Acad. Sci.* 42(9), 295–302.

Roberts, T. R. (1982a). Unculi (horny projections arising from single cells), an adaptive feature of the epidermis of ostariophysan fishes. *Zool. Scripta.* 11(1), 55–76.

Roberts, T. R. (1982b). A revision of the south and southeast Asian angler-catfishes (Chacidae). *Copeia* 1982(4), 895–901.

Roberts, T. R. (1982c). Revision of the Southeast Asian freshwater pufferfish genus *Chonerhinos* (Tetraodontidae), with descriptions of new species. *Proc. California Acad. Sci.* 43(1), 1–16.

Roberts, T. R. (1982d). Systematics and geographical distribution of the Asian silurid catfish genus *Wallago*, with a key to the species. *Copeia* 1982(4), 890–894.

Roberts, T. R. (1984a). Cobitidae. In *Check-list of the Freshwater Fishes of Africa* (J. Daget, J.-P. Gosse, and D. F. E. Thys van den Audenaerde, Eds.), p. 343. CLOFFA I, ORSTOM, Paris.

Roberts, T. R. (1984b). Hepsetidae. In *Check-list of the Freshwater Fishes of Africa* (J. Daget, J.-P. Gosse, and D. F. E. Thys van den Audenaerde, Eds.), pp. 138–139. CLOFFA I, ORSTOM, Paris.

Roberts, T. R. (1984c). Skeletal anatomy and classification of the neotenic Asian salmoniform superfamily Salangoidea (icefishes or noodlefishes). *Proc. California Acad. Sci.* 43(13), 179–220.

Roberts, T. R. (1986a). *Danionella translucida*, a new genus and species of cyprinid fish from Burma, one of the smallest living vertebrates. *Environ. Biol. Fishes* 16, 231–241.

Roberts, T. R. (1986b). Systematic review of the Mastacembelidae or spiny eels of Burma and Thailand, with description of two new species of *Macrognathus*. *Jpn. J. Ichthyol.* 33(2), 95–109.

Roberts, T. R. (1986c). Tetraodontidae. In *Check-list of the Freshwater Fishes of Africa* (J. Daget, J.-P. Gosse, and D. F. E. Thys van den Audenaerde, Eds.), pp. 434–436. CLOFFA II, ORSTOM, Paris.

Roberts, T. R. (1989a). The freshwater fishes of western Borneo (Kalimantan Barat, Indonesia). *Mem. California Acad. Sci.* 14.

Roberts, T. R. (1989b). Systematic revision and description of new species of suckermouth catfishes (*Chiloglanis*, Mochokidae) from Cameroon. *Proc. California Acad. Sci.* 46(6), 151–178.

Roberts, T. R. (1990). Mimicry of prey by fin-eating fishes of the African characoid genus *Eugnathichthys* (Pisces: Distichodidae). *Ichthyol. Exploration Freshwaters* 1(1), 23–31.

Roberts, T. R. (1992a). Systematic revision of the Old World freshwater fish family Notopteridae. *Ichthyol. Exploration Freshwater* 2(4), 361–383.

Roberts, T. R. (1992b). Systematic revision of the Southeast Asian anabantoid fish genus *Osphronemus* with descriptions of two new species. *Ichthyol. Exploration Freshwaters* 2(4), 351–360.

Roberts, T. R. (1994). *Osphronemus exodon*, a new species of giant gouramy with extraordinary dentition from the Mekong. *Nat. History Bull. Siam Soc.* 42, 67–77.

Roberts, T. R. (1998). Systematic observations on tropical Asian medakas or ricefishes of the genus *Oryzias*, with descriptions of four new species. *Ichthyol. Res.* 45(3), 213–224.

Roberts, T. R., and Karnasuta, J. (1987). *Dasyatis laosensis*, a new whiptailed stingray (family Dasyatidae) from the Mekong River of Laos and Thailand. *Environ. Biol. Fishes* 20(3), 161–167.

Roberts, T. R., and Kottelat, M. (1993). Revision of the Southeast Asian freshwater fish family Gyrinocheilidae. *Ichthyol. Exploration Freshwaters* 4(4), 375–383.

Roberts, T. R., and Kottelat, M. (1994). The Indo-Pacific tigerperches, with a new species from the Mekong basin (Pisces: Coiidae). *Ichthyol. Exploration Freshwaters* 5(3), 257–266.

Roberts, T. R., and Vidthayanon, C. (1991). Systematic revision of the Asian catfish family Pangasiidae, with biological observations and descriptions of three new species. *Proc. Acad. Nat. Sci. Philadelphia* 143, 97–144.

Robins, C. R., and Ray, G. C. (1986). *Atlantic Coast Fishes*. Houghton Mifflin, Boston.

Robins, C. R., Bailey, R. M., Bond, C. E., Brooker, J. R., Lachner, E. A., Lea, R. N., and Scott, W. B. (1991). *Common and Scientific Names of Fishes from the United States and Canada*, 5th ed., Special Publication No. 20. American Fisheries Society, Bethesda, MD.

Robison, H. W., and Buchanan, T. M. (1989). *Fishes of Arkansas*. Univ. of Arkansas Press, Fayetteville.

Rodriguez, C. M. (1997). Phylogenetic analysis of the tribe Poeciliinia (Cyprinodontiformes: Poeciliidae). *Copeia* **1977**(4), 663–679.

Rohde, F. C. (1980). *Prosopium*. In *Atlas of North American Freshwater Fishes* (D. S. Lee *et al.*, Eds.), pp. 97–102. North Carolina State Museum of Natural History, Raleigh.

Rohde, F. C., and Arndt, R. G. (1987). Two new species of pygmy sunfishes (Elassomatidae, *Elassoma*) from the Carolinas. *Proc. Acad. Nat. Sci. Philadelphia* **139**, 65–85.

Romand, R. (1992). Cyprinodontidae. In *Faune des Poissons d'eaux Douces et Saumâtres de l'Afrique de l'Ouest* (C. Lévêque, D. Paugy, and G. G. Teugels, Eds.), Vol. 2, pp. 586–654. ORSTOM, Paris.

Romero, A. (1985a). Ontogenetic change in phototactic responses of surface and cave populations of Astyanax fasciatus (Pisces: Characidae). *Copeia* **1984**(4), 1004–1011.

Romero, A. (1985b). Can evolution regress? *Natl. Speleol. Soc. Bull.* **47**(2), 86–88.

Romero, A. (1998a). Threatened fishes of the world: Amblyopsis rosae (Eigenmann, 1842) (Amblyopsidae). *Environ. Biol. Fishes* **52**(4), 434.

Romero, A. (1998b). Threatened fishes of the world: Typhlichthys subterraneus (Girard, 1860) (Amblyopsidae). *Environ. Biol. Fishes* **53**, 74.

Romero, A. (1998c). Threatened fishes of the world: Speoplatyrhinus poulsoni Cooper and Kuehne, 1974 (Amblyopsidae). *Environ. Biol. Fishes* **53**, 293–294.

Romero, A., and Bennis, L. (1998). Threatened fishes of the world: Amblyopsis spelaea DeKay, 1842 (Amblyopsidae). *Environ. Biol. Fishes* **51**, 420.

Romero, A., and McLeran, A. (2000). Threatened fishes of the world: Stygichthys typhlops Brittan & Böhlke 1965 (Characidae). *Environ. Biol. Fishes* **57**(3), 270.

Rosen, D. E. (1962). Comments on the relationships of the North American cave fishes of the family Amblyopsidae. *Am. Museum Novitates* **2109**, 1–35.

Rosen, D. E. (1973a). Suborder Cyprinodontoidei. In *Fishes of the Western North Atlantic*, Memoir

Sears Foundation for Marine Research No. 1, part 6, pp. 229–262.

Rosen, D. E. (1973b). Interrelationships of higher euteleostean fishes. *J. Linnean Soc. Zool.* **53**(Suppl. 1), 397–513.

Rosen, D. E. (1974). Phylogeny and zoogeography of salmoniform fishes and relationships of *Lepidogalaxias salamandroides*. *Bull. Am. Museum Nat. History* **153**(2), 265–326.

Rosen, D. E. (1975). A vicariance model of Caribbean biogeography. *Syst. Zool.* **24**(4), 431–464.

Rosen, D. E. (1978). Vicariant patterns and historical explanation in biogeography. *Syst. Zool.* **27**(2), 159–188.

Rosen, D. E. (1984). Zeiforms as primitive plectognath fishes. *Am. Museum Novitates* **2782**, 1–38.

Rosen, D. E. (1985). An essay on euteleostean classification. *Am. Museum Novitates* **2828**, 1–57.

Rosen, D. E., and Bailey, R. M. (1963). The poeciliid fishes (Cyprinodontiformes), their structure, zoogeography, and systematics. *Bull. Am. Museum Nat. History* **126**(1), 1–176.

Rosen, D. E., and Greenwood, P. H. (1976). A fourth Neotropical species of synbranchid eel and the phylogeny and systematics of synbranchiform fishes. *Bull. Am. Museum Nat. History* **157**(1), 1–70.

Rosen, D. E., and Parenti, L. R. (1981). Relationships of *Oryzias*, and the groups of atherinomorph fishes. *Am. Museum Novitates* **2719**, 1–25.

Rosenqvist, G. (1990). Male mate choice and female–female competition for mates in the pipefish *Nerophis ophidion*. *Anim. Behav.* **39**, 1110–1115.

Ross, S. W., and Burgess, G. H. (1980). *Dasyatis sabine* (Lesueur) Atlantic stingray. In *Atlas of North American Freshwater Fishes* (D. S. Lee *et al.*, Eds.), p. 37. North Carolina State Museum of Natural History, Raleigh.

Rostlund, E. (1952). *Freshwater Fish and Fishing in Native North America*, Publication in Geography No. 9. Univ. of California, Berkley.

Rowland, S. J. (1989). Aspects of the history and fishery of the Murray cod, *Maccullochella peeli* (Mitchell) (Percichthyidae). *Proc. Linnean Soc. New South Wales* **111**(3), 201–213.

Rowland, S. J. (1993). *Maccullochella ikei*, an endangered species of freshwater cod (Pisces: Percichthyidae) from the Clarence River system, NSW, and *M. peelii mariensis*, a new subspecies from the Mary River system, Qld. *Rec. Austr. Museum* **45**, 121–145.

Rowland, S. J. (1996). Threatened fishes of the world: *Maccullochella ikei* Rowland, 1985 (Percichthyidae). *Environ. Biol. Fishes* **46**, 350.

Rowland, S. J. (1998a). Aspects of the reproductive biology of Murray Cod, *Maccullochella peelii*

peelii. *Proc. Linnean Soc. New South Wales* **120**, 147–162.

Rowland, S. J. (1998b). Age and growth of the Australian freshwater fish Murray Cod, *Maccullochella peelii peelii*. *Proc. Linnean Soc. New South Wales* **120**, 164–180.

Rowland, S. J., Allan, G. L., Hollis, M., and Pontifex, T. (1995). Production of the Australian freshwater silver perch, *Bidyanus bidyanus* (Mitchell), at two densities in earthen ponds. *Aquaculture* **130**, 317–328.

Ruiz, V. H., and Berra, T. M. (1994). Fishes of the high Biobio River of south-central Chile with notes on diet and speculations on the origin of the ichthyofauna. *Ichthyol. Exploration Freshwaters* **5**, 5–18.

Runcorn, S. K. (Ed.) (1962). *Continental Drift*. Academic Press, New York.

Rundle, H. D., Nagel, L., Boughman, J. W., and Schluter, D. (2000). Natural selection and parallel speciation in sympatric sticklebacks. *Science* **287**, 306–308.

Russell, B. C. (1987). Clarification of the use of the family name Synodontidae in the Myctophiformes and Siluriformes. *Copeia* **1987**(2), 513–515.

Russell, D. F., Wilkens, L. A., and Moss, F. (1999). Use of behavioural stochastic resonance by paddle fish for feeding. *Nature* **402**, 291–294.

Saeed, B., and Ivantsoff, W. (1991). *Kalyptatherina*, the first telmatherinid genus known outside of Sulawesi. *Ichthyol. Exploration Freshwaters* **2**(3), 227–238.

Saeed, B., Ivantsoff, W., and Allen, G. R. (1989). Taxonomic revision of the Family Pseudomugilidae (Order Atheriniformes). *Austr. J. Mar. Freshwater Res.* **40**(6), 719–787.

Saeed, B., Ivantsoff, W., and Crowley, L. E. L. M. (1994). Systematic relationships of atheriniform families within Division I of the series Atherinomorpha (Acanthopterygii) with relevant historical perspectives. *J. Ichthyol.* **34**(9), 27–72.

Safina, C. (1995). The world's imperiled fish. *Sci. Am.* **273**(5), 46–53.

Safina, C. (1997). *Song for the Blue Ocean*. Holt, New York.

Sagua, V. O. (1979). Observations on the food and feeding habits of the African electric catfish, *Malapterurus electricus* (Gmelin). *J. Fish Biol.* **15**, 61–69.

Sagua, V. O. (1987). On a new species of electric catfish from Kainji, Niger, with some observations on its biology. *J. Fish Biol.* **30**, 75–89.

Sakai, H., Jeon, S.-R., Tsujii, H., and Iwata, A. (1996). An electrophoretic study of genetic differentiation in Korean *Odontobutis*. *Korean J. Limnol.* **29**(1), 1–7.

Sanford, C. P. J. (1990). The phylogenetic relationships of salmonoid fishes. *Bull. Br. Museum Nat. History Zool.* **65**(2), 145–153.

Sanger, A. C. (1984). Description of a new species of *Gadopsis* from Victoria. *Proc. R. Soc. Victoria* **96**, 93–97.

Sanger, A. C. (1990). Aspects of the life history of the two-spined blackfish, *Gadopsis bispinosus*, in King Parrot Creek, Victoria. *Proc. R. Soc. Victoria* **102**, 89–96.

Sasaki, K. (1989). Phylogeny of the family Sciaenidae, with notes on its zoogeography (Teleostei, Perciformes). *Mem. Faculty Fish. Hokkaido Univ.* **36**(1/2), 1–137.

Sato, T. (1986). A brood parasitic catfish of mouth-brooding cichlid fishes in Lake Tanganyika. *Nature* **323**, 58–59.

Saul, W. G. (1975). An ecological study of fishes at a site in upper Amazonian Ecuador. *Proc. Acad. Nat. Sci. Philadelphia* **127**, 93–134.

Sawada, Y. (1982). Phylogeny and zoogeography of the superfamily Cobitoidea (Cyprinoidei, Cypriniformes). *Mem. Faculty Fish. Hokkaido Univ.* **28**, 65–223.

Sazima, I. (1983). Scale-eating in characoids and other fishes. *Environ. Biol. Fishes* **9**(2), 87–101.

Sazima, I., and Guimaraes, S. de A. (1987). Scavenging on human corpes as a source for stories about man-eating piranhas. *Environ. Biol. Fishes* **20**, 75–77.

Sazima, I., and Machado, F. A. (1990). Underwater observations of piranhas in western Brazil. *Environ. Biol. Fishes* **28**, 17–31.

Schaefer, S. A. (1987). Osteology of *Hypostomus plecostomus* (Linneaus), with a phylogenetic analysis of the loricariid subfamilies. *Contrib. Sci. Nat. History Museum Los Angeles County* **394**, 1–31.

Schaefer, S. A. (1990). Anatomy and relationships of the scoloplacid catfishes. *Proc. Acad. Nat. Sci. Philadelphia* **142**, 167–210.

Schaefer, S. A. (1991). Phylogenetic analysis of the loricariid subfamily Hypoptopomatinae (Pisces: Siluroidei: Loricariidae), with comments on generic diagnoses and geographic distribution. *Zool. J. Linnean Soc.* **102**, 1–41.

Schaefer, S. A. (1996). *Nannoptopoma*, a new genus of loricariid catfishes (Siluriformes: Loricariidae) from the Amazon and Orinoco River basins. *Copeia* **1996**(4), 913–926.

Schaefer, S. A. (1997). The Neotropical cascudinhos: Systematics and biogeography of the *Otocinclus* catfishes (Siluriformes: Loricariidae). *Proc. Acad. Nat. Sci. Philadelphia* **148**, 1–120.

Schaefer, S. A. (1998). Conflict and resolution: Impact of new taxa on phylogenetic studies of the Neotropical Cascudinhos (Siluroidei: Loricariidae). In *Phylogeny and Classification of*

Neotropical Fishes (L. R. Malabara, R. E. Reis, R. P. Vari, C. A. S. Lucena, and Z. M. S. Lucena, Eds.), pp. 375–400. Museu de Ciências e Tecnologia da PUCRS, Porto Alegre, Brazil.

Schaefer, S. A., and Lauder, G. V. (1986). Historical transformation of functional design: Evolutionary morphology of feeding mechanisms in loricarioid catfishes. *Syst. Zool.* **35**(4), 489–508.

Schaefer, S. A., and Lauder, G. V. (1996). Testing historical hypotheses of morphological change: Biomechanical decoupling in loricarioid catfishes. *Evolution* **50**(4), 1661–1675.

Schaefer, S. A., Weitzman, S. H., and Britski, H. A. (1989). Review of the neotropical catfish genus *Scoloplax* (Pisces: Loricarioidea Scoloplacidae) with comments on reductive characters in phylogenetic analysis. *Proc. Acad. Nat. Sci. Philadelphia* **141**, 181–211.

Scheel, J. J. (1990). *Atlas of Killifishes of the Old World.* T. F. H., Neptune City, NJ.

Schmid, H., and Randall, J. E. (1997). First record of the tripletail, *Lobotes surinamensis* (Pisces: Lobotidae), from the Red Sea. *Fauna Saudi Arabia* **16**, 353–355.

Schmid, T. H., Ehrhart, L. M., and Snelson, F. F., Jr. (1988). Notes on the occurrence of rays (Elasmobranchii, Batoidea) in the Indian River lagoon system, Florida. *Florida Scientist* **51**(2), 121–128.

Schmitt, R. J., and Coyer, J. A. (1982). The foraging ecology of sympatric surfperch congeners (*Embiotoca*): Importance of foraging behavior in prey size selection. *Oecologia* **55**, 369–378.

Schultz, J. (1989). Origins and relationships of unisexual poeciliids. In *Ecology and Evolution of Livebearing Fishes* (G. K. Meffe and F. F. Snelson, Jr., Eds.), pp. 69–87. Prentice Hall, Englewood Cliffs, NJ.

Schultz, L. P. (1944). The catfishes of Venezuela, with a description of thirty-eight new forms. *Proc. U.S. Natl. Museum* **94**, 173–338.

Schultz, L. P. (1946). A revision of the genera of mullets, fishes of the family Mugilidae, with descriptions of three new genera. *Proc. U.S. Natl. Museum* **96**, 377–395.

Schultz, L. P. (1948). A revision of six subfamilies of atherine fishes, with descriptions of new genera and species. *Proc. U.S. Natl. Museum* **98**, 1–48.

Schultze, H.-P. (1994). Comparison of hypotheses on the relationships of sarcopterygians. *Syst. Biol.* **43**, 155–173.

Schultze, H.-P., and Wiley, E. O. (1984). The neopterygian *Amia* as a living fossil. In *Living Fossils* (N. Eldredge and S. M. Stanley, Eds.), pp. 153–159. Springer-Verlag, New York.

Schwarzhans, W. (1993). A comparative morphological treatise of recent and fossil otoliths of the family Sciaenidae (Perciformes). In *Piscium Catalogus: Part Otolithi Piscium*, Vol. 1. Verlag Dr. Friedrich Pfeil, Munich.

Schwassmann, H. O. (1976). Ecology and taxonomic status of diferent geographical populations of *Gymnorhamphichthys hypostomus* Ellis (Pisces, Cypriniformes, Gymnotoidei). *Biotropica*, 25–40.

Schwassmann, H. O. (1984). Species of *Steatogenys* Boulenger (Pisces, Gymnotiformes, Hypopomidae). *Boletim Museu Paraense Emilio Goeldi Zool.* **1**(1), 97–114.

Schwassmann, H. O. (1989). *Gymnorhamphichthys rosamariae*, a new species of knifefish (Rhamphichthyidae, Gymnotiformes) from the Upper Rio Negro, Brazil. *Stud. Neotropical Fauna Environ.* **24**, 157–167.

Schwassmann, H. O., and Kruger, L. (1965). Experimental analysis of the visual system of the four-eyed fish *Anapleps microlepis*. *Vision Res.* **5**, 269–281.

Scientific American (Readings from) (1973). *Continents Adrift.* Freeman, San Francisco.

Scott, G. K., Fletcher, G. L., and Davis, P. L. (1986). Fish antifreeze proteins: Recent gene evolution. *Can. J. Fish. Aquat. Sci.* **43**, 1028–1034.

Scott, T. D., Glover, C. J. M., and Southcott, R. V. (1974). *The Marine and Freshwater Fishes of South Australia.* South Australia Government Printer, Adelaide.

Scott, W. B., and Crossman, E. J. (1973). Freshwater fishes of Canada, Bulletin No. 184. Fisheries Research Board of Canada, Ottawa.

Secor, D. H., and Waldman, J. R. (1999). Historical abundance of Delaware Bay Atlantic sturgeon and potential rate of recovery. In *Life in the Slow Lane: Ecology and Conservation of Long-Lived Marine Animals* (J. A. Musick, Ed.), Symposium No. 23, pp. 203–216. American Fisheries Society, Bethesda, MD.

Seegers, L. (1996). The fishes of the Lake Rukwa drainage. *Koninklijk Museum Midden-Afrika Tervuren België Ann. Zool. Wetenschapen* **278**, 1–408.

Seehausen, O. J., Van Alphen, J. M., Witte, F. (1997). Cichlid fish diversity threatened by eutrophication that curbs sexual selection. *Science* **277**, 1808–1811.

Sideleva, V. G. (1980). Structural features of the seismosensory system of freshwater sculpins and Baikal oil-fishes (Cottidae and Comephoridae) in connection with a pelagic mode of life. *J. Ichthyol.* **20**(6), 119–124.

Siebert, D. J. (1997). Notes on the anatomy and relationships of *Sundasalanx* Roberts (Teleostei, Clupeidae), with descriptions of four new species from Borneo. *Bull. Nat. History Museum London (Zool.)* **63**(1), 13–26.

Silas, E. C. (1959). On the natural distribution of the Indian cyprinodont fish *Hoaraichthys setnai* Kulkarni. *J. Mar. Biol. Assoc. India* 1(2), 256.

Silfvergrip, A. M. C. (1992). *Zungaro*, a senior synonym of *Paulicea* (Teleostei: Pimelodidae). *Ichthyol. Exploration Freshwaters* 3(4), 305–310.

Silfvergrip, A. M. C. (1996). A systematic revision of the neotropical catfish genus *Rhamdia* (Teleostei, Pimelodidae). Ph.D. thesis, Department of Vertebrate Zoology, Swedish Museum of Natural History, Stockholm.

Simons, A. M. (1992). Phylogenetic relationships of the *Boleosoma* species group (Percidae: *Etheostoma*). In *Systematics, Historical Ecology, and North American Freshwater Fishes* (R. L. Mayden, Ed.), pp. 268–292. Stanford Univ. Press, Stanford, CA.

Simons, A. M., and Mayden, R. L. (1999) Phylogenetic relationships of North American cyprinids and assessment of homology of the open posterior myodome. *Copeia* 1999(1), 13–21.

Simpson, D. G., and Gunter, G. (1956). Notes on habitats, systematic characters and life histories of Texas salt water cyprinodontes. *Tulane Stud. Zool.* 4(4), 115–134.

Skelton, P. H. (1984). A systematic revision of species of the catfish genus *Amphilius* (Siluroidei, Amphiliidae) from east and southern Africa. *Ann. Cape Province Museum* 16(3), 41–71.

Skelton, P. H. (1989). Descriptions of two new species of west African amphiliid catfishes (Siluroidei, Amphiliidae). Ichthyological Bulletin of the J. L. B. Smith Institute of Ichthyology, Special Publication No. 48.

Skelton, P. H. (1992). Amphiliidae. In *Faune des Poissons d'eaux Douces et Saumâtres de l'Afrique de l'Ouest* (C. Lévêque, D. Paugy, and G. G. Teugels, Eds.), Vol. 2, pp. 450–467. ORSTOM, Paris.

Skelton, P. (1993). *A Complete Guide to the Freshwater Fishes of Southern Africa*. Southern Book, Halfway House, South Africa.

Skelton, P. H., and Teugels, G. G. (1989). Amphiliidae. In *Check-list of the Freshwater Fishes of Africa* (J. Daget, J.-P. Gosse, and D. F. E. Thys van den Audenaerde, Eds.), pp. 54–65. CLOFFA II, ORSTOM, Paris.

Smith, A. G. (1977). *Mesozoic and Cenozoic Paleocontinental Maps*. Cambridge Univ. Press, Cambridge, UK.

Smith, C. L. (1985). *The Inland Fishes of New York State*. New York State Department of Environmnetal Conservation, Albany.

Smith, D. G. (1989a). Order Anguilliformes: Family Anguillidae. In *Fishes of the Western North Atlantic. Part 9, Vol. 1. Orders Anguilliformes and Saccopharyngiformes* (E. B. Böhlke, Ed.), pp. 25–47. Sears Foundation for Marine Research, Memoir (Yale University), New Haven, CT.

Smith, D. G. (1989b). Family Anguillidae: Leptocephali. In *Fishes of the Western North Atlantic. Part 9, Vol. 2. Leptocephali* (E. B. Böhlke, Ed.), pp. 898–899. Sears Foundation for Marine Research, Memoir (Yale University), New Haven, CT.

Smith, G. R. (1966). Distribution and evolution of the North American catostomid fishes of the subgenus *Pantosteus*, genus *Catostomus*. *Miscellaneous Publ. Museum Zool. Univ. Michigan* 129, 1–133.

Smith, G. R. (1992). Phylogeny and biogeography of the Catostomidae, freshwater fishes of North America and Asia. In *Systematics, Historical Ecology, and North American Freshwater Fishes* (R. L. Mayden, Ed.), pp. 778–826. Stanford Univ. Press, Stanford, CA.

Smith, G. R., and Hoehn, R. K. (1971). Phenetic and cladistic studies of biochemical and morphological characteristics of *Catostomus. Syst. Zool.* 20(3), 282–297.

Smith, G. R., and Lundberg, J. G. (1972). The Sand Draw fish fauna. *Bull. Am. Museum Nat. History* 148, 40–54.

Smith, G. R., and Stearley, R. F. (1989). The classification and scientific names of rainbow and cutthroat trouts. *Fisheries* 14(1), 4–10.

Smith, G. R., and Todd, T. N. (1984). Evolution of species flocks of fishes in north temperate lakes. In *Evolution of Fish Species Flocks* (A. A. Echelle and I. Kornfield, Eds.), pp. 45–68. Univ. of Maine Press, Orono.

Smith, H. M. (1936). The archer fish. *Nat. History* 38, 3–11.

Smith, H. M. (1945). The freshwater fishes of Siam, or Thailand. *Bull. U.S. Natl. Museum* 188.

Smith, J. L. B. (1950). Two noteworthy non-marine fishes from South Africa. *Ann. Magazine Nat. History* 12(3), 705–710.

Smith, M. L., and Miller, R. R. (1987). *Allotoca goslinei*, a new species of goodeid fish from Jalisco, Mexico. *Copeia* 1987(3), 610–616.

Smith, M. L., Cavender, T. M., and Miller, R. R. (1975). Climatic and biogeographic significance of a fish fauna from the late Pliocene–early Pleistocene of the Lake Chapala basin (Jalisco, Mexico). Studies on Cenozoic Paleontology and Stratigraphy, Papers on Paleontology No. 12, pp. 29–39. University of Michigan, Ann Arbor.

Smith, M. L., Song, J., and Miller, R. R. (1984). Redescription, variation, and zoogeography of the Mexican darter *Etheostoma pottsi* (Pisces: Percidae). *Southwestern Nat.* 29(4), 395–402.

Smith-Vaniz, W. F., Collette, B. B., and Luckhurst, B. E. (1999). Fishes of Bermuda: History, zoogeography, annotated checklist, and identification keys, Special Publication No. 4. American Society of Ichthyologists and Herpetologists, Lawrence, KS.

Snelson, F. F. (1991). Review of "Phylogenetic Studies of North American Minnows, with Emphasis on the Genus *Cyprinella* (Teleostei: Cyprinidae)" by R. L. Mayden. *Copeia* **1991**, 258–260.

Soares-Porto, L. M. (1998). Monophyly and inter-relationships of the Centromochlinae (Siluriformes: Auchenipteridae). In *Phylogeny and Classification of Neotropical Fishes* (L. R. Malabarba, R. E. Reis, R. P. Vari, Z. M. S. Lucena, and C. A. S. Lucena, Eds.), pp. 331–350. EDIPUCRS, Porto Alegre, Brazil.

Soares-Porto, L. M., Walsh, S. J., Nico, L. G., and Netto, J. M. (1999). A new species of *Gelanoglanis* from the Orinoco and Amazon river basins, with comments on miniaturization within the genus (Siluriformes: Auchenipteridae: Centromochlinae). *Ichthyol. Exploration Freshwaters* **10**(1), 63–72.

Soltz, D. L., and Naiman, R. J. (1978). The natural history of native fishes in The Death Valley System. *Nat. History Museum Los Angeles County Sci. Ser.* **30**, 1–76.

Song, C. B., Near, T. J., and Page, L. M. (1998). Phylogenetic relations among percid fishes as inferred from mitochondrial cytochrome b DNA sequence data. *Mol. Phylogenet. Evol.* **10**(3), 343–353.

Soto, C. S., and Noakes, D. L. G. (1994). Color and gender in the hermaphroditic fish *Rivulus marmoratus* Poey (Teleostei: Rivulidae). *Ichthyol. Exploration Freshwaters* **5**, 79–90.

Springer, V. G. (1982). Pacific plate biogeography, with special reference to shorefishes. *Smithsonian Contrib. Zool.* **367**, 1–182.

Starnes, W. C., and Schindler, I. (1993). Comments on the genus *Apareiodon* Eigenmann (Characiformes: Parodontidae) with the description of a new species from the Gran Sabana region of eastern Venezuela. *Copeia* **1993**(3), 754–762.

Stauffer, J. R., Jr., LoVullo, T. J., and McKaye, K. R. (1993). Three new sand-dwelling cichlids from Lake Malawi, Africa, with a discussion of the status of the genus *Copadichromis* (Teleostei: Cichlidae). *Copeia* **1993**(4), 1017–1027.

Stauffer, J. R., Jr., Bowers, N. J., Kellogg, K. A., and McKaye, K. R. (1997). A revision of the blue-black *Pseudotropheus zebra* (Teleostei: Cichlidae) complex from Lake Malawi, Africa, with a description of a new genus and 10 new species. *Proc. Acad. Nat. Sci. Philadelphia* **148**, 189–230.

Stead, D. G. (1906). *Fishes of Australia.* William Brooks, Sydney.

Stearley, R. F., and Smith, G. R. (1993). Phylogeny of the Pacific trouts and salmons, *Onchorhynchus*, and genera of the family Salmonidae. *Trans. Am. Fish. Soc.* **122**(1), 1–33.

Steindachner, F. (1879a). Fisch-fauna des Magdalenen-Stromes. *Denkschriften Akad. Wissenschaften. Wien* **39**, 19–78, Plates 1–15.

Steindachner, F. (1879b). Über einige neue und seltene fisch-arten aus den k. k. zool. Museen zu Wien, Stuttgart und Warschau. *Denkschriften Akad. Wissenschaften. Wien* **41**, 1–52, Plates 1–9.

Steindachner, F. (1882a). Beiträge zur Kenntniss der flussfische Südamerika's. II. *Denkschriften Akad. Wissenschaften. Wien* **43**, 103–146, Plates 1–7.

Steindachner, F. (1882b). Beiträge zur Kenntniss der flussfische Südamerika's. III. *Denkschriften Akad. Wissenschaften. Wien* **44**, 1–18, Plates 1–5.

Steindachner, F. (1883). Beiträge zur Kenntniss der flussfische Südamerika's. IV. *Denkschriften Akad. Wissenschaften. Wien* **46**, 1–44, Plates 1–7.

Steindachner, F. (1915). Beiträge zur Kenntniss der flussfische Südamerika's. V. *Denkschriften Akad. Wissenschaften. Wien* **93**, 1–92, Plates 1–13.

Sterba, G. (1966). *Freshwater Fishes of the World*, rev. ed. Studio Vista, London.

Sterba, G. (Ed.) (1986). *The Aquarium Encyclopedia.* MIT Press, Cambridge, MA.

Stevenson, H. M. (1976). *Vertebrates of Florida.* Univ. Presses of Florida, Gainesville.

Stewart, D. J. (1985). A review of the South American catfish tribe Hoplomyzontini (Pisces, Aspredinidae), with descriptions of new species from Ecuador. *Fieldiana Zool.* **25**, 1–19.

Stewart, D. J. (1986a). Revision of *Pimelodina* and description of a new genus and species from the Peruvian Amazon (Pisces: Pimelodidae). *Copeia* **1986**(3), 653–672.

Stewart, D. J. (1986b). A new pimelodid catfish from the deep-river channel of the Rio Napo, eastern Ecuador (Pisces: Pimelodidae). *Proc. Acad. Nat. Sci. Philadelphia* **138**(1), 46–52.

Steyskal, G. C. (1980). The grammar of family-group names as exemplified by those of fishes. *Proc. Biol. Soc. Washington* **93**(1), 168–177.

Stiassny, M. L. J. (1987). Cichlid familial intrarelationships and the placement of the neotropical genus *Cichla* (Perciformes, Labroidei). *J. Nat. History* **21**, 1311–1331.

Stiassny, M. L. J. (1990a). Notes on the anatomy and relationships of the bedotiid fishes of Madagascar, with a taxonomic revision of the genus *Rheocles* (Atherinomorpha: Bedotiidae). *Am. Museum Novitates* **2979**, 1–33.

Stiassny, M. L. J. (1990b). *Tylochromis*, relationships and the phylogenetic status of the African Cichlidae. *Am. Museum Novitates* 2993, 1–14.

Stiassny, M. L. J. (1991). Phylogenetic intrarelationships of the family Cichlidae: An overview. In *Cichlid Fishes, Behaviour, Ecology, and Evolution* (M. H. A. Keenleyde, Ed.), pp. 1–35. Chapman & Hall, London.

Stiassny, M. L. J. (1992). Phylogenetic analysis and the role of systematics in the biodiversity crisis. In *Systematics, Ecology, and the Biodiversity Crisis* (N. Eldredge, Ed.), pp. 109–120. Columbia Univ. Press, New York.

Stiassny, M. L. J. (1993). What are grey mullets? *Bull. Mar. Sci.* 52(1), 197–219.

Stiassny, M. L. J. (1996). An overview of freshwater biodiversity: With some lessons from African fishes. *Fisheries* 21(9), 7–13.

Stiassny, M. L. J., and Gerstner, C. L. (1992). The parental care behavior of *Paratilapia polleni* (Perciformes, Labroidei), a phylogenetically primitive cichlid from Madagascar, with a discussion of the evolution of maternal care in the family Cichlidae. *Environ. Biol. Fishes* 34, 219–233.

Stiassny, M. L. J., and Jensen, J. S. (1987). Labroid intrarelationships revisited: Morphological complexity, key innovations, and the study of comparative diversity. *Bull. Museum Comp. Zool.* 151(5), 269–319.

Stiassny, M. L. J., and Meyer, A. (1999). Cichlids of the rift lakes. *Sci. Am.* 280(2), 64–69.

Stiassny, M. L. J., and Mezey, J. G. (1993). Egg attachment systems in the family Cichlidae (Perciformes: Labroidei), with some comments on their significance for phylogenetic studies. *Am. Museum Novitates* 3058, 1–11.

Stiassny, M. L. J., and Raminosoa, N. (1994). The fishes of the inland waters of Madagascar. In *Biological Diversity of African Fresh- and Brackish Water Fishes* (G. G. Teugels, J.-F. Guégan, and J.-J. Albaret, Eds.), pp. 133–148. Annales Musée Royal de L'Afrique Centrale, Zoologiques, Belgique.

Stiassny, M. L. J., and Reinthal, P. N. (1992). Description of a new species of *Rheocles* (Atherinomorpha, Bedotiidae) from the Nosivolo Tributary, Mangoro River, Eastern Malagasy Republic. *Am. Museum Novitates* 3031, 1–8.

Stoddard, P. K. (1999). Predation enhances complexity in the evolution of electric fish signals. *Nature* 400, 254–255.

Sturmbauer, C., Verheyen, E., Rüber, L., and Meyer, A. (1997). Phylogeographic patterns in populations of cichlid fishes from rocky habitats in Lake Tanganyika. In *Molecular Systematics of Fishes* (T. D. Kocher and C. A. Stepien, Eds.), pp. 97–111. Academic Press, San Diego.

Su, J., Hardy, G. S., and Tyler, J. C. (1986). A new generic name for *Anchisomus multistriatus* Richardson 1854 (Tetraodontidae), with notes on its toxicity and pufferfish biting behavior. *Rec. Western Austr. Museum* 13(1), 101–120.

Sültmann, H., and Mayer, W. E. (1997). Reconstruction of cichlid fish phylogeny using nuclear DNA markers. In *Molecular Systematics of Fishes* (T. D. Kocher and C. A. Stepien, Eds.), pp. 39–51. Academic Press, San Diego.

Suttkus, R. D. (1961). Additional information about blind catfishes from Texas. *Southwestern Nat.* 6, 55–64.

Suttkus, R. D. (1963). Order Lepisostei. In *Fishes of the Western North Atlantic*, Memoir of the Sears Foundation for Marine Research No. 1, pt. 3, pp. 61–88. Sears Foundation for Marine Research, New Haven, CT.

Sverliji, S. B., Schenke, R. L. D., López, H. L., and Ros, A. E. (1998). *Peces del Río Uruguay*. Publicaciones de la comision Administradora del Rio Uruguai, Uruguay.

Svetovidov, A. N. (1962). Gadiformes. *Fauna of the U.S.S.R.* 9(4) [Translation of the 1948 Russian edition]. National Science Foundation, Israel Program for Scientific Translation.

Swanson, C., Cech, J. J., Jr., and Piedrahita, R. H. (1996). *Mosquitofish Biology, Culture, and Use in Mosquite Control*. Mosquito and Vector Control Association of California, Elk Grove.

Swofford, D. L., Branson, B. A., and Sievert, G. A. (1980). Genetic differentiation of cavefish populations (Amblyopsidae). *Isozyme Bull.* 13, 109–110.

Talwar, P. K., and Jhingran, A. G. (1992). *Inland Fishes of India and Adjacent Countries*, 2 vols. Balkema, Rotterdam.

Tan, T. H. T., and Ng, P. K. L. (1996). Catfishes of the *Ompok leiacanthus* (Bleeker, 1853) species group (Teleostei: Siluridae) from southeast Asia, with description of a new species. *Reffles Bull. Zool.* 44(2), 531–542.

Tarling, D., and Tarling, M. (1975). *Continental Drift*. Anchor, Garden City, NY.

Tarp, F. H. (1952). A revision of the family Embiotocidae (the surfperches). *California Department Fish Game Fish Bull.* 88, 1–99.

Taylor, D. S. (1992). Diet of the killifish *Rivulus marmoratus* collected from land crab burrows, with further ecological notes. *Environ. Biol. Fishes* 33, 389–393.

Taylor, D. S., Davis, W. P., and Turner, B. J. (1995). *Rivulus marmoratus*: Ecology of distributional patterns in Florida and the central Indian River lagoon. *Bull. Mar. Sci.* 57(1), 202–207.

Taylor, J. N., Courtenay, W. R., Jr., and McCann, J. A. (1984). Known impacts of exotic fishes in the continental United States. In *Distribution, Biology, and Management of Exotic Fishes* (W. R. Courtenay, Jr., and J. R. Stauffer, Jr., Eds), pp. 322–373. Johns Hopkins Univ. Press, Baltimore.

Taylor, W. R. (1964). Fishes of Arnhem Land. *Rec. American–Australian Sci. Expedition Arnhem Land* 4, 45–307.

Taylor, W. R. (1969). A revision of the catfish genus *Noturus* Rafinesque, with an analysis of higher groups in the Ictaluridae. *Bull. U.S. Natl. Museum* 282, 1–315.

Taylor, W. R. (1986). Ariidae. In *Check-list of the Freshwater Fishes of Africa* (J. Daget, J.-P. Gosse, and D. F. E. Thys van den Audenaerde, Eds.), pp. 153–159. CLOFFA II, ORSTOM, Paris.

Tesch, F.-W. (1977). *The Eel: Biology and Management of Anguillid Eels.* Chapman & Hall, London.

Teugels, G. G. (1986). Clariidae. In *Check-list of the Freshwater Fishes of Africa* (J. Daget, J.-P. Gosse, and D. F. E. Thys van den Audenaerde, Eds.), pp. 66–101. CLOFFA II, ORSTOM, Paris.

Teugels, G. G. (1990a). Denticipitidae. In *Faune des Poissons d'eaux Douces et Saumâtres de l'Afrique de l'Ouest* (C. Lévêque, D. Paugy, and G. G. Teugels, Eds.), Vol. 1, pp. 95–97. ORSTOM, Paris.

Teugels, G. G. (1990b). Pantodontidae. In *Faune des Poissons d'eaux Douces et Saumâtres de l'Afrique de l'Ouest* (C. Lévêque, D. Paugy, and G. G. Teugels, Eds.), Vol. 1, pp. 116–118. ORSTOM, Paris.

Teugels, G. G. (1992a). Clariidae. In *Faune des Poissons d'eaux Douces et Saumâtres de l'Afrique de l'Ouest* (C. Lévêque, D. Paugy, and G. G. Teugels, Eds.), Vol. 2, pp. 468–495. ORSTOM, Paris.

Teugels, G. G. (1992b). Malapteruridae. In *Faune des Poissons d'eaux Douces et Saumâtres de l'Afrique de l'Ouest* (C. Lévêque, D. Paugy, and G. G. Teugels, Eds.), Vol. 2, pp. 496–499. ORSTOM, Paris.

Teugels, G. G. (1992c). Nandidae. In *Faune des Poissons d'eaux Douces et Saumâtres de l'Afrique de l'Ouest* (C. Lévêque, D. Paugy, and G. G. Teugels, Eds.), Vol. 2, pp. 710–713. ORSTOM, Paris.

Teugels, G. G. (1992d). Channidae. In *Faune des Poissons d'eaux Douces et Saumâtres de l'Afrique de l'Ouest* (C. Lévêque, D. Paugy, and G. G. Teugels, Eds.), Vol. 2, pp. 655–658. ORSTOM, Paris.

Teugels, G. G. (1996). Taxonomy, phylogeny, and biogeography of catfishes (Ostariophysi, Siluroidei): An overview. *Aquat. Living Resourc.* 9, 9–34.

Teugels, G. G., and Daget, J. (1984). *Parachanna* nom. nov. for the African snake-heads and rehabilitation of *Parachanna insignis* (Sauvage, 1884) (Pisces, Channidae). *Cybium* 8(4), 1–7.

Teugels, G. G., and Thys Van Den Audenaerde, D. F. E. (1992). Cichlidae. In *Faune des Poissons d'eaux Douces et Saumâtres de l'Afrique de l'Ouest* (C. Lévêque, D. Paugy, and G. G. Teugels, Eds.), Vol. 2, pp. 714–779. ORSTOM, Paris.

Teugels, G. G., Breine, J. J., and Thys van den Audenaerde, D. F. E. (1986). Channidae. In *Check-list of the Freshwater Fishes of Africa* (J. Daget, J.-P. Gosse, and D. F. E. Thys van den Audenaerde, Eds.), pp. 288–290. CLOFFA II, ORSTOM, Paris.

Teugels, G. G., Risch, L. M., De Voss, L., and Thys Van Den Audenaerde, D. F. E. (1991). Generic review of the African bagrid genera *Auchenoglanis* and *Parauchenoglanis* with description of a new genus. *J. Nat. History (London)* 25, 499–517.

Thomerson, J. E., and Thorson, T. B. (1977). The bull shark, *Carcharhinus leucas*, from the upper Mississippi River, near Alton, Illinois. *Copeia* 1977(1), 166–168.

Thompson, K. W. (1979). Cytotaxonomy of forty-one species of neotropical Cichlidae. *Copeia* 1979(4), 679–691.

Thomson, D. A., Findley, L., and Kerstitch, A. N. (1979). *Reef Fishes of the Sea of Cortez.* Wiley, New York.

Thomson, J. M. (1964). A bibliography of systematic references of the grey mullets (Mugilidae), Technical Paper No. 16. CSIRO Division of Fisheries and Oceanography, Melbourne.

Thomson, J. M. (1986). Mugilidae. In *Check-list of the Freshwater Fishes of Africa* (J. Daget, J.-P. Gosse, and D. F. E. Thys van den Audenaerde, Eds.), pp. 344–349. CLOFFA II, ORSTOM, Paris.

Thomson, J. M. (1996). Grey mullets. In *Freshwater Fishes of South-Eastern Australia* (R. M. McDowall, Ed.), pp. 191–197. Reed, Sydney.

Thorson, T. B. (Ed.) (1976). *Investigations of the Ichthyofauna of Nicaraguan Lakes.* School of Life Sciences, University of Nebraska, Lincoln.

Thorson, T. B., and Gerst, J. W. (1972). Comparison of some parameters of serum and uterine fluid of pregnant, viviparous sharks *Carcharhinus leucus* and serum of the near-term young. *Comp. Biochem. Physiol. A* 42, 33–40.

Thorson, T. B., Cowan, C. M., and Watson, D. E. (1967). *Potamotrygon* spp. Elasmobranchs with low urea content. *Science* 158, 375–377.

Thorson, T. B., Wotton, R. M., and Georgi, T. D. (1978). Rectal gland of freshwater stingrays, *Potamotrygon* spp. (Chondrichthyes: Potamotrygonidae). *Biol. Bull.* **154**(3), 508–516.

Thys van den Audenaerde, D. F. E. (1961a). Existence de deux races géographiques distinctes chez *Phractolaemus ansorgei* Boulenger, 1901 (Pisces, Clupeiformes). *Bull. Séances l'Académie R. Sci. d'Outremer* 7(2), 222–251.

Thys van den Audenaerde, D. F. E. (1961b). L'anatomie de *Phractolaemus ansorgei* Blgr et la position systématique des Phractolaemidae. *Ann. Musée R. l' Afrique Centrale* 103, 99–167.

Thys van den Audenaerde, D. F. E. (1984). Phractolaemidae. In *Check-list of the Freshwater Fishes of Africa* (J. Daget, J.-P. Gosse, and D. F. E. Thys van den Audenaerde, Eds.), pp. 136–137. CLOFFA I, ORSTOM, Paris.

Thys van den Audenaerde, D. F. E., and Breine, J. (1986). Nandidae. In *Check-list of the Freshwater Fishes of Africa* (J. Daget, J.-P. Gosse, and D. F. E. Thys van den Audenaerde, Eds.), pp. 342–343. CLOFFA II, ORSTOM, Paris.

Tinbergen, N. (1952). The curious behavior of the sticklebacks. *Sci. Am.* **187**, 22–26.

Todd, J. H. (1971). The chemical language of fishes. *Sci. Am.* **224**, 99–108.

Toledo-Piza, M., Menezes, N. A., and Mendes dos Santos, G. (1999). Revision of the neotroical fish genus *Hydrolycus* (Ostariophysi: Cynodontinae) with the description of two new species. *Ichthyol. Exploration Freshwaters* 10(3), 255–280.

Tominaga, Y. (1968). Internal morphology, mutual relationships and systematic position of the fishes belonging to the family Pempheridae. *Jpn. J. Ichthyol.* 15(2), 43–95.

Tortonese, E. (1986a). Cyprinodontidae. In *Fishes of the North-Eastern Atlantic and the Mediterranean* (P. J. P. Whitehead, M.-L. Bauchot, J.-C. Hureau, J. Nielsen, and E. Tortonese, Eds.), Vol. 2, pp. 623–626. UNESCO, Paris.

Tortonese, E. (1986b). Moronidae. In *Fishes of the North-Eastern Atlantic and the Mediterranean* (P. J. P. Whitehead, M.-L. Bauchot, J.-C. Hureau, J. Nielsen, and E. Tortonese, Eds.), Vol. 2, pp. 793–796. UNESCO, Paris.

Tortonese, E. (1986c). Apogonidae. In *Fishes of the North-Eastern Atlantic and the Mediterranean* (P. J. P. Whitehead, M.-L. Bauchot, J.-C. Hureau, J. Nielsen, and E. Tortonese, Eds.), Vol. 2, pp. 803–809. UNESCO, Paris.

Tortonese, E. (1986d). Tetradontidae. In *Fishes of the North-Eastern Atlantic and the Mediterranean* (P. J. P. Whitehead, M.-L. Bauchot, J.-C. Hureau, J. Nielsen, and E. Tortonese, Eds.), Vol. 2, pp. 1341–1345. UNESCO, Paris.

Trautman, M. B. (1981). *The Fishes of Ohio*, rev. ed. Ohio State Univ. Press, Columbus.

Travers, R. A. (1984a). A review of the Mastacembeloidei, a suborder of synbranchiform teleost fishes. Part I: Anatomical descriptions. *Bull. Br. Museum Nat. History Zool.* 46(1), 1–133.

Travers, R. A. (1984b). A review of the Mastacembeloidei, a suborder of synbranchiform teleost fishes. Part II: Phylogenetic analysis. *Bull. Br. Museum Nat. History Zool.* 47(2), 83–150.

Travers, R. A. (1988). Diagnosis of a new African mastacembelid spiny eel genus *Aethiomastacembelus* gen. nov. (Mastacembeloidei: Synbranchiformes). *Cybium* 12(3), 255–257.

Travers, R. A. (1992a). *Caecomastacembelus taiaensis* and *Aethiomastacembelus praensis*, two new species of mastacembelid spiny-eels from West Africa. *Ichthyol. Exploration Freshwaters* 2(4), 331–340.

Travers, R. A. (1992b). Mastacembelidae. In *Faune des Poissons d'eaux Douces et Saumâtres de l'Afrique de l'Ouest* (C. Lévêque, D. Paugy, and G. G. Teugels, Eds.), Vol. 2, pp. 848–857. ORSTOM, Paris.

Travers, R. A., Eynikel, G., Thy van den Audenaerde, D. F. E. (1986). Mastacembelidae. In *Check-list of the Freshwater Fishes of Africa* (J. Daget, J.-P. Gosse, and D. F. E. Thys van den Audenaerde, Eds.), pp. 415–427. CLOFFA II, ORSTOM, Paris.

Trewavas, E. (1942). The cichlid fishes of Syria and Palestine. *Ann. Magazine Nat. History* 9(11th series), 526–536.

Trewavas, E. (1973). On the cichlid fishes of the genus *Pelmatochromis* with proposal of a new genus for *P. congicus*; On the relationship between *Pelmatochromis* and *Tilapia* and the recognition of *Sarotherodon* as a distinct genus. *Bull. Br. Museum Nat. History Zool.* 25(1), 1–26.

Trewavas, E. (1977). The sciaenid fishes (croakers or drums) of the Indo-West Pacific. *Trans. Zool. Soc. London* 33, 253–541.

Trewavas, E. (1981). Nomenclature of tilapias of southern Africa. *J. Limnol. Soc. South Africa* 7(1), 42.

Tringali, M. D., Bert, T. M., and Seyoum, S. (1999). Genetic identification of centropomine fishes. *Trans. Am. Fish. Soc.* **128**(3), 446–458.

Triques, M. L. (1996). *Iracema caiana*, new genus and species of electrogenic Neotropical freshwater fish (Rhamphichthyidae: Gymnotiformes: Ostariophysi: Actinopterygii). *Rev. Francaise d' Aquariol.* 23(3/4), 91–92.

Tsukamoto, K., Nakai, I., and Tesch, W.-V. (1998). Do all freshwater eels migrate? *Nature* **396**, 635.

Tucker, D. W. (1959). A new solution to the Atlantic eel problem. *Nature* 183, 495–501.

Turner, B. J., and Liu, R. K. (1977). Extensive inter-specific genetic compatibility in the New World killifish genus *Cyprinodon*. *Copeia* 1977(2), 259–269.

Turner, B. J., Davis, W. P., and Taylor, D. S. (1992). Abundant males in populations of a selfing hermaphrodite fish, *Rivulus marmoratus*, from some Belize cays. *J. Fish Biol.* 40, 307–310.

Turner, C. L. (1946). A contribution to the taxonomy and zoogeography of the goodeid fishes. *Occasional Papers Museum Zool. Univ. Michigan* 495, 1–13.

Turner, L. M. (1886). Contributions to the natural history of Alaska, Arctic Ser. Publ. Signal Service, U.S. Army, No. 2. Government Printing Office, Washington, DC.

Turner, R. W., Maler, L., and Burrows, M. (Eds.) (1999). Electroception and electrocommunication. *J. Exp. Biol.* 202(10), 1167–1458.

Tyler, J. C. (1964). A diagnosis of the two species of South American puffer fishes (Tetraodontidae, Plectognathi) of the genus *Colomesus*. *Proc. Acad. Nat. Sci. Philadelphia* 116, 119–148.

Tyler, J. C. (1980). Osteology, phylogeny, and high classification of the fishes of the order Plectognathi (Tetraodontiformes), NOAA technical report. *Natl. Mar. Fish. Service Circular* 434, 1–422.

Tyler, J. C., Johnson, G. D., Nakamura, I., and Collette, B. B. (1989). Morphology of *Luvarus imperialis* (Luvaridae), with a phylogenetic analysis of the Acanthuroidei (Pisces). *Smithsonian Contrib. Zool.* 485, 1–78.

Udvardy, M. D. F. (1969). *Dynamic Zoogeography*. Van Nostrand–Reinhold, New York.

Unmack, P., and Brumley, C. (1991). Initial observations on spawning and conservation status of redfinned blue-eye (*Scaturiginichthys vermeilipinnis*). *Fishes Sahul (J. Austr. New Guinea Fishes Assoc.)* 6(4), 282–284.

Uwa, H., and Parenti, L. R. (1988). Morphometric and merisitc variation in ricefishes, genus *Oryzias*: A comparison with cytogenetic data. *Jpn. J. Ichthyol.* 35(2), 159–166.

Uwa, H., Wang, R.-F., and Chen, Y.-R. (1988). Karyotypes and geographical distribution of ricefishes from Yunnan, southwestern China. *Jpn. J. Ichthyol.* 35(3), 332–340.

Uyeno, T., and Miller, R. R. (1962). Relationships of *Empetrichthys erdisi*, a Pliocene cyprinodontid fish from California, with remarks on the Fundulinae and Cyprinodontinae. *Copeia* 1962(3), 520–532.

Uyeno, T., and Smith, G. R. (1972). Tetraploid origin of the karyotype of catostomid fishes. *Science* 175, 644–646.

Uyeno, T., Miller, R. R., and Fitzsimons, J. M. (1983). Karyology of the cyprinodontoid fishes of the Mexican family Goodeidae. *Copeia* 1983(2), 497–510.

Van Oijen, M. J. P., Snoeks, J., Skelton, P. H., Maréchal, C., and Teugels, G. G. (1991). *Haplochromis*. In *Check-list of the Freshwater Fishes of Africa* (J. Daget, J.-P. Gosse, G. G. Teugels, and D. F. E. Thys van den Audenaerde, Eds.), pp. 100–184. CLOFFA IV, ORSTOM, Paris.

Vari, R. P. (1977). Notes on the characoid subfamily Iguanodectinae, with a description of a new species. *Am. Museum Novitates* 2612, 1–6.

Vari, R. P. (1978). The therapon perches (Percoidei, Teraponidae). A cladistic analysis and taxonomic revision. *Bull. Am. Museum Nat. History* 159(5), 175–340.

Vari, R. P. (1979). Anatomy, relationships and classification of the families Citharinidae and Distichodontidae (Pisces, Characoidea). *Bull. Br. Museum Nat. History Zool.* 36(2), 261–344.

Vari, R. P. (1982a). Systematics of the neotropical characoid genus *Curimatopsis* (Pisces: Characoidei). *Smithsonian Contrib. Zool.* 373, 1–28.

Vari, R. P. (1982b). Fishes of the Western North Atlantic. The seahorses. Subfamily Hippocampinae. *Sears Foundation Mar. Res. Mem.* 1(8), 173–189.

Vari, R. P. (1983). Phylogenetic relationships of the families Curimatidae, Prochilodontidae, Anostomidae, and Chilodontidae (Pisces: Characiformes). *Smithsonian Contrib. Zool.* 378, 1–60.

Vari, R. P. (1984). Systematics of the neotropical characiform genus *Potamorhina* (Pisces: Characiformes). *Smithsonian Contrib. Zool.* 400, 1–36.

Vari, R. P. (1988). The Curimatidae, a lowland neotropical fish family (Pisces: Characiformes): Distribution, endemism, and phylogenetic biogeography. In *Proceedings of a Workshop on Neotropical Distribution Patterns* (W. R. Heyer and P. E. Vanzolini, Eds.), pp. 343–377. Academia Brasileira de Ciências, Rio de Janeiro.

Vari, R. P. (1989a). A phylogenetic study of the neotropical characiform family Curimatidae (Pisces: Ostariophysi). *Smithsonian Contrib. Zool.* 471, 1–71.

Vari. R. P. (1989b). Systematics of the neotropical characiform genus *Curimata* Bosc (Pisces: Characiformes). *Smithsonian Contrib. Zool.* 474, 1–63.

Vari, R. P. (1989c). Systematics of the neotropical characiform genus *Psectrogaster* Eigenmann and Eigenmann (Pisces: Characiformes). *Smithsonian Contrib. Zool.* 481, 1–43.

Vari, R. P. (1989d). Systematics of the neotropical characiform genus *Pseudocurimata* Fernández-Yépez (Pisces: Ostariophysi). *Smithsonian Contrib. Zool.* **490**, 1–28.

Vari, R. P. (1991). Systematics of the neotropical characiform genus *Steindachnerina* Flower (Pisces: Ostariophysi). *Smithsonian Contrib. Zool.* **507**, 1–118.

Vari, R. P. (1992a). Systematics of the neotropical characiform genus *Cyphocharax* Fowler (Pisces: Ostariophysi). *Smithsonian Contrib. Zool.* **529**, 1–137.

Vari, R. P. (1992b). Systematics of the neotropical characiform genus *Curimatella* Eigenmann and Eigenmann (Pisces: Ostariophysi), with summary comments on the Curimatidae. *Smithsonian Contrib. Zool.* **533**, 1–48.

Vari, R. P. (1992c). Redescription of *Mesopristes elongatus* (Guichenot, 1866), an endemic Malagasy fish species (Pisces, Terapontidae). *Am. Museum Novitates* **3039**, 1–7.

Vari, R. P. (1995). The neotropical fish family Ctenoluciidae (Teleostei: Ostariophysi: Characiformes): Supra- and intrafamilial phylogenetic relationships, with a revisionary study. *Smithsonian Contrib. Zool.* **564**, 1–97.

Vari, R. P. (1998). Higher level phylogenetic concepts within Characiforms (Ostariophysi): A historical review. In *Phylogeny and Classification of Neotropical Fishes* (L. R. Malabarba, R. E. Reis, R. P. Vari, Z. M. S. Lucena, and C. A. S. Lucena, Eds.), pp. 111–121. EDIPUCRS, Porto Alegre, Brazil.

Vari, R. P., and Barriga S. R. (1990). *Cyphocharax pantostictos*, a new species (Pisces: Ostariophysi: Characiformes: Curimatidae) from the western portions of the Amazon basin. *Proc. Biol. Soc. Washington* **103**(3), 550–557.

Vari, R. P., and Blackledge, T. A. (1996). New curimatid, *Cyphocharax laticlavius* (Ostariophysi, Characiformes), from Amazonian Ecuador, with a major range extension for *C. gouldingi*. *Copeia* **1996**(1), 109–113.

Vari, R. P., and Ferraris, C. J., Jr. (1988). The neotropical catfish genus *Epapterus* Cope (Siluriformes: Auchenipteridae): A reappraisal. *Proc. Biol. Soc. Washington* **111**(4), 992–1007.

Vari, R. P., and Goulding, M. (1985). A new species of *Bivibranchia* (Pisces: Characiformes) from the Amazon River basin. *Proc. Biol. Soc. Washington* **98**(4), 1054–1061.

Vari, R. P., and Harold, A. S. (1998). The genus *Creagrutus* (Teleostei: Characiformes: Characidae): Monophyly, relationships, and undetected diversity. In *Phylogeny and Classification of Neotropical Fishes* (L. R. Malabarba, R. E. Reis, R. P. Vari, Z. M. S.

Lucena, and C. A. S. Lucena, Eds.), pp. 245–260. EDIPUCRS, Porto Alegre, Brazil.

Vari, R. P., and Ortega, H. (1986). The catfishes of the neotropical family Helogenidae (Ostariophysi: Siluroidei). *Smithsonian Contrib. Zool.* **442**, 1–20.

Vari, R. P., and Ortega, H. (1997). A new Chilodus species from southeastern Peru (Ostariophysi: Characiformes: Chilodontidae): Description, phylogenetic discusion, and comments on the distribution of other chilodontids. *Ichthyol. Exploration Freshwaters* **8**(1), 71–80.

Vari, R. P., and Raredon, S. J. (1991). The genus *Schizodon* (Teleostei: Ostariophysi: Anostomidae) in Venezuela, a reappraisal. *Proc. Biol. Soc. Washington* **104**(1), 12–22.

Vari, R. P., and Reis, R. (1995). *Curimata acutirostris*, a new fish (Teleostei: Characiformes: Curimatidae) from the rio Araguaia, Brazil: Description and phylogenetic relationships. *Ichthyol. Exploration Freshwaters* **6**(4), 297–304.

Vari, R. P., and Weitzman, S. H. (1990). A review of the phylogenetic biogeography of the freshwater fishes of South America. In *Vertebrates in the Tropics* (G. Peters and R. Hutterer, Eds.), pp. 381–393. Museum Alexander Koening, Bonn.

Vari, R. P., and Williams, A. M. (1987). Headstanders of the neotropical anostomid genus *Abramites* (Pisces: Characiformes: Anostomidae). *Proc. Biol. Soc. Washington* **100**(1), 89–103.

Vari, R. P., Castro, R. M. C., and Raredon, S. J. (1995). The neotropical fish family Chilodontidae (Teleostei: Characiformes): A phylogenetic study and revision of *Caenotropus* Günther. *Smithsonian Contrib. Zool.* **577**, 1–32.

Vidthayanon, C. (1995). *Odontobutis aurarmus*, a new species of odontobutid goby from wetlands of the Mekong basin, Thailand. *Ichthyol. Exploration Freshwater* **6**(3), 235–242.

Vinton, K. W., and Stickler, W. H. (1941). The carnero: A fish parasite of man and possibly of other mammals. *Am. J. Surg.* **54**, 511–519.

Vladykov, V. D. (1984). Petromyzonidae. In *Fishes of the North-Eastern Atlantic and the Mediterranean* (P. J. P. Whitehead, M.-L. Bauchot, J.-C. Hureau, J. Nielsen, and E. Tortonese, Eds.), Vol. 1, pp. 64–67. UNESCO, Paris.

Vladykov, V. D., and Kott, E. (1982). Comments on Reeve M. Bailey's view of lamprey systematics. *Can. J. Fish. Aquat. Sci.* **39**, 1215–1217.

von der Emde, G., Schwarz, S., Gomez, L., Budelli, R., and Grant, K. (1998). Electric fish measure distance in the dark. *Nature* **395**, 890–894.

Vreven, E. J., and Teugels, G. G. (1996). Description of a new mastacembelid species

(Synbranchiformes; Mastacembelidae) from the Zaïre River basin in Africa. *Copeia* **1996**(1), 130–139.

Vreven, E. J., and Teugels, G. G. (1997). *Aethiomastacembelus traversi*, a new spiny-eel from the Zaïre River basin, Africa (Synbranchiformes: Mastacembelidae). *Ichthyol. Exploration Freshwaters* 8(1), 81–87.

Wainwright, P. C., and Lauder, G. V. (1992). The evolution of feeding biology in sunfishes (Centrarchidae). In *Systematics, Historical Ecology, and North American Freshwater Fishes* (R. L. Mayden, Ed.), pp. 472–491. Stanford Univ. Press, Stanford, CA.

Waite, E. R. (1902). New records or recurrences of rare fishes from eastern Australia. *Rec. Austr. Museum* 4(7), 263–273, Plates xli–xliii.

Waite, E. R. (1904a). A review of the eleotrids of New South Wales. *Rec. Austr. Museum* 5(5), 277–286, Plates xxxiv–xxxvi.

Waite, E. R. (1904b). The breeding habits of the fighting fish (*Betta pugnax*, Cantor). *Rec. Austr. Museum* 5(5), 293–295, Plate xxxviii.

Waite, E. R. (1923). *The Fishes of South Australia*. Government Printer, Adelaide, South Australia.

Waldman, J. R. (1986). Systematics of *Morone* (Pisces: Moronidae), with notes on the lower percoids. Ph.D. thesis, The City University of New York, New York.

Walker, B. W. (1959). The timely grunion. *Nat. History* 68(6), 302–434.

Wallace, A. R. (1860). On the zoological geography of the Malay Archipelago. *J. Proc. Linnean Soc. London* 4, 172–184.

Wallace, A. R. (1876). *The Geographical Distribution of Animals*, 2 vols. Macmillian, London. (Reprint 1962, Hafner, New York).

Wallis, G. P., Judge, K. F., Bland, J., Waters, J., and Berra, T. M. (2001). Genetic diversity in New Zealand galaxiid fish: A test of a biogeographic hypothesis. *J. Biogeog.* (In press.)

Walsh, S. J. (1990). A systematic revision of the neotropical catfish family Ageneiosodae (Teleostei: Ostariophysi: Siluriformes). Ph.D. dissertation, University of Florida, Gainesville.

Walsh, S. J., and Burr, B. M. (1984). Life history of the banded pygmy sunfish, *Elassoma zonatum* Jordan (Pisces, Centrarchidae) in western Kentucky. *Bull. Alabama Museum Nat. History* 8, 31–52.

Walsh, S. J., and Gilbert, C. R. (1995). New species of troglobitic catfish of the genus, *Prietella* (Siluriformes: Ictaluridae) from Northeastern México. *Copeia* **1995**(4), 850–861.

Wantanabe, M. (1972). First record of the gobioid fish, *Rhyacichthys aspro*, from Formosa. *Jpn. J. Ichthyol.* 19, 120–124.

Warren, M. L., Jr. (1992). Variation of the spotted sunfish, *Lepomis punctatus* complex (Centrarchidae): Meristics, morphometrics, pigmentation and species limits. *Bull. Alabama Museum Nat. History* 12, 1–56.

Washington, B. B., Eschmeyer, W. N., and Howe, K. M. (1984). Scorpaeniformes: Relationships. In *Ontogeny and Systematica of Fishes* (H. G. Moser *et al.*, Eds.), Special Publication No. 1, pp. 438–447. American Society of Ichthyologists and Herpetologists, Lawrence, KS.

Waters, J. M. (1966). Aspects of the phylogeny, biogeography and taxonomy of galaxioid fishes. Ph.D. thesis, University of Tasmania, Hobart.

Waters, J. M., and Burridge, C. P. (1999). Extreme intraspecific mitochondrial DNA sequence divergence in *Galaxias maculatus* (Osteichthys: Galaxiidae), one of the world's most widespread freshwater fish. *Mol. Phylogenet. Evol.* 11(1), 1–12.

Waters, J. M., and Cambray, J. A. (1997). Intraspecific phylogeography of the Cape galaxias from South Africa: Evidence from mitochondrial DNA sequences. *J. Fish Biol.* 50, 1329–1338.

Waters, J. M., and Wallis, G. P. (2001). Mitochondrial DNA phylogenetics of the *Galaxias vulgaris* complex from South Island, New Zealand: Rapid radiation of a species flock. *J. Fish Biol.* (In press.)

Waters, J. M., and White, R. W. G. (1997). Molecular phylogeny and biogeography of the Tasmanian and New Zealand mudfishes (Salmoniformes: Galaxiidae). *Austr. J. Zool.* 45, 39–48.

Waters, J., López, A., and Wallis, G. (2000). Molecular phylogenetics and biogeography of galaxiid fishes (Osteichthys: Galaxiidae): Dispersal, vicariance, and the position of *Lepidogalaxias salamandroides*. *System. Biol.* 49(4), 777–795.

Watson, R. E. (1995). Review of the freshwater goby genus *Cotylopus* (Teleostei: Gobiidae: Sicydiinae). *Ichthyol. Exploration Freshwaters* 6(1), 61–70.

Watson, R. E. (1996a). A review of *Stiphodon* from New Guinea and adjacent regions, with descriptions of five new species (Teleostei: Gobiidae: Sicydinnar). *Rev. Française d'Aquariol.* 23(3/4), 113–132.

Watson, R. E. (1996b). Revision of the subgenus *Awaous (Chonophorus)* (Teleostei: Gobiidae). *Ichthyol. Exploration Freshwaters* 7(1), 1–18.

Watson, R. E. (1998). *Stiphodon martenstyni*, a new species of freshwater goby from Sri Lanka (Teleostei: Gobiidae: Sicydiini). *J. South Asian Nat. History* 3(1), 69–78.

Watson, R. E. (1999a). Two new subgenera of *Sicyopus*, with a redescription of *Sicyopus zosterophorum* (Teleostei: Gobioidei: Sicydiinae). *Aqua J. Ichthyol. Aquat. Biol.* 3(3), 93–104.

Watson, R. E. (1999b). *Stiphodon hydoreibatus*, a new species of freshwater goby from Samoa (Teleostei: Gobiidae). *Ichthyol. Exploration Freshwaters* 10(1), 89–95.

Watson, R. E., and Horsthemke, H. (1995). Revision of *Euctenogobius*, a monotypic subgenus of *Awaous*, with discussion of its natural history (Teleostei: Gobiidae). *Rev. Française d'Aquariol.* 22(3/4), 83–92.

Watson, R. E., and Pöllabauer, C. (1998). A new genus and species of freshwater goby from New Caledonia with a complete lateral line (Pisces: Teleostei: Gobioidei). *Senckenbergiana Biol.* 77(2), 147–153.

Watson, R. E., Allen, G. R., and Kottelat, M. (1998). A review of *Stiphodon* from Halmahera and Irian Jaya, Indonesia, with descriptions of two new species (Teleostei: Gobiidae). *Ichthyol. Exploration Freshwaters* 9(3), 293–304.

Weber, M. (1910). A new case of parental care among fishes. *Proc. Section Sci. Koninklijke Adademie Wetenschappen Amsterdam* 13, 583–587.

Weber, M. (1913). Süsswasserfische aus Niederländisch süd-und nord-Neu-Guinea. In *Nova Guinea. Résultats de L'Expédition Scientifique Néerlandaise à la Nouvelle-Guinée en 1907 et 1909. Zoologie,* Vol. 9, pp. 513–613, Plates 12–14. Brill, Leiden.

Weber, M., and DeBeaufort, L. F. (1916). *The Fishes of the Indo-Australian Archipelago. III. Ostariophysi: II Cyprinoidea, Apodes, Synbranchi.* Brill, Leiden.

Weber, M., and DeBeaufort, L. F. (1922). *Fishes of the Indo-Australian Archipelago, IV: Heteromi, Solenichthyes, Synentognathi, Percesoces, Labyrinthici, Microcyprini.* Brill, Leiden.

Weber, M., and DeBeaufort, L. F. (1929). *Fishes of the Indo-Australian Archipelago, V: Anacanthini, Allotriognathi, Heterosomata, Berycomorphi, Percomorphi.* Brill, Leiden.

Weber, M., and DeBeaufort, L. F. (1931). *Fishes of the Indo-Australian Archipelago. VI. Perciformes (continued).* Brill, Leiden.

Weber, M., and DeBeaufort, L. F. (1936). *Fishes of the Indo-Australian Archipelago, VII: Perciformes (Continued).* Brill, Leiden.

Weber, M., and DeBeaufort, L. F. (1953). *The Fishes of the Indo-Australian Archipelago. X. Gobioidea.* Brill, Leiden. (This volume authored by F. P. Koumans)

Webster, D. (1998). The Orinoco. *Natl. Geographic* 193(4), 2–31.

Wegener, A. (1966). *The Origin of Continents and Oceans.* Dover, New York. (Reprint of 4th ed., 1929)

Wei, Q., Ke, F., Zhang, J., Zhuang, P., Luo, J., Zhou, R., and Yang, W. (1997). Biology, fisheries, and conservation of sturgeons and paddlefish in China. *Environ. Biol. Fishes* 48, 241–255.

Weitzman, S. (1954). The osteology and relationships of the South American characid fishes of the subfamily, Gasteropelecinae. *Stanford Ichthyol. Bull.* 4(4), 212–263.

Weitzman, S. (1960). Further notes on the relationships and classification of the South American fishes of the subfamily Gasteropelecinae. *Stanford Ichthyol. Bull.* 7(4), 217–239.

Weitzman, S. H. (1964). Osteology and relationships of South American characid fishes of subfamily Lebiasininae and Erythrininae with special reference to subtribe Nannostomia. *Proc. U.S. Natl. Museum* 116(3499), 127–170.

Weitzman, S. H. (1966). Review of South American characid fishes of subtribe Nannostomia. *Proc. U.S. Natl. Museum* 119(3538), 1–56.

Weitzman, S. H. (1978). Three new species of fishes of the genus *Nannostomus* from the Brazilian states of Para and Amazonas (Teleostei: Lebiasinidae). *Smithsonian Contrib. Zool.* 263, 1–14.

Weitzman, S. H., and Cobb, J. S. (1975). A revision of the South American fishes of the genus *Nannostomus* Gunther (family Lebiasinidae). *Smithsonian Contrib. Zool.* 186, 1–36.

Weitzman, S. H., and Fink, S. V. (1985). Xenurobryconin phylogeny and putative pheromone pumps in glandulocaudine fishes (Teleostei: Characidae). *Smithsonian Contrib. Zool.* 421, 1–121.

Weitzman, S. H., and Fink, W. L. (1983). Relationships of the neon tetras, a group of South American freshwater fishes (Teleostei, Characidae), with comments on the phylogeny of new world characiformes. *Bull. Museum Comp. Zool.* 150(6), 339–395.

Weitzman, S. H., and Géry, J. (1981). The relationships of the South American pygmy characoid fishes of the genus *Elachocharax*, with a description of *Elachocharax junki* (Teleostei: Characidae). *Proc. Biol. Soc. Washington* 93(4), 887–913.

Weitzman, S. H., and Kanazawa, R. H. (1976). *Ammocryptocharax elegans*, a new genus and species of riffle-inhabiting characoid fish (Teleostei: Characidae) from South America. *Proc. Biol. Soc. Washington* 89, 325–346.

Weitzman, S. H., and Malabarba, L. R. (1998). Perspectives about the phylogeny and classification of the Characidae (Teleostei: Characiformes). In *Phylogeny and Classification of Neotropical Fishes* (L. R. Malabarba, R. E. Reis, R. P. Vari, Z. M. S. Lucena, and C. A. S. Lucena, Eds.), pp. 161–233. EDIPUCRS, Porto Alegre, Brazil.

Weitzman, S. H., and Malabarba, L. R. (1999). Systematics of *Spintherobolus* (Teleostei: Characidae: Cheirodontinae) from eastern Brazil. *Ichthyol. Exploration Freshwaters* 10(1), 1–43.

Weitzman, S. H., and Menezes, N. A. (1998). Relationships of the tribes and genera of Glandulocaudinae (Ostariophysi: Characiformes: Characidae) with a description of a new genus, *Chrysobrycon*. In *Phylogeny and Classification of Neotropical Fishes* (L. R. Malabarba, R. E. Reis, R. P. Vari, Z. M. S. Lucena, and C. A. S. Lucena, Eds.), pp. 171–192. EDIPUCRS, Porto Alegre, Brazil.

Weitzman, S. H., and Palmer, L. (1996). Do freshwater hatchetfishes really fly? *Tropical Fish Hobbyist* 45(1), 195–206.

Weitzman, S. H., and Vari, R. P. (1987). Two new species and a new genus of miniature characid fishes (Teleostei: Characiformes) from northern South America. *Proc. Biol. Soc. Washington* 100(3), 640–652.

Weitzman, S. H., and Vari, R. P. (1988). Miniaturization in South American freshwater fishes; An overview and discussion. *Proc. Biol. Soc. Washington* 101(2), 444–465.

Weitzman, S. H., and Vari, R. P. (1994). Characins and their allies. In *Encyclopedia of Fishes* (J. P. Paxton and W. N. Eschmeyer, Eds.), pp. 101–105. Academic Press, San Diego.

Weitzman, S. H., and Weitzman, M. (1982). Biogeography and evolutionary diversification in neotropical freshwater fishes with comments on the refuge theory. In *Biological Diversification in the Tropics* (G. T. Prance, Ed.), pp. 403–422. Columbia Univ. Press, New York.

Weitzman, S. H., Menezes, N. A., and Britski, H. A. (1986). *Nematocharax venustus*, a new genus and species of fish from the Rio Jequitinhonha, Minas Gerais, Brazil (Teleostei: Characidae). *Proc. Biol. Soc. Washington* 99(2), 335–346.

Weitzman, S. H., Menezes, N. A., and Weitzman, M. J. (1988). Phylogenetic biogeography of the Glandulocaudini (Teleostei: Characiformes, Characidae) with comments on the distributions of other freshwater fishes in eastern and southeastern Brazil. In *Proceedings of a Workshop on Neotropical Distribution Patterns* (W. R. Heyer and P. E. Vanzolini, Eds.), pp. 379–427. Academia Brasileira de Ciências, Rio de Janeiro.

Werns, S., and Howland, H. C. (1976). Size and allometry of the saccular air bladder of *Gnathonemus petersi* (Pisces: Mormyridae): Implications for hearing. *Copeia* 1976, 200–202.

Wheeler, A. (1975). *Fishes of the World*. Macmillan, New York.

Wheeler, A., and Baddokway, A. (1981). The generic nomenclature of the marine catfishes usually referred to the genus *Arius* (Osteichthyes-Siluriformes). *J. Nat. History* 15, 769–773.

White, B. N. (1985). Evolutionary relationships of the Atherinopsinae (Pisces: Atherinidae). *Los Angeles County Museum Nat. History Contrib. Sci.* 368, 1–20.

White, B. N. (1986). The isthmian link, antitropicality and American biogeography: distributional history of the Atherinopsinae (Pisces: Atherinidae). *Syst. Zool.* 35(2), 176–194.

Whitehead, P. J. P. (1984a). Engraulidae. In *Fishes of the North-Eastern Atlantic and the Mediterranean* (P. J. P. Whitehead, M.-L. Bauchot, J.-C. Hureau, J. Nielsen, and E. Tortonese, Eds.), Vol. 1, pp. 282–283. UNESCO, Paris.

Whitehead, P. J. P. (1984b). Clupeidae. In *Fishes of the North-Eastern Atlantic and the Mediterranean* (P. J. P. Whitehead, M.-L. Bauchot, J.-C. Hureau, J. Nielsen, and E. Tortonese, Eds.), Vol. 1, pp. 268–281. UNESCO, Paris.

Whitehead, P. J. P. (1985a). King herring: His place among the clupeoids. *Can. J. Fish. Aquat. Sci.* 42, 3–20.

Whitehead, P. J. P. (1985b). FAO species catalogue. Vol. 7: Clupeoid fishes of the world (suborder Clupeoidei). Part 1. Chirocentridae, Clupeidae and Pristigasteridae. *FAO Fish. Synopsis (125)* 7(Pt. 1), 1–303.

Whitehead, P. J. P., Nelson, G. J., and Wongratana, T. (1988). FAO species catalogue. Vol. 7: Clupeoid fishes of the world (suborder Clupeoidei). Part 2. Engraulidae. *FAO Fish. Synopsis (125)* 7(Pt. 2), 305–579.

Whitmore, T. C. (Ed.) (1981). *Wallace's Line and Plate Tectonics*. Clarendon, Oxford.

Wiest, F. C. (1995). The specialized locomotory apparatus of the freshwater hatchetfish family Gasteropelecidae. *J. Zool. (London)* 1995(236), 571–592.

Wildekamp, R. H., Romand, R., and Scheel, J. J. (1986). Cyprinodontidae. In *Check-list of the Freshwater Fishes of Africa* (J. Daget, J.-P. Gosse, and D. F. E. Thys van den Audenaerde, Eds.), pp. 165–276. CLOFFA II, ORSTOM, Paris.

Wiley, E. O. (1976). The phylogeny and biogeography of fossil and recent gars (Actinopterygii: Lepisosteidae), Miscellaneous Publication No. 64, pp. 1–111. University of Kansas Museum of Natural History, Lawrence.

Wiley, E. O. (1980). *Lepisosteus osseus* Linneaus longnose gar. In *Atlas of North American Freshwater Fishes* (D. S. Lee et al., Eds.), p. 49. North Carolina State Museum of Natural History, Raleigh.

Wiley, E. O. (1986). A study of the evolutionary relationships of *Fundulus* topminnows (Teleostei: Fundulidae). *Am. Zool.* 26, 121–130.

Wiley, E. O. (1988a). Vicariance biogeography. *Annu. Rev. Ecol. Syst.* **19**, 513–542.

Wiley, E. O. (1988b). Parsimony analysis and vicariance biogeography. *Syst. Zool.* **37**, 271–290.

Wiley, E. O. (1992). Phylogenetic relationships of the Percidae (Teleostei: Perciformes): A preliminary hypothesis. In *Systematics, Historical Ecology, and North American Freshwater Fishes* (R. L. Mayden, Ed.), pp. 247–267. Stanford Univ. Press, Stanford, CA.

Wiley, E. O., and Collette, B. B. (1970). Breeding tubercles and contact organs in fishes: Their occurrence, structure, and significance. *Bull. Am. Museum Nat. History* **143**(3), 143–216.

Wiley, E. O., and Hagen, R. H. (1997). Mitochondrial DNA sequence variation among the sand darters (Percidae: Teleostei). In *Molecular Systematics of Fishes* (T. D. Kocher and C. A. Stepien, Eds.), pp. 75–96. Academic Press, San Diego.

Wiley, E. O., and Schultze, H.-P. (1984). Family Lepisosteidae (gars) as living fossils. In *Living Fossils* (N. Eldredge and S. M. Stanley, Eds.), pp. 160–165. Springer-Verlag, New York.

Wilimovsky, N. J. (1956). *Protoscaphirhynchus squamosus*, a new sturgeon from the Upper Cretaceous of Montana. *J. Paleontol.* **30**(5), 1205–1208.

Williams, G. C., and Koehn, R. K. (1984). Population genetics of North Atlantic catadromous eels (*Anguilla*). In *Evolutionary Genetics of Fishes* (B. J. Turner, Ed.), pp. 530–560. Plenum, New York.

Williams, J. D., and Burgess, G. H. (1999). A new species of bass, *Micropterus cataractae* (Teleostei: Centrarchidae), from the Apalachicola River basis in Alabama, Florida, and Georgia. *Bull. Florida Museum Nat. History* **42**(2), 80–114.

Williams, J. D., and Clemmer, G. H. (1991). *Scaphirhynchus suttkusi*, a new sturgeon (Pisces: Acipenseridae) from the Mobile Basin of Alabama and Mississippi. *Bull. Alabama Museum Nat. History* **10**, 17–31.

Williams, R. R. G. (1987). The phylogenetic relationships of the salmoniform fishes based on the suspensorium and its muscles. Ph.D. thesis, Department of Zoology, University of Alberta, Edmonton, Canada.

Williams, R. R. G. (1996). Jaw muscles and suspensoria in the Aplochitonidae (Teleostei: Salmoniformes) and their possible phylogenetic significance. *Mar. Freshwater Res.* **47**, 913–917.

Williams, R. R. G. (1997). Bones and muscles of the suspensorium in the Galaxioids and *Lepidogalaxias salamandroides* (Teleostei: Osmeriformes) and their phylogenetic significance. *Rec. Austr. Museum* **49**, 139–166.

Williamson, D. F., Benz, G. W., and Hoover, C. M. (Eds.) (1999). *Proceedings of the Symposium on the Harvest, Trade and Conservation of North American Paddlefish and Sturgeon.* TRAFFIC North America and World Wildlife Fund, Washington, DC.

Williamson, J. H., Carmichael, G. J., Graves, K. G., Simco, B. A., and Tomasso, J. R. (1993). Centrarchids. In *Culture of Nonsalmonid Freshwater Fishes* (R. R. Stickney, Ed.), 2nd ed., pp. 145–197. CRC Press, Boca Raton, FL.

Willson, M. F., Gende, S. M., and Marston, B. H. (1998). Fishes and the forest. *Bioscience* **48**(6), 455–462.

Wilson, M. V. H. (1984). Osteology of the Palaeocene teleost *Esox tiemani*. *Palaeontology (London)* **27**(3), 597–608.

Wilson, M. V. H., and Li, G.-Q. (1999). Osteology and systematic position of the Eocene salmonid †*Eosalmo driftwoodensis* Wilson from western North America. *Zool. J. Linnean Soc.* **124**, 279–311.

Wilson, M. V. H., and Veilleux, P. (1982). Comparative osteology and relationships of the Umbridae (Pisces: Salmoniformes). *Zool. J. Linnean Soc.* **76**, 321–352.

Wilson, M. V. H., and Williams, R. R. G. (1991). New Paleocene genus and species of smelt (Teleostei: Osmeridae) from freshwater deposits of the Paskapoo formation of Alberta, Canada, and comments on osmerid phylogeny. *J. Vertebr. Paleontol.* **11**(4), 434–451.

Wilson, M. V. H., and Williams, R. R. G. (1992). Phylogenetic, biogeographic, and ecological significance of early fossil records of North American freshwater teleostean fishes. In *Systematics, Historical Ecology, and North American Freshwater Fishes* (R. L. Mayden, Ed.), pp. 224–244. Stanford Univ. Press, Stanford, CA.

Wilson, M. V. H., Brinkman, D. B., and Neuman, A. G. (1992). Cretaceous Esocoidei (Teleostei): Early radiation of the pikes in North American fresh waters. *J. Paleontol.* **66**(5), 839–846.

Wimberger, Reis, R. E., and Thornton, K. R. (1998). Mitochondrial phylogenetics, biogeography, and evolution of parental care and mating systems in *Gymnogeophagus* (Perciformes: Cichlidae). In *Phylogeny and Classification of Neotropical Fishes* (L. R. Malabarba, R. E. Reis, R. P. Vari, Z. M. S. Lucena, and C. A. S. Lucena, Eds.), pp. 509–518. EDIPUCRS, Porto Alegre, Brazil.

Winterbottom, R. (1974). The familial phylogeny of the Tetraodontiformes (Acanthopterygii: Pisces) as evidenced by their comparative myology. *Smithsonial Contrib. Zool.* **155**, 1–201.

Winterbottom, R. (1980). Systematics, osteology, and phylogenetic relationships of fishes of the ostariophysan subfamily Anostominae (Characoidei, Anostomidae). *Life Sci. Contrib. R. Ontario Museum* 123, 1–112.

Winterbottom, R. (1993a). Search for the gobioid sister group (Actinopterygii: Percomorpha). *Bull. Mar. Sci.* 52(1), 395–414.

Winterbottom, R. (1993b). Myological evidence for the phylogeny of recent genera of surgeonfishes (Percomorpha, Acanthuridae), with comments on the Acanthuroidei. *Copeia* 1993(1), 21–39.

Winterbottom, R., and Emery, A. R. (1981). A new genus and two new species of gobiid fishes (Perciformes) from the Chagos Archipelago, central Indian Ocean. *Environ. Biol. Fishes* 6(2), 139–149.

Winterbottom, R., and Emery, A. R. (1986). Review of the gobioid fishes of the Chagos Archipelago, central Indian Ocean. *Life Sci. Contrib. R. Ontario Museum* 142, 1–82.

Winterbottom, R., and McLennan, D. A. (1993). Cladogram versatility: Evolution and biogeography of acanthuroid fishes. *Evolution* 47(5), 1557–1571.

Wood, G. L. (1972). *The Guinness Book of Animal Facts and Feats*. Guinness, Middlesex, UK.

Wood, R. M. (1996). Phylogenetic systematics of the darter subgenus *Nothonotus* (Teleostei: Percidae). *Copeia* 1996(2), 300–318.

Wood, R. M., and Mayden, R. L. (1997). Phylogenetic relationships among selected darter subgenera (Teleostei: Percidae) as inferred from analysis of allozymes. *Copeia* 1997(2), 265–274.

Wood, R. M., and Raley, M. E. (2000). Cytochrome b sequence variation in the crystal darter *Crystallaria asprella* (Actinopterygii: Percidae). *Copeia* 2000(1), 20–26.

Woods, L. P., and Inger, R. F. (1957). The cave, spring, and swamp fishes of the family Amblyopsidae of central and eastern United States. *Am. Midland Nat.* 58(1), 232–256.

Wootton, R. J. (1976). *The Biology of the Sticklebacks*. Academic Press, London.

Wootton, R. J. (1984). *A Functional Biology of the Sticklebacks*. Croom Helm, London.

Wright, K. (1999). Pupfish in peril. *Discover* 20(7), 42–44.

Wu, C. H. (1984). Electric fish and the discovery of animal electricity. *Am. Sci.* 72, 598–607.

Wydoski, R. S., and Whitney, R. R. (1979). *Inland Fishes of Washington*. Univ. of Washington Press, Seattle.

Xu-Friedman, M. A., and Hopkins, C. D. (1999). Central mechanisms of temporal analysis in the knollenorgan pathway of mormyrid electric fish. *J. Exp. Biol.* 202(10), 1311–1318.

Yabe, M. (1985). Comparative osteology and myology of the superfamily Cottoidea (Pisces: Scorpaeniformes), and its phylogenetic classification. *Mem. Faculty Fish. Hokkaido Univ.* 32(1), 1–130.

Yamamoto, Y., and Jeffery, W. R. (2000). Central role for the lens in cave fish eye degeneration. *Science* 289, 631–633.

Yazdani, G. M. (1972). A new genus and species of fish from India. *J. Bombay Nat. History Soc.* 69(1), 134–135.

Yazdani, G. M. (1976). A new family of mastacembeloid fish from India. *J. Bombay Nat. History Soc.* 73, 166–170.

Yazdani, G. M. (1978). Adaptive radiation in mastacembeloid fishes. *Bull. Zool. Surv. India* 1(3), 279–290.

Zahl, P. A. (1978). The four-eyed fish sees all. *Natl. Geographic* 153, 390–394.

Zahl, P. A., McLaughlin, J. A., and Gomprecht, R. J. (1977). Visual versatility and feeding of the four-eyed fishes, *Anableps*. *Copeia* 1977(4), 791–793.

Zakaria-Ismail, M. (1994). Zoogeography and biodiversity of the freshwater fishes of Southeast Asia. *Hydrobiologia* 285, 41–48.

Zardoya, R., and Meyer, A. (1997). The complete DNA sequence of the mitochondrial genome of a "living fossil" (*Latimera chalumnae*). *Genetics* 146, 995–1010.

Zimmer, C. (2000). In search of vertebrate origins: Beyond brain and bone. *Science* 287, 1576–1579.

Ziuganov, V. V. (1991). The family Gasterosteidae of world fish fauna. In *Fauna of USSR, Fishes*, Vol. 5, No. 1. Nauka, Leningrad. (In Russian)

Subject Index

Alphabetical List of Families and Page Numbers